COMPUTATIONAL AND EXPERIMENTAL ANALYSIS OF FUNCTIONAL MATERIALS

AAP Research Notes on Polymer Engineering Science and Technology

COMPUTATIONAL AND EXPERIMENTAL ANALYSIS OF FUNCTIONAL MATERIALS

Edited by

Oleksandr V. Reshetnyak, DSc
Gennady E. Zaikov, DSc

APPLE ACADEMIC PRESS

Apple Academic Press Inc.
3333 Mistwell Crescent
Oakville, ON L6L 0A2 Canada

Apple Academic Press Inc.
9 Spinnaker Way
Waretown, NJ 08758 USA

International Standard Book Number-13: 978-1-77188-342-9 (Hardcover)
International Standard Book Number-13: 978-1-315-36635-7 (CRC Press/Taylor & Francis eBook)
International Standard Book Number-13: 978-1-77188-343-6 (eBook)

Library and Archives Canada Cataloguing in Publication

Computational and experimental analysis of functional materials / edited by Oleksandr Reshetnyak, DSc, Gennady E. Zaikov, DSc.

(AAP research notes on polymer engineering science and technology series)
Includes bibliographical references and index.
Issued in print and electronic formats.
ISBN 978-1-77188-342-9 (hardcover).--ISBN978-1-315-36635-7 (PDF)

1. Polyanilines--Synthesis. 2. Polymeric composites. I. Zaikov, G. E. (Gennadii Efremovich), 1935-, editor
II. Reshetnyak, Oleksandr, author, editor III. Series: AAP research notes on polymer engineering science and technology series

TP1180.P552C64 2017 668.9 C2016-907864-7 C2016-907865-5

Library of Congress Cataloging-in-Publication Data

Names: Reshetnyak, Oleksandr, editor. | Zaikov, G. E. (Gennadiæi Efremovich), 1935- editor.
Title: Computational and experimental analysis of functional materials / editors, Oleksandr Reshetnyak, DSc, Gennady E. Zaikov, DSc.
Description: Toronto : Apple Academic Press, [2017] | Includes bibliographical references and index.
Identifiers: LCCN 2016054503 (print) | LCCN 2016057620 (ebook) | ISBN 9781771883429 (hardcover)
| ISBN 9781315366357 (ebook)
Subjects: LCSH: Polyanilines. | Conducting polymers. | Materials science.
Classification: LCC TP1180.P552 C66 2017 (print) | LCC TP1180.P552 (ebook) | DDC 620.1/9204297--dc23
LC record available at https://lccn.loc.gov/2016054503

Apple Academic Press also publishes its books in a variety of electronic formats. Some content that appears in print may not be available in electronic format. For information about Apple Academic Press products, visit our website at **www.appleacademicpress.com** and the CRC Press website at **www.crcpress.com**

This book is dedicated to the memory of
Professor Eugen P. Koval'chuk, DSc
(1939–2012)

ABOUT THE EDITORS

Oleksandr V. Reshetnyak, DSc

Oleksandr Reshetnyak, DSc, is Head of the Department of Physical and Colloid Chemistry of Ivan Franko National University of Lviv (Lviv, Ukraine). He is the author of two monographs and more than 150 publications in various journals and conference proceedings. His scientific activity is in the fields of physical chemistry of nanosystems, conductive polymers and its composites, electrochemistry of organic compounds, and chemi- and electrochemiluminescence. He is a member of the Editorial Board of the *Visnyk Lviv University Journal* (*Series: Chemistry*) (Lviv, Ukraine) as well as a member of the Editorial Board of the bimonthly scientific and technical international journal *Materials Science*. He is also an active member of the Taras Shevchenko Scientific Society (Lviv, Ukraine).

Contact address:
6 Kyryla & Mefodiya Str., Lviv, 79005, Ukraine
Tel: (+38) (032) 2600–397
Fax: (+38) (032) 2600–397
E-mail: reshetniak@franko.lviv.ua; Olexandr.Reshetnyak@gmail.com.

Gennady E. Zaikov, DSc

Gennady Zaikov, DSc, is Head of the Department of the Kinetics of Chemical & Biological Processes at the N. M. Emanuel Institute of Biochemical Physics at the Russian Academy of Sciences (Moscow, Russia) and Professor at Moscow State Academy of Fine Chemical Technology (Russia) as well as Professor at Kazan National Research Technological University (Kazan, Russia). He is also a prolific author, researcher, and lecturer.

Contact address:
4 Kosygin Str., Moscow, 119991, Russia
Tel: 7 (495) 1374–101
Fax: 7 (495) 9397–320
E-mail: chembio@sky.chph.ras.ru

CONTENTS

LIST OF CONTRIBUTORS

Olena I. Aksimentyeva
Department of Physical and Colloid Chemistry, Faculty of Chemistry of the Ivan Franko National University of Lviv; Doctor of Science in Chemical Sciences, Chief Scientific Officer. *E-mail: aksimen@ukr.net*

Liliya I. Bazylyak
Department of Physical Chemistry of Fossil Fuels, Institute of Physical–Organic Chemistry and Coal Chemistry named after L. M. Lytvynenko, National Academy of Sciences of Ukraine; Candidate of Science (Ph.D.) in Chemical Sciences, Senior Researcher. *E-mail: bazyljak.l.i@nas.gov.ua*

Jerzy Błażejowski
Department of Physical Chemistry, Faculty of Chemistry of the University of Gdańsk; Doctor of Science in Chemical Sciences, Professor, Head of the Department. *E-mail: jerzy.blazejowski@ug.edu.pl*

Lidiya M. Boichyshyn
Department of Physical and Colloid Chemistry, Faculty of Chemistry of the Ivan Franko National University of Lviv; Candidate of Science (Ph.D.) in Chemical Sciences, Docent (Associate Professor). *E-mail: lboichyshyn@yahoo.com*

Mar'yana V. Buzhans'ka
Department of Chemistry and Physics, Lviv Academy of Commerce; Candidate of Science (Ph.D.) in Chemical Sciences, Docent (Associate Professor)

Ivanna I. Demchyna
Department of Physical and Colloid Chemistry, Faculty of Chemistry of the Ivan Franko National University of Lviv; Master in Chemistry, Postgraduate Student

Yuliia A. Hnizdiukh
Department of Physical and Colloid Chemistry, Faculty of Chemistry of the Ivan Franko National University of Lviv; Master in Chemistry, Postgraduate Student. *E-mail: yulya_hnisdyuch@ukr.net*

Yulia Yu. Horbenko
Department of Physical and Colloid Chemistry, Faculty of Chemistry of the Ivan Franko National University of Lviv; Master in Biology and Biochemistry, Junior Researcher. *E-mail: y-bilka@ukr.net*

Oksana I. Konopelnyk
Department of General Physics, Faculty of Physics of the Ivan Franko National University of Lviv; Candidate of Science (Ph.D.) in Physico–Mathematical Sciences, Docent (Associate Professor)

Yaroslav S. Kovalyshyn
Department of Physical and Colloid Chemistry, Faculty of Chemistry of the Ivan Franko National University of Lviv; Candidate of Science (Ph.D.) in Chemical Sciences, Docent (Associate Professor). *E-mail: kovalyshyn@yahoo.com*

Andriy I. Krupak
Department of Physical and Colloid Chemistry, Faculty of Chemistry of the Ivan Franko National University of Lviv; Master in Chemistry, Junior Researcher

Viktoria M. Makogon
Department of Physical and Colloid Chemistry, Faculty of Chemistry of the Ivan Franko National University of Lviv; Master in Chemistry, Postgraduate Student

Galyna V. Martynyuk
Department of Methodology of Teaching of Physics and Chemistry, State Humanitarian University of Rivne; Candidate of Science (Ph.D.) in Chemical Sciences, Docent (Associate Professor). *E-mail: galmart@ukr.net*

Iryna Ye. Opaynych
Department of Physical and Colloid Chemistry, Faculty of Chemistry of the Ivan Franko National University of Lviv; Candidate of Science (Ph.D.) in Chemical Sciences, Senior Researcher

Bogdan B. Ostapovych
Department of Physical and Colloid Chemistry, Faculty of Chemistry of the Ivan Franko National University of Lviv; Candidate of Science (Ph.D.) in Chemical Sciences, Docent (Associate Professor). *E-mail: bohdanostapovych@ukr.net*

Dmytro O. Poliovyi
Department of Biology and General Ecology, Natural and Engineering Faculty of Kremenetski Regional Humanitarian Pedagogical Institute named after Taras Shevchenko; Candidate of Science (Ph.D.) in Chemical Sciences, Docent (Associate Professor)

Oleksandr V. Reshetnyak
Department of Physical and Colloid Chemistry, Faculty of Chemistry of the Ivan Franko National University of Lviv; Doctor of Science in Chemical Sciences, Professor, Head of the Department. *E-mail: reshetniak@franko.lviv.ua*

Mykhaylo M. Yatsyshyn
Department of Physical and Colloid Chemistry, Faculty of Chemistry of the Ivan Franko National University of Lviv; Candidate of Science (Ph.D.) in Chemical Sciences, Docent (Associate Professor). *E-mail: m_yatsyshyn@franko.lviv.ua*

Oksana M. Yevchuk
Department of Physical and Colloid Chemistry, Faculty of Chemistry of the Ivan Franko National University of Lviv; Bachelor in Chemistry, Ph.D. getter

Gennady E. Zaikov
Kinetics of Chemical & Biological Processes Department of Institute of Biochemical Physics named after N. N. Emanuel, Russian Academy of Sciences D.Sc. in Chemical Sciences, Head of the Department

Viktor P. Zakordonskiy
Department of Physical and Colloid Chemistry, Faculty of Chemistry of the Ivan Franko National University of Lviv; Candidate of Science (Ph.D.) in Chemical Sciences, Senior Researcher. *E-mail: zakordonskiy@franko.lviv.ua*

LIST OF ABBREVIATIONS

AA	Al–alloys
AAc	acrylic acid
AAr	aminoarenes
AcN	acetonitrile
ABS	acrylonitrile–butadiene–styrene
AFM	atomic force microscopy
AMA	amorphous metal alloys
An	aniline
$An–BF_3$	aniline doped by tetrafluorborate acid
AnSi	*N*–(3–trimethoxysilylpropyl)aniline
ANSA	aminonaphthylsulphoacid
AO	atomic orbitals
APS	ammonium persulphate
ASA	acetyl salicylic acid
Au–NPs	nanoparticles functionalized by gold
BP	benzoyl peroxide
BSA	bovine serum albumin
C_F	Faraday capacitance
CCC	cerium conversion coatings
CcFs	coconut fibers
CeAFs	cellulose acetate nanofibers
CFs	curauá fiber
CL	chemiluminescence
CNT	carbon nanotubes
CR	contrast
CSA	camphorosulfonic acid
Cu–NPs	nanoparticles functionalized by copper
CVA	cyclic voltammetry
DBSA	dodecylbenzenesulfonic acid
DCA	dichloric–acetic acid
D_{eff}	effective diffusion coefficient
DEL	double electric layer
DMFA	dimethylformamide
DN	donor number

DNNSA	dinonylnaphthalenesulfonic acid
DMI	1,3–dimethyl–2–imidazolidinone
DMSO	dimethylsulfoxide
DSC	differential scanning calorimetry
DTA	differential thermal analysis
DTG	differential mass loss curves
E	emeraldine
EB	emeraldine base
ECe	ethyl cellulose
EChM	electrochromic materials
EChP	electrochromic polymers
ECL	electrochemiluminescence
ECP	electroconductive polymers
EDOT	3,4–ethylenedioxythiophene
EES	electrochemical energy storage
EPD	electrophoretic monochrome display
ER	epoxy resin
ES	emeraldine salt
EPD	electrophoretic monochrome display
EPO/PAn	epoxy–polyaniline composites
FEP	fluorinated ethylene propylene copolymer
FSA	ferrocenesulfonic acid
FTIR	Fourier transform infrared spectroscopy
HDPE	high-density polyethylene
HOMO	high occupied molecular orbital
HRP	horseradish peroxidase
ICPs	intrinsically conducting polymers
IDPD	integral procedural decomposition temperature
IPN	interpenetrating polymer networks
ITO	indium–tin–oxide
IR	infrared (spectrum, spectroscopy)
JFs	jute fibers
k_s	heterogeneous constant of a charge transport
L	leucoemeraldine
LB	leucoemeraldine base
LB films	*Langmuir–Blodgett* films
LC	liquid crystals
LDPE	low-density polyethylene
LECs	light-emitting electrochemical cells
LEDs	light-emitting diodes

LGS	lignosulfonate
Lum	luminol
LUMO	low unoccupied molecular orbital
λ_{max}	maximum absorption/transmission
LS	leucoemeraldine salt
MDTIR	multiply disturbed total internal reflection
MMT	montmorrilonite
MPD	N–methylpyrrolidone
MS	mass spectrometry
MWCNT	multiwalled carbon nanotubes
NAn	nitroaniline
NASA	naphthylaminosulfonic acid
NaSS	Na salt of styrene sulfonic acid
ND	nanodiamond
NM	natural minerals
NMP	N–methylpyrrolidinone
NSA	2–naphthalenesulfonic acid
P	pernigraniline
PA	polyacetylene
PAA	polyacrylic acid
PACC	protective anticorrosive coating
PAn	polyaniline
$PAn\text{–}BF_3$	polyaniline doped by tetrafluorborate acid
PAn/CeFs	polyaniline–cellulose fibers
PAn/SF	polyaniline/silk fibroin composite fibers
PAr	polyarenes
PAAr	polyaminoarene
PAN	polyacrylonitrile
PAPh	polyaminophenols
PAT	polyaminothiazole
PBMA	polybuthylmethacrylate
Pd–NPs	nanoparticles functionalized by palladium
PnB	pernigraniline base
PnS	pernigraniline salt
PEDOT	poly–3,4–ethylenedioxythiophene
PES	polyester
PET	polyethylene terephtalate
PI	polyimide
PLm	polyluminol
PmAPh	poly–*m*–aminophenol

PMA	polymetacrylic acid
PMMA	polymethyl methacrylate
PoAPh	poly–*o*–aminophenol
PoA	poly–*o*–anisidine
PoEA	poly–*o*–ethoxyaniline
PoMA	poly–*o*–methoxyaniline
PoT	poly–*o*–toluidine
PT	polythiophene
PPh	polyphenylene
PpPh	poly–*p*–phenylene
PpPhV	poly–*p*–phenylene–vinylene
PP	polypropylene
PPy	polypyrrol
PS	polystyrene
PSF	polysulfone film
PSS	poly(sodium styrenesulfonate)
PSSA	polystyrenesulfonic acid
PSSMA	copolymer of styrenesulfonate and maleic acid
PTFE	poly(tetrafluoroethylene)
PU	polyurethane
PVA	polyvinyl alcohol
PVC	polyvinylcarbazole
PVDF	polyvinylidenefluoride
PVN	polyvinylnaphthalene
REM	rare-earth metals
RF	radio frequency
R_e	electronic resistance
R_i	ionic resistance
R_s	resistance of electrolyte solution
SAW	surface acoustic waves
SCE	saturated calomel electrode
Si–Gl/PAn	silica–glauconite/polyaniline composite
SPSF	sulfonated polysulfones
SSA	sulfosalicylic acid
StMA	copolymer of styrene with maleic anhydride
TG	integrated mass loss curves
TGA	differential thermal gravimetric analysis
TMA	thermal mechanical analysis
TSA	toluene sulfonic acid
TMSPA	*N*–(3–(trimethoxysylil)–propyl)aniline

p–TSA	*p*–toluene sulfonic acid
τ_D	diffusion time
UV	ultraviolet (spectrum, spectroscopy)
WE	working electrodes
XRD	*X*-ray diffraction analysis

SUPPORTING INSTITUTIONS

- Department of Physical and Colloid Chemistry, Faculty of Chemistry of the Ivan Franko National University of Lviv (*Kyryla i Mefodiya Str. 6, 79005 Lviv, Ukraine);*
- Department of General Physics, Faculty of Physics of the Ivan Franko National University of Lviv *(Kyryla I Mefodiya Str. 8, 79005 Lviv, Ukraine);*
- Kinetics of Chemical and Biological Processes Department, Institute of Biochemical Physics named after N. N. Emanuel, Russian Academy of Scien*ces (Kosygin Str. 4, 119991 Moscow, Russian Ferderation);*
- Department of Physical Chemistry, Faculty of Chemistry, University of Gdańsk *(J. Sobieskiego Str., 18, 80-952 Gdańsk, Poland);*
- Kremenetskyi Regional Humanitarian Pedagogical Institute named after Taras Shevchenko *(Lyceum Str. 1, 47003 Kremenets, Ternopil region, Ukraine);*
- Lviv Academy of Commerce *(Samchuka Str. 9, 79011 Lviv, Ukraine);*
- Department of Physical Chemistry of Fossil Fuels, Institute of Physical–Organic Chemistry and Coal Chemistry named after L. M. Lytvynenko, National Academy of Sciences of Ukraine (*Naukova Str. 3a, 79053 Lviv, Ukraine);*
- Department of Methodology of Teaching of Physics and Chemistry, State Humanitarian University of Rivne (*Stepan Bandera Str. 12, 33000 Rivne, Ukraine).*

PREFACE

The main attention in this presented collection of the scientific papers is paid to the recent theoretical and practical advances in conductive polymers and nanocomposites. It also coveres the different branches of knowledge: from the obtaining of nanoparticles with the use of new synthetic techniques to application of the obtained nanomaterials and nanocomposites in different fields of industry.

During the last two decades, conductive polymers were the object of intensive research and development in the academic world and also in the chemicals and electronics industry worldwide. In the 2003–2013 period only, over 5000 articles and more than 35 chapters of books were dedicated to the polyaniline and to the composites on its basis and were published in the scientific journals of the American Chemical Society, Elsevier, and John Wiley & Sons Publishing. However, the books dedicated solely to the polyaniline and its composites are practically absent, which makes this Book unique among the modern scientific books.

This book of scientific papers represents by itself the result of the perma-systematical and methodical investigations and also scientific accumulations of the scientific team of the Department of Physical and Colloid Chemistry (Chemical Faculty) of Ivan Franko National University of Lviv in the field of the synthesis, investigation of the properties and practical application of the electroactive polymers and composites on their basis, which have been performed within the scientific school "Physical chemistry of polymers" initiated by Professor Eugen P. Koval'chuk over 25 years ago. Under his guidance two theses for a Doctor's as well as over 10 Ph.D. theses were defended, and the results of the investigations were published in more than 200 articles in scientific journals. The research in this area are still continuing.

This book comprehends the questions both of the synthesis of polyaniline by different methods, under different conditions, for various applications, and also the studies of its properties by a wide range of the modern physical–chemical methods. The advantages of this book are comprehensive analysis of experimental results from the point of view of the correlations in the triad "Synthesis Conditions–Structure–Physical and Chemical Properties". Some chapters contain the integrated results, which will be very valuable for the potential purchaser as a referenced data.

The mechanism of the initial stages of oxidative polymerization of aniline and its derivatives with the formation of conductive polymers with the system of conjugated π-bonds are analyzed in *Chapter 1*. It is shown that both during chemical and electrochemical polymerization the process is initiated by the cation-radicals of the initial monomer, which then recombine in accordance with the type "head to tail". The obtained dimers represent by themselves the structural units of the future polymeric chain, and then they can take part in similar chain of the transformations. In the process of polymeric chain propagation, the molecules with varying polymerization degree can participate, because their reactivity is determined by the presence of the end amino-groups, and, therefore the oxidative polymerization is typical polyaddition process. On example of nitro- and oxymethyl- derivatives it was evaluated an impact and position of the substituent in the aromatic ring on the reactivity of the aniline derivatives in the reaction of oxidative polymerization. On the basis of spectral studies and elemental analysis it was determined the most likely structures of the obtained polymeric anilines, including the polyluminol. It was shown that in the case of polyanilines obtaining by chemical method regardless of the nature of the initial monomer the final product represents by itself a form of emeraldine salt.

Chapter 2 is dedicated to the research of the morphology of polyaniline's films electrochemically deposited on the surface of Al-based amorphous metal alloys. The films of polyaniline have been synthesized by potentiodynamic oxidation of aniline at the electrodes of amorphous metal alloys of the composition $Al_{87}Ni_8(REM)_5$, where REM ≡ Y, Ce, Gd and Dy, as well as polycristalline aluminum. The process of electrochemical oxidation of aniline on amorphous metal alloys-electrodes was analyzed and compared with the process of aniline's oxidation on the polycrystalline Al-electrode. It was established, that the difference in the form of cyclic voltammogramms is conditioned by the presence of amorphic components in the amorphous metal alloys composition, which cause of different resistance of surface oxide films on the working electrodes. With the use of X-ray and IR-spectral analysis it was showed, that the structure of polyaniline's films on the surface of the Al-electrode and the $Al_{87}Ni_8(REM)_5$ electrodes is amorphous-crystalline. At this, the polyaniline deposited on $Al_{87}Ni_8(REM)_5$ electrodes has a higher degree of crystallinity. An analysis of the images of scanning electron microscopy showed, that on the surface of the working electrodes of amorphous alloys $Al_{87}Ni_8(REM)_5$ the polyaniline's films have spongy and porous branched morphology.

Polymers of polyvinyl series (namely, polyaminoarenes) are considered as the perspective materials for electrochromic devices from the point of view of physical chemistry of electrooptic phenomena in *Chapter 3*. The main regularities of polyaminoarenes films formation on transparent semi-conducting surfaces have been studied based on the results of electrochemical, optical and spectroelectrochemical investigations. The relationship between the electrochemical and also electrooptical characteristics and the structure of polymers was established. On a basis of determined diffusion coefficients and heterogeneous constants of a charge transport in different electrolytic media it was shown, that the structure of a polymeric chain has the decisive effect on the electrochromic material efficiency. Some possible methods of electrochromic characteristics improvement for polyaniline and its derivatives have been proposed.

The properties of Langmuir films of polyaniline and also its derivatives (namely, poly-o-methoxyani-line, poly-o-toluidine), as well as compositions of these polymers with polymethyl methacrylate and copolymer of styrene and maleic anhydride have been investigated and described in *Chapter 4*. Monomo-lecular Langmuir films were obtained on water surface by their application both from the individual and from the mixed solutions of polymers in organic solvents (tetrahydrofuran, chloroform). It is shown, that the initial surface concentration of the substance and the nature of a solvent used for the deposition of a monolayer has an impact on the magnitude of the surface pressure of polyaniline monolayers. Dependence of the effective area of monomer link ($S_{eff.}$) on the composition of binary mixture has been studied on a basis of isotherms of surface pressure of binary compositions. It was discovered both of $S_{eff.}$ magnitude increasing and decreasing compared to the total values of the areas for individual components. Such behavior of Langmuir films can be conditioned due to the conformational changes undergone by macromolecules into composite monolayers. During an investigation of monomolecular films under conditions of compression–expansion it was determined that the polyanilines can form both equilibrium and non-equilibrium monolayers. The polymer "memory" effect is developed in a case when the monolayers of polyanilines are applied on water surface.

The mechanism of formation and physical chemical properties of epoxy-polyanilines composites obtained using both the curing agent and also the electroconductive filler of polyaniline doped by tetra-fluorborate acid has been studied in *Chapter 5*. The factors having an impact on curing mechanism and properties of composites were analyzed. It was shown that the combination of properties both of curing agent and electroconductive component into one complex polyaniline– tetrafluorborate acid gives the possibility essentially to simplify the conductive epoxy polymer obtaining, to receive the polymer–polymeric composites characterizing by electric conductivity 10^{-6}–10^{-4} S/cm at relatively low content of complex polyaniline– tetrafluorborate acid; at the same time, such composites are characterized by high thermal and physical mechanical properties which makes it possible to use them for producing of antistatic coatings, electroconductive adhesives, electrodes, etc.

The temperature dependences of optical absorption (transmittance) spectra of the conducting polymer films—polyaniline, poly-o-toluidine, poly-o-anisidine, poly-o-aminophenol doped by sulfuric acid, potassium ferricyanide and silver nanoparticles have been represented in *Chapter 6.* It has shown that temperature increasing causes a change in optical transmittance (absorption) spectra of polymers in all temperature range. However, characteristics of thermo-optical changes or thermochromic effects depend on polymer nature and doping conditions. Changes in parameters of the optical spectra were connected with conformation of polymer chain and modification of electron properties of conducting polymers under temperature action.

The basic aspects of the electroactive polymers using in general as well as the polyaniline as classical representative of these polymers in particular under design of chemo- and biosensors are considered in *Chapter 7.* On example of typical experimental and industrial samples it was analyzed the architecture and performance properties of chemosensors by resistive, amperometric, potentiometric, voltamametric, optical and gravimetric types, their advantages and disadvantages at the analysis of various inorganic and organic substrates. It's shown the prospects of application of nanostructured polyaniline and composites on its basis with nano-dispersed metallic, oxide and mineral fillers at the design of sensor devices. An application of polyaniline in biosensorics is considered in a separate way; in particular, the methods of biological components immobilization on the polymer-modified substrates have been analyzed.

The main aspects of aluminum and aluminum alloys protection by corrosion coatings based on various forms of polyaniline were considered in *Chapter 8.* The advantages and disadvantages of modern chemical and electrochemical methods of application of polyaniline protective coatings on aluminum-containing substrates were analyzed as well as the relationship between the conditions of application of protective layers and their protective properties was shown. The modern approaches to the use of polyaniline in protective corrosion coatings, such as doping by polymeric acids, the formation of double-strand polyaniline complexes, the use of polyaniline as the pigment filler in paint coating, as well as a major component of various polymer–polymer or polymer–inorganic hybrid composites were considered. Proposed today mechanisms of protective action of protonated and deprotonated forms of polyaniline were in detail analyzed concerning to the corrosion of aluminum-containing substrates by forming of the protective oxide or salt passivation layers, as well as the inhibition of redox processes by polyaniline doping anions was considered. Special attention has been paid into the negative role of the intermetallic surface inclusions concerning to the initiation of corrosion and to the use of conversion cerium coatings to eliminate of this problem.

The synthesis conditions and physico–chemical properties of conductive composites of conjugated po-lyaminoarenes (polyaniline, polyorthotoluidine, polyorthomethoxy–aniline in polymer matrixes of polyme-thylmethacrylate, polyvinyl alcohol, polybuthylmethacrylate, polyacrylic acid and polymetacrylic acid are described in *Chapter 9*. It was found that the dependence of specific conductivity on the conducting polymer content has a percolation character with extremely low "threshold". It was established a connection between electrical, mechanical and thermomechanical properties of polymer–polymer composites studied. On the basis of study optical changes in film composites under gas action the method of obtaining the flexible color indicators for express control of ammonia content in gas environment has been developed.

Chapter 10 is dedicated to the investigation and analysis of the sources of the electrochemi-luminescence during the electrochemical synthesis of polyaniline and copolymers of aniline and luminol by different composition at high electrode potentials. It was proposed that the generation of low intensity of luminosity in acidic aqueous solutions at the high electrode potentials is related with irreversible destruction of the polyanilines in the form of pernigraniline as a result of their interaction with free radical intermediates of oxidation of solvent (water) and anions of the base electrolyte. It was shown, that the obtained copolymers exhibit the electrochemiluminescent activity in aqueous alkaline solutions. The feature of electrochemiluminescence of copolymers is the presence of two waves of the appearance of luminosity; that's why it was suggested, that the source of the electrochemiluminescence is a direct electrochemical oxidation (the first wave) and chemical oxidation by free radicals of electrochemically generated intermediates (the so-called second wave) of luminolic fragments of copolymeric chain.

The structure and physical chemical properties of the synthesized polymer–inorganic composites by intercalation type based on polyaniline/copolymers of aniline and nitroaniline with xerogel of vanadium pentaoxide are described in *Chapter 11*. The electrical conductivity of synthesized composites was studied and they were tested as cathode materials of lithium chemical power sources. It was shown that the use of copolymers of aniline with its nitro derivatives improves the electrical characteristics of lithium chemical power sources based on hybrid polymer–xerogel composites by introducing of the electroactive nitro groups into the structure of the polymer.

The subject of *Chapter 12* is the modification of large-scale (films, sheets, tapes, threads, fibers, etc.) non-conductive material-substrates of different polymeric nature by polyaniline layer and also obtaining of homogeneous polyaniline / polymeric composites. The methods of preparation of non-conductive polymeric substrates, and also the basic techniques of applying of the polyaniline layers or films on such substrates, namely the technique from the solution in situ, from the gaseous phase, by polymerization method in plasma, by forming of polyaniline's films on the surface of polymeric substrates from its solutions and suspensions, by preparation of polyaniline / polymeric composites using the methods of co-dissolving, by making of melts as well as by combined methods are discussed. The basic stages of the processes have been analyzed, the physico–chemical properties and morphology of the obtained in each case composites were considered, the examples of composite materials based on polyaniline and nonconducting massive polymers of natural, artificial and synthetic origin are demonstrated. It is shown that due to its physical and chemical properties such materials are promising for applications as sensitive sensory materials, anticorrosive, anti-static, electrostatic coating materials for shielding of electromagnetic waves, elements of organic optoelectronic devices, diaphragms, artificial muscles, electroconductive fabrics, biologically active substrates, etc.

One of the main tasks of modern physical and chemical investigations is the development of scientific principles and new approaches to the creation of polymeric and composite materials with given functional properties, namely electrical, optical, magnetic, an ability to the charge storage and its transfer, the catalysis of a number of reactions, sensorics and etc. The actual problem of the science and technology at the present step is an investigation of the mechanism of formation and the development of new nanosized composite materials based on polymers doped or filled by inorganic clusters, in particular carbonic (graphene, fullerene, nanotubes), silicium (silicum (IV) oxide, nanocrystals of silicon, porous silicon, silicum carbide) as well as by the compounds of transition metals, in particular by ferrum-containing ones, namely clusters of ferrum and its oxide, magnetite, the complexes of ferrum and etc. Due to the nanostructures of such composites (the size of the particles consists from some units to dozens nm) they have unique magnetic, spectral and electrochemical properties. General scientific problem at the development of high-dispersed and film composites of the polymers with inorganic nanoparticles is understanding of the character of components interaction. For this purpose, it is necessary to investigate the physical chemistry both of initial compounds and of the obtained composites; this gives the possibility to control the parameters of the synthesized materials. This knowledge can be obtained via the complex investigations of crystalline, molecular, supramolecular, and electronic structure, the dispersion degree of the particles, their distribution, determination of the relationship between the structure and magnetic, electrical and optical properties of nanomaterials. The investigations presented in *Chapter 13* are dedicated to the solution of the above-said tasks.

In recent decades, researchers actively developed the various hybrid mineral–polymeric composites, among which special attention is given to the materials based on polyaniline and natural minerals, which combine the properties both of the polymer and natural minerals showing the synergism effect. Combination of properties of the polyaniline and inorganic substances of micro-, and especially of nanosized dispersion degree leads to the formation of composites with physical and chemical properties that are much better compared to individual properties both of polymer and inorganic mineral. In many cases the thermal stability of polyaniline into obtained composites is increased, and the composites acquire electromagnetic, magnetic, catalytic, adsorptive and other properties. *Chapter 14* is dedicated to the investigations of the structure and thermal stability of silica-glauconite / polyaniline composite.

The results of the unique experimental investigations and original methodology of the description of physical–chemical and electrochemical phenomena at the interface surfaces are organically combined in the presented book. It is shown an influence of such phenomena on the applied aspects of the polyaniline and nanocomposites on its basis applications. The last fact advantageously distinguishes the presented book of the scientific papers among existent range of the works dedicated to such subject matter at the present time.

This book will be useful first of all for scientists and post-graduate students who are engaged in physical chemistry of conductive polymers in the whole and polyaniline particularly. It will be helpful also for professors teaching courses dedicated to the materials science, chemistry of high-molecular compounds, physical–chemical properties of conductive polymers and its applications in electronics, power sources, sensors, anticorrosive coatings, etc.

This presented work has been performed under the partial financial support of the scientific grants of the Ministry of Education and Science of Ukraine (state registration numbers 0112U001283, 0112U001294, 0112U001295, 0113U003055, 0115U003262 and 0115U003263).

—Olexandr Reshetnyak
Lviv, July 2016

CHAPTER 1

POLYANILINES: THE ROLE OF PARTICLES OF RADICAL NATURE IN OBTAINING OF POLYMERS/COPOLYMERS WITH A SYSTEM OF CONJUGATED π-BONDS

O. V. RESHETNYAK[1], M. M. YATSYSHYN[1], and L. I. BAZYLYAK[2]

[1]*Department of Physical and Colloid Chemistry, Faculty of Chemistry, Ivan Franko National University of Lviv, 6 Kyryla & Mefodia Str., Lviv 79005, Ukraine*

[2]*Department of Physical Chemistry of Fossil Fuels, L. M. Lytvynenko Institute of Physical–Organic Chemistry and Coal Chemistry, National Academy of Sciences of Ukraine, 3a Naukova Str., Lviv 79053, Ukraine*

Corresponding author: reshetniak@franko.lviv.ua

CONTENTS

ABSTRACT

The mechanism of the initial stages of oxidative polymerization of aniline (**An**) and its derivatives with the formation of conductive polymers with the system of conjugated π-bonds has been analyzed. It is shown that both during chemical (in the presence of peroxydisulfate anions) and electrochemical polymerization, the process is initiated by the cation-radicals of the initial monomer, which then recombine in accordance with the type "head to tail." The obtained dimers represent by themselves the structural units of the future polymeric chain, and then they can take part in similar chain of the transformations. In the process of polymeric chain propagation, the molecules with varying polymerization degree can participate, because their reactivity is determined by the presence of the end amino-groups, and, therefore the oxidative polymerization is typical polyaddition process. On example of nitro- and oxymethyl- derivatives, the impact and position of the substituent in the aromatic ring on the reactivity of the aniline derivatives in the reaction of oxidative polymerization were evaluated. On the basis of spectral studies and elemental analysis, the most likely structures of the obtained polymeric anilines, including the polyluminol, were determined. It was shown that in the case of polyanilines (PAn) obtained by chemical method, regardless of the nature of the initial monomer the final product represents by itself a form of emeraldine salt.

1.1 INTRODUCTION

The discovery of polymers possessing own electroconductivity due to the presence of the system of alternant σ- and π-bonds results in the occurrence and the development of new fundamental areas of researches in chemistry, physics, materials science, etc.[1–3] Electroconductive polymers (ECPs) combine high flexibility and plasticity, typical for the polymers with high electroconductivity whose value may approach to the conductivity of metals. This property of ECPs has led them to be often called as "synthetic metals." Among the large number of modern organic and inorganic materials, the ECPs are valued as "strategic" materials that have become the objects of intense researches in the laboratories of the world's leading scientific centers and major industrial corporations.[4]

Polyaniline (PAn) and polymers based on its derivatives occupy an especial place among the ECP. Due to the special electronic spatial structure of aniline and its derivatives, their polymers have a wide range of physical and chemical properties, such as the ability to the reversible oxidation–reduction, long-continued resistance to not only moisture or air, but also so much more aggressive media. The combination of low prime cost with ease of synthesis makes such materials indispensable in molecular electronics, biochemistry, medicine, chemical current sources, etc. The color of **PAn** depends on the pH value and on the applied potential, and can vary from light yellow (at pH = 1 and the electrode potential $E = -0.2$ V) to purple-red (pH = 4 and $E = +1.4$ V), passing through yellow, green, dark-blue, black, and many their tints.[5,6] This property of **PAn** was the basis for its use as pH indicators,[5,7] in electrochromic[6,8–11] and electroluminescent[12,13] devices. PAn layers possess good corrosion properties during protection of steel,[14–17] zinc,[18] copper,[19] aluminum, and its alloys (**Chapter 8**). Thin layers of **PAn** are used in molecular electronics,[2,20] as detectors and dosimeters of γ-radiation[21] in the manufacture of various types of sensor devices for determining pH,[22,23] viruses,[24] and also important analytes such as ammonia,[25] H_2O_2,[26] NO_2,[27] CO,[28] HCN,[29] H_2,[30] CH_4,[31] glucose,[32] carbamide,[33] etc. PAn and its derivatives are used as cathode materials in chemical sources of electrical energy[34,35] and electrochemical supercapacitors.[36–39]

The polymerization of aniline is very easily carried out chemically in aqueous acidic solutions using as oxidizing agents the peroxide compounds (in particular, peroxydisulfates,[40–42] hydrogen peroxide,[7,43] benzoyl peroxide[44]), the ions in higher oxidation level (Fe^{3+}, Ce^{4+}, $Cr_2O_7^{2-}$, IO_3^-, ClO_2^-, and VO_3^-),[45–50] the oxides of transition metals (V_2O_5,[51] WO_3,[52] MnO_2, and PbO_2[50]) and enzymes,[53–56] etc. PAn layers on the surface of metals or semiconductors were also obtained by the method of vacuum deposition[57,58] or by the polymerization in plasma.[59,61] However, often the method of electrochemical polymerization is used due to the opportunity to get thin polymeric layers on the surface of the conductive substrate. It has been reported about obtaining of **PAn** on the platinum electrode during its polarization under potentiodynamic[62] or galvanostatic[63] conditions in both aqueous[62–64] and organic[65] media. In all cases, the structure and properties of the polymerizates differ significantly, mainly due to varying degrees of protonation of nitrogen atoms in the polymeric chain. This difference leads to the volatility in the conductivity of the synthesized samples in a wide range[66] from 10^{-15} to 10 $S{\cdot}cm^{-1}$.

With the use of IR-, UV–Vis, ESR, NMR, Raman, and X-ray photoelectron spectroscopy methods, the existence of three forms of **PAn** was confirmed, when the chain is a sequence: (1) only reduced (which contain only benzene rings)

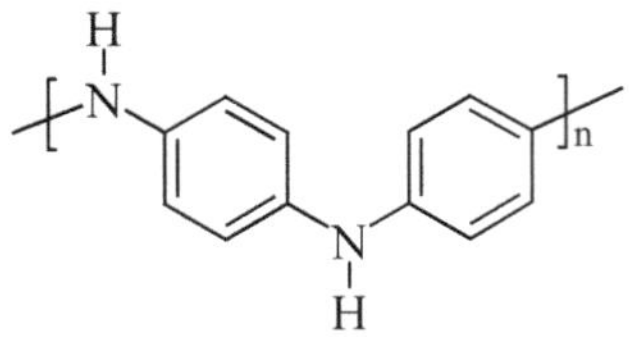

dimer fragments, or, so-called, leucoemeraldine, ***L***; (2) only oxidized (which contain only quinoid fragments)

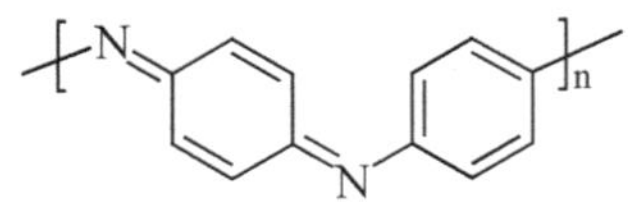

dimer segments, or, so-called, pernigraniline, ***Pn***; (3) reduced and oxidized dimer links, which are alternated, or, so-called, emeraldine, ***E***.[67–70] In addition, the forms of **PAn** may vary by degree of the protonation of nitrogen in the main chain.

MacDiarmid and coworkers,[71] by measuring the potential value of the platinum electrode during the chemical oxidative polymerization of aniline in aqueous acidic solutions $(NH_4)_2S_2O_8$, found that during the process it changes from +0.4 V (initial value) to +0.66 V, and then to +0.75 V (immediately and after 2 min of adding the oxidant in the reaction mixture, respectively). After the 10-min period, the potential is decreased again to the +0.47 V. All fragments of PAn, which was formed at the potential +0.75 V, are oxidized, that is, the polymer is in the form of pernigraniline, while the product formed at +0.43 V is the emeraldine. This led the authors to the conclusion that the main product of the aniline polymerization is emeraldine that can be formed by at least in two ways: as a result of the pernigraniline reduction and oxidation of aniline by pernigraniline. According to experimental data, the more likely is the second way.

Mechanism of the aniline formation has been the subject of many studies. Formation of the structure of PAn proceeds through the stages of the oxidation of aniline (an electron detachment) with the following recombination of radical particles–intermediates that leads to an elongation of the polymer chain. It is proposed that the oxidation of aniline molecule is single-electron process whereby the cation-radical is formed,[72] that was confirmed by ESR spectra during the electro-oxidative deposition of **PAn** in 1.0 M H_2SO_4 aqueous solution at the platinum electrode under scanning potential range from −0.1 to +0.85 V.[73] In this case, the strong dependence of the intensity of the ESR signal of PAn on electrode potential was observed in the range +(0.2–0.8) V. Possible ways to convert the aniline's cation-radicals were the subject of the discussion in many papers,[5,70,74–77] but the proposed schemes of the transformations of aniline are imperfect since the existence of various tautomeric forms of particles–intermediates and the optimality of the proposed routes of further transformations were not confirmed. Moreover, with the exception of the MacDiarmid's and A. Epstein's article,[78] the chemical mechanism of the initiating of aniline's polymerization was almost not considered. Therefore, the mechanism of oxidative polymerization of aniline and their derivatives in aqueous solutions, particularly in the presence of peroxydisulfate anions was analyzed by us in detail, based on the results of electrochemical studies and quantum mechanical calculations.

1.2 THE MECHANISM OF OXIDATIVE POLYMERIZATION OF ANILINE IN AQUEOUS SOLUTIONS

1.2.1 ANALYSIS OF THE INITIAL STAGES OF OXIDATION OF ANILINE

Cyclical voltammogram of aniline obtained in acidic aqueous solution at a platinum electrode is shown in Figure 1.1. The first maximum of oxidation current at +0.22 V, the height of which increases with each successive cycle of the potential scanning was attributed to the oxidation of neutral forms (i.e., ***L*** and ***E***) of PAn to cation-radicals, namely $\boldsymbol{L}^{+\bullet}$ and $\boldsymbol{E}^{+\bullet}$, respectively.[79] According to the results of *ref.*,[40] this maximum corresponds to the transition into the redox couple "leucoemeraldine ↔ emeraldine." The second maximum at +0.8 V corresponds to the transformation of emeraldine into pernigraniline, which is accompanied by the transfer of one electron and by the detachment of two ions[73] H^+. Transformation of emeraldine ↔ pernigraniline in the acetonitrile can also occur as a result of chemical oxidation of emeraldine by pyridine or by superoxide-ion $O_2^{-\bullet}$ (in the presence of dissolved molecular oxygen in solution).[80] The attempts of the interpretation of the third (middle) peak were made in the *ref.*,[81,82] under which it can be attributed to secondary transformations of PAn that result in the reduction of conductivity of PAn's layer on the electrode surface. In particular, such secondary transformations can be cross-linking of polymeric chains through nitrenium cation,[81] or the oxidation of products of partial degradation of **PAn**, which is formed at the polarization of the electrode under potentials over than +1.2 V.[83] Growth of the heights of maxima of oxidation current during each subsequent scan cycle of potential suggests that they are associated with the conversion of polymeric aniline, which is formed on the electrode.

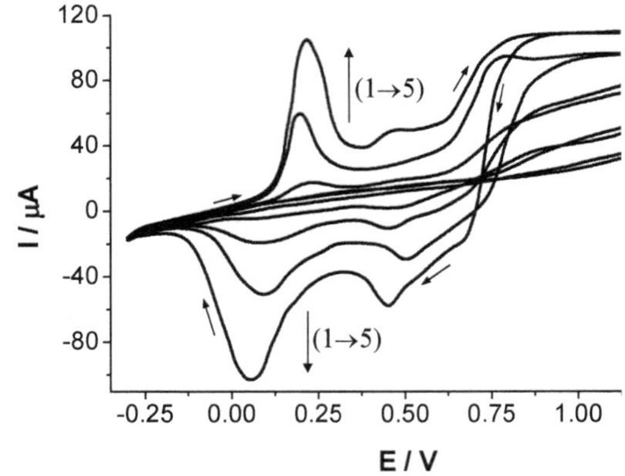

FIGURE 1.1 *Cyclical voltammogram of platinum electrode in 0.1 M solution of aniline during the first five cycles of the potential scanning (base electrolyte is 1 M HCl aqueous solution; $s_E = 20\ mV \cdot c^{-1}$).

*Here and hereinafter, in the chapter, all electrode potentials are presented relatively saturated Ag/AgCl electrode.

According to the conventional scheme (Fig. 1.2), the oxidation of aromatic amines occurs through several stages, which cover the transfer of two electrons and the detachment of two H^+ ions.[72] This scheme can be considered only preliminary, because it does not detail the mechanism of electron transfer, and does not estimate the likelihood of the proceeding of process accordingly with each two possible ways of its occurrence. There are two possibilities for the detachment of the second electron. The first is that possible direct electron transfer (the oxidation of cation-radical ***II*** to doubly charged cation ***III***), while in the second stage, that precedes the deprotonation of cation-radical to the radical ***IV***.

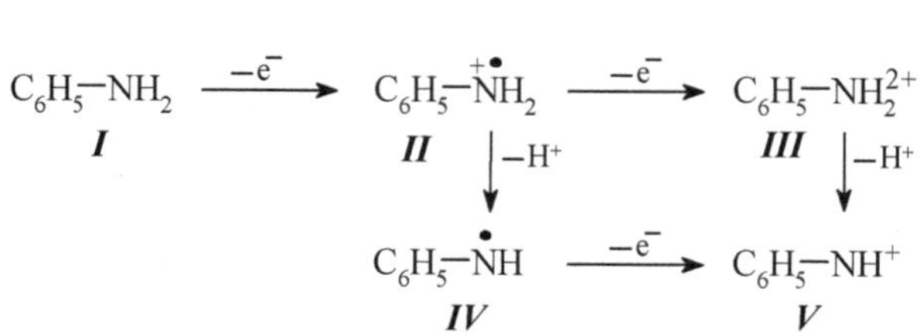

FIGURE 1.2 The sequence of transformations of aniline during its oxidative polymerization according to *ref.*[72]

Moreover, given scheme ignores also the possibility of two ways of one-electron oxidation of original aniline.[84] According to the first, presented in the scheme, the electron detachment occurs from 2*s* level of nitrogen atom leading to the formation of cation-radical ***II***.

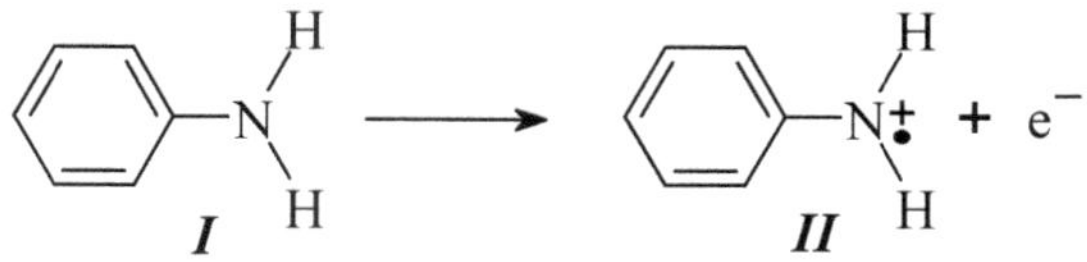

(1.1)

At the same time, one-electron oxidation of aniline's molecules is also possible, which will be accompanied by heterolytic breaking of N–H bond

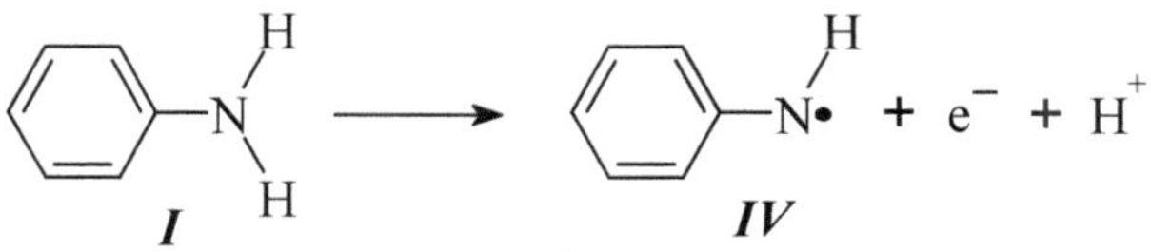

(1.2)

Also, it is necessary to take into account the possibility of isomerization of cation-radical ***II*** and radical ***IV***, which proceeds according to the following schemes:

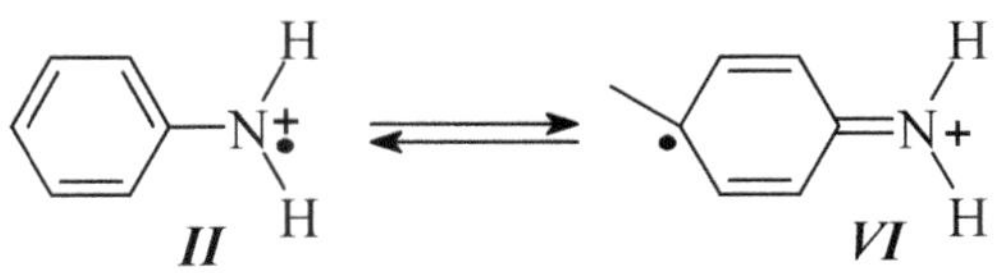

(1.3)

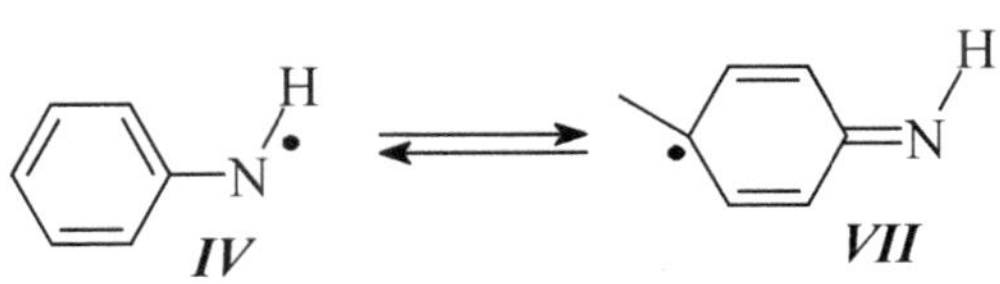

(1.4)

The results of quantum chemical calculations, which have been done for the particles **I–VII**,[84] showed, an increase of the oxidation level in sequence aniline → cation-radical → dication is accompanied by an increase both of total energy E_{tot} of respective particles (from −103,518 to −92,653 kJ·mol^{-1}) and its binding E_{bin} (from −6210 to −4131 kJ·mol^{-1}) as well as electronic components E_{el} (from −421,384 to −403,648 kJ·mol^{-1}). At this, the energy of the highest occupied molecular orbital (**HOMO**) E_{HOMO} regularly is decreased for the same sequence, since as the oxidation of particles is complicated with increasing of their oxidation level. In addition, it was concluded about almost practically identical probability of the existence of benzenoid cation-radical ***II*** and isomeric to it cation-radical of quinoid type ***VI***, since they are characterized by practically almost identical energy characteristics.

TABLE 1.1 Electron Occupancy of Atomic Orbitals of the Nitrogen Atom.

Particle		Atomic orbitals of the nitrogen			
		2s	p_x	p_y	p_z
I	α	1.427	1.103	1.041	1.841
II	α	0.580	0.513	0.485	0.938
	β	0.608	0.493	0.464	0.344
III	α	0.579	0.525	0.472	0.502
	β	0.586	0.527	0.474	0.545
IV	α	0.804	0.686	0.543	0.875
	β	0.799	0.673	0.512	0.170
V	α	0.760	0.741	0.523	0.346
	β	0.760	0.741	0.523	0.346
VI	α	0.605	0.532	0.485	0.918
	β	0.583	0.486	0.464	0.351
VII	α	0.809	0.692	0.549	0.867
	β	0.796	0.649	0.512	0.194

Let us separately analyze the oxidation process of radical ***IV***. E. Geniés and M. Łapkowski[72] claim that this neutral radical easily can be oxidized to the particle ***V***, since as redox potential of the process ***IV*** → ***V*** is lower compared with the transition ***I*** → ***II***. However, this statement does not agree with the results of quantum chemical calculations. Given the fact that the electron transfer during the oxidation of the particles takes place from **HOMO**, then the comparison of values E_{HOMO} for aniline and particle ***IV*** (−8.2135 against −9.2442 eV) shows, that the transformation ***I*** → ***II*** is more probable compared to the ***IV*** → ***V*** process.[84] Therefore, at the excess of aniline in the reaction mixture in the beginning, its full primary oxidation takes place and then possible further conversion of monomeric particles that leads to the formation of the polymer.

Analysis of the electron occupancy of atomic orbitals (**AO**) of nitrogen atom, conducted in accordance with the Mulliken's method[85] shows the significant delocalization of electrons in a molecule of aniline and intermediates of its oxidation.[84] In particular, on the 2*s* **AO** of nitrogen into molecule of aniline (***I***) is 1.427 electrons instead of two, while on the p_z **AO** is 1.841 electrons instead of one (Table 1.1). This permits to consider the electronic structure as *sp*-hybridization, since 0.574 electrons move from 2*s* to 2*p* level. This transition takes place mainly on $2p_z$ level due to the same type of symmetry (axial) for *s*- and p_z types of orbitals. It should be noted, that the redistribution of the electron density in the molecule of aniline causes the delocalization of electrons, so even the hydrogen atom in *ortho*-position of benzene ring carries an excess positive charge (to 0.108 of elementary charge). In the particles–intermediates of the oxidation process, the delocalization of the electron density is increased and the charge on the same hydrogen atom exceeds 0.199 of elementary charge. The strongest change of the charge is observed for the nitrogen atoms, in particular, from 0.051 (neutral molecule of aniline ***I***) to 0.464 and 0.795 of elementary charge (cation-radical ***II*** and dication ***III***, respectively).

1.2.2 ANALYSIS OF THE STAGE OF RECOMBINATION OF INITIAL RADICAL PARTICLES

The next stage of the transformations during an oxidative polymerization of aniline is the formation of dimers as the result of recombination of initial radical products of the aniline oxidation, namely particles ***II***, ***IV***, ***VI***, and ***VII***. It should be noted, that the radical of imine type (***IV***), can take part in the chain transfer to another molecule of aniline as a result of hydrogen atom detachment from it, the most probable in *para*-position (Eq 1.5) which can also takes part in recombination reactions.

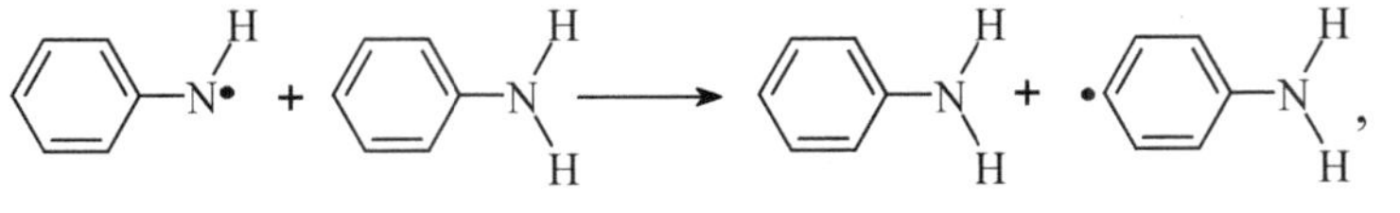
(1.5)

Moreover, it can be assumed, that the isomerization of initial radical particles (Eq 1.3 and 1.4) is a sequence of the processes analogous to the Eq 1.5, with the following rearrangement of the formed particle into the intermediate of quinoid type.

Also, it should be noted that with the use of the methods of mass- and ion-spectrometry, the existence of two isomeric ions of aniline with a positive charges on the nitrogen or carbon atoms was revealed.[86] This means that the protonation of aniline can occur not only at the place of the nitrogen atom but also carbon atoms of the benzene ring. Quantum chemical calculations performed on *DFT* and *MP*4-levels confirmed the possibility of the proton addition to the carbon atom in the *para*-position of the molecule of aniline.[87] Conducted at this analysis on the basis of the orbital Fukui indices did not give the clear answer to what is protonation of nucleophilic centers—carbon or nitrogen atom in the *para*-position prevails—because the probability of both processes are virtually identical. Therefore, in our opinion, the oxidation of aniline in theory is possible also in the place of carbon atom in the *para*-position of the benzene ring that will occur in accordance with the scheme.[88]

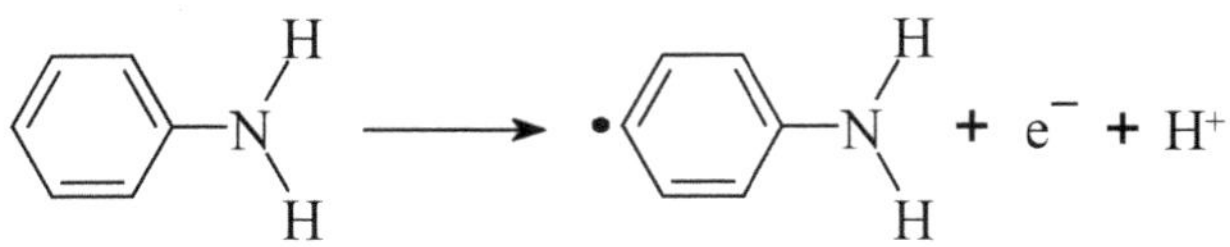

(1.6)

At this, in the scheme of the polymerization transformations of aniline is superfluous the stage of isomerization of nitrene-radical, and the formation of dimer particles will occur through direct recombination of nitrene and *para*-aminobenzene radicals. However, since the existence of the latter particle experimentally was not confirmed, the possibility of the process (1.6) we have not taken into account.

In order to standardize, the possible processes of recombination can be classified as the interaction of type "head to head" (*h–h*), "tail to tail" (*t–t*) or cross-connection of type "head to tail" (*h–t*) and "tail to head" (*t–h*). In accordance with the general principles of thermodynamics, the most probable will be the structures that have minimal energy. Therefore, the main condition for the formation of bond between the particles, starting from aniline and ending by polymer, is the less total energy of the forming particle compared with the sum of the energies of initial fragments, which, be combined, actually form the particle–product, that is,

$$E_s < \sum E_{f,i} \tag{1.7}$$

where E_s is the energy of the compound formation as a result of recombination; $E_{f,i}$ is the energy of initial fragments (radicals or cation-radicals).

According to the results presented in Table 1.2, the ratio (1.6) in the case of recombination of cation-radicals is not satisfied as to the total energy of the particle and binding energy regardless of the recombination type of particles with each other (direct, namely *h–h* and *t–t* or cross, type *h–t* and *t–h*). This result may be due to electrostatic repulsion between of the same charged particles. Exactly by this interaction can be explained the big values of total energy for dimers formed as a result of the interaction of "head to head." In the case of uncharged radical, it is observed a clear dependence of E_{tot} and E_{bin} on the type of linkage of initial radicals: the highest values are observed for dimers formed by connection type "head to tail." Moreover, the energy of dimers of type "head to tail" and "tail to head," which are formed from the cross-connections, is also different; energetically more favorable is the latter structure, as evidenced by the corresponding values of E_{tot}, $E_{bin,}$ and E_{el}. Therefore, based on quantum chemical calculations, it can be predicted that the recombination will easily take place in the case of initial radicals and their linkage will occur as a result of the cross-interaction. The recombination of the type "tail to head" is more energetically favorable compared with the interaction of "head to tail." E. Geniés and M. Łapkowski[70] suggested that the formation of **PAn** is the result of recombination of cation-radicals according to the type "head to tail" with the simultaneous deprotonation.

$$2\ C_6H_5 - NH_2^{+\bullet} \longrightarrow C_6H_5 - NH - C_6H_4 - NH_2 + 2\ H^+ \qquad (1.8)$$

However, the minimal values of energy do not agree with the uncharged dimers (Table 1.2). Therefore, the formation of structure

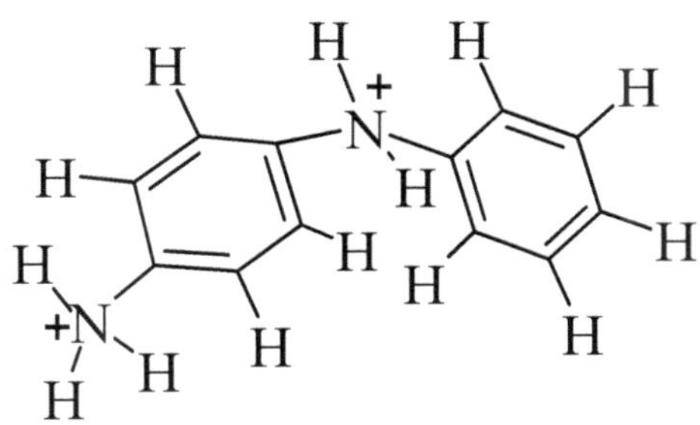

which is a product of recombination of cation-radicals (*II* + *VI*) of type "head to tail" and the structural unit of future polymeric chain looks more likely.

The proposed mechanism of oxidative polymerization of aniline describes only the main stages of this process, based on the fact that it is developed in accordance with the most likely scheme of the transformations, namely: monomer → dimer → tetramer → ... → oligomer → ... etc. and with the formation, as a result, of the conductive polymer having the unbranched structure. It should be noted at once, that such a mechanism is implemented in highly acidic (pH < 2) medium. At higher pH values, the interaction between the monomeric and di- or trimeric intermediates, as theoretically was shown by J. Stejskal et al.,[89] could lead to the formation in a reaction mixture along with the polymer also low molecular oligomeric products, as evidenced by the fact that the yield of the polymeric product is always much lower than 100%. In addition, as a result of intramolecular oxidative cyclization, the formation of branched tetramers of aniline such as pseudomauveine,[89] which is well-known by-product of the oxidation of aniline,[90] is also possible which, however, may also participate in the PAn's chain propagation.[91] The formation of these and other similar products, which is occurred at pH = 2.5–10.0 and/or weak oxidant is analyzed in detail in *ref.*[92,93]

TABLE 1.2 Energy Changes in the Result of the Recombination of Initial Radicals.

Initial particles	Total energy of starting particles, kJ·mole−1			Type of dimer	Energy of dimers, kJ·mole−1		
	$-E_{tot}$,	$-E_{bin}$,	$-E_{el}$,		$-E_{tot}$	$-E_{bin}$	$-E_{el}$
II + *II*	188,057	10,941	812,863	*t–t* (*VI* + *VI*)	187,595	10,479	1,206,770
				h–h (*II* + *II*)	205,452	10,907	1,182,070
				h–t (*II* + *VI*)	205,402	10,781	1,204,690
IV + *IV*	186,349	11,759	777,434	*h–h* (*IV* + *IV*)	186,417	11,828	1,153,140
				t–t (*VII* + *VII*)	186,530	11,858	1,110,970
				h–t (*IV* + *VII*)	204,334	11,916	1,152,380
II + *IV*	187,203	11,350	795,148	*h–t* (*IV* + *VII*)	205,038	11,518	1,174,790
				t–h (*VI* + *II*)	205,042	11,527	1,182,140

1.2.3 AN INTERACTION ANILINE—PEROXYDISULFATE

Chemical method of the **PAn**'s synthesis, particularly as a result of oxidation of the initial monomer by peroxydisulfate in acidic aqueous solutions is probably the most common. Moreover, since the processes of chemical and electrochemical oxidation of aniline are occurred almost identically, and the obtained products are neared both in structure and physical–chemical properties, it is likely that the radical intermediates which are formed during the electrochemical and chemical processes are also identical. Therefore, the chemical oxidation of aniline with peroxydisulfate ions has been analyzed separately.[84]

It is well known that the aromatic amines in combination with the peroxides form a redox system. The first stage of the process is the formation of complex "**PAn** – peroxydisulfate anion," the decomposition of which further leads to the generation into a system of radical nature particles, which respectively initiate further polymerization.[94] The possibility of the formation of complex was investigated by us using the method of cyclic voltammetry. In particular, the influence of the peroxydisulfate admixtures on the process of electrochemical oxidation of aniline has been analyzed. According to the cyclical voltammograms of aqueous solutions of aniline, which are shown in Figure 1.3, after adding $(NH_4)_2S_2O_8$, the position of the first maximum of oxidation (−(0.2–0.24) V) virtually is unchanged compared to the cyclical voltammograms in the absence of peroxydisulfate ions. The height of this maximum increases with the increase in the concentration of peroxydisulfate because it accelerates the formation of **PAn**, which was recorded visually.

At the same time, an impact of the aniline's additives on the reducing process of $S_2O_8^{2-}$ anions is much appreciable. In this case (Fig. 1.4) for the process, it was observed the shift of potential toward more positive potentials. In particular, at the addition of 0.01 M of aniline solution to 0.01 M $(NH_4)_2S_2O_8$ solution, the reduction potential of $S_2O_8^{2-}$ anions is shifted from −1.47 to −1.05 V. But, an increase of the concentration of aniline to 0.02 M does not lead to further significant shift of the reduction potential of the $S_2O_8^{2-}$ anions. The highest rate of reduction was observed at addition of $2 \cdot 10^{-3}$ M aniline solution when the magnitude of the oxidation current reaches a value 41 μA. This effect may be due to the catalytic influence of small amounts of aniline, while a further increase of its content can inhibit the active centers on the surface of the electrode.

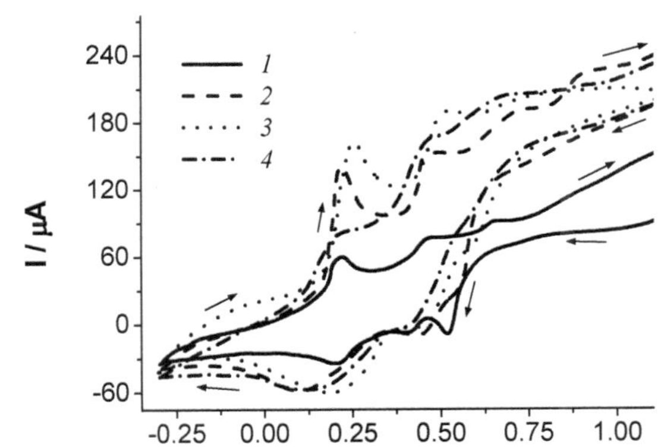

FIGURE 1.3 Cyclical voltammograms of platinum electrode in 0.1 M solution of aniline + 1 M aqueous solution of HCl in the presence of $(NH_4)_2S_2O_8$ at the concentration, M: 0.0 (1); 0.01 (2); 0.02 (3), and 0.04 (4) (s_E=20 mV·s^{-1}).

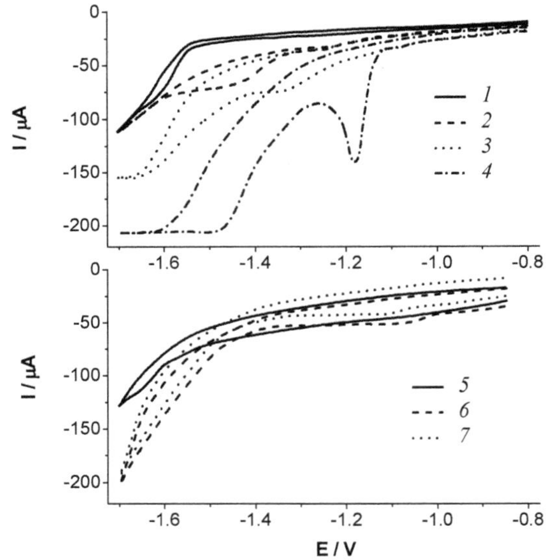

FIGURE 1.4 Cyclical voltammograms of platinum electrode in aqueous solution of aniline + 0.01 *M* $(NH_4)_2S_2O_8$ + 1 M KCl at the concentration of aniline, *M*: 0.0 (1); $1\cdot10^{-3}$ (2); $2\cdot10^{-3}$ (3); $5\cdot10^{-3}$ (4); $8\cdot10^{-3}$ (5); $1\cdot10^{-2}$ (6) та $2\cdot10^{-2}$ (7) (s_E=20 mV·s^{-1}).

Maximal shift of the potential is observed at equimolar (1:1) ratio of aniline and peroxydisulfate, so, obviously this is the stoichiometric ratio of the components in the intermediate reactive complex. For such systems, the donor–acceptor nature of this complex is the most likely. It is obvious that in a couple "aniline–peroxydisulfate" the donor is aniline, while the acceptor of the electrons is peroxydisulfate (its peroxide group). The values calculated by us of total energy and binding energy for the complex **An**–$S_2O_8^{2-}$ were lower on 89.25 and 89.41 kJ·mol^{-1}, respectively, compared with isolated initial particles, and the value of the enthalpy of complex formation[84] $\Delta_f H$ was −93.65 kJ·mol^{-1}. Since the values of the all energies are decreased with the formation of the complex, it shows its energy advantage.

The decomposition of complex An-$S_2O_8^{2-}$

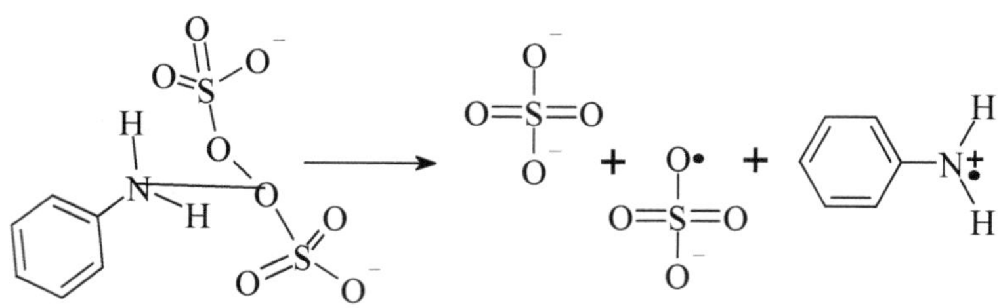

(1.9)

leads to the formation of aniline's cation-radical, $SO_4^{\bullet-}$ anion-radical and SO_4^{2-} anion. Further, the particles of radical nature can enter into the secondary chemical transformations. In our opinion, the anion-radicals can recombine both with the cation-radicals of aniline

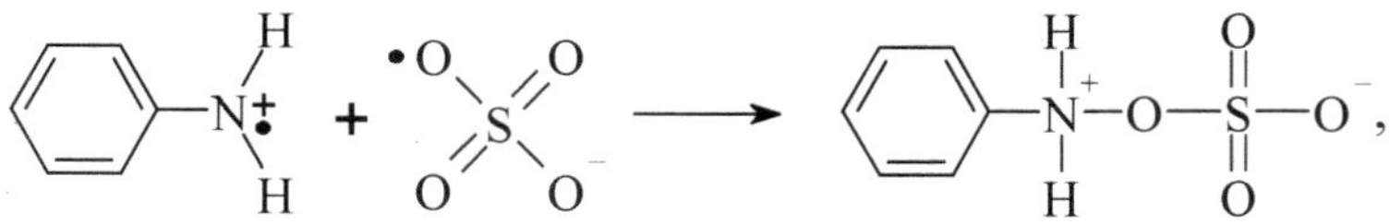

(1.10)

and also to oxidize the neutral aniline to radical ***IV***.

$$C_6H_5 - NH_2 + SO_4^{\bullet-} \longrightarrow C_6H_5 - \dot{N}H + HSO_4^- \tag{1.11}$$

At the same time, it is possible that the simultaneous occurrence of these two processes, resulting in the formation the dianiline sulfate, which will next be decomposed into aniline's radical and hydrosulfoaniline:

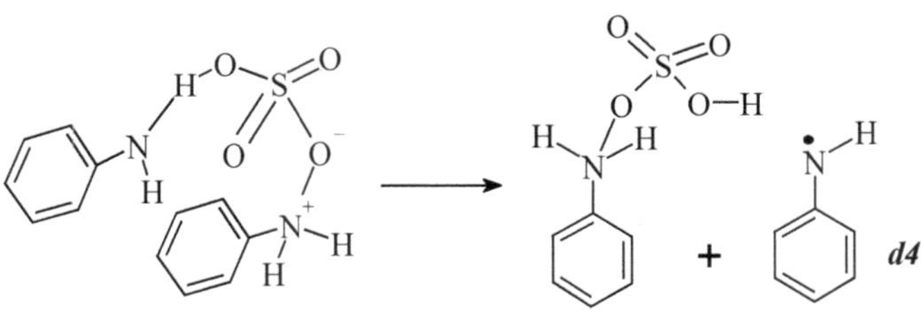

(1.12)

In accordance with S. H. Glarum and J. H. Marshall,[73] during the electrochemical oxidation of aniline in the range of electrode potentials +(0.05–2.00) V it was found the greatest intensity of the ESR signal, which indicates the presence of paramagnetic particles. In this case, the ESR spectrum contained only one absorption line with the splitting factor equal to 2.00270 ± 0.00005. It was assumed that this signal corresponds to the product of one-electron oxidation of aniline. Therefore, the assumption about the formation of particle ***IV*** during the chemical oxidation of aniline is quite reasonable.

1.3 AN IMPACT OF THE SUBSTITUENT'S NATURE ON MECHANISM OF THE SYNTHESIS, STRUCTURE, AND PROPERTIES OF POLYANILINES

In view of the above-proposed mechanism of polymerization, the significant effect on the rate of the process and the structure of products will have the nature and position of the substituent in benzene ring of aniline. Therefore, we have studied the process of polymerization of nitro-[84,94] and oxymethyl[95,96] derivatives of aniline. This choice was caused primarily by the difference in electronic properties of the substituents, since as NO_2-group has strong electron-acceptor properties, whereas the group CH_3O-group possesses electron-donor properties as well as by voluminosity of the substituents and, therefore, by possible creation of spatial difficulties during the polymerization process at the stage of the radicals recombination.

1.3.1 COPOLYMERS OF ANILINE AND NITROANILINES

An attempt of chemical synthesis of polymeric nitroanilines was performed in 1 M HCl aqueous solutions, using the $(NH_4)_2S_2O_8$ as oxidant.[94] Despite the fact that the color change of reaction mixture clearly pointed to the interaction between the components of the system, but the polymer formation even after 24 h of stirring the reaction mixture was not observed. In view of the solubility of the products of interaction, it was concluded that the polymerization process is stopped at the stage of the formation of dimers. Moreover, it was found that the introduction of *ortho*-nitroaniline in the reaction mixture during the chemical synthesis of aniline significantly reduces the yield of polymeric product. At this, the results presented in Table 1.3 show that the yield of polymeric product the smaller, the higher is content of the nitro-derivative in the initial polymerization mixture and thus nitro-aniline derivatives inhibit the radical chain process. It can be assumed that in this case, the initial cation-radical of nitroaniline reacts with the same particle on the place of oxygen atom of nitro-group. As a result, poorly reactive dimer of nitroaniline is formed, unable for further polyaddition reactions, and therefore the polymerization process terminates at the stage of the dimers formation.

TABLE 1.3 The Yield of **PAn** and of Copolymer Aniline: *o*-Nitroaniline During Their Chemical Synthesis Depending on Molar Ratio of Monomers (*r*) into Reactive Mix at the Use of $(NH_4)_2S_2O_8$ as Oxidant (the Duration of the Synthesis was 24 h), and also the Conductivity of the Obtained Samples.

***r*(aniline:*o*-nitroaniline)**	10:0 (**PAn**)	9:1	8:2	7:3	6:4	5:5
Yield of the product/%	75.2	66.9	63.6	53.3	50.3	51.1
Conductivity/*S·cm*$^{-1}$	$3.02\cdot10^{-3}$	$1.06\cdot10^{-3}$	$1.39\cdot10^{-3}$	$3.35\cdot10^{-4}$	$2.25\cdot10^{-4}$	$1.09\cdot10^{-5}$

In order to interpret further of the obtained results, the copolymerization of aniline with various nitroanilines was studied by cyclic voltammetry. It turned out that unlike aniline (Fig. 1.1) for which there are three peaks of currents oxidation, for nitroaniline only one weak peak is characteristic in the vicinity of +0.2 V (Fig. 1.5), and then the working electrode scanning in the range of potentials (−0.3) − (+1.3) does not result in the formation of polynitroanilines on the electrode surface. At the same time, the significant changes was not observed on the cyclical voltammograms of acidic aqueous solutions of aniline at the introduction into them of additives of *o*-, *m*- or *p*-nitroanilines (Fig. 1.6), which may be evidence of the similarity of chemical or electrochemical oxidation of aniline and nitroanilines.

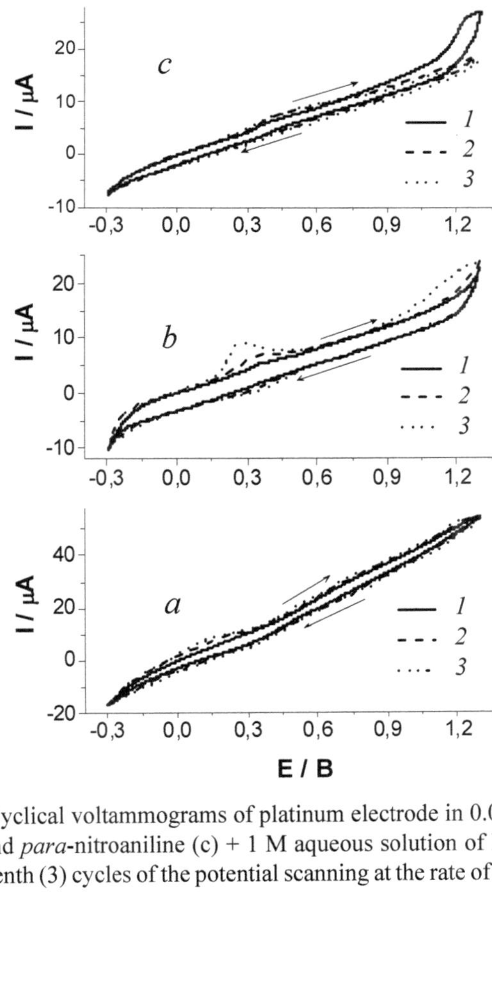

FIGURE 1.5 Cyclical voltammograms of platinum electrode in 0.01 M solutions of *ortho*- (a), *meta*- (b), and *para*-nitroaniline (c) + 1 M aqueous solution of HCl during the first (1), second (2), and tenth (3) cycles of the potential scanning at the rate of scanning s_E=20 mV·s^{-1}.

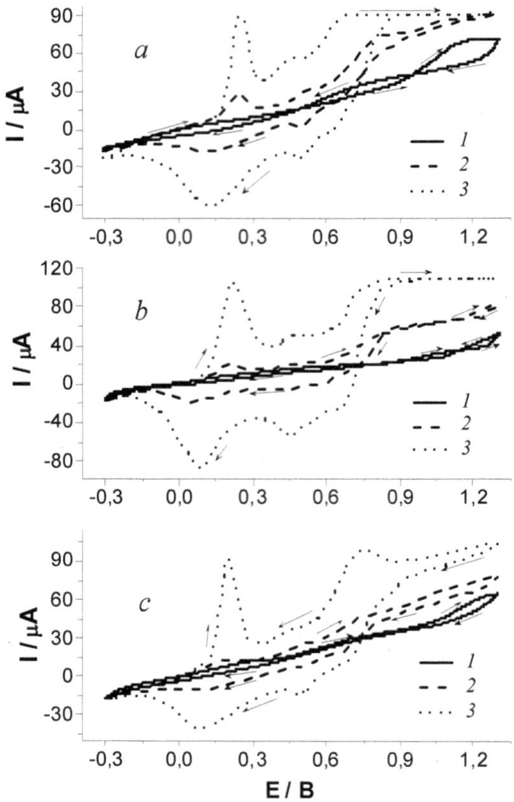

FIGURE 1.6 Cyclical voltammograms of platinum electrode in 0.01 M solutions of aniline + 1 M aqueous solution of HCl with the additives of 0.01 M solutions of *ortho*- (a), *meta*- (b), and *para*-nitroaniline (c) during the first (1), third (2), and fifth (3) cycles of the potential scanning (s_E=20 mV·s^{-1}).

1.3.1.1 SPECTRAL PROPERTIES AND THE STRUCTURE OF POLYANILINE AND COPOLYMERS OF ANILINE WITH ITS NITRO-DERIVATIVES

The structure of the synthesized samples of PAn and copolymers of aniline with *o*-nitroaniline was analyzed on the basis of their spectral properties.[94] FTIR-spectra of the investigated samples, pressed into tablets with KBr, were recorded in the spectral range 4400–400 cm^{-1} using the spectrophotometer BRUKER IFS 66. For obtaining Raman spectra, the spectrophotometer BRUKER FRA106 was used with cooled liquid nitrogen, MCT detector and Nd: YAG laser (λ = 1064 *nm*) as the excitation source.

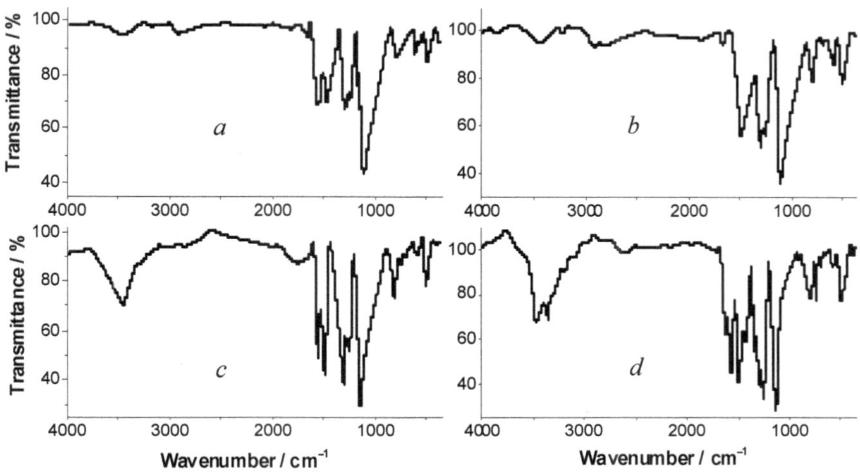

FIGURE 1.7 FTIR spectra of samples of polyaniline (a) and copolymers of aniline and *o*-nitroaniline at molar ratio of the monomers 9:1 (b), 8:2 (c), and 7:3 (d).

Analysis of IR spectrum of PAn obtained by chemical method (Fig. 1.7a), testified that it represents by itself the oxidized form of emeraldine. This conclusion is based on the presence of absorption bands in the spectrum corresponding to the valence stretching vibrations of bonds C=N (bands at 3434 and 1659 cm^{-1}), $^{+}$N–H (2921 and 1291–1238 cm^{-1}) and C–N= (1103 and 4963 cm^{-1}),[97] typical for this modification of **PAn**. The presence of >C=N^{+} group in the structure of **PAn** was determined based on the Raman spectrum (Fig. 1.8), where there is a clear splitted band at 2200–2100 cm^{-1}. Wide band in the range of 3500–2300 cm^{-1} (Fig. 1.8) can be attributed to stretching vibration of unoxidized group >C=N–. In addition, the band in the same area can generate the vibrations of C–N group, and also quaternizated nitrogen atom, whose structure is similar to unoxidized electrically neutral atom. This allows us to propose the following approximate formula for the synthesized sample of PAn

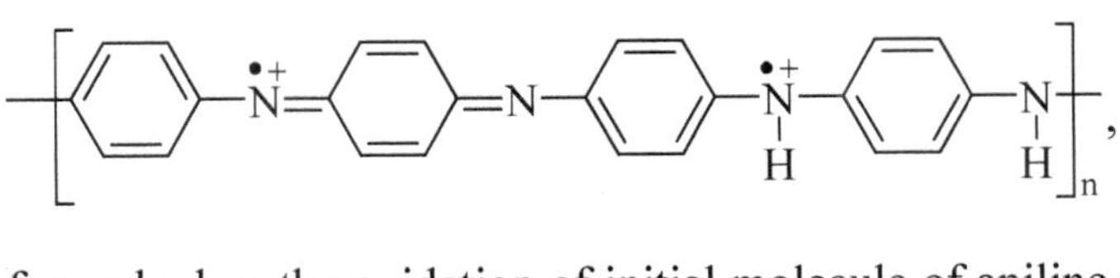

which is formed when the oxidation of initial molecule of aniline proceeds as the result of electron transfer from the 2*s* sublevel of the nitrogen atom.

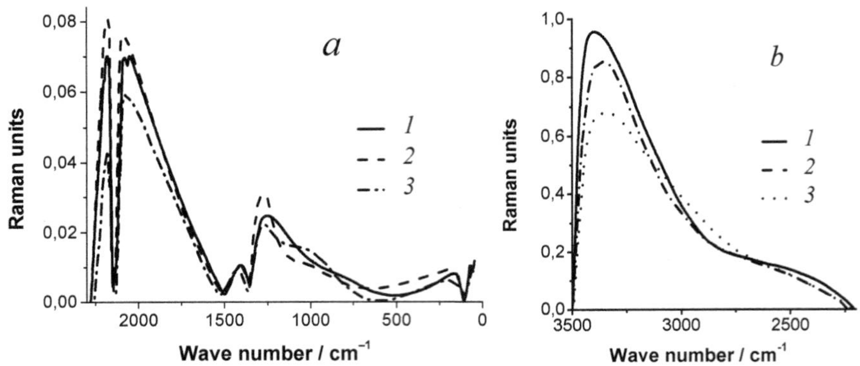

FIGURE 1.8 Raman spectra of polyaniline (1) and copolymers of aniline and *o*-nitroaniline at molar ratio of monomers 8:2 (2) and 7:3 (3) in the spectral areas 3500–2250 (a) and 2250–2270 cm^{-1} (b), respectively.

Analysis of the spectra of copolymerization products obtained at different molar ratios (*r*) aniline : *o*-nitroaniline in the reaction mixture (Fig. 1.7b–d) is complicated due to partial overlapping of the absorption bands corresponding to nitro-group (ν_{as} = 1650–1500 cm^{-1} and ν_s = 1370–1250 cm^{-1}), with the bands corresponding to valence stretching vibrations of C–N bond (at 1360–1250 cm^{-1}) and deformation vibrations of (–HN^+–) group (1600–1500 cm^{-1}). The main differences in the spectra of copolymerization products are observed in the range of wave numbers from 1800 to 1200 cm^{-1} and in the high-frequency region 3000–2750 cm^{-1}. Polymerization of the initial mixture at r = 7:3 leads to the appearance in spectrum of the product the band at 1626 cm^{-1}, which can be attributed to asymmetric stretching vibrations of NO_2 group. At the same time, for all the investigated ratios of the components in the spectrum, it is evident that band at 616–582 cm^{-1} is associated with the presence of SO_4^{2-} ions in the product. The form and the intensities ratio of the adsorption bands at 3442–3476 and 2900–300 cm^{-1} are varied depending on the initial composition of the reaction mixture. So, for the ratio of 7:3, the band at ~3400 cm^{-1} is splitted into two maxima at 3476.4 and 3353.9 cm^{-1}. Since the frequencies of the –HN– ta –HN^+ groups vibrations correspond to these bands, it indicates the presence both of oxidized and unoxidized links in a polymer chain.

Comparison of Raman spectra (Fig. 1.8) allows us to follow the trend of the changes in the oxidation level of polymer on place of N=C group, using a decrease in the intensity of the absorption band at 2173 cm^{-1} and increase of the bandwidth with a maximum at 2060 cm^{-1}. At the same time, the intensity of the band at 1282 cm^{-1}, which corresponds to asymmetric vibrations of >C–N< group increases, suggesting the possibility of some decrease of oxidation level of products of compatible polymerization of aniline and *o*-nitroaniline compared to homopolymer of aniline. In spectra both of **PAn** and copolymer, the band at 1432–1488 cm^{-1} is observed, which is associated with the vibrations of N=N group.[67] The presence of N–N bonds in the structure of polymer indicates the possibility of recombination of the type "head to head" at the stage of the dimers formation. The possibility of such units' connection in the polymer chain was also considered in *ref.*[67] Thus, according to IR spectroscopic studies the fragment of polymeric chain containing the nitroaniline link can be described as follows:

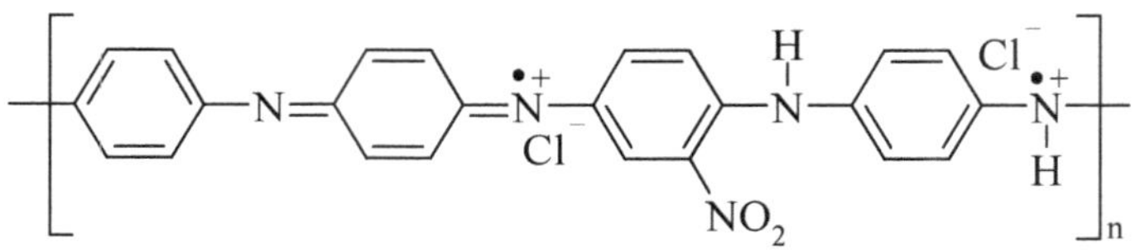

(1.13)

Exactly this structure was used by us to interpret the data of element analysis for the calculation of theoretical composition of copolymer.

It is well known that the neutral PAn, no matter in which of the three forms it is existed, is the typical insulator with a size of an conductivity ~10^{-10} $S\cdot cm^{-1}$, and its transformation into the high-conductive state (doping) can be made by chemical or electrochemical ways, otherwise acidic or oxidative.[2] With the use of the method of impedance spectroscopy, two components of the conductivity of PAns were found, namely electronic and ionic.[98] In order to explain the electroconductivity in terms of electronic component when the charge is transferred by the electrons of the conduction band of **PAn**, the notion of "*solitons*" and "*polarons*" is used.[1] However, in our view, the spread of the soliton hypothesis proposed for the processes for polyacetylene doping, on the other, including the PAn, conductive polymers, are not sufficiently justified, as this virtual structures is the result of the *Schrödinger* equation solution. At the same time, the polarons and bipolarons, whose formation is a high probable at the initial stages of the polymerization process, can be interpreted as a cation-radical center or positively charged center in the singlet state in the polymer chain with a conjugated system of π-bonds, which are often used in chemistry. The increase of electronic conductivity of PAn, which is determined by the number of paramagnetic centers on the chain (polarons or bipolarons), can be achieved during its oxidation. On the other hand, the conductivity of **PAn** increases also as a result of the processing of PAn base with strong acids, (so-called acidic doping with the formation of the salt of appropriate forms of **PAn**) due to the quaternization of nitrogen atom of imine group as a result of donor–acceptor interaction with H^+ ions. In this case, the charge will be transferred mainly by counterions. In turn, ionic component of the conductivity associated with such macro-characteristics of polymer as the porosity, crystallinity degree, orientation of macro-chains, globularity, or fibrillation of the structure of polymer. Moreover, when the result of oxidation of **PAn** is the formation of polarons or bipolarons, then the charge on them is compensated by the ions, which are transferred from the volume of the electrolyte solution through the pores of the surface or due to the deprotonation of imine groups of the polymer and, thus two components of the conductivity of polymers with the system of conjugated π-bonds are interrelated.

An increase of the number of unpaired electrons results in the growth of electronic component of the conductivity for the all forms of PAn. At the same time, the ion component increases with the increase of the number of charges on the chain, which are localized on the quaternized atoms of nitrogen. However, given the possibility of acidic doping and the fact that the polarons is too short-lived particles, it must be concluded that the structure (1.13) cannot exist long. Moreover, according to data presented in Table 1.3, the conductivity of synthesized samples does not reach the values 1–5 $S\cdot cm^{-1}$ that corresponds to 50th% degree of the acidic doping.[99] Therefore, the structure of synthesized samples likely can be described by the formula

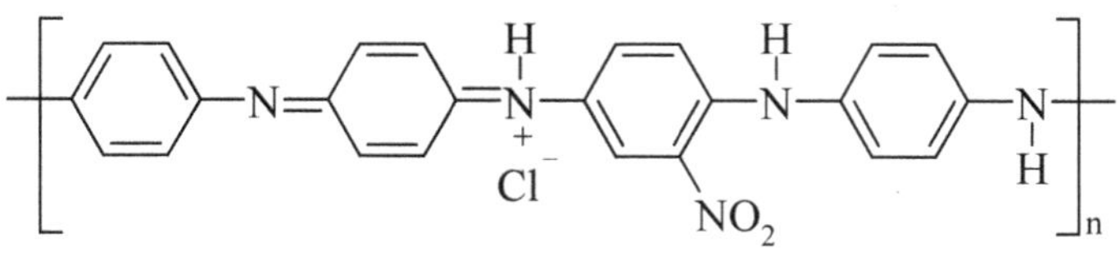

(1.14)

which corresponds to the H^+-doped form of emeraldine, namely emeraldine salt with the doping level of polymer 25%, resulting from the values of the conductivity of obtained polycondensates.

1.3.1.2 ELEMENT ANALYSIS AND THERMOGRAVIMETRY OF THE SYNTHESIZED COPOLYMERS

The results of the element analysis of the obtained samples are represented in Table 1.4. It should be noted at once, that the experimental values of the carbon content were on 1–6 *mass.*% lesser compared with the theoretical values calculated in accordance with the above-proposed structure. The content of chlorine and nitrogen was undervalued compared with hydrogen the content of which, in turn, was somewhat overvalued. In addition, it was found, that the total content of the elements (C, H, and N) in the sample was only 85 and 95% for the copolymer and **PAn**, respectively. However, similar results also were obtained by other researchers.[100,101] According to T. Kobayashi et al.[100] these experimental data can be explained by the presence of water in the polymer even after its drying under vacuum, the content of which can be ranged from 2 up to 17 *mass.%*. So, if to carry out the **PAn**'s synthesis under special conditions, then total content of the elements can reach 97%.[102] In addition, the results presented in Table 1.4 show that both the PAn and the all samples of copolymers contain sulfur. As a result of electrostatic binding of HSO_4^{2-} anions (the product of the reduction of peroxydisulfate, which was used for the synthesis as an oxidizing agent), to remove these anions from the samples by repeated washing with water is practically impossible. So, the anions HSO_4^{2-} thus have the function of doping agent (counter-ion) of **PAn**. On the other hand, it is also possible chemical binding of the sulfur in the structure of the polymer. As it was shown above, at the initial stages of chemical synthesis of **PAn** after the destruction of the An-$S_2O_8^{2-}$ complex, which is accompanied by the transfer of an electron from the aniline's molecule to peroxide group, the recombination of radical intermediates, namely cation-radicals of aniline and sulfate-anion radicals is possible (Eq 1.11) with the formation of N–O bond. The possibility of N–O bond formation confirms the presence into IR-spectrum of the polymerization product the absorption bands in the range 1302–1238 cm^{-1}.

So, if to take into account the analytically determined sulfur, then the total content of the components reaches about ~100%. At this, the excess of hydrogen can be attributed to the presence in synthesized samples of ammonium ions and also water, which was confirmed by the results of the thermogravimetry studies. As shown on Figure 1.9, a slight decrease in mass of the sample (~1.5 *mass.%*) is observed in the temperature range 45–105 °C and related, apparently, with the removal of physically bound water. These data are in good agreement with the results of differential scanning calorimetry,[103,104] according to which the endothermic peak at 30–120 °C corresponds exactly to this process (desorption of water). During heating of the sample from 30 to 500 °C the most significant decrease of mass (~ 35 *mass.%*) was observed in the temperature range 105–250 °C. Spectral analysis of gaseous products of destruction of polymer sample with $r = 7{:}3$ showed that for the basic product characteristic are the absorption bands with maxima at 2359 and 2344 cm^{-1}. Somewhat less intense bands are observed at 668 and 1370–1320 cm^{-1} (Fig. 1.10). If the first two bands can be attributed to stretching vibration of N=C and C=O groups, the following two asymmetric valence vibrations correspond to NO_2 group and an amino-group in aromatic amines.[97] Also, somewhat intense band at 668 cm^{-1} can also be associated with the deformation vibrations of N–H group.

TABLE 1.4 The Results of Element Analysis of Synthesized Samples of Polyaniline and its Copolymers with *o*-Nitroaniline (Analyzer GA 1108, Carlo Erbo).

Sample		Composition, mass.%					
		C	H	N	S	SO_4^{2-}	Σ
Theoretical data							
Polyaniline (emeraldine form)		62.74	4.13	12.20	6.97	20.91	99.98
Copolymer	9:1	57.01	3.76	13.86	6.33	19.00	99.97
(molar ratio	8:2	57.01	3.76	13.86	6.33	19.00	99.97
aniline:*o*-nitroaniline)	7:3	57.01	3.76	13.86	6.33	19.00	99.97
	6:4	57.01	3.76	13.86	6.33	19.00	99.97
Experimental data							
Polyaniline (emeraldine form)		53.05	4.46	10.23	4.35	13.04	85.13
Copolymer	9:1	56.15	4.32	11.58	3.97	11.82	83.87
(molar ratio	8:2	56.00	4.19	12.77	3.81	11.43	84.59
aniline : *o*-nitroaniline)	7:3	54.23	4.16	13.77	4.16	10.52	82.67
	6:4	55.75	4.09	14.21	2.71	8.13	82.18

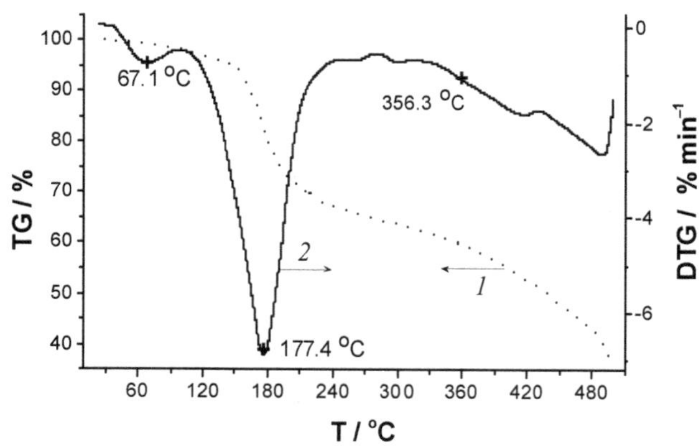

FIGURE 1.9 Integral (1) and differential (2) derivatogram of the sample of copolymer of aniline and *o*-nitroaniline (r = 7:3) (microbalance NETZSCH TG29; atmosphere is Ar; scanning rate of the temperature 15 K·min^{-1}).

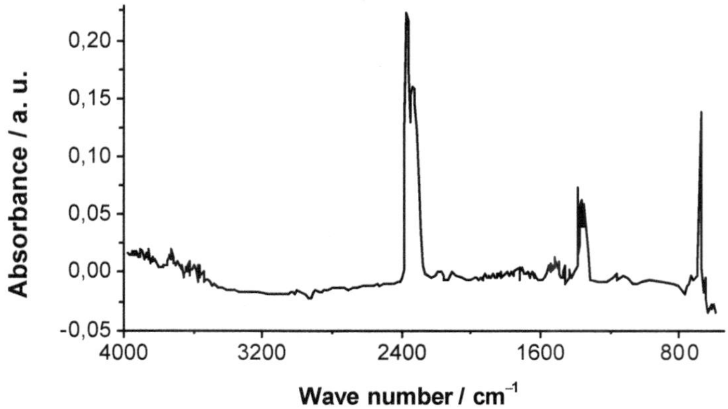

FIGURE 1.10 FTIR absorption spectrum of gaseous products of thermolysis of the sample of copolymer of aniline and *o*-nitroaniline ($r = 7{:}3$) 1364 s after the start of thermolysis.

Electrochemical destruction of PAn was studied in detail by D. E. Stilwell and S.-M. Park.[83] They confirmed fixed earlier[100] fact of the benzoquinone accumulation during the destruction of **PAn**. At the same time, the final product of the destruction depends on the conditions of the process and the nature of the initial sample. In the presence of water, the oxidative hydrolysis of PAn takes place, the main products of which are the quinone and NH_3:

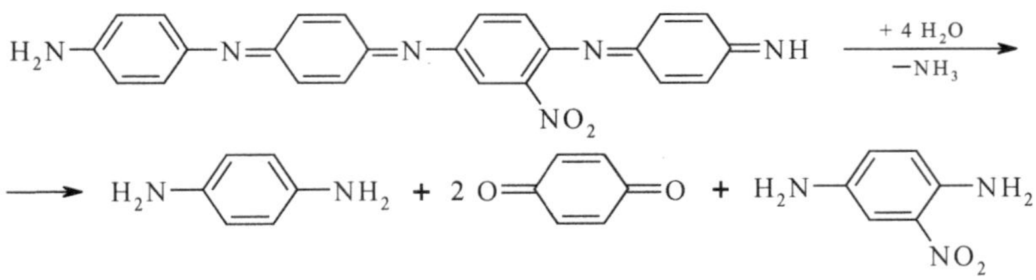

(1.15)

The oxidative degradation of the copolymer, in contrast to electrochemical destruction, passing through another sequence of transformations, which are different from the proposed in *ref.*[83] First, we took into account the presence of oxidant's particles in the sample. In our opinion, during the thermolysis the particles of radical nature may be generated due to the thermal decomposition of adduct, which is formed in reaction (1.10):

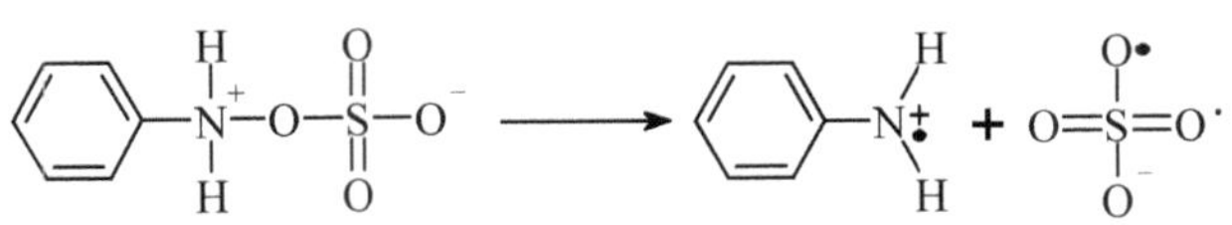

(1.16)

Sulfate-anion radicals may further oxidize partially oxidized polymer (emeraldine form) to pernigraniline:

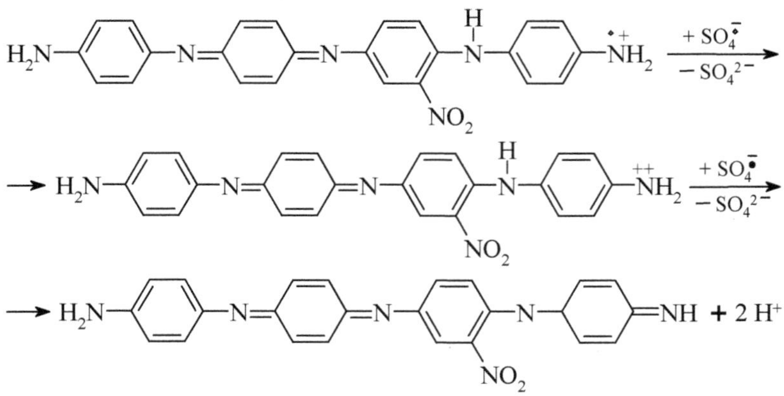

(1.17)

If the emeraldine is oxidized by cation-radicals of aniline, then in addition to pernigraniline also the aniline is formed:

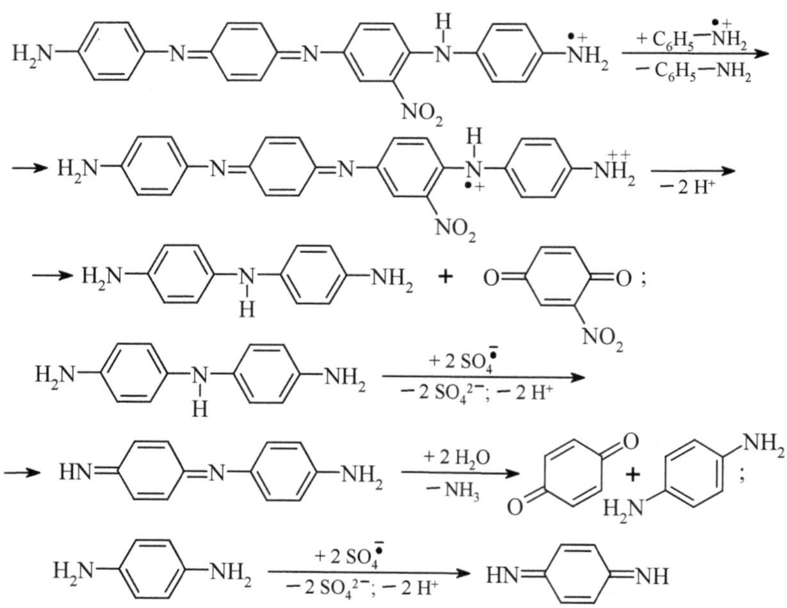

(1.18)

1.3.2 SYNTHESIS AND PHYSICO–CHEMICAL PROPERTIES OF POLYANISIDINES

The researches of various anilines' derivatives are far from the completion. For example, among the all methoxyanilines, the polymerization process only of its *ortho*-isomer is sufficiently thoroughly investigated.[105–107] Therefore, to identify the relationships between the conditions of synthesis and the properties of polymers on the one hand, and the electronic structure of initial monomers on the other hand, we have investigated the oxidative polymerization of isomeric anisidines and the structure of obtained polymers.[96]

1.3.2.1 CYCLICAL VOLTAMMETRY OF ISOMERIC ANISIDINES

Quantify to estimate the processes of the electron transfer from the electroactive particle to electrode allows the cyclic voltammetry. As shown on Figure 1.11, among the three isomeric anisidines, the most easily is oxidized *ortho*-isomer, while the oxidation of *meta*-derivative is substantially difficult: the potential corresponding to the first oxidation wave is shifted from +0.21 V for *ortho*-anisidine to +0.52 V for *meta*-anisidine. Such impact of the position of substituent is true for different concentrations of electroactive monomers, and for different scanning rates of the potential electrode. Starting from 0.01 M initial concentration of monomer, the oxidation of *ortho*-anisidine takes place in several stages, as shown by the presence of three peaks of the oxidation current on the cyclic voltammogram (Fig. 1.11a), similarly as in the case of aniline (Fig. 1.1). Therefore, the first maximum of oxidation current at +0.21 V corresponds to the transition "leucoemeraldine → emeraldine," the second oxidation current at +0.58 V corresponds to oxidation to the pernigraniline form, and the middle oxidation current is associated with the oxidation of products of partial destruction of the polymer or with the cross-linking of the polymer chains.[80] With each next cycle of the potential scanning ,the values of redox currents increase, which indicates the accumulation of the polymer and its subsequent conversion on the electrode surface,[70] so, the anisidine unlike nitro-derivatives of aniline capable to form of homopolymers. Moreover, an oxidation of poly-*ortho*-anisidine is reversible, since the difference between the maxima highs and reduction of polymer oxidation is negligible, which is evidence of the formation of electroconductive polymer with the system of conjugated π-bonds.

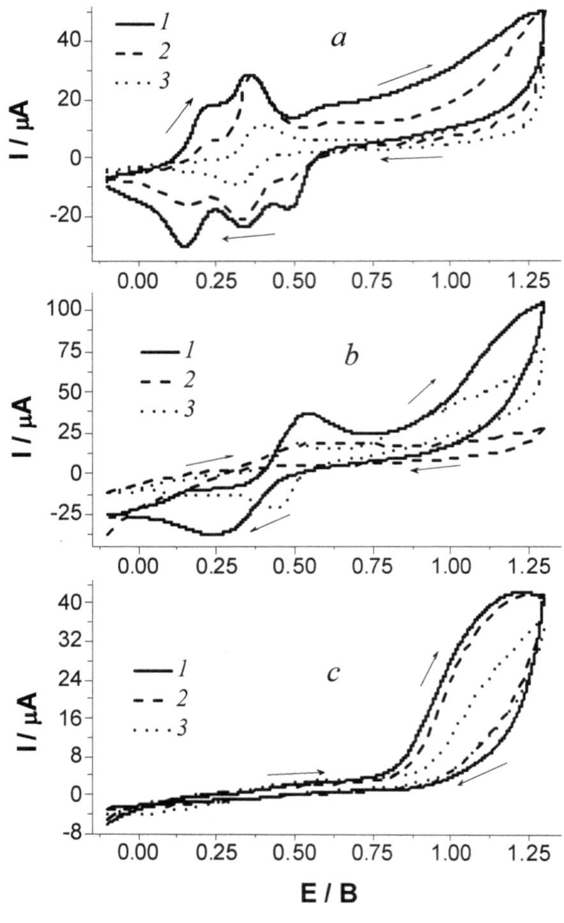

FIGURE 1.11 Cyclical voltammograms of 0.1 (1), 0.05 (2) and 0.01 M (3) solutions of *ortho*- (a), *para*- (b), and *meta*-anisidines (c) in 0.2 M aqueous solution of HCl during the first cycle of the potential scanning (s_E=70 mV·s^{-1}).

In accordance with the value of the first potential of oxidation (Table 1.5), the isomeric anisidines can be placed in the following consequence:

meta-anisidine > *para*-anisidine > *ortho*-anisidine.

At the same time, electron-donor properties of anisidine depend on the localization of electron density on the nitrogen atom of electrically amine-group. The results of the calculations of energy characteristics and the electronic structure of isomeric anisidines and aniline (Table 1.5) conducted using the MOPAC 93 (semi-empirical PM3 level) showed that the values of electronic energy and the energy of molecules binding as well as the enthalpy of formation for the consequence *meta* < *para* < *ortho*-isomer are decreased. In other words, the potential of oxidation increases with the increasing of thermodynamic stability of the anisidine isomers. At the same time, the increasing of HOMO energy and, therefore, reducing of the negative charge on the nitrogen atom of the amino-group indicate the difficulty of the electron transfer process.

TABLE 1.5 An Impact of the Electronic Structure on the Position of the First Wave of Oxidation and Energy Characteristics of Aniline and Isomeric Anisidines.

Compound	$E_{ox}^{(I)}$ **/V**	**$-E_{bin}$/kJ·mol–1**	**$-E_{el}$/kJ·mol–1**
Aniline	0.18	6185.87	407,436.01
ortho-Anisidine	0.21	7757.39	645,210.92
para-Anisidine	0.44	7763.21	645,553.56
meta-Anisidine	0.52	7682.19	647,565.70
Compound	E_{HOMO}/eV	**$\Delta_f H$/kJ·mol^{-1}**	**Charge on atom *N/e***
Aniline	−8.0678	−107.10	0.051
ortho-Anisidine	−8.5662	−63.26	0.077
para-Anisidine	−8.3841	−69.08	0.073
meta-Anisidine	−8.6422	−71.85	0.072

However, an unambiguous relationship between the value of oxidative potential of anisidines and the position of oxymethyl substituent in the aromatic ring of the initial monomer is not observed. Therefore, it can be concluded that the difference between the electrochemical properties of anisidines is determined not only by the redistribution of electron density between the substituents and benzene ring, but also the orientation of molecules in the adsorption layer that is associated with the geometry of the monomer's molecule. Obviously, the electroactive NH_2 groups of *ortho*-isomer of anisidine are oriented profitable compared with *meta*- and *para*-isomers, primarily because of the possibility of interaction with the surface also oxygen of oxymethyl group that promotes the electron transfer from the adsorbed particles on the electrode.

1.3.2.2 THE STRUCTURE OF POLYMERIC ANISIDINES

In accordance with the above-discussed mechanism of aniline's oxidation[84,88] the first stage of the process is the transfer of an electron from 2*s*-orbital of the nitrogen atom with the formation of a cation-radical, which in the presence of base *B* can be accompanied by the deprotonation of amine-groups:

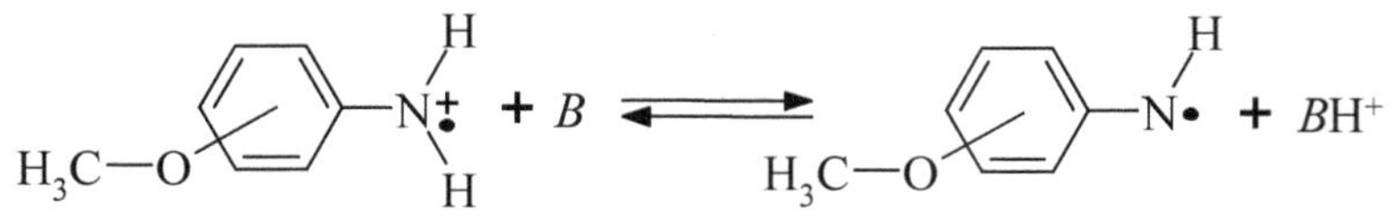
(1.19)

Presented in the Table 1.6, results of quantum chemical calculations of energy parameters of initial radical particles that arise as a result of the oxidation of isomeric anisidines, allow to conclude that according to Eq 1.19, the formation of nitrenium-radicals is advantageous compared with the existence of the corresponding to them cation-radicals. In particular, for the all anisidines heat of formation, electron energy and the energy of binding of nitrenium-radicals are lower compared to cation-radicals, that is, an equilibrium of the reaction (1.19) is shifted to the right.

TABLE 1.6 The Enthalpy of Formation, Energy of Binding, and the Electron Energy Of Intermediates of the Anisidine Oxidation.

Particle	$\Delta_f H$/kJ·mol^{-1}	$-E_{bin}$/kJ·mol^{-1}	$-E_{el}$/kJ·mol^{-1}
$o\text{-}CH_3O-C_6H_4-\overset{+}{N}H$	101.53	302.96	22,993.87
$p\text{-}CH_3O-C_6H_4-\overset{+}{N}H$	105.76	324.02	21,250.69
$m\text{-}CH_3O-C_6H_4-\overset{+}{N}H$	109.99	330.25	20,647.92
$o\text{-}CH_3O-C_6H_4-\overset{+\bullet}{N}H$	708.28	706.90	1190.73
$p\text{-}CH_3O-C_6H_4-\overset{+\bullet}{N}H$	712.47	712.51	613.03
$m\text{-}CH_3O-C_6H_4-\overset{+\bullet}{N}H$	732.73	732.73	17.96

Further recombination of nitrenium-radicals leads to the formation of N–N bond, which must be present in the structure of the polymer. This fact is confirmed by the results of IR-spectroscopy (Table 1.7) of chemically synthesized in 1 M aqueous HCl solutions samples of polyanisidines ($(NH_4)_2S_2O_8$ was used as an oxidant). The band at 1461 cm^{-1} which formerly was observed by S. Patil et al.[107] for electrochemically synthesized samples of poly-*ortho*-anisidine corresponds to the stretching vibrations of N–N bond. In addition, the band at ~1490 cm^{-1} which is associated with the existence of quinoid structure indicates the deeper oxidation of poly-*ortho*- and poly-*meta*-anisidine chains during the chemical synthesis of the polymer.

TABLE 1.7 Analysis of the Absorption Bands of IR-Spectra of Isomeric Anisidines.

Poly-*para*-anisidine		**Poly-*ortho*-anisidine**		**Poly-*meta*-anisidine**	
ω/cm⁻¹	**Vibrations**	**ω/cm⁻¹**	**Vibrations**	**ω/cm⁻¹**	**Vibrations**
3403	$\nu_{N\text{-}H}$	3448	$\nu_{N\text{-}H}$	3424	$\nu_{N\text{-}H}$
3220	$\nu_{N\text{-}H}$	3204	$\nu_{N\text{-}H}$	2937	ν_{-CH_3}; ν_{-CH_2-}
2038	ν_{-NH^+}	1579	$\nu_{N=N}$	2834	ν_{-O-CH_3}
1608	$\nu_{N=N}$; ν_{-N-H}	1488	ν_{C-N^+}	2038	ν_{-NH^+}
1564	$\nu_{C=C}$; ν_{N-H}	1461	ν_{N-N}	1606	$\nu_{N=N}$; γ_{-N-H}
1510	γ_{-N-H}	1255	ν_{-C-O-}; $\nu_{-C-O-CH_3}$	1578	$\nu_{C=C}$; $\nu_{N=N}$
1347	$\nu_{-C-N^{\bullet+}}$	1205	$\nu_{C-N=}$	1493	$\nu_{C=N-}$; ν_{N-H}
1289	γ_{N-N}	1172	ν_{C-N}	1464	ν_{N-H}
1251	ν_{-C-O-}; $\nu_{-C-O-CH_3}$	1179	ν_{C-N}; ν_{N-N}	1254	ν_{-C-O-}; $\nu_{-C-O-CH_3}$
1172	ν_{-C-N}; ν_{N-N}	1046	ν_{C-N}	1207	$\nu_{C-N=}$
1030	$\nu_{-C-N<}$	1017	γ_{C-C}; $\nu_{N=N}$	1158	ν_{C-N}; ν_{N-N}
826	$\gamma_{=C-H}$	828	$\gamma_{=C-H}$	1039	$\nu_{-C-N<}$
520	γ_{C-H}; γ_{N-N}; $\gamma_{C-N=}$	583	ν_{-C-S}	832	$\gamma_{=C-H}$
		547	γ_{C-H}; ν_{N^+-H};	688	γ_{C-C}; γ_{C-H}; γ_{C-N}
		451	ν_{C-N}	580	ν_{-C-S}

The results, which are shown in Table 1.7, demonstrate the similarity of IR-spectra of isomeric polyanisidines. However, for the spectra of poly-*para*-anisidine typical are some significant differences. In particular, only for poly-*para*-anisidine the band at 1347 cm^{-1} is observed, showing the formation of polaron $C{-}N^{\bullet+}$, as it was shown for **PAn**, doped with the sulfate-ions.[108] On the other hand, the spectrum of poly-*para*-anisidine is somewhat simpler. So, it is not observed of the absorption bands in the range of wave numbers 1493–1461 cm^{-1} indicating the absence of the N–N bonds in this polymer.

The spectra of proton magnetic resonance (NMR) of synthesized samples of polyanisidines are also quite similar to each other (Fig. 1.12). However, unlike aniline,[109] the signal from the protons of benzene ring at 7.4–7.7 ppm is weak, while the proton signal of the oxymethyl group at 2.3–2.7 ppm is enough intense. However, this fact was also recorded for the polymer of toluidine.[110]

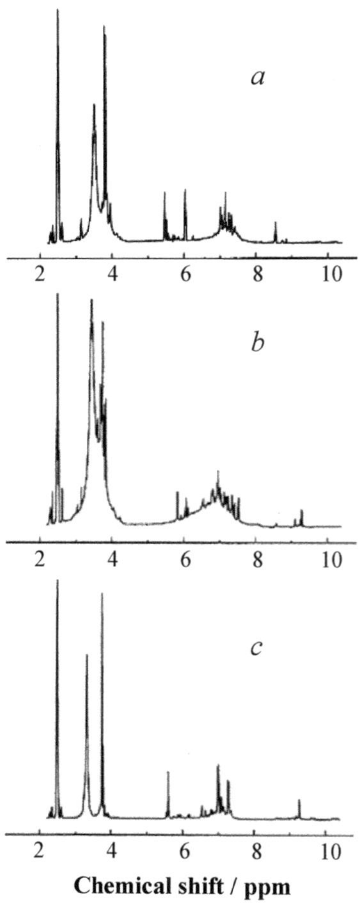

FIGURE 1.12 NMR-spectra of synthesized samples of polymeric *ortho-* (a), *para-* (b), and *meta*-anisidine (c) (400 MHz; spectrometer Varian Whity 500 Plus; solvent is deuterated **DMSO**; external standard is $Si(CH_3)_4$).

The reason for differences in spectral characteristics of polyanisidines is a different mechanism of their formation. The linkage of initial radicals of the type "head to tail" can occur only when the isomerization of nitrenium-radical into quinoid structure takes place. If the isomerization of *ortho-* and *meta*-anisidines easily proceeds due to the stabilization of the quinoid structure at the expense of the conjugation of π-bond of the cycle with the unpaired electron pairs of oxygen of the CH_3O group

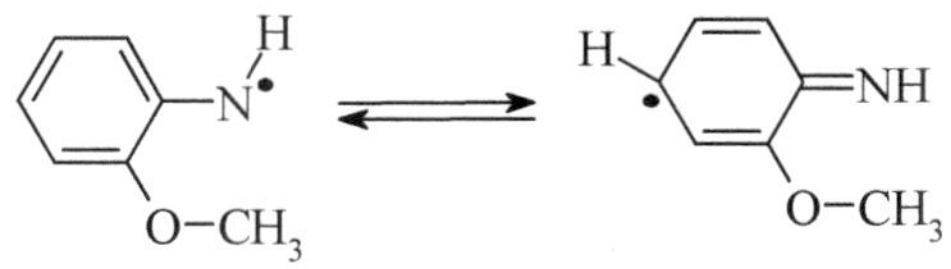

(1.20)

then for *para*-isomer this process

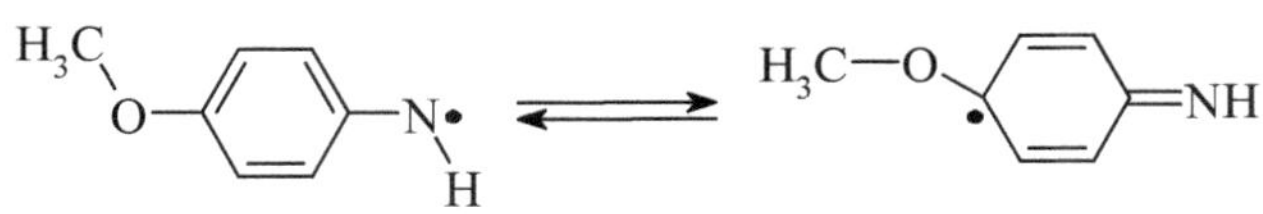

(1.21)

is complicated, and therefore, the formation of the dimer by the interaction of the type "head to head" with the formation of N–N bond, makes further growth of the chain more likely impossible. Moreover, in the case of recombination of "head to tail" type

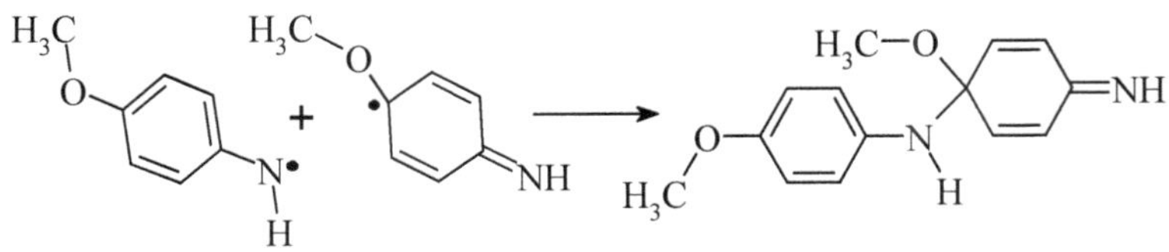

(1.22)

the conjugation in the main chain is broken and thus the conductivity of the polymer is decreased and becomes impossible the re-isomerization of dimer. Therefore, in our opinion, the isomerization of the nitrenium-radical of *para*-anisidine will be occurred differently

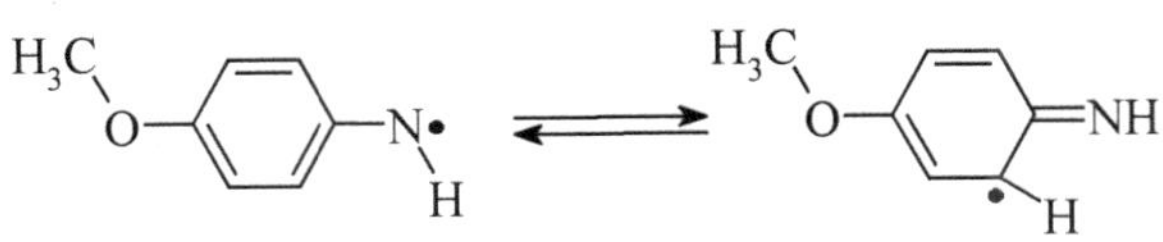

(1.23)

which removes all the previous disagreements. Therefore, the formation of polymer after interaction of initial and isomerized radicals

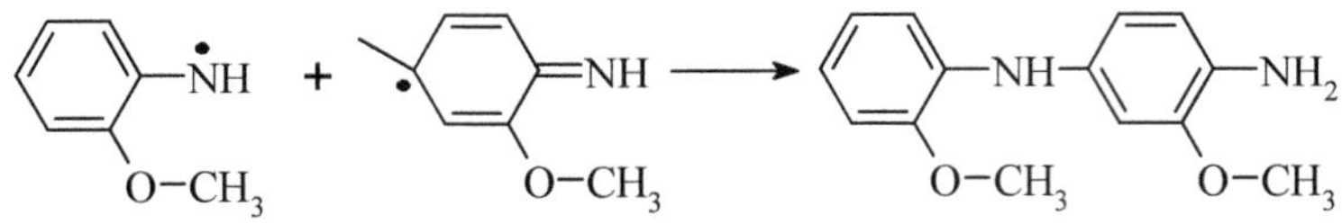

(1.24)

is possible due to the re-isomerization of the formed particle into the structure of benzenoid type with the following its oxidation (according to Eq 1.1) and further interaction with another particle of the radical nature. For example, the trimer formation can be as a result of the recombination of oxidized dimeric particle with isomerized initial radical of anisidine,

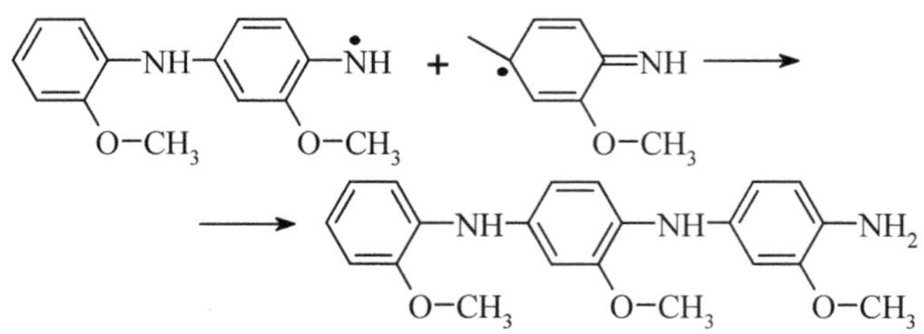

(1.25)

and tetramer as a result of the recombination of two oxidized dimer particles or trimer and initial radicals, similarly as in the case of oxidation of aniline or nitro-anilines.[84,94] This propagation of polymer chain is the result of recombination of oligomeric and monomeric particles, leading to a rapid depletion of the monomer, as it was revealed during the oxidation of aniline with peroxydisulfate in aqueous solutions.[111] Leucoemeraldine (***L***), which is formed in the first cycle of the potential scanning next more consistently is oxidized to emeraldine salt (***ES***) and pernigraniline (***P***):

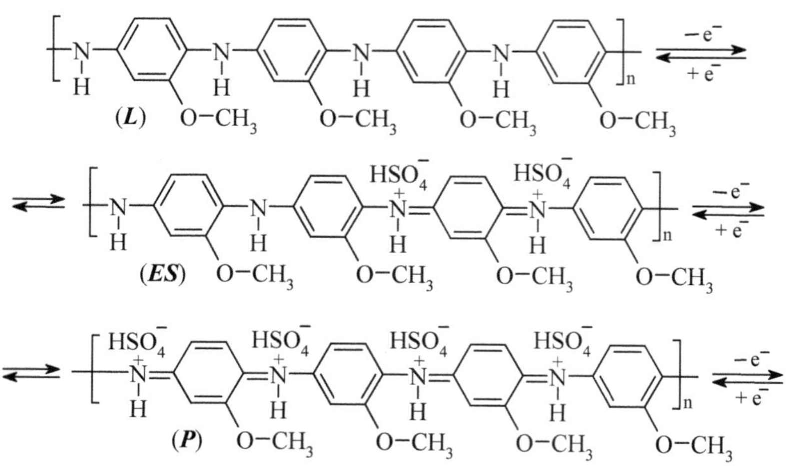

(1.26)

As it can be seen from the data presented in Table 1.8, experimentally found values of the hydrogen content in the synthesized sample of poly-*ortho*-anisidine exceed the theoretical value and vice versa the content of carbon is understated, which may evidence that chemically synthesized poly-*ortho*-anisidine is not as emeraldine base but emeraldine salt (***ES***) or pernigraniline. Such transformations were observed, in particular, for the photoelectrochemical response of **PAn** intercalated in the film of the ethylphosphoric acid.[112] However, there is also another possibility, related with the transformation of initial radicals without the reisomerization of the products of the recombination. In this case, the quinoid structure is remained unchanged and, as a result, the structure of the oxazyne type is formed

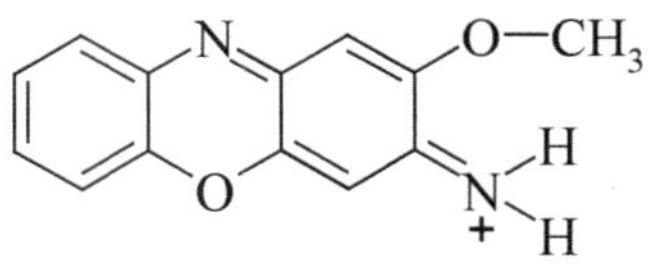

(1.27)

The possibility of such structure formation was showed by J. Widera et al.[106] for the oxidation of *ortho*-anisidine on platinum electrode. The content of carbon in the structure of oxazyne type is less than the leucoemeraldine form of poly-*ortho*-anisidine explaining the reduced value of the experimentally found carbon content in the polymer sample (56.03 *mass.%*, Table 1.8) compared with the theoretical (58.21 *mass.%*).

TABLE 1.8 The Results of Element Analysis of Synthesized Samples of Polyanisidines (Analyzer GA 1108, Carlo Erbo).

Sample	Composition/*mass.%*					
	C	N	H	SO_4^{2-}	O	Σ
Theoretical data (emeraldine form)						
Polyanisidine	58.21	9.70	4.64	16.46	11.05	100.00
Experimental data						
Poly-*ortho*-anisidine	56.03	8.00	5.01	13.00	11.05	93.09
Poly-*meta*-anisidine	58.40	7.90	4.45	12.10	11.05	93.90
Poly-*para*-anisidine	62.49	8.22	4.49	11.70	11.05	97.95

1.4 CHEMICAL SYNTHESIS OF POLYLUMINOL

The solubility of **PAn** in water and organic solvents is very low, causing some difficulties during the research of these materials as well as in the design of different kinds of devices on their basis. Therefore, the search for new materials based on PAn with improved physical and chemical properties continues. Two the most important directions of the researches today are the obtaining of polymers based on derivatives of aniline, considered above, and the synthesis of copolymers of aniline with heterocyclic compounds, such as pyrrol,[113] luminol,[114,115] etc. From this perspective, the luminol, thanks to its structure as heterocycle and benzene ring, bound to the amino group, corresponds to both criteria. However, despite the wide range of the researches of polyluminol (**PLm**), many questions regarding the reaction mechanism with its participation remained unclear. In particular, the polymerization of luminol was studied only in terms of its electrochemical oxidation when the polymer was delayed on the electrode in the form of a thin layer.

The electrochemical obtaining of the polyluminol has been studied in detail for the first time by G.-F. Zhang and H.-Y. Chen.[116] They obtained the polymer in acidic aqueous solutions in potentiodynamic mode, by 50-fold potential scanning of gold electrode ranging from −0.2 to +1.2 V. Today, in addition to the gold electrode,[114–119] the electropolymerization of luminol was made on the traditional electrode materials such as ITO,[117,120] glass–carbon,[116–119,121] platinum.[1,86,122] Depending on the electrode material the polyluminol can be obtained by scanning of the electrode potential also in a narrower range of potentials (e.g., 0.0–(+0.6) V on glass–carbon in aqueous H_2SO_4 solution with pH ~1.5).[121] At the same time, it is showed that the graphite electrode can be modified with a layer of electrically **PLm** also in weakly acidic medium at pH = 6[121] which has significant advantages during the immobilization of enzymes on such a surface to create the biosensors.

Since the luminol can be considered as a derivative of aniline, then the proposed mechanism of its electrochemical polymerization[116] is similar to the mechanism of the synthesis of **PAn**. This fact was confirmed by enhanced surface of Raman spectra of the **PLm**, which was by evidence of the transformation of the amine-groups into imide ones during oxidative polymerization of luminol, while N−N group of hydrazide ring remained intact,[117] or were deprotonated passing into N=N fragment.[116] Therefore, by analogy with **PAn**, the structure of the reduced form of **PLm** can be represented as a set of dimer fragments (excluding possible protonation of imine-groups):

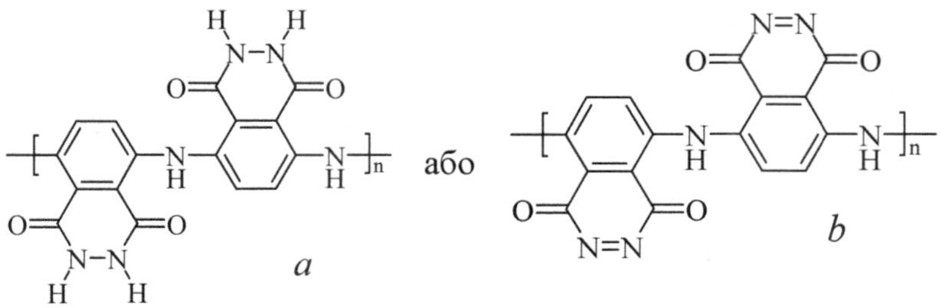 (1.28)

while the oxidized form of which is as follows:

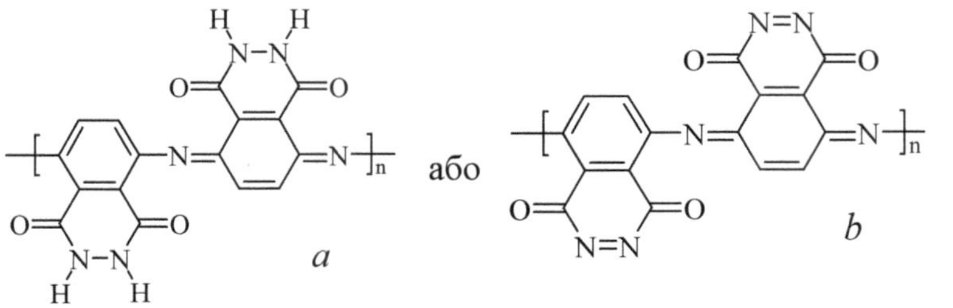 (1.29)

Another confirmation of the structure of polyluminol was the results of chemiluminescent researches. Since the synthesized polymer retained the fluorescent properties, it was by evidence that the heteroatomic cycle of luminol was kept during the oxidative polymerization. Moreover, it was shown that via copolymerization of luminol and aniline (molar ratio of monomers 1:(40–60)) the polymer which, unlike **PAn**, due to the presence in the structure of the polymer hydrazide group of luminol is electroactive and stable in alkaline medium.[114,115]

As it was already noted, the synthesis of **PLm** was carried out only electrochemically. However, given the similarity of **PLm** and **PAn**, logically to expect that the polyluminol can be also obtained chemically, which was conducted using a mixed aqueous–organic medium and suitable oxidant.

1.4.1 CYCLICAL VOLTAMMETRY OF 5-AMINE-2,3-DIHYDRO-1,4-PHTHALAZINEDIONE

Due to low solubility of inorganic oxidants in organic solvents, and luminol in water, chemical synthesis of **PLm** was conducted only in mixed aqueous–organic solvents. To estimate the possibility of polyluminol obtaining in such media, where organic component was **DMSO**, we have used the method of cyclic voltammometry.[123] As shown on Figure 1.13, at electrode potential scanning in the range 0–(+1.3) V the maximum of current on cyclical voltammogram is observed at potentials +(0.92–1.04) V. This maximum of current obviously is referred to the oxidation of luminol upon place of the amine-group since as potential of the aniline's oxidation consists of +1.13 V.[124] Similar results were also obtained by the authors of ref.[114,115] who have noted that as a result of lower potential of the luminol's oxidation versus aniline the oxidation process of last is suppressed during the electrochemical copolymerization of these monomers at their molar ratio from 1:1 to 1:4. Each successive scanning of the potential leads to the depression of the oxidation currents due to the screening of the surface of electrode by the layer precipitated on its surface of insoluble product, indicating the non-conductivity of the formed film of polyluminol. Another evidence of the fact that the formed polymer represents by itself the dielectric is absence on the cyclical voltammograms of current peaks responsible for the formation of polarons and bipolarons due to electrochemical doping (for **PAn** at +0.2 and +0.8 V).[73] The height of the current maxima is increased with the speed sweep potential increasing (Fig. 1.14). This result points to the fact that the speed of the process is limited by the diffusion of particles of electroactive material.

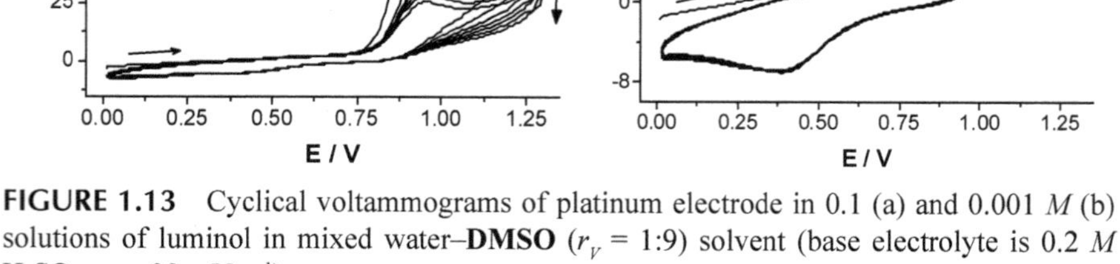

FIGURE 1.13 Cyclical voltammograms of platinum electrode in 0.1 (a) and 0.001 *M* (b) solutions of luminol in mixed water–**DMSO** (r_V = 1:9) solvent (base electrolyte is 0.2 *M* H_2SO_4; s_E = 30 mV·s^{-1}).

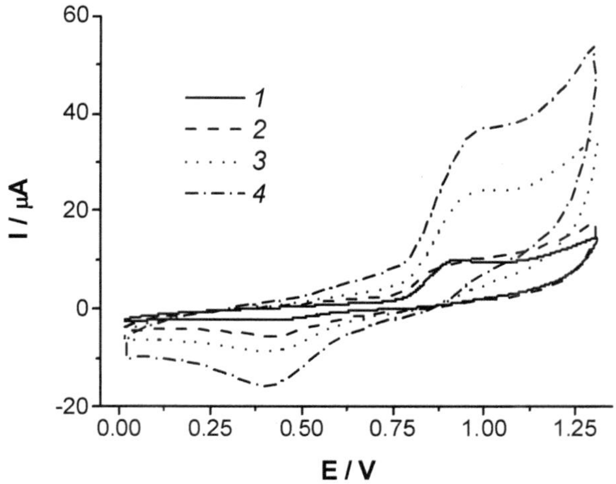

FIGURE 1.14 Cyclical voltammograms of platinum electrode in 0.05 *M* solution of luminol in mixed water–**DMSO** solvent (r_V = 1:9) at the values s_E, mV·s^{-1}: 10 (1); 30 (2); 50 (3) тa 100 (4) (base electrolyte is 0.2 *M* H_2SO_4).

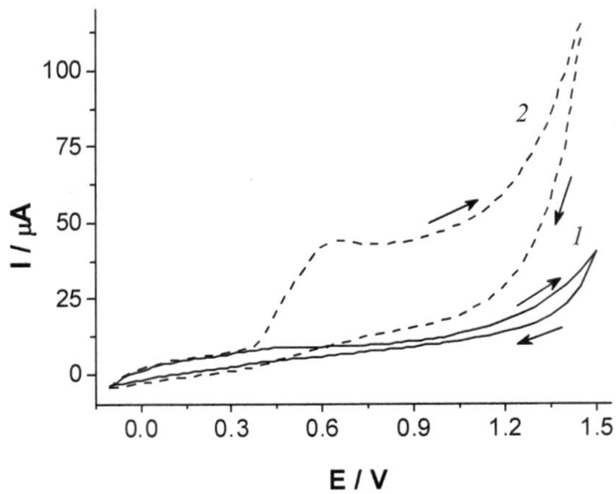

FIGURE 1.15 Cyclical voltammogram of platinum electrode in borate buffer solution (pH = 10.0) in the absence of (1) and after addition (2) of 5·10^{-5} M solution of luminol (s_E =10 mV·s^{-1}).

Studying the electrochemical oxidation of luminol (8.4·10^{-4} M) in 0.5 M aqueous solution of H_2SO_4 on the glass–carbon electrode G.-F. Zhang and H.-Y. Chen[116] received several other results, due to using of another electrolyte and electrode material. They observed two additional mild anodic peaks at +332 and +552 mV and the corresponding to them two cathode current peaks at +223 and +467 mV, the heights of which gradually increase with the increasing of number of cycles. These peaks corresponded, according to the authors' opinion, to the oxidation and reduction of the formed polyluminol's film. However, the attributing of both these current highs to the redox processes of polyluminol film is seemed to us somewhat incorrect. In particular, the second maximum the most plausible is associated with redox processes involving the imide groups of luminol, as indicated by cyclic voltammogram of luminol in aqueous alkaline solutions (Fig. 1.15).

1.4.2 CHEMICAL POLYMERIZATION OF LUMINOL

Based on electrochemical measurements, it can be concluded that at low values of the electrode potential the luminol is oxidized by hydrazine group, while at the high at the place of the amine-group. So, for the chemical synthesis two oxidants, namely $(NH_4)_2S_2O_8$, the standard values of the redox potential of which is 2.05 V, and KIO_3, the standard potential of which is twice smaller and is 1085 V were selected. It can be hoped that the oxidation of various functional groups differently are displayed on the process of the condensation of luminol rings and the structure of the condensation product. As the standard values of the redox potentials are given relatively to the hydrogen electrode, and the oxidation potentials on cyclic voltammograms relatively saturated Ag/AgCl electrode (E = +0.26 V), it is clear that the redox potential of KIO_3 is insufficient for the oxidation of luminol in the place of amine-group.

The possibility of luminol polymerization by chemical oxidation with the elimination of the product of reaction as separate phase has been investigated by us in a mixed aqueous–organic medium, when the dimethylsulfoxide (**DMSO**), dimethylformamide (**DMFA**) or N-methylpyrrolidone (**MPD**) were used as organic components. It was found, that the optimal volume ratio (r_V) of the components in a mixed solvent organic component: water was 9:1. At this, the yield of the product of polymerization of luminol does not depend on the nature of the solvent, and is determined by the ratio monomer: solvent (Table 1.9). The lowest yield of the polycondensate was observed at the use of KIO_3 as an oxidant. An increase of the excess of $(NH_4)_2S_2O_8$ over luminol actually does not affect the yield, while an increase of excess amounts of monomer leads to a significant increase of the quantity of the obtained product.

TABLE 1.9 Dependence of the Yield of the Luminol's Oxidation Products on the Conditions of Synthesis (r_V (Water:Organic Component) = 1:9; the Temperature of Reacting Mix is 20°C; the Duration of the Reaction 24 h).[123]

Organic solvent	C(Lum)/M	C(oxidant)/M	Yield/%
PLm I (Oxidant is KIO_3)			
DMSO	0.22	0.11	5.7
PLm II (Oxidant is $(NH_4)_2S_2O_8$)			
MPD	0.22	0.11	26.5
DMFA	0.22	0.11	26.9
DMSO	0.22	0.11	26.9
DMSO	0.22	0.22	27.5
DMSO	0.22	0.33	27.8
DMSO	0.44	0.22	41.6
DMSO	0.55	0.22	42.5

1.4.3 SPECTRAL PROPERTIES OF POLYCONDENSATES OF LUMINOL

Color of the obtained final products of oxidative polymerization of luminol depends on the nature of oxidant. At the use of KIO_3 the black polymer (PLm I) is formed, and in the presence of $(NH_4)_2S_2O_8$ the obtained polymer had a green color (PLm II) like to PAn, which, along with the results of spectral studies is an evidence of significant differences in polymerization mechanism and product's structure depending on the nature of the used oxidant.

IR-spectra of luminol and samples of synthesized polyluminols are shown on Figure 1.16. Spectra of **PLm** (spectra *b–d*) similar to the spectra of luminol (spectrum *a*), and the most essential changes are observed in the high-frequency (2900–3500 cm^{-1}) area, where there are the bands of amine-group: the bands at 3331, 3420, and 3473 cm^{-1} correspond to asymmetric and symmetric vibrations of amine-groups of aromatic amines.[125] The first and the third from these bands are characteristic for the spectra of the samples **PLm II**, synthesized in **DMSO** and **MPD** (Fig. 1.16, the spectra *b* and *c*). The band at 3420 cm^{-1} which corresponds to the vibration of N–H bonds is absent in the spectra of samples **PLm II**, indicating the oxidation of the amine-group with peroxydisulfate during the synthesis. Available in a spectrum of luminol the bands of deformation and torsion vibrations of NH_2 group at 1628 and 492 cm^{-1}, respectively, are also absent in the spectrum of the polymer, which once again confirms the oxidation of luminol in a place of amine-group. In the spectra of polyluminols also available the azogroup –N=N–, the vibration frequency of which centered in the vicinity of 1588 cm^{-1}. Valence vibrations ν_{C-H} of benzene ring are observed at 3164–2911 and 3167–2920 cm^{-1} in monomer and polymer, respectively. Three bands of vibrations in the area 1627–1754 cm^{-1} correspond to the vibrations of amide-group.[97] An absence of the bands in the area 1610–1550 and at 1400 cm^{-1} is agreed with the assumption that the carboxylates are absent in the structure of polyluminol.

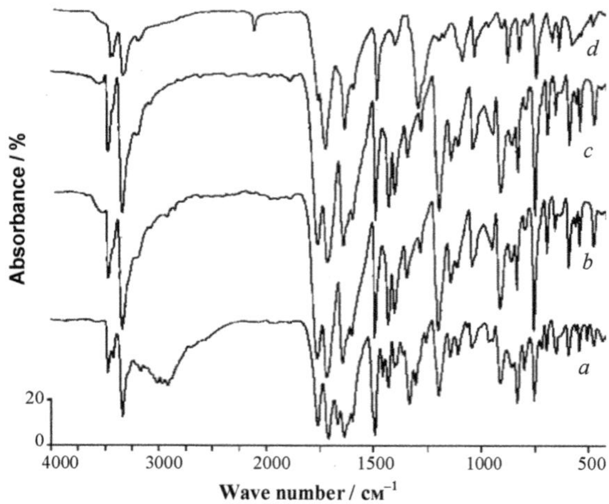

FIGURE 1.16 IR absorption spectra of the initial luminol (a), **PLm II** (synthesized in mixed water–**DMSO** (b), and water–**MPD** (c) solvent) and **PLm I** (d).

Replacement of oxidant $(NH_4)_2S_2O_8$ on KIO_3 leads to the change of the structure of product that is displayed on the IR-spectrum of polyluminol obtained under these conditions. The form of absorption bands in the range 3200–3500 cm^{-1} (Fig. 1.16, the spectrum *d*) which is similar to luminol (the spectrum *a*) indicates the tenure of the amine-group during the oxidation of luminol by potassium iodate. At the same time, the new absorption band at 2123 and 1289 cm^{-1} can be attributed to the vibrations of –N=C=O group and amine-group of aromatic amines, respectively.

Luminol and **PLm** can be represented as analogs of aniline and polymeric anilines. However, unlike **PAn**, in the molecule of polyluminol no bands at 1500 and 1600 cm^{-1} which were assigned by the authors of *ref.*[26] to structural fragments.

and

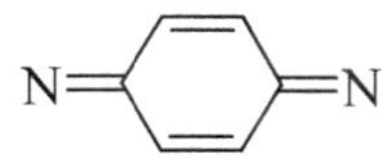

In addition, in spectra of the all synthesized samples of polyluminols absent the absorption bands, the vibration frequencies of which are in the vicinity of 1610 and 1400 cm^{-1}. Therefore, it can be assumed, that the final product of the luminol's oxidation both with $(NH_4)_2S_2O_8$ and KIO_3 is not aminophthalate (emitter of chemiluminescence), and, therefore, heteroatomic ring of luminol is not oxidized during the polymerization.

Initial luminol (Fig. 1.17a) and synthesized samples of **PLm** (Fig. 1.17b and 1.18) were also characterized by Raman spectroscopy. It was found that the intensity of the Raman bands for the samples **PLm II** on two orders and for the sample **PLm I** on three orders lower compared of with the spectrum of luminol. This is obviously due to the strong absorption of dark green and black synthesized polymer samples (**PLm I** and **PLm II**, respectively). Compared with luminol the band at 3100 cm^{-1} which corresponds to the valence vibrations of N–H bond of the aromatic amines significantly is expanded, although the position of maximum (3089 cm^{-1}) remained unchanged. This not can be said for the spectrum of **PLm I**, where this maximum stretched in the range from 3046 to 3250 cm^{-1} and reminds the band for the PAn (Fig. 1.8). Position of doublet band at 1749 (intensive) and 1705 cm^{-1} (low intensive) corresponding to valence vibrations of carbonyl group is little varied during the transition from luminol to **PLm II**. At the same time, for polyluminol obtained in the presence of KIO_3, $\nu_{c=o}$ is shifted to 1775 cm^{-1} due to the different surroundings of the functional group in both cases. The bands at 1628, 1589, and 647 cm^{-1} (intense) corresponding to the vibrations of benzene ring are observed in the spectra of the all synthesized samples.

References data on the Raman spectra of electroconducting polymers in general and **PLm** in particular are rather poor. The Raman spectra enhanced with the surface of electrochemically synthesized on Au-electrode polyluminol were of low quality,[116] but major changes were recorded for the band of the stretching vibration $\nu_{N\text{-}H}$, the intensity of which for polyluminol is much higher than similar intensity for the same band of luminol. High-quality spectra of (1′,2′-dicarboxy)ethylbenzotriazole on the copper surface were obtained by the authors of *ref.*[127] Here, existing distinct bands of the stretching vibrations of the carboxyl group, benzene ring, and a clear band of deformation vibrations $\delta_{N\text{–}H}$ at 1162 cm^{-1}, which into obtained by us spectra was centered at 1174 cm^{-1} for luminol, ranging from 1185 to 1187 cm^{-1} for **PLm II**, and was absent in the case of **PLm I**. In the Raman spectrum of luminol and polyluminols available low-intensive band at 1245 cm^{-1}, which can be attributed, by analogy with **PAn**,[128] to the deformation vibrations $\delta_{C\text{–}N}$. In spectrum of polyluminols the bands that correspond to the presence in the structure of the C=N bonds were not found. Obviously, this is related to the absence of quinoid type structures, the formation of which is possible due to oxidation of polyluminol. This is a serious argument in favor of unparticipation of polyluminol chains in deep oxidation with the formation of structures such as emeraldine or pernigraniline, as it takes place in the case of chemical condensation of anilines.[126-129]

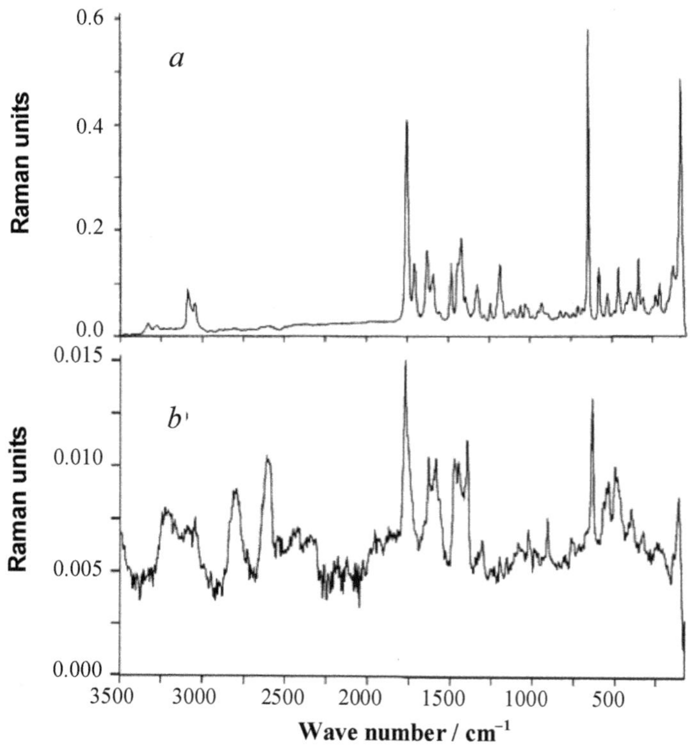

FIGURE 1.17 Raman spectra of the starting luminol (a) and **PLm I** (b).

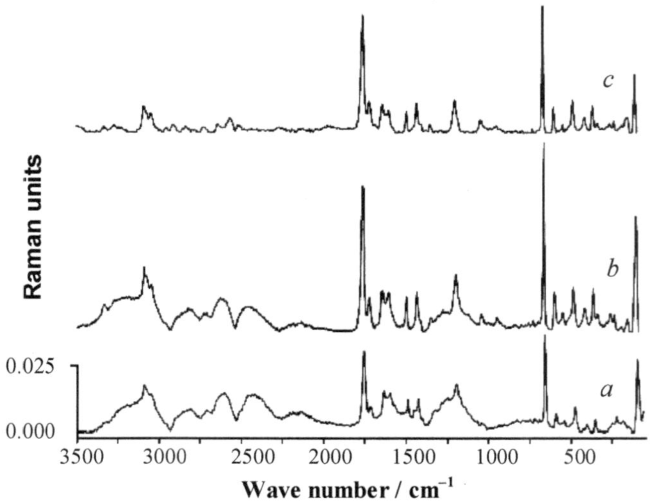

FIGURE 1.18 Raman spectra of the **PLm II** synthesized in mixed water–organic ($r_V = 1{:}9$, organic component: **DMSO** (a), **DMFA** (b), and **MPD** (c)) solvent.

1.4.4 THE STRUCTURE OF POLYCONDENSATES OF LUMINOL

The data of element analysis of polyluminols are shown in Table 1.10. Characteristically, that the content of carbon, nitrogen and hydrogen is little differed for the luminols obtained by oxidation with $(NH_4)_2S_2O_8$, while at the use of KIO_3 the element composition of samples is significantly different. The interpretation of these differences in chemical composition is based on various possible mechanisms of oxidation in these two cases. Oxidation of luminol with KIO_3 takes place on amide groups which are easier oxidated. The fact that the content of nitrogen is 5% less, it can be postulated the loss of nitrogen atom by each monomer link.

Therefore, by use of KIO_3 as oxidant agent, the dimer of type

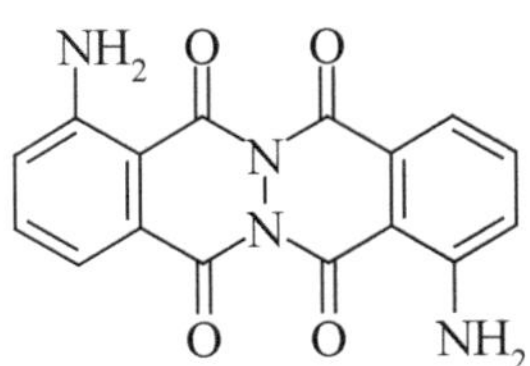

is formed.

At this, the N–N bond is formed by connection of two residues of the luminol's molecules, each containing one nitrogen atom. The calculation of element composition for this structure gives the following results: 59.62% C; 17.39% N; 3.10% H, and 19.88% O. A good agreement of the results of analysis with the above- proposed structure takes place for nitrogen and hydrogen, however the carbon content exceeded on 5%. The products of KIO_3 reduction may be the iodine atoms, so they can be part of the product of luminol oxidation. On the possibility of formation of complex compound of iodine with nitrates or quaternary ammonium salts, particularly was pointed by A. Wells.[130] The structure of the product then can be represented by the following formula

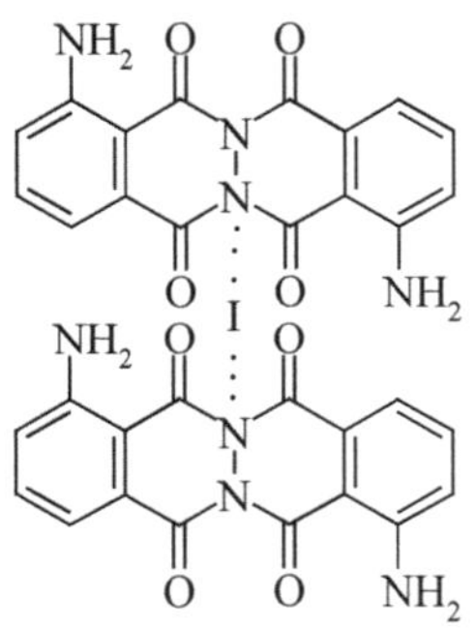

(1.30)

Calculation of chemical composition based on the structure (1.30) gave the following results: 49.74% C; 14.51% N; 2.85% H; 16.58% O; 16.32% I, and, so element composition in near to the experimentally found (Table 1.10). In favor of this structure indicates the presence of the absorption band at 567 cm^{-1} and IR bands at 526 and 581 cm^{-1} in Raman spectra. Therefore, we can assume that the most likely structure of the luminol oxidation with KIO_3 (**PLm I**) product corresponds to depicted formula (1.30).

Oxidation of luminol with peroxydisulfate takes place on the place of the amino-group of luminol and the structure of the formed polycondensate contains the monomer links that are almost indistinguishable from the original structure of luminol. In favor of this, there are data of the element analysis, IR, and Raman spectra of the products of oxidation. By analogy with the PAn for polyluminol synthesized in strong acidic medium, it was proposed the structure (1.28, *a*), which corresponds to fully reduced form of polyluminol.[116] However, it was not considered that the amide groups can be also oxidized, indirect evidence of which is, for example, the oxidation of luminol in the presence of montmorillonite[131,132] containing the Fe^{3+} ions.

TABLE 1.10 Data of Element Analysis of Luminol and Synthesized Samples of Polyluminol (Analyzer GA 1108, Carlo Erbo, Italy).

Sample	Composition, %			
	C	H	N	O
Theoretical				
Luminol (Lum)	54.24	3.96	23.73	18.08
Experimental				
Lum	54.24	3.94	23.61	18.22*
PLm I	54.25	3.88	23.66	18.21*
PLm II				
DMSO; C(Lum)/C(Ox) = 1:1	53.80	3.84	23.40	18.96*
DMSO; C(Lum)/C(Ox) = 2:1	54.17	3.83	23.48	18.52*
MPD; C(Lum)/C(Ox) = 2:1	47.39	3.08	18.38	31.19*

*Note: The values calculated based on found experimentally total content C, H, and N.

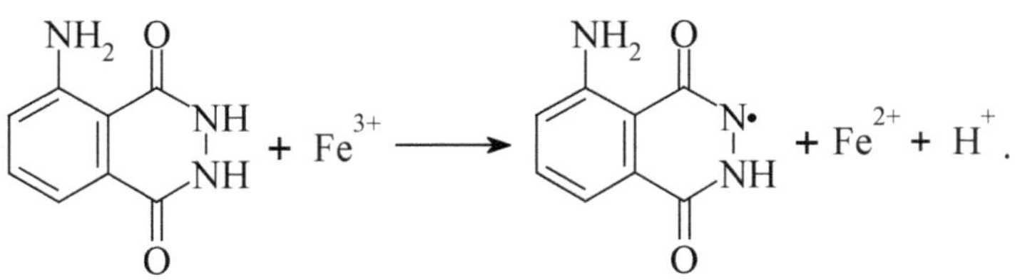

(1.31)

Further electrochemical oxidation of the intermediate radical leads to the formation of azoquinone.[133] Studying the electrochemical oxidation of an aqueous solution of luminol in an acidic medium, it was suggested that in the area of low potential reversibly is formed exactly this compound[116]

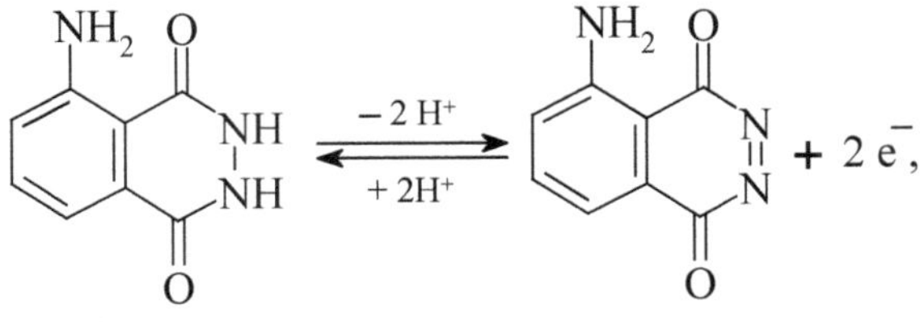

(1.32)

and the oxidation takes place on amine-group at higher potentials. So, again similar to the analogy with the condensation of aniline, it can be suggested that chemically synthesized polyluminol does not contain the hydrazine –NHNH– group, but azogroup –N=N in heteroatomic ring and the structure of the link will correspond to the formula (1.28, *b*). The hydrogen content in such a structure should be much smaller. Calculated for this structure content of the hydrogen (2.11%) is much smaller compared with the experimental values 3.83–3.88% for polymer and 3.95% for luminol. If to assume that the amide group in the luminol is not changed during the oxidation with $(NH_4)_2S_2O_8$ and bridge atoms of the nitrogen are protonated, then the content of hydrogen will be near to the experimentally determined. Because it can be argued that the oxidation of hydrazine groups of luminol in the presence of $(NH_4)_2S_2O_8$ does not proceed and then the polymer structure is similar (1.28, *a*). In the presented formula of the structural fragment of polyluminol, the authors of *ref.*[116] do not exclude the possibility of an existence of protonated atoms of nitrogen.

An indirect confirmation of the conclusions about the mechanism of the formation of polymerizates, namely that the oxidation of luminol in mixed aqueous–organic solvents proceeds primarily not by hydrazine group (resulting in the formation of the emitter radiation 3−aminophthalate), but by amine-group with the formation of the polymer, can be the results of chemiluminescent researches in reaction systems, in which the polycondensate of luminol was obtained. In systems 0.22 M **Lum** + 0.11 M $(NH_4)_2S_2O_8$ slight luminescence was observed only in a mixed water–**MPD** solvent, while in the water–**DMSO** and water–**DMFA** solutions the generation of luminescence was not fixed. Adding to the mix besides $(NH_4)_2S_2O_8$ of hydrogen peroxide leads to the intensification of luminescence in the presence of **MPD** and before its appearance in the presence of **DMSO** and **DMFA**. However, the intensity of radiation even in the presence of H_2O_2 is very small (on the 2–3 orders lower compared to the intensity of chemiluminescence of luminol under optimal conditions), that is more obvious, if we take into account on 1–2 order higher concentration of luminol during the synthesis of its polymeric form compared to concentrations, that are used in chemiluminescent techniques.

1.4.5 THERMAL STABILITY OF CONDENSATES OF LUMINOL

We have also studied the thermal stability of polyluminols synthesized under different conditions and compared to the properties of the original luminol. As it can be seen from the integral and differential curves of thermal analysis shown on Figure 1.19, the less thermally stable is monomeric luminol, for the sample of which the first maximum on **DTG**-curve, which is associated with destruction on the hydrazine group with the nitrogen elimination is observed at the destruction temperature (T_D) 256.5°C, while for the most thermally stable **PLm II** (synthesized in **MPD**) T_D = 334.2 °C (against 280.2 °C for **PLm II** synthesized in **DMSO**).

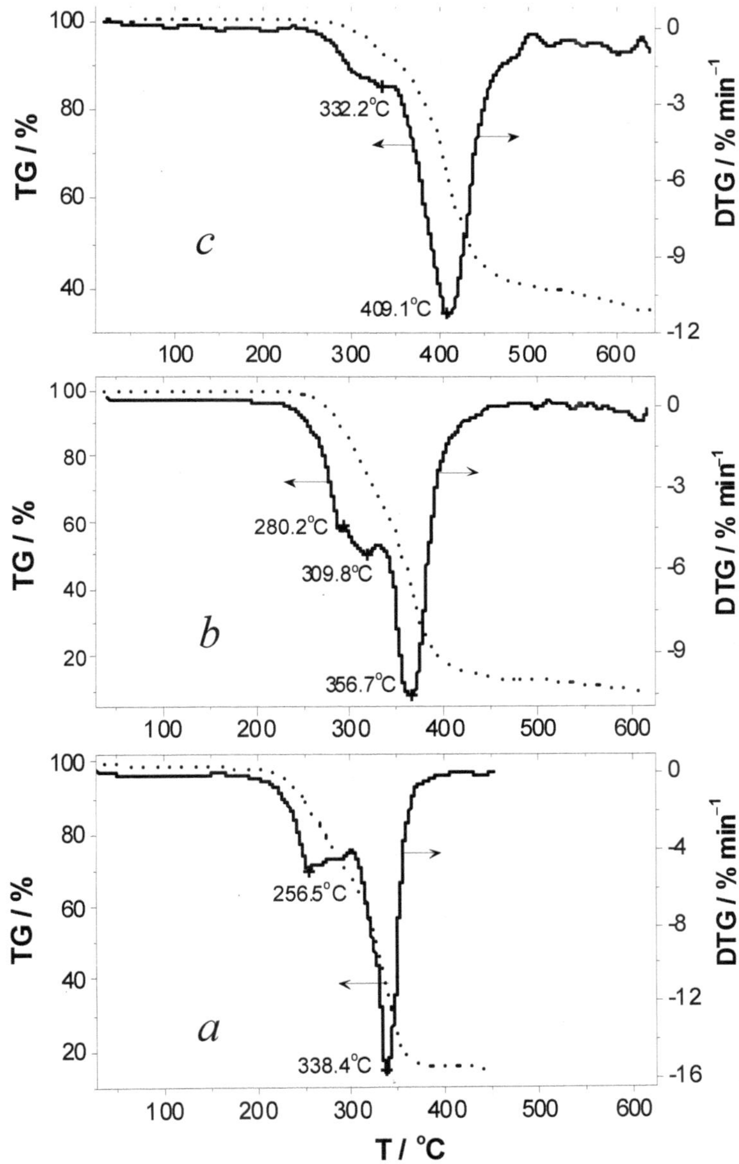

FIGURE 1.19 Integral and differential derivatograms of initial luminol (a) and **PLm II** synthesized in mixed water–**DMSO** (b) and water–**MPD** (c) solvent (atmosphere is Ar; microbalance NETZSCH TG29; s_T = 10 °C·min^{-1}).

Another derivatogram has been obtained for polyluminol synthesized with the use of KIO_3 as oxidant (Fig. 1.20). Substantial differences in the nature of **DTG** curves of **PLm I** and **PLm II** once again affirm the significant difference in the structure of the obtained samples. Differential thermogram of sample **PLm I** (Fig. 1.20) is much "richer" as to the number of available highs, indicating not only the much more complex mechanism of thermolysis of the structure (1.30), but also the entire difference between the products of **PLm I** and **PLm II**. In particular, if the final losses of mass both of luminol and samples **PLm II (DMSO)** make up more than 85%, for the samples of **PLm II (MPD)** and **PLm I** this value consists of only ~55%.

Sample **PLm I** is characterized by significantly lower thermal stability even compared to monomeric luminol. In particular, the first maximum on the **DTG**-curve related, obviously, with the loss of mass by the iodine sample, is observed at 62.8 °C and a maximal rate of the loss of mass for the sample **PLm I** is observed at $T_{D,max}$ = 282.5 °C. In comparison, for luminol and samples **PLm II** synthesized in **DMSO** and **MPD**, $T_{D,max}$ is 338.4 °C, 356.7 °C, and 409.1 °C, respectively.

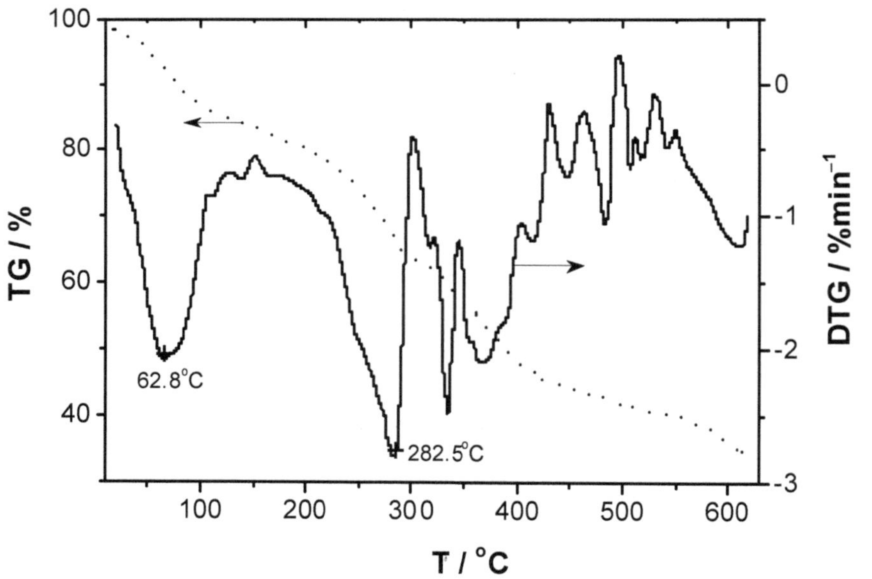

FIGURE 1.20 Integral and differential derivatograms of the sample **PLm I**.

1.5 CONCLUSIONS

Analysis of the above-presented results shows that the synthesis of polymeric aniline and its derivatives is a combination of the processes of electron transfer with the formation of initial particles in the doublet state with their subsequent isomerization, deprotonation, and recombination. The initial stages of oxidative polymerization of aniline and its derivatives (at pH < 2) both during chemical and electrochemical synthetic methods, is likely the sequence of following transformations:

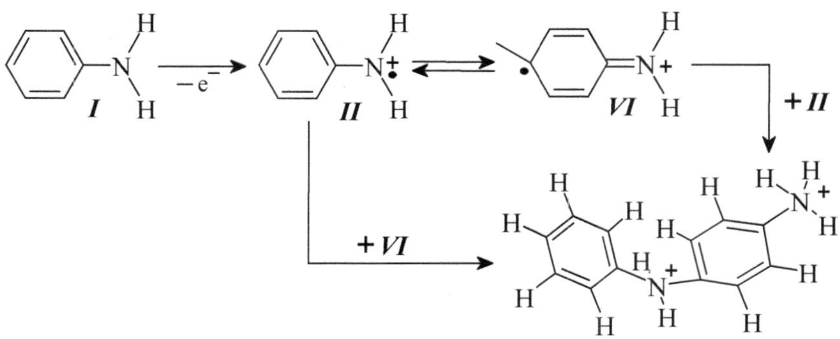

The first stage of the process is the formation of cation-radical of aniline as a result of the detachment of the electron from the 2*s*-level of nitrogen atom. If at the electrochemical synthesis, this process takes place in one-step directly on the electrode, then in the case of chemical synthesis the intermediate of the process is the complex of initial monomer with an oxidant (e.g., peroxydisulfate anion). Formed cation-radicals of aniline by electrochemical oxidation of the monomer or the decomposition of complex monomer–oxidant enter into the secondary reactions of recombination, the most probably of "head to tail" type, which is possible as a result of the isomerization of starting cation-radical of benzenoid type into its quinoid form. The resulting dimers are the structural units of the future polymer chain, because next may take part in such a chain of transformations to form tetrameters, octamers, ..., oligomers, and finally of the polymer product.

So, based on the proposed mechanism, the oxidative polymerization of anilines can be defined in accordance with *ref.*[134] as a process of polyaddition. At the same time, since the composition of links of the main polymeric chain and the molecules of original polymer are not identical, the same process can be classified as the condensation polymerization. However, it should be noted that in this case not simple molecules are eliminated (according to the definition presented in *ref.*[135]) but the H^+ ion.

By spectroscopic and element analysis it was found, that regardless of the nature of the original monomer the most likely structure of PAns obtained by chemical way is the form of emeraldine salt, positive charge in which is compensating by the anions of the reaction medium. At the same time, during the electrochemical synthesis the form of PAns will be determine by the final value of the electrode potential at which the polymer deposition was occurred. However, introduction of a substituent in the aromatic ring significantly alters the reactivity of the monomer in oxidative polycondensation reactions compared with aniline. So, the presence of strong electron-acceptor nitro-group makes it impossible to obtain the polymer products both during chemical and electrochemical synthesis regardless of its position in the aromatic ring. At the same time, the reactivity of the aniline's derivatives with electron-donor substituents substantially increases. Separately, it should be noted, that an important factor that will determine the reactivity of the aniline's monomers is the position of substituent in benzene cycle that will affect not only the distribution of electron density in the original molecule or the possibility of isomerization intermediates oxidation, but also on the orientation of molecules in the adsorption layer during the electrochemical polycondensation.

KEYWORDS

- **aniline**
- **polymerization mechanism**
- **oxidative polycondensation**
- **polyaddition**
- **quantum chemical calculations**
- **thermal properties**
- **luminol**

REFERENCES

1. Heeger, A. J. Semiconducting and Metallic Polymers: The Fourth Generation of Polymeric Materials. *Curr. Appl. Phys.* **2001,** *1,* 247–267.
2. MacDiarmid, A. G. "Synthetic Metals": A Novel Role of Organic Polymers. *Curr. Appl. Phys.* **2001,** *1,* 269–279.
3. Shirakawa, H. The Discovery of Polyacetylene Film the Dawning of an Era of Conducting Polymers. *Curr. Appl. Phys.* **2001,** *1,* 281–286.
4. Croce, F.; Panero, S.; Passerini, S.; Scrosati, B. The Role of Conductive Polymers in Advanced Electrochemical Technology. *Electrochim. Acta.* **1994,** *39,* 255–263.
5. Batich, C. D.; Laitinen, H. A.; Zhou, H. C. Chromatic Changes in Polyaniline Films. *J. Electrochem. Soc.* **1990,** *137,* 883–885.
6. Mortimer, R. J.; Dyer, A. L.; Reynolds, J. R. Electrochromic Organic and Polymeric Materials for Display Applications. *Displays*. **2006,** *27,* 2–18.
7. Dutta, D.; Sarma, T. K.; Chowdhury, D.; Chattopadhyay, A. A Polyaniline-Containing Filter Paper That Acts as a Sensor, Acid, Base, and Endpoint Indicator and also Filters Acids and Bases. *J. Colloid Interf. Sci.* **2005,** *283,* 153–159.
8. Jelle, B. P.; Hagen, G. Performance of an Electrochromic Window Based on Polyaniline, Prussian Blue and Tungsten Oxide. *Solar Energy Mater. Solar Cells.* **1999,** *58,* 277–286.
9. Somani, P.; Mandale, A. B.; Radhakrishnan, S. Study and Development of Conducting Polymer-Based Electrochromic Display Devices. *Acta Mater.* **2000,** *48,* 2859–2871.
10. Mortimer, R. J. Electrochromic Materials. *Chem. Soc. Rev.* **1997,** *26,* 147–156.
11. Lacroix, J. C.; Kanazawa, K. K.; Diaz, A. Polyaniline: A Very Fast Electrochromic Material. *J. Electrochem. Soc.* **1989,** *136,* 1308–1313.
12. Carpi, F.; De Rossi, D. Colours from Electroactive Polymers: Electrochromic, Electroluminescent and Laser Devices Based on Organic Materials. *Opt. Laser Technol.* **2006,** *38,* 292–305.
13. Yang, Y.; Heeger, A. J. Polyaniline as a Transparent Electrode for Polymer Light-Emitting Diodes: Lower Operating Voltage and Higher Efficiency. *Appl. Phys. Lett.* **1994,** *64,* 1245–1247.
14. Sazou, D.; Kourouzidou, M.; Pavlidou, E. Potentiodynamic and Potentiostatic Deposition of Polyaniline on Stainless Steel: Electrochemical and Structural Studies for a Potential Application to Corrosion Control. *Electrochim. Acta.* **2007,** *52,* 4385–4397.
15. Nicho, M. E.; Hu, H.; González-Rodriguez, J. G.; Salinas-Bravo, V. M. Protection of Stainless Steel by Polyaniline Films Against Corrosion in Aqueous Environments. *J. Appl. Electrochem.* **2006,** *36,* 153–160.
16. Cook, A.; Gabriel, A.; Siew, D.; Laycock, N. Corrosion Protection of Low Carbon Steel with Polyaniline: Passivation or Inhibition? *Curr. Appl. Phys.* **2004,** *4,* 133–136.
17. Santos, J. R. Jr.; Mattoso, L. H. C.; Motheo, A. J. Investigation of Corrosion Protection of Steel by Polyaniline Films. *Electrochim. Acta.* **1998,** *43,* 309–313.
18. Williams, G., Holness, R. J.; Worsley, D. A.; McMurray, H. N. Inhibition of Corrosion-Driven Organic Coating Delamination on Zinc by Polyaniline. *Electrochem. Commun.* **2004,** *6,* 549–555.
19. Özyılmaz, A. T.; Tüken, T.; Yazıcı, B.; Erbil, M. The Electrochemical Synthesis and Corrosion Performance of Polyaniline on Copper. *Prog. Org. Coat.* **2005,** *52,* 92–97.
20. Saxena, V.; Malhotra, B. D. Prospects of Conducting Polymers in Molecular Electronics. *Curr. Appl. Phys.* **2003,** *3,* 293–305.

21. Laranjeira, J. M. G.; Khoury, H. J.; de Azevedo, W. M.; de Vasconcelos, E. A.; da Silva, E. F. Jr. Polyaniline Nanofilms As a Monitoring Label and Dosimetric Device for Gamma Radiation. *Mater. Charact.* **2003,** *50,* 127–130.
22. Grummt, U. W.; Pron, A.; Zagorska, M.; Lefrant, S. Polyaniline Based Optical pH Sensor. *Anal. Chim. Acta.* **1997,** *357,* 253–259.
23. Talaie, A. Conducting Polymer Based pH detector: A New Outlook to pH Sensing Technology. *Polymer.* **1997,** *38,* 1145–1150.
24. Tahir, Z. M.; Alocilja, E. C.; Grooms, D. L. Polyaniline Synthesis and its Biosensor Application. *Biosens. Bioelectron.* **2005,** *20,* 1690–1695.
25. Timmer, B.; Olthuis, W.; van den Berg, A. Ammonia Sensors and Their Applications - A Review. *Sensor. Actuat. B–Chem.* **2005,** *107,* 666–677.
26. Michira, I.; Akinyeye, R.; Somerset, V.; Klink, M. J.; Sekota, M.; Al-Ahmed, A; Baker, P. G. L.; Iwuoha, E. Synthesis, Characterisation of Novel Polyaniline Nanomaterials and Application in Amperometric Biosensors. *Macromol. Symp.* **2007,** *255,* 57–69.
27. Do, J. S.; Chang, W. B. Amperometric Nitrogen Dioxide Gas Sensor Based on Pan/Au/Nafion® Prepared by Constant Current and Cyclic Voltammetry Methods. *Sensor. Actuat. B Chem.* **2004,** *101,* 97–106.
28. Ram, M. K.; Yavuz, Ö.; Lahsangah, V.; Aldissi, M. CO Gas Sensing from Ultrathin Nano-Composite Conducting Polymer Film. *Sensor. Actuat. B Chem.* **2005,** *106,* 750–757.
29. Li, J.; Petelenz, D.; Janata, J. Suspended Gate Field-Effect Transistor Sensitive to Gaseous Hydrogen Cyanide. *Electroanalysis.* **1993,** *5,* 791–794.
30. Conn, C.; Sestak, S.; Baker, A. T.; Unsworth, J. A Polyaniline-Based Selective Hydrogen Sensor. *Electroanalysis.* **1998,** *10,* 1137–1141.
31. Campos, M.; Bulhões, L. O. S.; Lindino, C. A. Gas-Sensitive Characteristics of Metal/Semiconductor Polymer Schottky Device. *Sensor. Actuat. A Phys.* **2000,** *87,* 67–71.
32. Pan, X.; Kan, J.; Yuan, L. Polyaniline Glucose Oxidase Biosensor Prepared with Template Process. *Sensor. Actuat. B Chem.* **2004,** *102,* 325–330.
33. Luo, Y.-C.; Do, J.-S. Urea Biosensor Based on PANi(urease)-Nafion®/Au Composite Electrode. *Biosens. Bioelectron.* **2004,** *20,* 15–23.
34. MacDiarmid, A. G.; Yang, L. S.; Huang, W. S.; Humphrey, B. D. Polyaniline: Electrochemistry and Application to Rechargeable Batteries. *Synth. Met.* **1987,** *18,* 393–398.
35. Novák, P.; Müller, K.; Santhanam, K. S. V.; Haas, O. Electrochemically Active Polymers for Rechargeable Batteries. *Chem. Rev.* **1997,** *97,* 207–281.
36. Fan, L. Z.; Hu, Y. S.; Maier J.; Adelhelm, P.; Smarsly, B.; Antonietti, M. High Electroactivity of Polyaniline in Supercapacitors by Using a Hierarchically Porous Carbon Monolith as a Support. *Adv. Funct. Mater.* **2007,** *17,* 3083–3087.
37. Nadagouda, M. N.; Varma, R. S. Green Approach to Bulk and Template-Free Synthesis of Thermally Stable Reduced Polyaniline Nanofibers for Capacitor Applications. *Green Chem.* **2007,** *9,* 632–637.
38. Jang, J.; Bae J.; Choi, M.; Yoon, S. H. Fabrication and Characterization of Polyaniline Coated Carbon Nanofiber for Supercapacitor. *Carbon.* **2005,** *43,* 2730–2736.
39. Ryu, K. S.; Lee, Y.; Han, K. S.; Park, Y. J.; Kang, M. G.; Park, N. G.; Chang, S. H. Electrochemical Supercapacitor Based on Polyaniline Doped with Lithium Salt and Active Carbon Electrodes. *Solid State Ionics.* **2004,** *175,* 765–768.
40. MacDiarmid, A. G.; Epstein, A. J. Polyanilines: A Novel Class of Conducting Polymers. Faraday Discuss. *Chem. Soc.* **1989,** *88,* 317–332.

41. Chiang, J. C.; MacDiarmid, A. G. 'Polyaniline': Protonic Acid Doping of the Emeraldine Form to the Metallic Regime. *Synth. Met.* **1986,** *13,* 193–205.
42. MacDiarmid, A. G.; Chiang, J. C.; Richter, A. F.; Epstein, A. J. Polyaniline: A New Concept in Conducting Polymers. *Synth. Met.* **1987,** *18,* 285–290.
43. Sun, Z.; Geng, Y.; Li, J.; Jing, X.; Wang, F. Chemical Polymerization of Aniline with Hydrogen Peroxide as Oxidant. *Synth. Met.* **1997,** *84,* 99–100.
44. Ram, M. S.; Palaniappan, S. Benzoyl Peroxide Oxidation Route to Polyaniline Salt and its Use as Catalyst in the Esterification Reaction. *J. Mol. Catal. A Chem.* **2003,** *201,* 289–296.
45. Armes, S. P.; Gottesfeld, S.; Beery, J. G.; Garzon, F.; Agnew, S. F.. Conducting Polymer-Colloidal Silica Composites. *Polymer*. **1991,** *32,* 2325–2330.
46. Yan, H.; Toshima, N. Chemical Preparation of Polyaniline and its Derivatives by Using Cerium(IV) Sulphate. *Synth. Met.* **1995,** *69,* 151–152.
47. Ayad, M. M.; Shenashin, M. A. Polyaniline Film Deposition from the Oxidative Polymerization of Aniline Using $K_2Cr_2O_7$. *Euro. Polym. J.* **2004,** *40,* 197–202.
48. Chowdhury, P.; Saha, B. Potassium Iodate-Initiated Polymerization of Aniline. *J. Appl. Polym. Sci.* **2007,** *103,* 1626–1631.
49. Li, X. X.; Li, X. W. Oxidative Polymerization of Aniline Using $NaClO_2$ as an Oxidant. *Mater. Lett.* **2007,** *61,* 2011–2014.
50. Ballav, N. High-Conducting Polyaniline via Oxidative Polymerization of Aniline by MnO_2, PbO_2 and NH_4VO_3. *Mater. Lett.* **2004,** *58,* 3257–3260.
51. Kuwabata, S.; Idzu, T.; Martin, R. C.; Yoneyama, H. Charge-Discharge Properties of Composite Films of Polyaniline and Crystalline V_2O_5 Particles. *J. Electrochem. Soc.* **1998,** *145,* 2707–2710.
52. Bernard, M. C.; Hugot-Le Goff, A.; Zeng, W. Elaboration and Study of a PANI/PAMPS/WO_3 All Solid-State Electrochromic Device. *Electrochim. Acta.* **1998,** *44,* 781–796.
53. Cruz-Silva, R.; Romero-García, J.; Angulo-Sánchez, J. L.; Flores-Loyola, E.; Farias, M. H.; Castillon, F. F.; Diaz, J. A. Comparative Study of Polyaniline Cast Films Prepared from Enzymatically and Chemically Synthesized Polyaniline. *Polymer.* **2004,** *45,* 4711–4717.
54. Cruz-Silva, R.; Romero-García, J.; Angulo-Sánchez, J. L.; Ledezma-Perez, A.; Arias-Marin, E.; Moggio, I.; Flores-Loyola, E. Template-Free Enzymatic Synthesis of Electrically Conducting Polyaniline Using Soybean Peroxidase. *Euro. Polym. J.* **2005,** *41,* 1129–1135.
55. Shen, Y.; Sun, J.; Wu, J.; Zhou, Q. Synthesis and Characterization of Water-Soluble Conducting Polyaniline by Enzyme Catalysis. *J. Appl. Polym. Sci.* **2005,** *96,* 814–817.
56. Rumbau, V.; Pomposo, J. A.; Alduncin, J. A.; Grande, H,; Mecerreyes, D.; Ochoteco, E. A New Bifunctional Template for the Enzymatic Synthesis of Conducting Polyaniline. *Enzyme Microb. Tech.* **2007,** *40,* 1412–1421.
57. Traore, M. K.; Stevenson, W. T. K.; McCormick, B. J.; Dorey, R. C.; Wen, S.; Meyers, D. Thermal Analysis of Polyaniline Part I. Thermal Degradation of HCl-Doped Emeraldine Base. *Synth. Met.* **1991,** *40,* 137–153.
58. Ivanov, V. F.; Gribkova, O. L.; Nekrasov, A. A.; Vannikov, A. V. Comparative Spectroelectrochemical Investigation of Vacuum Evaporated and Electrochemically Synthesized Electrochromic Polyaniline Films AgI. *J. Electroanal. Chem.* **1994,** *372,* 57–61.
59. Karakişla, M.; Saçak, M.; Akbulut, U. Conductive Polyaniline/Poly(Methyl Methacrylate) Films Obtained by Electropolymerization. *J. Appl. Polym. Sci.* **1996,** *59,* 1347–1354.

60. Lin, Y.; Yasuda, H. Effect of Plasma Polymer Deposition Methods on Copper Corrosion Protection. *J. Appl. Polym. Sci.* **1996,** *60,* 543–555.
61. Gong, X.; Dai, L.; Mau, A. W. H.; Griesser, H. J. Plasma-Polymerized Polyaniline Films: Synthesis and Characterization. *J. Polym. Sci. Pol. Chem.* **1998,** *36,* 633–643.
62. Boschi, T.; Di Vona, M. L.; Tagliatesta, P.; Pistoia, G. Behaviour of Polyaniline Electrodes in Aqueous and Organic Solutions. *J. Power Sources.* **1988,** *24,* 185–193.
63. Taguchi, S.; Tanaka, T. Fibrous Polyaniline as Positive Active Material in Lithium Secondary Batteries. *J. Power Sources.* **1987,** *20,* 249–252.
64. Diaz, A. F.; Logan, J. A. Electroactive Polyaniline Films. *J. Electroanal. Chem.* **1980,** *111,* 111–114.
65. Osaka, T.; Ogano, S.; Naoi, K. Electroactive Polyaniline Deposit from a Nonaqueous Solution. *J. Electrochem. Soc.* **1988,** *135,* 539–540.
66. Geniès, E. M.; Boyle, A.; Łapkowski, M.; Tsintavis, C. Polyaniline: A Historical Survey. *Synth. Met.* **1990,** *36,* 139–182.
67. Ohsaka, T.; Ohnuki, Y.; Oyama, N.; Katagiri, G.; Kamisako, K. IR Absorption Spectroscopic Identification of Electroactive and Electroinactive Polyaniline Films Prepared by the Electrochemical Polymerization of Aniline. *J. Electroanal. Chem.* **1984,** *161,* 399–405.
68. Syed, A. A.; Dinesan, M. K. Review: Polyaniline - A Novel Polymeric Material. *Talanta.* **1991,** *38,* 815–837.
69. Kaplan, S.; Conwell, E. M.; Richter, A. F.; MacDiarmid, A. G. Ring Flips as a Probe of the Structure of Polyanilines. *Macromolecules.* **1989,** *22,* 1669–1675.
70. Łapkowski, M.; Geniés, E. M. Evidence of Two Kinds of Spin in Polyaniline from In Situ EPR and Electrochemistry: Influence of the Electrolyte Composition. *J. Electroanal. Chem.* **1990,** *279,* 157–168.
71. Manohar, S. K.; MacDiarmid, A. G.; Epstein, A. J. Polyaniline: Pernigraniline, an Isolable Intermediate in the Conventional Chemical Synthesis of Emeraldine. *Synth. Met.* **1991,** *41,* 711–714.
72. Łapkowski, M.; Geniés, E. M. Spectroelectrochemical Evidence for an Intermediate in the Electropolymerization of Aniline. *J. Electroanal. Chem.* **1987,** *236,* 189–197.
73. Glarum, S. H.; Marshall, J. H. The In Situ ESR and Electrochemical Behavior of Poly(Aniline) Electrode Films. *J. Electrochem. Soc.* **1987,** *134,* 2160–2165.
74. La Croix, J. C.; Diaz, A. F. Electrolyte Effects on the Switching Reaction of Polyaniline. *J. Electrochem. Soc.* **1988,** *135,* 1457–1463.
75. Rudzinski, W. E.; Lozano, L.; Walker, M. The Effects of PH on the Polyaniline Switching Reaction. *J. Electrochem. Soc.* **1990,** *137,* 3132–3136.
76. Habib, M. A.; Maheswari, S. P. Electrochromism of Polyaniline: An *In Situ* FTIR Study. *J. Electrochem. Soc.* **1989,** *136,* 1050–1053.
77. Shim, Y. B.; Won, M. S.; Park, S. M. Electrochemistry of Conductive Polymers VIII. *J. Electrochem. Soc.* **1990,** *137,* 538–544.
78. MacDiarmid, A. G.; Epstein, A. J. The Polyanilines: Potential Technology Based on New Chemistry and New Properties. In *Science and Applications of Conducting Polymers,* Proceedings of Sixth Europhysics Industrial Workshop, Lofthus, Norway, May 1990; Salaneck, W.R., Clark, D. T., Samuelsen, E. J., Eds.; Adam Hilder, IOP Publishing Ltd.: Bristol, 1991; pp 117–127.
79. Boudreaux, D. S.; Chance, R. R.; Wolf, J. F.; Shacklette, L. W.; Bredas, J. L.; Themans, B.; Andre, J. M.; Silbey, R. Theoretical Studies on Polyaniline. *J. Chem. Phys.* **1986,** *85,* 4584–4590.

80. Pekmez, N.; Pekmez, K.; Yıldız, A. Electrochemical Behavior of Polyaniline Films in Acetonitrile. *J. Electroanal. Chem.* **1994,** *370,* 223–229.
81. Cattarin, S.; Doubova, L.; Mengoli, G.; Zotti, G. Electrosynthesis and Properties of Ring-Substituted Polyanilines. *Electrochim. Acta.* **1988,** *33,* 1077–1084.
82. Geniès, E. M.; Łapkowski, M.; Penneau, J. F. Cyclic Voltammetry of Polyaniline: Interpretation of the Middle Peak. *J. Electroanal. Chem.* **1988,** *249,* 97–107.
83. Stilwell, D. E.; Park, S. M. Electrochemistry of Conductive Polymers. *J. Electrochem. Soc.* **1988,** *135,* 2491–2502.
84. Koval'chuk, E. P.; Whittingham, S.; Skolozdra, O. M.; Zavalij, P. Y.; Zavaliy, I. Yu.; Reshetnyak, O. V.; Seledets, M. Co-polymers of Aniline and Nitroanilines. Part I. Mechanism of Aniline Oxidation Polycondensation. *Mater. Chem. Phys.* **2001,** *69,* 154–162.
85. Pilar, F. L. *Elementary Quantum Chemistry*, 2nd ed.; McGraw-Hill Publ.: New York, 1990.
86. Karpas, Z.; Berant, Z.; Stimac, R. M. An Ion Mobility Spectrometry/Mass Spectrometry (IMS/MS) Study of the Site of Protonation in Anilines. *Struct. Chem.* **1990,** *1,* 201–204.
87. Russo, N.; Toscano, M.; Grand, A.; Mineva, T. Proton Affinity and Protonation Sites of Aniline. Energetic Behavior and Density Functional Reactivity Indices. *J. Phys. Chem. A.* **2000,** *104,* 4017–4021.
88. Koval'chuk, E. P.; Reshetnyak, O. V.; Błażejowski, J. Protonation-Extraction of Hydrogen Ions During Oxidative Condensation of Aromatic Amines. First Russian-Ukrainian-Polish Conference on Molecular Interaction. School of Physical Organic Chemistry: Book of Abstracts. Gdańsk, June 10–16, 2001. Zakład Poligrafii Fundacji Rozwoju Universytetu Gdańskiego: Sopot, 2001; pp 111–112.
89. Ćirić-Marjanović, G.; Trchová, M.; Stejskal, J. Theoretical Study of the Oxidative Polymerization of Aniline with Peroxydisulfate: Tetramer Formation. *Int. J. Quantum Chem.* **2008,** *108,* 318–333.
90. Hedayatullah, M. Oxidation of Primary Aromatic Amines - Review. *Bull. Soc. Chim. France.* **1972,** 7, 2957–2974.
91. Ćirić-Marjanović, G.; Trchová, M.; Stejskal, J. MNDO-PM3 Study of the Early Stages of the Chemical Oxidative Polymerization of Aniline. *Collect. Czech. Chem. Commun.* **2006,** *71,* 1407–1426.
92. Sapurina, I.; Stejskal, J. The Mechanism of the Oxidative Polymerization of Aniline and the Formation of Supramolecular Polyaniline Structures. *Polym. Int.* **2008,** *57,* 1295–1325.
93. Ćirić-Marjanović, G. Recent Advances in Polyaniline Research: Polymerization Mechanisms, Structural Aspects, Properties and Applications. *Synth. Met.* **2013,** *177,* 1–47.
94. Koval'chuk, E. P.; Whittingham, S.; Skolozdra, O. M.; Zavalji, P. Y.; Zavaliy, I. Y.; Reshetnyak, O.V.; Blazejowski, J. Copolymer Aniline and Ortho-Nitroaniline. Part II. Physicochemical Properties. *Mater. Chem. Phys.* **2001,** *70,* 38–48.
95. Koval'chuk, E. P.; Stratan, N. V.; Reshetnjak, O. V.; Whittingham, M. S. In *Synthesis and Properties of Polyanisidines*, XIVth International Symposium on the Reactivity of Solids: Program and Abstracts, August 27–31, 2000; Officina Press: Szeged, Budapest, Hungary, 2000; pp 151.
96. Koval'chuk, E. P.; Stratan, N. V.; Reshetnyak, O. V.; Blazejowski, J.; Whittingham, M.S. Synthesis and Properties of the Polyanisidines. *Solid State Ionics.* **2001,** *142,* 217–224.

97. Nakanishi, K. *Infrared Absorption Spectroscopy, Practical.* Holden-Day, Inc.: San Francisco, 1963.
98. Roßberg, K.; Paasch, G.; Dunsch, L.; Ludwig, S. The Influence of Porosity and the Nature of the Charge Storage Capacitance on the Impedance Behaviour of Electropolymerized Polyaniline Films. *J. Electroanal. Chem.* **1998,** *443,* 49–62.
99. Chiang, J. C.; MacDiarmid, A. G. 'Polyaniline': Protonic Acid Doping of the Emeraldine Form to the Metallic Regime. *Synth. Met.* **1986,** *13,* 193–205.
100. Kobayashi, T.; Yoneyama, H.; Tamura, H. Oxidative Degradation Pathway of Polyaniline Film Electrodes. *J. Electroanal. Chem.* **1984,** *177,* 293–297.
101. Mengoli, G.; Munari, M. T.; Bianco, P.; Musiani, M. M. Anodic Synthesis of Polyaniline Coatings onto Fe Sheets. *J. Appl. Polym. Sci.* **1981,** *26,* 4247–4257.
102. Macdiarmid, A. G.; Mu, S. L.; Somasiri, N. L. D.; Wu, W. Electrochemical Characteristics of "Polyaniline" Cathodes and Anodes in Aqueous Electrolytes. *Mol. Cryst. Liq. Cryst.* **1985,** *121,* 187–190.
103. Tsocheva, O.; Zlatkov, T.; Terlemezyan, L. Thermoanalytical Studies of Polyaniline 'Emeraldine Base'. *J. Therm. Anal. Calorim.* **1998,** *53,* 895–904.
104. Anand, J.; Palaniappan, S.; Sathyanarayana, D. N. Spectral, Thermal, and Electrical Properties of Poly(***o***- and *m*-toluidine)-Polystyrene Blends Prepared by Emulsion Pathway. *J. Polym. Sci. Pol. Chem.* **1998,** *36,* 2291–2299.
105. Gonçalves, D.; Matvienko, B.; Bulhões, L. O. S. Electrochromism of Poly(o-methoxyaniline) Films Electrochemically Obtained in Aqueous Medium. *J. Electroanal. Chem.* **1994,** *371,* 267–271.
106. Widera, J.; Grochala, W.; Jackowska, K.; Bukowska, J. Electrooxidation of o-Methoxyaniline as Studied by Electrochemical and SERS Methods. *Synth. Met.* **1997,** *89,* 29–37.
107. Patil, S.; Mahajan, J. R.; More, M. A.; Patil, P. P. Electrochemical Synthesis of Poly(o-methoxyaniline) Thin Films: Effect of Post Treatment. *Mater. Chem. Phys.* **1999,** *58,* 31–36.
108. Bernard, M. C.; de Torresi, S. C.; Hugot-Le Goff, A. *In Situ* Raman Study of Sulfonate-Doped Polyaniline. *Electrochim. Acta.* **1999,** *44,* 1989–1997.
109. Naudin, E.; Gouérec, P.; Bélanger, D. Electrochemical Preparation and Characterization in Non-Aqueous Electrolyte of Polyaniline Electrochemically Prepared from an Anilinium Salt. *J. Electroanal. Chem.* **1998,** *459,* 1–7.
110. Yang, C. H. Electrochemical Polymerization of Aniline and Toluidines on a Thermally Prepared Pt Electrode. *J. Electroanal. Chem.* **1998,** *459,* 71–89.
111. MacDiarmid, A. G.; Epstein, A. J. New Developments in the Synthesis and Doping of Polyacetylene and Polyaniline. In *Conjugated Polymeric Materials: Opportunities in Electronics, Optoelectronics, Molecular Electronics*, Proceedings of the NATO Advances Research Workshop on Conjugated Polymeric Materials: Opportunities in Electronics, Optoelectronics, and Molecular Electronics, Mons, Belgium, September 3–8, 1989; Brédas, J. L., Chance, R. R., Eds.; Kluwer Academic Publishers: Dordrecht, 1990; pp 53–63.
112. Maia, D. J.; das Neves, S.; Alves, O. L.; De Paoli, M. A. Photoelectrochemical Measurements of Polyaniline Growth in a Layered Material. *Electrochim. Acta.* **1999,** *44,* 1945–1952.
113. Sari, B.; Talu, M. Electrochemical Copolymerization of Pyrrole and Aniline. *Synth. Met.* **1998,** *94,* 221–227.

114. Ferreira, V.; Cascalheira, A. C.; Abrantes, L. M. Electrochemical Copolymerisation of Luminol with Aniline: A New Route for the Preparation of Self-Doped Polyanilines. *Electrochim. Acta.* **2008,** *53,* 3803–3811.
115. Ferreira, V.; Cascalheira, A. C.; Abrantes, L. M. Electrochemical Preparation and Characterisation of Poly(Luminol–Aniline) Films. *Thin Solid Films.* **2008,** *516,* 3996–4001.
116. Zhang, G. F.; Chen, H. Y. Studies of Polyluminol Modified Electrode and Its Application in Electrochemiluminescence Analysis with Flow System. *Analyt. Chim. Acta.* **2000,** 419, 25–31.
117. Chen, S. M.; Lin, K. C. The Electrocatalytic Properties of Biological Molecules Using Polymerized Luminol Film-Modified Electrodes. *J. Electroanal. Chem.* **2002,** *523,* 93–105.
118. Chang, Y. T.; Lin, K. C.; Chen, S. M. Preparation, Characterization and Electrocatalytic Properties of Poly(Luminol) and Polyoxometalate Hybrid Film Modified Electrodes. *Electrochim. Acta.* **2005,** *51,* 450–461.
119. Lin, K. C.; Chen, S. M. Reversible Cyclic Voltammetry of the NADH/NAD^+ Redox System On Hybrid Poly(luminol)/FAD Film Modified Electrodes. *J. Electroanal. Chem.* 2006, *589,* 52–59.
120. Wang, C. H.; Chen, S. M.; Wang, C. M. Co-Immobilization Of Polymeric Luminol, Iron(II) Tris(5-aminophenanthroline) and Glucose Oxidised at an Electrode Surface, and its Application as a Glucose Optrode. *Analyst.* **2002,** *127,* 1507–1511.
121. Sassolas, A.; Blum, L. J.; Leca-Bouvier, B. D. Electrogeneration of Polyluminol and Chemiluminescence for New Disposable Reagentless Optical Sensors. *Anal. Bioanal. Chem.* **2008,** *390,* 865–871.
122. Mendonça, T. P.; Moraes, S. R.; Motheo, A. J. Influence of the Synthesis Parameters on the Polyluminol Properties. *Mol. Cryst. Liq. Cryst.* **2006,** *447,* 383–391.
123. Koval'chuk, E. P.; Grynchyshyn, I. V.; Reshetnyak, O. V.; Błażejowski, J. Oxidative Condensation and Chemiluminescence of 5-Amino-2,3-Dihydro-1,4-Phtalazinedione. *Euro. Polym. J.* **2005,** *41,* 1315–1325.
124. Malinauskas, A.; Holze, R. An *In Situ* UV–Vis Spectroelectrochemical Investigation of the Initial Stages in the Electrooxidation of Selected Ring- and Nitrogen-Alkyl substituted Anilines. *Electrochim. Acta.* **1999,** *44,* 2613–2623.
125. Alpert, N. L.; Keiser, W. E.; Szymański, H. A. *IR: Theory and Practice of Infrared Spectroscopy*; Plenum Press: New York, 1970; pp 388.
126. De Azevedo, W. M.; de Souza, J. M.; de Melo, J. V. Semi-Interpenetrating Polymer Networks Based on Polyaniline and Polyvinyl Alcohol-Glutaraldehyde. *Synth. Met.* **1999,** *100,* 241–248.
127. Schweinsberg, D. P.; Bottle, S. E.; Otieno-Alego, V.; Notoya, T. A Near-Infrared FT-Raman (SERS) and Electrochemical Study of the Synergistic Effect of 1-[(1′,2′-Dicarboxy)Ethyl]-Benzotriazole and KI on the Dissolution of Copper in Aerated Sulfuric Acid. *J. Appl. Electrochem.* **1997,** *27,* 161–168.
128. Baibarac, M.; Cochet, M.; Łapkowski, M.; Mihut, L.; Lefrant, S.; Baltog, I. SERS Spectra of Polyaniline Thin Films Deposited on Rough Ag, Au and Cu. Polymer Film Thickness and Roughness Parameter Dependence of SERS Spectra. *Synth. Met.* **1988,** *96,* 63–70.
129. Hatchett, D. W.; Josowicz, M.; Janata, J. Comparison of Chemically and Electrochemically Synthesized Polyaniline Films. *J. Electrochem. Soc.* **1999,** 146, 4535–4538.
130. Wells, A. F. *Structural Inorganic Chemistry,* 5th ed.; Clarendon Press: Oxford, 1987.

131. Ouyang, C. S.; Wang, C. M. Clay-Enhanced Electrochemiluminescence and Its Application in the Determination of Glucose. *J. Electrochem. Soc.* **1998,** *145,* 2654–2659.
132. Ouyang, C. S.; Wang, C. M. Electrochemical Characterization of the Clay-Enhanced Luminol ECL Reaction. *J. Electroanal. Chem.* **1999,** *474,* 82–88.
133. Chen, S. M.; Lin, K. C. The Electrocatalytic Properties of Biological Molecules Using Polymerized Luminol Film-Modified Electrodes. *J. Electroanal. Chem.* **2002,** *523,* 93–105.
134. Jenkins, A. D.; Kratochvíl, P.; Stepto, R. F. T.; Suter, U. W. Glossary of Basic Terms in Polymer Science (IUPAC Recommendations 1996). *Pure Appl. Chem.* **1996,** *68,* 2287–2311.
135. Purple Book: IUPAC Compendium of Macromolecular Nomenclature. Blackwell Scientific Publications: Oxford, 1991; pp 18.

CHAPTER 2

MORPHOLOGY OF POLYANILINE'S FILMS ELECTROCHEMICALLY DEPOSITED ON THE SURFACE OF AL-BASED AMORPHOUS METAL ALLOYS

M. M. YATSYSHYN, L. M. BOICHYSHYN, I. I. DEMCHYNA, and YU. A. HNIZDIUKH

Department of Physical and Colloid Chemistry, Faculty of Chemistry, Ivan Franko National University of Lviv, 6 Kyryla & Mefodia Str., Lviv 79005, Ukraine

Corresponding author: reshetniak@franko.lviv.ua

CONTENTS

ABSTRACT

The films of polyaniline (**PAn**) have been synthesized by potentiodynamic oxidation of aniline (**An**) in aqueous solution of 0.5 M H_2SO_4 at the electrodes of amorphous metal alloys (**AMA**) of the composition $Al_{87}Ni_8(REM)_5$, where REM≡Y, Ce, Gd, and Dy, as well as polycristalline aluminum. The process of electrochemical oxidation of **An** on **AMA**-electrodes was analyzed and compared with the process of **An**'s oxidation on the polycrystalline Al-electrode. It was established that the difference in the form of cyclic voltammogramms is conditioned by the presence of amorphic components in the **AMA** composition, which cause of different resistance of surface oxide films on the working electrodes (**WE**). With the use of X-ray and IR-spectral analysis it was showed that the structure of **PAn**'s films on the surface of the Al-electrode and the $Al_{87}Ni_8(REM)_5$ electrodes is amorphous–crystalline. At this, the **PAn** deposited on $Al_{87}Ni_8(REM)_5$ electrodes has a higher degree of crystallinity. An analysis of the images of scanning electron microscopy (**SEM**) showed that on the surface of the **WE**s of amorphous alloys $Al_{87}Ni_8(REM)_5$ the **PAn**'s films have spongy and porous branched morphology. An analysis of spectra of energy variance of electrons confirmed the presence on the surface of the electrodes of **PAn**, and showed the presence of impurities in the polymeric film of metal sulfates, which are part of the **WE**s.

2.1 INTRODUCTION

Electroconductive polymers (**ECP**) are relatively new class of polymers,[1–5] the most famous representative among which is the polyaniline (**PAn**). Easy methods of synthesis, the possibility of acquiring through the mechanism of doping/dedoping of various forms, unique physicochemical properties among which important are high electrical conductivity, chemical sensitivity, multicoloured electrochromism, catalytic properties, limited solubility, chemical and thermal resistance, high adhesiveness to the surfaces of different nature etc., make the **PAn** and composites on its basis by important materials in modern tehnologies.[1, 2, 4–9]

The most common usable methods after chemical synthesis are the electrochemical methods of **PAn** obtaining, namely galvanostatic method (GS), potentiostatic method (PS), and potentiodynamic method (PD).[10–14] Galvanostatic polymerization of aniline (**An**) makes easy to control the properties of the **PAn**'s films, namely molecular weight of polymer and the thickness of deposited coatings. Potentiostatic mode of polymerization in turn permits to control by the reactivity of electrochemically active intermediates of the starting monomer's oxidation during reaction. Potentiodynamic polymerization (cyclic voltammetry), which is carried out under cyclic scanning of the potential electrode allows to control in real time both of **An** oxidation and redox conversions of deposited **PAn**. Cyclic voltammetry is actively used for the determination of mechanisms of **An** oxidation, redox reactions of **PAn**,[15] mechanisms of ion exchange in films of **PAn**,[16] as well as for the researches of the film's stability,[17] dispersions and **PAn**'s composites,[18] capacitive characteristics of **PAn**'s films,[19–23] electrical activity of **PAn**'s films,[23] and others.

Electrochemical methods are used for obtaining of polymeric coatings free from the oxidants directly on the metals surface. In addition to noble metals (Pd, Pt, Au), which are used the most often, by the working electrodes (**WE**) can serve also active metals Mg, Al, Ti, Cr, Fe, Ni, Cu, Zn, Ag, In, Pb),[17, 2538] the alloys based on Fe including also stainless steels, such as trademark SS 304,[39,40] SS 316L,[41,42] etc., the aluminum alloys (**AA**), including polycrystalline **AA** 1100,[43] **AA** 2024 43-T3,[44,45] **AA** 2024-T6,[46] **AA** 3004,[47] **AA** 5182,[48] **AA** 7075[49] and amorphous alloys,[50–53] magnesium alloy by trademark AZ91D,[54] the alloy $Co_{67}Cr_{29}W_4$,[55] and others. Sometimes the films of **PAn** are synthesized on metal oxide surfaces such as IrO_x,[56] SnO_2,[57] PbO_2,[58] $In_2O_3 \times SnO_2$ (ITO).[59]

By choosing the electrochemical method and the conditions of the **An**'s oxidation on the surface of **WE** it can be obtained the **PAn**'s film with different structure of macromolecular chains (from linear to branched), different morphology from micro- or nano-wires to micro- or nano-tubes and also different topography of the surface from smooth films to the layers with developed surface.[60–63] The morphology of electrochemically deposited **PAn**'s films on the surfaces of electrodes is influenced by various factors, including the method of electrochemical oxidation of **An**,[10–13, 47, 64] current density,[61] velocity potential sweep in the first and subsequent cycles,[65] the boundary of the potential sweep[39] and the value of applied potential,[39] duration of polymerization and conductivity of electrode,[66] nature of anione of acid-electrolyte,[67] nature,[12, 68] temperature,[68] pH of medium,[64] the presence of oxide film on the surface of **WE**,[69] others. Films of **PAn**, deposited by PD have more developed surface than obtained by other electrochemical methods.[49, 70, 71]

An important feature of **ECP**, which makes their use, is a phenomenon of the proceeding of redox transformations in macromolecules of **ECP** under the action of potential, current, or chemical factors. Another feature of films of **ECP**, which is important in their application, is their morphology and structure. These characteristics come to the fore, when **PAn** is used in chemo- and bio-sensors (since determine the size of response of **PAn** films at detecting of various substances-analytes),[72] in the manufacture of supercapacitors (increasing of their capacity),[19, 73, 74] electrochemical energy sources (reducing the polarization losses),[23, 75] in catalytic coatings (increasing the effectiveness of catalysis),[76–80] in nano- and micro-drives (artificial muscles) (increasing an angle of their deformation),[81,82] more. The morphology of **PAn**'s layers is also crucial under their application in modern electronic technologies. The large number of active centers per unit of mass of the polymer accessible to oxidation/reduction reaction as well as the porosity of the **PAn**'s structures allow rapid diffusion of ions in polymer networks that are the most important requirements for the construction of a high-energy batteries, supercapacitors, transducers of chemosensors and bio-sensors, nano- and micro-drives (artificial muscles),[23] etc. However, smooth and adhesive, chemically and thermally stable, impermeable to H_2O and small ions such Cl^-, films of **PAn** also have broad prospects for their use, including a protective anticorrosion coatings of active metals and alloys on their basis.[36, 49, 83]

Surface condition of **WE**, especially from the active metals, has an important significance for the deposition of **PAn**'s film. Usually, in literature data it is not described the condition of **WE** surfaces, such as cleanness of the surface treatment, the presence of oxide films, the composition and structure of alloys of different nature, but there is only way of the surface preparation **WE**. The clarification of an impact of the condition and composition of oxide films on potentiodynamic oxidation of An is an important task because it allows the use of active metals without additional costs for surface preparation, and in many cases enhances the adhesion of electrochemically deposited films of **ECP**. Suitable materials for **WE**, which are characterized with listed above properties, can serve the samples of AMAs based on aluminum.[50–52] Therefore, important is the study of potentiodynamic deposition of **PAn**'s films on amorphous metal surfaces of **WE** due to the possibilities of their future use.

2.2 EXPERIMENTAL

2.2.1 MATERIALS

An (99.5%) of the "*Aldrich*" company was distilled in a vacuum. Solutions of sulfuric acid (H_2SO_4) were prepared from the standard titrimetric substance of Cherkasy State Plant of Chemicals Production. Distilled water was used as the solvent. Ethanol was distilled under normal conditions.

Samples of **AMA** based on aluminum of composition $Al_{87}Ni_8Y_5$ (AlNiY-electrode), $Al_{87}Ni_8Ce_5$ (AlNiCe-electrode), $Al_{87}Ni_8Gd_5$ (AlNiGd-electrode),

$Al_{87}Ni_8Dy_5$ (AlNiDy-electrode), as well as polycrystalline aluminum (Al-electrode, purity 99.995%) in the form of plates with the thickness ~40 μm, size ~2.0 × 0.2 cm and active surface ~0.2 cm^2 were used as the **WE**.

2.2.2 ELECTROCHEMICAL DEPOSITION OF PAn's FILMS

In the ribbon **AMA** obtained by the flow turning method there are two sides, namely contact (adjacent to the cooling drum) with developed (defective) surface and the external (which is in contact with the atmosphere of helium) having a smooth surface.[51] In the presented work, the films of **PAn** were deposited simultaneously on both sides of the **AMA** samples, which were used as the **WE**s. **WE**s were previously washed with ethanol and were air-dried for 5 min. Electrodeposition of **PAn** was performed from air-free argon for 10 min 0.25 M aqueous solution of An in 0.5 M H_2SO_4 simultaneously on both sides of the electrodes at the speed of a potential sweep 50 mV/s within −200−1200 mV. The films of **PAn** on the **WE**s were formed for 75 cycles of the sweep potential. Electrodes with the coated films of **PAn** were washed with distilled water and dried at room temperature.

2.2.3 INSTRUMENTAL METHODS

The deposition of **PAn**'s films was conducted by PD at the facility for electrochemical and electrochemiluminescent researches CVA-1 accordingly to three-electrodic scheme with Ag/AgCl reference electrode by EVL-1M4 mark. All values of the electrode potentials are regarded as for this reference electrode. Cyclic voltammogramms were recorded on a personal computer. Platinum plate (99.9%) with the size of 1 × 1 cm was used as the antielectrode.

The structure of the synthesized **PAn** films was studied using X-ray diffraction (XRD) and Fourier-transform infrared spectral (**FTIR**) analysis. The difractograms of the samples were received with the use of diffractometer by DRON-3 mark (*Cu–Kα* radiation, λ = 1.54060 Å). **FTIR** spectra of the samples were recorded with the use of spectrophotometer NICOLET IS 10 in reflection mode. XRD and **FTIR** analysis of **PAn**'s films was performed directly on the surface of **WE**.

To study the morphology of obtained **PAn**'s layers (scanning electron microscope (**SEM**)-images) on **WE** and energy dispersive X-ray (EDX) analysis of composition of films it was used the SEM-microanalyzer SELMA by PEMMA-102–02 mark.

2.3 RESULTS AND DISCUSSION

2.3.1 POTENTIODYNAMIC OXIDATION OF ANILINE

Analysis of potentiodynamic curves of **An**'s oxidation and redox transformations of **PAn** on the surface of AlNiY-electrode[50] showed the difference in a form of a peak for oxidation of **An** and cyclic voltammogramms on a peak of **An**'s oxidation and curves of redox transformations of **PAn** on Pt, Au, Al, and other electrodes.[17, 49, 70] Therefore, to characterize and understand the processes occurring under potentiodynamic oxidation of **An** on the electrodes with the Al-based **AMA** we have conducted researches on potentiodynamic oxidation of **An** on polycrystalline Al-electrode. Cyclical voltammograms depicted in Figure 2.1 obtained by oxidation of **An** in its 0.25 M aqueous solution of 0.5 M H_2SO_4. The first branch of the anode cycle within potentials 0–1200 mV is slightly dissolving of Al-electrode.[36] As shown in Figure 2.1a, characteristic high-current peak, just as on Pt-electrode,[17, 50, 52] which is associated with the electrochemical oxidation of **An**, is not observed on the cyclical voltammogramms due to the presence of oxide film (Al_2O_3) on the surface of Al-electrode, which prevents the oxidation of **An**. At the prolonged (for 25 cycles) potential (*E*) scanning of Al-electrode, gradually dissolving of the oxide film or the formation of pittings in it takes place, where the oxidation of a small number of **An** is occured.[50] Clearly to establish the potential of **An**'s oxidation is difficult due to very small current. However, at detailed study of cyclical voltammogramms it can be said that this takes place at E ≈ 1000 mV. For 25 cycle on the cathodic branch of cyclical voltammogramm at *E*≈ 460 mV there is a low-current peak of the reduction of **PAn**, or rather transformaton of its pernigraniline form into emeraldine, which at *E* ≈ 200 mV is reduced deeper, namely to leucoemeraldine form of **PAn**. On the next anodic branch of cyclical voltammogramm a peak at *E* ≈ 250 mV is observed, which corresponds to oxidative transformation leucoemeraldine/emeraldine **(L/E)** and as well as a peak at *E* ≈ 650 mV, which corresponds to the oxidation of emeraldine to pernigraniline *E/Pn* (see Fig. 2.1a).[15, 36, 84] In subsequent cycles, the values of anodic and cathodic current peaks are increased (see Fig. 2.1). At the anode potential sweep to ~1000 mV on the **WE** the **PAn** is formed as the pernigraniline salt (**PnS**), which is one of the main forms of **PAn**. And at the cathodic potential sweep to ~ −200 mV the **PAn** is formed as leucoemeraldine salt (**LS**), which is also a form of **PAn**.[1, 2, 4, 5, 15, 36]

Increase of the number of cycles leads to the displacement of the potential of the first anodic peak (**L/E**) in the region of higher potentials (see Fig. 2.1a), which is a sign of some growing of the electrical resistance of **PAn**'s film. The difference between the values of potentials of the first anodic peaks of the 30th and the 75th cycles is ~250 mV. Figure 2.1b demonstrates the presence of the induction period (parallel to the abscissas plot of the curve) with the duration ~650 s.[17] The duration of the induction period was determined only taking into account the time of anodic potential sweep. An availability of the induction period is caused by the dissolution of the oxide film on the surface of Al-electrode. The ascending branch of the curve shows that the growth of the film of **PAn** on the Al-electrode begins actively to occur after the end of the induction period. The potentials of the first anodic peak grow stronger than that of the second anodic peak (see Fig. 2.1b). Overall, the form of cyclical voltammogramms of the redox transformations of **PAn** on Al-electrode is similar to that described in ref.[36]

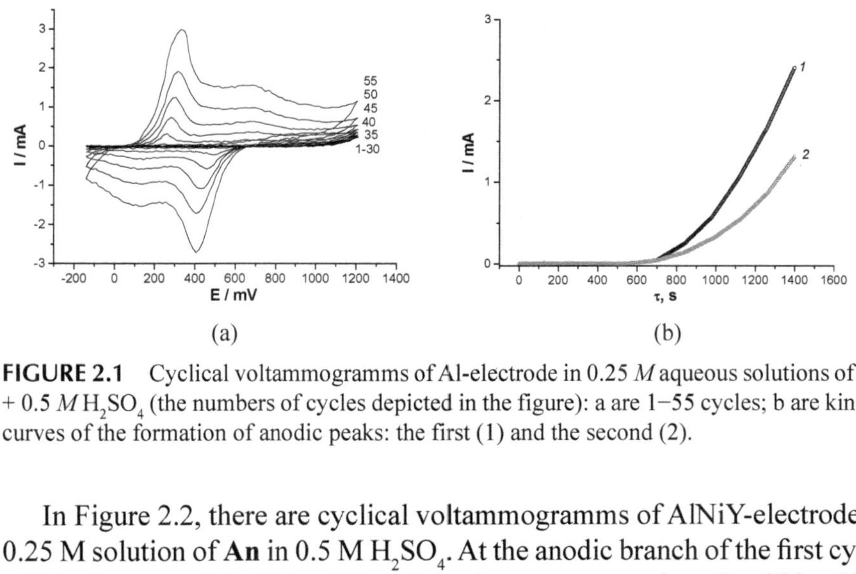

FIGURE 2.1 Cyclical voltammogramms of Al-electrode in 0.25 *M* aqueous solutions of **An** + 0.5 *M* H_2SO_4 (the numbers of cycles depicted in the figure): a are 1–55 cycles; b are kinetic curves of the formation of anodic peaks: the first (1) and the second (2).

In Figure 2.2, there are cyclical voltammogramms of AlNiY-electrode in 0.25 M solution of **An** in 0.5 M H_2SO_4. At the anodic branch of the first cycle within the potentials 220 – 470 mV the low-current peak at $E = 280$ mV of AlNiY-electrode dissolution is observed. Then the passivation of **WE** takes place. Peak, which is responsible for oxidation of **An**, is not observed on the anodic branches of 1–40 cycles (see Fig. 2.2a). At the cathodic branch of the ~ 40 cycle of the potential sweep at $E \approx 300$ mV the cathodic peak is observed that as in the case of Al-electrode, corresponds to the reduction of **Pn** to **E** form of **PAn**. After this peak at $E \approx 50$ mV the reduction process of **E** to **L** is started. At the anodic potential scanning at $E \approx 400$ mV the peak is formed, which corresponds to the oxidation of **L** in **E** (see Fig. 2.2a). Active formation of **PAn** on AlNiY-electrode is started after ~800 s of anodic scanning potential and is described by a sharp rise on kinetic curve (see Fig. 2.2b). Currents of the second anodic peak grow more intense than that of the first anodic peak (see Fig. 2.2 b).

Figure 2.2c shows the cyclical voltammogramms of AlNiCe-electrode. For 45 scanning cycles of the potential the obvious signs of **An**'s oxidation are not traced. An analysis of the cathodic branch of cyclical voltammogramm of the 45 cycle shows that the low-current peak at $E \approx 220$ mV corresponds to reduction of PNAn. Already in the 46 cycle on the cyclical voltammogram, two peaks of **PAn** oxidation at $E \approx 620$ and 1050 mV are traced. Form of the red/ox peaks on cyclical voltammogramms is differed from the peaks on the cyclical voltammogramms of Al-electrode (see Fig. 2.2a). Cathode peak at $E \approx 200$ mV corresponds to the reduction of **PAn** obtained in the anode process. This peak with increasing of the number of cycles is shifted to the cathode side and in 75 cycle, its potential is equal to 25 mV. Continuous scanning of the potential leads to higher anodic and cathodic currents peaks (see Fig. 2.2d). The character of anodic branches of cyclical voltammogramms shows that the oxidation of **L** into **E** and **E** into **Pn** is similar to this process on other **WE**. The difference between the values of the potentials (ΔE) of the first anodic peaks of 51 and 75 cycles consists of ~301 mV. For the kinetic curve (see Fig. 2.2d) inherent an existence of the induction period. Active formation of **PAn** begins after achievement of 1100 s; then, sharp rise is appeared on the curves. The currents of the first anodic peak are increased stronger than the currents of the second peak (see Fig. 2.2d).

Figure 2.2e shows the cyclical voltammogramms of AlNiGd-electrode. For 50 cycle on cyclical voltammogramm at $E \approx 450$ mV, the peak of **PAn** reduction is observed (see Fig. 2.2e). At the anode branch of 51 cycle two peaks at potentials ~580 and ~1080 mV are observed. Continuous scanning of the potentials leads to higher currents redox peaks of **PAn** and the shift of their anode potential in the anode side and cathode potential in the cathode side. The difference between the values of potentials (ΔE) of the first anodic peaks of 51 and 75 cycles consists of ~897 mV. Potential of the cathodic peak of 75 cycle is 130 mV. Kinetic curve of the anodic processes proceeding on AlNiGd-electrode is shown in Figure 2.2e. Apparently, the sharp rise on the kinetic curve is observed after 1500 s and is a sign of the start of active formation of **PAn** on the surface of **WE**. Currents of anodic peaks increase with the same intensity (see Fig. 2.2f).

Figure 2.2, *g* presents the cyclical voltammogramms of AlNiDy–electrode. At the anodic branch of the first cycle within the potentials 100 ÷ 300

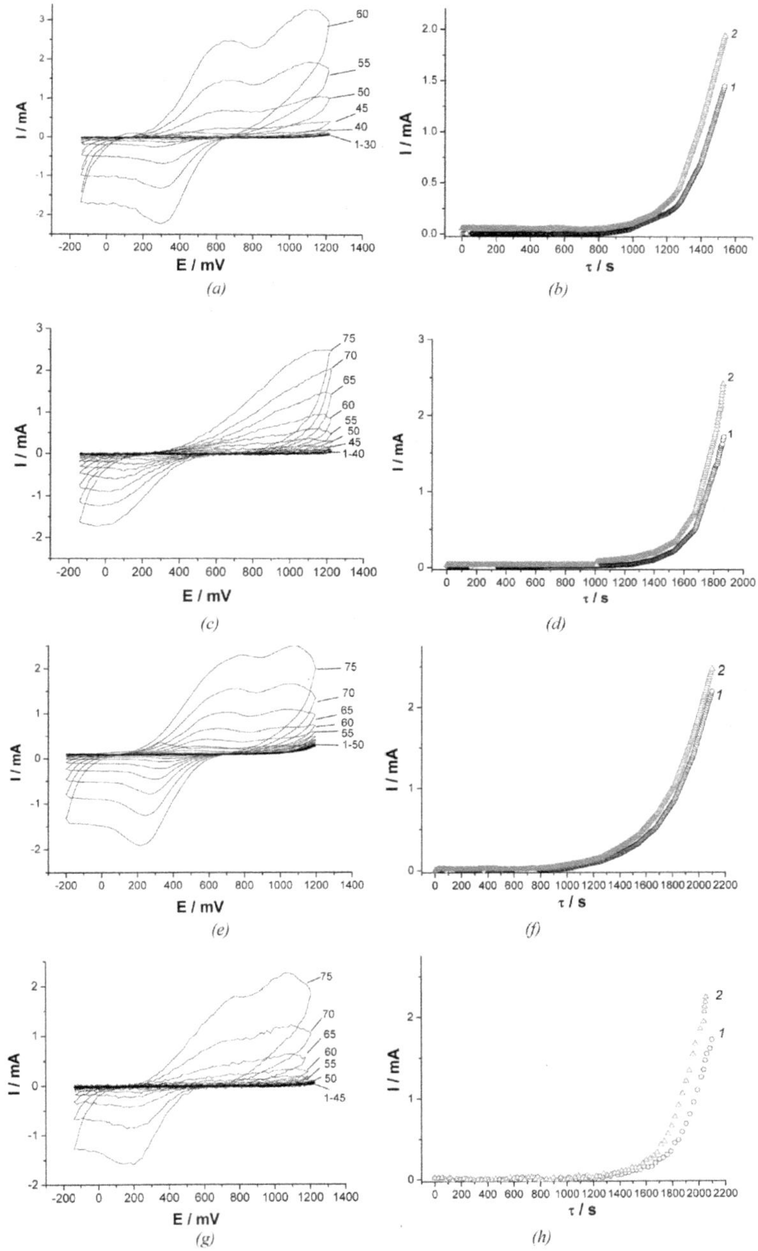

FIGURE 2.2 Cyclical voltammogramms of **WE**s in 0.25 M aqueous solutions of **An** + 0.5 M H_2SO_4 (the numbers of cycles depicted in the figure): a is AlNiY; c is AlNiCe; e is AlNiGd; g is AlNiDy and kinetic curves of the formation of anodic peaks on the electrodes; b is AlNiY; d is AlNiCe; f is AlNiGd; h is AlNiDy.

mV the low–current peak is observed at $E = 280$ *mV*, which corresponds to the dissolution of AlNiDy–electrode. Then the passivation of **WE** takes place. The pick, which corresponds to the **An** oxidation on the anodic branches of 1 – 54 cycles is not observed (*see* Figure 2.2, *g*). At long potential sweep the cathodic peak at $E \approx 190$ *mV* is formed on the 54 cycle. At the anodic branch of 55 cycle two distinct peaks with anodic potentials ~790 and ~1080 *mV* are formed. These red / ox peaks correspond to conversion of **PAn** like to Al–electrode. As shown on Figure 2.2, *h* the start of active formation of **PAn**'s film begins after 1150 *s* of the completion of the induction period. The currents of the first anodic peak grow stronger than the currents of the second peak (*see* Figure 2.2, *h*).

Different duration of the induction period, typical for the beginning of active formation of **PAn**'s film on each of the investigated **WE**, caused by the different properties of the passivation surface oxide films and the duration of their dissolution.[17, 85] There is an obvious fact that the **An**'s molecules on the surface of **WE** do not inhibit the process of surface oxide films dissolution.[86]

For red/ox transformations of **PAn** on the Al-electrode is characteristic higher reversibility (small difference of the potentials $\Delta E'$ between conjugated anodic and cathodic peaks on the cyclical voltammogramms) compared with **AMA**-electrodes (see Table 2.1). At the transition from AlNiY, AlNiGd to AlNiDy-electrode the reversibility of the red/ox process conversion of **PAn** is increased, the value ΔE is decreased (see Table 2.1). Exclusion has the AlNiCe-electrode: the reversibility red/ox conversion in macromolecule of **PAn** is significantly lower than on other electrodes of the **AMA**.

TABLE 2.1 Parameters of Potentiodynamic Oxidation of **An** and Redox Transformations of **PAn** on the Investigated Electrodes (the 45th Cycle).

WE	Potentials ± 1.0/mV						ΔE/mV
	Oxidation	Redox transformations of PAn				ΔE/mV	
	An	Anodic peaks		Cathode peak			
Al	~1000	270	660	82	–	280	250
AlNiY	~1050	700	1100	220	–	880	350
AlNiCe	~1050	620	1185	210	–	887	301
AlNiGd	~1050	750	1083	190	–	897	224
AlNiDy	~1050	780	1080	180	–	900	161

Notes: *ΔE (reversibility) was determined as the difference between of the first anodic and the second cathode peaks. **$\Delta E'$ was determined as the difference between the potentials of first anodic peaks for 50th and 75th cycle of potential scanning.

Analysis of the results of potentiodynamic oxidation of **An** and red/ox transformations of **PAn** on the the electrodes of Al and **AMA** shows that the existing difference in peaks' form of red/ox transformations of **PAn**, as well as in the kinetics of **PAn**'s deposition are due to surface condition of **WE**, which is determined by the influence amorphic component (Y, Ce, Gd, and Dy), which, obviously affects the higher resistance of the oxide film on the surface of **WE**. With an increasing of number of scanning cycles of potential the difference between the potential of the first anodic peak of 50 and 75 cycles is decreased in the series (see Table 2.2):

$$\text{AlNiY} > \text{AlNiCe} > \text{AlNiGd} > \text{AlNiDy}.$$

The obtained films of **PAn** on **WE** had dark green/black color characteristic for **PAn** in oxidized form **E/Pn**,[36] so in general, electrochemical oxidation reaction of An at the anode (**WE**) can be described by the scheme:[17, 87]

$$C_6H_5NH_2 \xrightarrow{-ne^-,-nH^+} -(C_6H_4\text{-}NH)_n - \cdot$$

The main peaks of electrochemical conversion of **PAn** find the anodic peaks at potential ~300 and ~750 mV, which correspond to the oxidation of **L** to partially oxidized **E** form, and the second peak corresponds to the oxidation of **E** to fully oxidized **Pn** form.[23]

2.3.2 STRUCTURAL STUDIES

X-ray diffraction analysis of PAn products on WEs. Figure 2.3 shows the difractograms of **PAn**, potentiodynamically deposited on Al and **AMA**-electrodes in the form of films. As shown from Figure 2.3, on the all difractograms against the background of halo within $2\theta = 17 - 28°$ there are two characteristic peaks of **PAn** at $2\theta = 20.4°$, which corresponds to EmB and at $2\theta = 24.1°$, which corresponds to EmS.[53, 88, 89] The peak centered at $2\theta = 20.4°$ can be ascribed to the periodicity parallel to the polymer chain, while the peak at $2\theta = 24.1°$ may be caused by the periodicity perpendicular to the polymer chains of **PAn**.[75, 90, 91]

The intensities of diffraction peaks of the **PAn**'s sample, formed on the Al-electrode (see Fig. 2.3, curve 1) are the lowest, and the intensities of diffraction peaks of **PAn** on the AlNiY-electrode are almost proportionate. The intensity of the diffraction peak of **PAn** at $2\theta = \sim 24.1°$ formed on the surface of AlNiCe-, AlNiGd-, and AlNDy-electrodes is higher than the intensity of peak at $2\theta = \sim 20.4°$ certifying the change ratio EmB/EmS toward EmS and its higher crystallinity (sharp and narrow peak). Characteristic peaks at $2\theta = \sim 24.1°$ on diffractograms of **PAn**'s samples on the AlNiCe-, AlNiGd-, and AlNDy-electrodes (see Fig. 2.3, difractograms 3, 4, and 5) is higher than the peaks on the Al- and AlNiY-electrodes (see Fig. 2.3, difractograms 1 and 2), which is indicative of a higher degree of crystallinity of **PAn** in films (see Table 2.2). The presence of two diffraction peaks in the background amorphous halo is a sign that the formed **PAn** has amorphous–crystalline structure in which coexist crystallites of EmS (metalloids) with the crystallites of unprotonated form of EmB.[36, 92] High intensity of the diffraction peaks at $2\theta = \sim 24.1°$ compared with the peaks at $2\theta = \sim 20.2°$ shows that the proportion of metalloid crystallites is too higher (see Table 2.2).

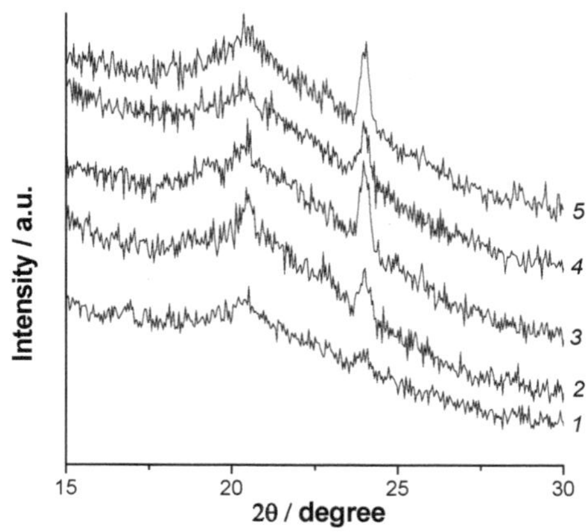

FIGURE 2.3 Difractograms of **PAn** on the surface of **WE**s: 1 is Al; 2 is AlNiY; 3 is AlNiCe; 4 is AlNiGd; 5 is AlNDy.

TABLE 2.2 Structural Parameters of **PAn** in the Films on the Surface of **WE.**

Sample	Crystallinity degree/% at		The ratio of intensities	
	$2\theta = 24.1$	$2\theta = 20.4$	$2\theta_{24.1}/2\theta_{20.4}$	I_{1484}/I_{1570}
Al	1.3	13.8	1.2	1.3
AlNiY	1.0	13.9	1.1	1.2
AlNiCe	2.3	18.1	2.0	1.3
AlNiDy	1.8	8.2	2.6	1.3
AlNiGd	1.3	16.8	3.0	1.2

FTIR spectra of PAn products on WEs. The **FTIR** spectra of reflection of **PAn**'s films deposited by PD on the surface of **WE** are shown in Figure 2.4 and the band assignments are collected in Table 2.3. **FTIR** spectra of **PAn**'s films on the surface of **WE** are, mainly, similar to the **FTIR** spectra of **PAn** synthesized in chemical[93–97] and electrochemical[44, 47, 90, 98–102] ways. The main characteristic bands characteristic for **PAn** are the following 3443, 3200, 3050, 2920, 2846, 1576, 1490, 1301–1299, 1150–1136 i 808–783 cm^{-1}. The locations of these characteristic peaks present a good agreement with the literature.[44, 47, 90, 93–102] The peak positions of various bonds obtained in these spectra (see Fig. 2.4) along with their bonds are given in Table 2.3.

TABLE 2.3 The Assignment and Frequency of the Peaks in the Region **FTIR** Spectra 1600–650 cm^{-1} of **PAn** Samples in Figure 2.4.

Sample	Interatomic bond/Vibrational freguensy, cm^{-1}							
	C=C quinoid ring	C=C benzoid ring	C–N	C–N–C	C=C quinoid ring	SO_4^{2-}	B–NH–Q or B–NH–B	C–H
Al	1564	1481	1296	1244	1136	1065	973	796
AlNiY	1564	1487	1301	1246	1149	1046	970	799
AlNiCe	1574	1487	1302	1247	1149	1070	972	799
AlNiGd	1574	1488	1301	1245	1149	1077	986	799
AlNiDy	1574	1489	1302	1244	1150	1073	972	799

Deformation bands at 3400–3200 cm^{-1} and 2950–2830 cm^{-1} correspond to valence vibrations of N–H groups of aromatic amines, and to valence vibrations of C–H groups of aromatic ring.[68, 103, 104] Absorption bands within 3400–2800 cm^{-1}, commonly are called "H-peaks," also attributed to the hydrogen bonds between the regularly placed **PAn** chains that form the amino (–NH–) and attaching proton imino (–NH^+–) groups.[94, 89] Bands within 3270 and 3200 cm^{-1} are also referred to intra- and inter-molecular hydrogen-bonded N–H stretching vibrations of the secondary amines.[105] The presence of a weak band at 2920 cm^{-1}, which corresponds to N–H modes may be a sign of crosslinking (structuring) of polymer chains[106] or the formation of fully reduced LEm and partially oxidized Em bonds at 2922 and 2923 cm^{-1} correspond to N–H modes with sharp intense bond edges and fully oxidized pernigraniline (PNAn) form of bond at 2922 cm^{-1} with low intense bond edge as shown in Figure 2.4.[102]

A weak peak on **FTIR** spectra of **PAn** on Al-electrode at 1733 cm^{-1} (see Fig. 2.4, the spectrum 1) obviously can be attributed to the absorptions of C=O group, which indicates that **PAn** in films is slightly overoxidized during the growth process of electropolymerization.[107] A similar band on the **FTIR** spectra of **PAn**'s films on the **AMA**-electrodes is not observed (see Fig. 2.4, the spectra 2–5). The weak band at ~1665 and an intense band at ~1407 cm^{-1} on **FTIR** spectra of **PAn** on **AMA**-electrodes are referred to cross-linking of **PAn**'s macromolecules.[108] The combination observed on the **FTIR** spectra of the absorption bands at 1665, 1407, and 687 cm^{-1} was attributed by the authors of ref.[109] to the first oligomeric products of oxidation of **An**.

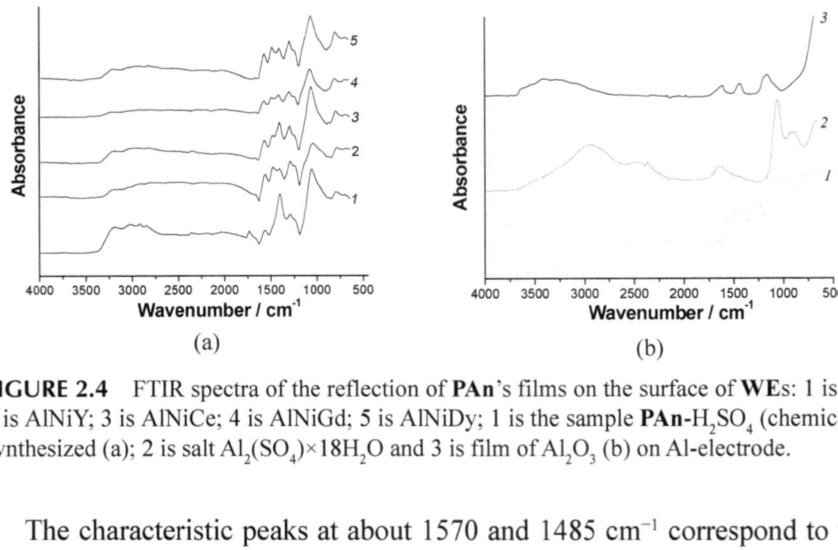

(a) (b)

FIGURE 2.4 FTIR spectra of the reflection of **PAn**'s films on the surface of **WE**s: 1 is Al; 2 is AlNiY; 3 is AlNiCe; 4 is AlNiGd; 5 is AlNiDy; 1 is the sample **PAn**-H_2SO_4 (chemically synthesized (a); 2 is salt $Al_2(SO_4) \times 18H_2O$ and 3 is film of Al_2O_3 (b) on Al-electrode.

The characteristic peaks at about 1570 and 1485 cm^{-1} correspond to the stretching vibrations of N=Q=N ring, N–B–N ring, respectively (where Q refers to the quinonic type rings and B refers to the benzenic type rings).[44, 47, 68, 89, 93–103, 105–107] An intensity of the "benzoid" band exceeds an intensity of the "quinoid" band, which is a sign of dominance of the benzoid groups in the structure of macromolecular chain.

On **FTIR** spectra of the samples at 1404–1410 cm^{-1} an intensive band exists (see Fig. 2.4a). The intensity of this band on the spectrum of **PAn** on Al- and AlNiCe-electrodes is high enough (see Fig. 2.4a, the spectra 1 and 3), and on the other **AMA**-electrodes (see Fig. 2.4a, the spectra 2, 4, 5) is commensurate with the intensity of the band at ~1490 cm^{-1}. Nature of band at 1404–1410 cm^{-1} (see Fig. 2.4a) requires a detailed study. In the works, in which there are **FTIR** spectra of **PAn** on the aluminum-containing surfaces, particularly AA 2024 T6 (sharp band at 1380 cm^{-1})[46] and **PAn**/pumice (acute low-intensity band at 1415 cm^{-1})[110] not identified. On **FTIR** spectra of **PAn** on the not aluminum-containing surfaces, such as, **PAn**/activated carbon (band at 1400 cm^{-1})[111] or **PAn**/dodecylbenzenesulfonic acid (weak sharp band at 1410 cm^{-1})[112] the nature of this band is also not determined.

The weak band on **FTIR** spectra of **PAn** at 1404 cm^{-1} authors of ref.[113] attributed to the stretching vibrations of –C–N and Q–B–Q group of semiquinoid ring of **PAn**. The peaks at 1403 and 1490 cm^{-1} authors of ref.[114] attributed to the C=C stretching vibrations of benzenoid group. Small peaks appeared at 1400 cm^{-1} authors of ref.[79, 105, 115] attributed to the aromatic C–C stretching vibrations of phenazine-like segments or branched structures in **PAn** oligomers. The pair of bands at 1443 and 1041 cm^{-1} on **FTIR** spectra of **PAn** authors of ref.[116, 117] attributed to the N–N single bond, which is the defect of the polymer chains. The intense sharp bands on Raman spectra of **PAn** at about 1651 and 1400 cm^{-1} authors of ref.[108] attributed to the cross-linking of **PAn**. Weak peak at 1400 cm^{-1} the authors of ref.[118] attributed to the C–N stretching in the neighborhood of a quinonoid ring. The intensity of the peak at 1404–1410 cm^{-1} also can correspond to the oxidized units (quinoid rings), and it very low due to low polymerization potential.[119]

The band at 1404–1410 cm^{-1} on **FTIR** spectra of **PAn** the authors of ref.[120, 121] attributed to stretching vibration of Al–O bond or stretching vibrations –C–N and Q–B–Q groups of semiquinoid ring of **PAn**, or which were associated with phenazine-like or branched structures in **PAn** oligomers.[120–122]

For identification of band at ~1407 cm^{-1} on **FTIR** spectra of **PAn**'s samples on **WE**, we have conducted an analysis of **FTIR** spectra of samples, namely, chemically synthesized sample in 0.5 M solution of H_2SO_4 (**PAn-**H_2SO_4), salt $Al_2(SO_4)_3 18H_2O$, and film of Al_2O_3 on Al-electrode (see Fig. 2.4b). As shown in Figure 2.4b, on the **FTIR** spectra of the sample **PAn-**H_2SO_4 in the spectrum 1 there are all characteristic bands peculiar for chemically synthesized **PAn** in the presence of H_2SO_4.[100, 115, 123, 124] On **FTIR** spectra of $Al_2(SO_4)_3 \times 18H_2O$ the sharp intense peak at 1062 cm^{-1} can correspond to SO_4^{2-} groups (see Fig. 2.4b, the spectrum 2).[100, 113, 123] On **FTIR** spectra of Al_2O_3 within the 1700–900 cm^{-1} there are three characteristic peaks (see Fig. 2.4b, the spectrum 3). The most intense is the peak at 1153 cm^{-1}, which apparently corresponds to the valence vibrations of Al–O bond.[121]

High intensity of the peak at 1407 cm^{-1} on **FTIR** spectra of the **PAn**'s sample, deposited on the Al-electrode (see Fig. 2.4a, the spectrum 1) may be caused by overlapping of the bands inherent to $Al_2(SO_4)_3$, which may be present in the **PAn**'s film, by the formation of branched structures or semiquinoid ring in macromolecules of **PAn**, or which were associated with phenazine-like structures in **PAn** oligomers. Obviously, the presence of the peak at 1407 cm^{-1} on **FTIR** spectra of **PAn**'s samples on **WE** may correspond to the above factors together.[114–117, 125, 129]

The intensity peak at about ~1299 cm^{-1} and shoulder at ~1247 cm^{-1} are attributed to C–N stretching vibration corresponding to $-NH^{+}-$ in protonic acid doped.[95, 104, 115] The intensity of the band at about ~1299 cm^{-1} indicative of the degree of **PAn**'s doping, that is the presence of EmS providing the conductivity of **PAn**. It is well known that the doping-protonation of **PAn** is occurred on imine nitrogen atoms of the macromolecules with the formation of semi-quinoid cation-radicals, which are so called polyarons.[94]

The form of band within ~1210–890 cm^{-1} with intensive peak at 1068 cm^{-1} on all **FTIR** spectra is complex and obviously has all three bands, namely band (shoulder) at ~1135 cm^{-1}, band at ~1068 cm^{-1}, and band (shoulder) at ~982 cm^{-1} (see Fig. 2.4a).[126] The weak bands at ~1135 and ~982 cm^{-1} can correspond to B–NH–Q or B–NH–B groups that are formed during the doping reaction.[104, 127] These characteristic absorption peaks at ~1135 cm^{-1} (corresponding to electron delocalization degree) and ~1244 cm^{-1} (corresponding to $-NH^{+}-$ in protonic acid doped **PAn**) (see Fig. 2.4a) became lower in intensity with decreasing acidity of the preparation media.[115] The authors of ref.[128] attributed of the intense peak at ~1144 cm^{-1} to the plane vibrations of C–H groups of the pernigraniline. At ~1135 cm^{-1} it can be occurred the band of Me–O bond in composites **PAn**/metal oxides[129] or C–O stretching vibrations can be found[130] at 1132 cm^{-1}. The bands at ~1135 and ~796 cm^{-1} can also correspond to the plane and out of plane deformation vibrations of C–H bond in benzene cycle, respectively, confirming the para-replacement in the structure of the molecules of synthesized **PAn**.[131–133] Low intensity band at ~796 cm^{-1} which can be traced on the **FTIR** spectra of samples is a unique key to identify the type of joining molecules of An in macromolecular chain of **PAn**.[95, 99, 103, 113, 134, 135] The presence of this band indicates 1.4-linkage of **An**'s molecules in macromolecules of **PAn**.[99, 131, 132, 133, 136]

After the finish of potentiodynamic deposition of **PAn**'s film on **WE** in 0.5 $mol \cdot l^{-1}$ H_2SO_4, **PAn** is in doped SO_4^{2-} state, namely in the form of EmS/PNAnS. This confirms also the peak at ~1068 cm^{-1}, which is a sign of the absorptions of SO_4^{2-} groups.[100, 113, 123]

On **FTIR** spectra of electrochemically deposited **PAn**'s films on **WE** from **AMA** it is observed the shift of the main characteristic bands (see Fig. 2.4a and Table 2.3) regarding to **PAn**'s film on Al-electrode (~1564 cm^{-1}) in the short-wave area of the spectrum for **PAn** on AlNiY-electrode to ~1564 and 1485 cm^{-1}, while the main characteristic bands of **PAn** in films on AlNiCe-, AlNiGd-, and AlNiDy-electrodes are "quinoid" bands at 1574 cm^{-1}, whereas for the "benzoid" band at ~1487 cm^{-1} a small shift is observed. More significant in relation to characteristic band at 1065 cm^{-1} of **PAn** on the Al-electrode is bathochromic displacement of this characteristic band for **PAn** on AlNiY-electrode (~1046 cm^{-1}) and hypsochrome displacement for the **PAn**'s films on AlNiCe (~1070 cm^{-1}), AlNiDy (~1073 cm^{-1}), and AlNiGd-electrode (~1076 cm^{-1}).

Based on difractogramms (see Fig. 2.3) it were calculated the crystallinity degrees of **PAn** in films on **WE**, and by **FTIR** spectra (see Fig. 2.4a) it were determined the ratio of the intensities of "quinoid" (~1484 cm^{-1}) and "benzoid" (~ 1570 cm^{-1}) peaks (see Table 2.2). As shown in Table 2.2, the crystallinity degree of **PAn** in the form of EmB ($2\theta = 24.1°$) is low, and the crystallinity degree of **PAn** in the form of EmS ($2\theta = 20.4°$) is much higher. During the electrochemical oxidation of **An** on **WE** the EmS and PNAnS are formed to a greater extent than the EmB. The ratio of the "quinoid" (~1484 cm^{-1}) and "benzoid" (~1570 cm^{-1}) intensities is greater than 1 confirming the formation predominantly of benzoid groups in the composition of **PAn**'s film on the investigated **WE**.

2.3.3 MORPHOLOGICAL PROPERTIES OF PAn's FILMS ON WORKING ELECTRODES

The surfaces of **WE** and **PAn**'s films on these electrodes were analyzed by SEM-microanalyzer (**SEM**-images). For the analysis the electrodes with the **PAn**'s films deposited at 75 cycles of the potential scanning were used. The analysis of the **PAn**'s films deposited only on the contact surfaces of **WE** has been done. Figure 2.5 shows the **SEM**-images of the contact surfaces of Al and, as an example of the **AMA**' surfaces, the surface of AlNiY-electrode.

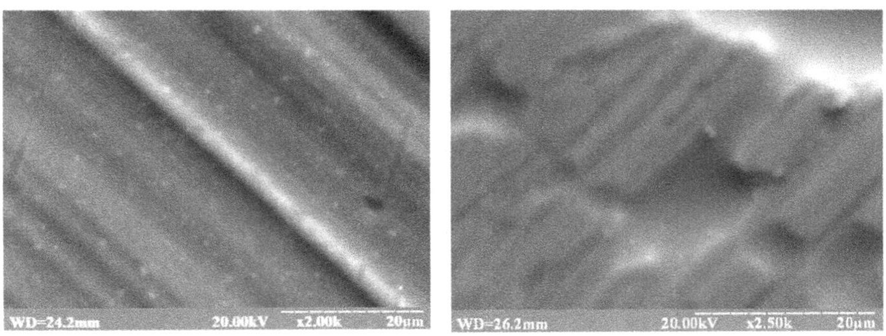

(a) (b)

FIGURE 2.5 SEM-images of the contact side of Al (a) and AlNiY (b) electrodes.

The surface of Al-electrode is smoother than that of AlNiY-electrode. The surfaces of AlNiY-electrode, as well as other **AMA**-electrodes are more developed due to the topography of the drum on which the tapes of **AMA** were formed. Ledges on the surface of **AMA**-electrodes are rough and can serve as defects, and the deepenings are smoother (see Fig. 2.5b). For oxide films on Al and on alloys on its basis inherent an inhomogeneity in thickness[17, 43, 69] to facilitate the formation of pittings in areas with thinner oxide film or in areas with the defects.

Figure 2.6 illustrates the **SEM**-images of the surfaces of **PAn**'s films with different magnifications, deposited from 0.25 M solution of An in 0.5 M H_2SO_4 on **WE**. From the **SEM**-images (see Fig. 2.6a, c, i, g, h) it is shown, that against the background of spongy, branched polymeric units on Al-electrode dark areas (deepenings) are traced (see Fig. 2.6a), and on the surface of **WE** from **AMA**, namely AlNiY, AlNiGd, and AlNiDy the cracks exist (see Fig. 2.6, c, g, h). **SEM**-image in Figure 2.6i certify uneven deposition of **PAn** on AlNiCe-electrode. A similar morphology of the film was obtained by the authors of ref.[137] at the anode polymerization of **An** on ITO and Au-electrodes.

(a)

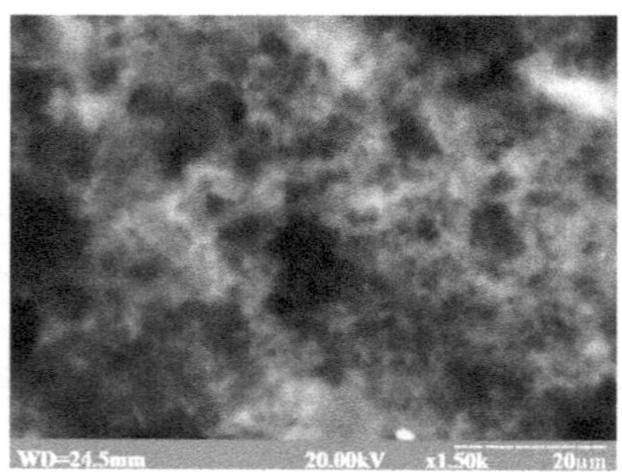

(b)

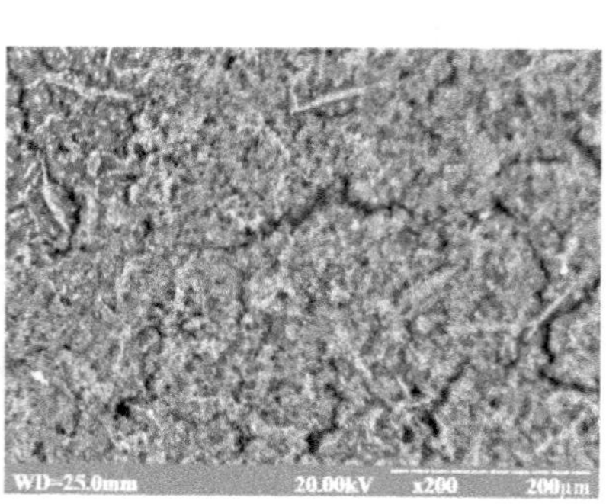

(c)

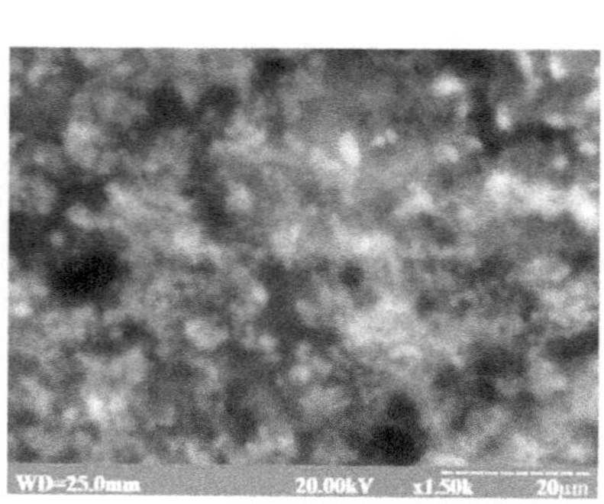

(d)

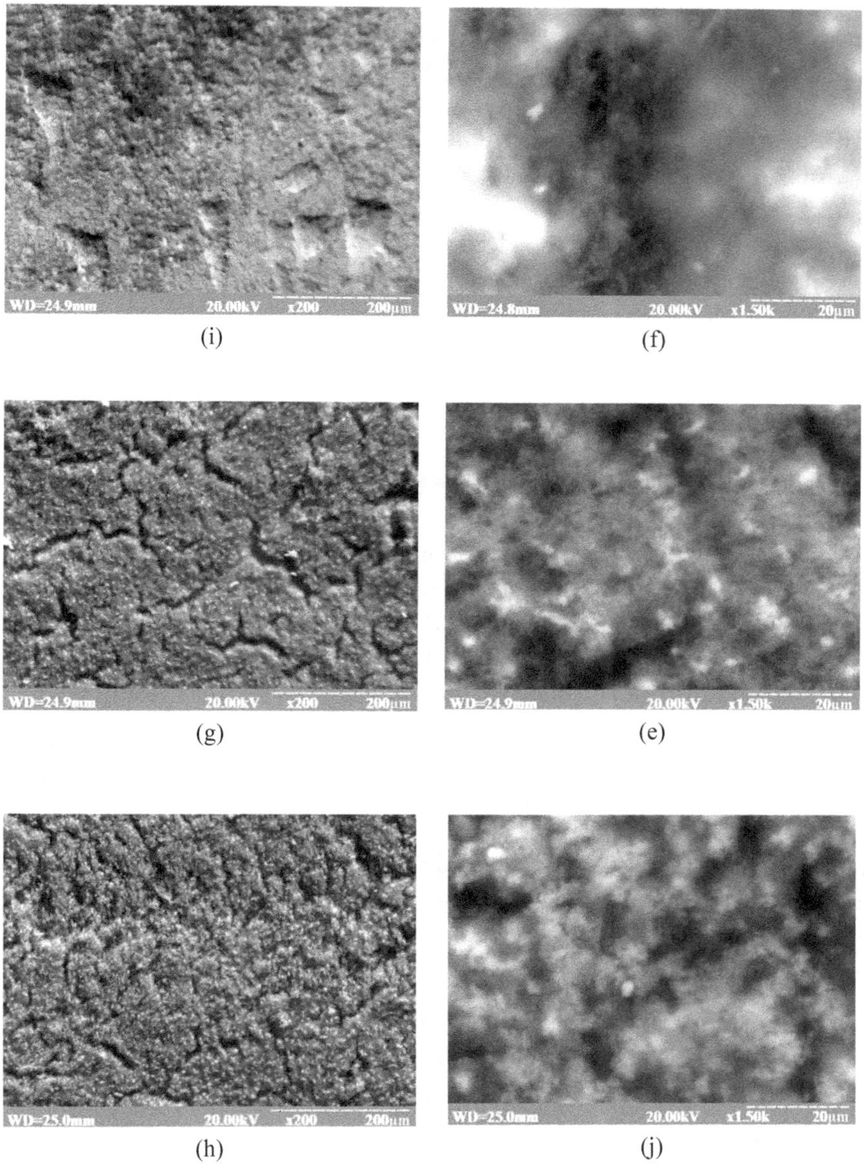

FIGURE 2.6 **SEM** images of the contact side of **WE** with deposited **PAn**'s films: a, b—Al, c, d—AlNiY, i, f—AlNiCe, g, e—AlNiGd, h, j—AlNiDy. a, c, i, g, h—magnification in 200 times. b, d, f, e, j—magnification in 1500 times.

As shown in Figure 2.6a, the **PAn**'s film on Al-electrode has the surround character and is more porous. **SEM**-images of higher magnifications (see Fig. 2.6, b, d, f, e, j show that the **PAn**'s film is uneven on the thickness formation. For the film, formation characteristic is the formation of spatial pungent structures. The **PAn**'s films deposited via potentiodynamic oxidation of **An** from the solutions of H_2SO_4 on Al and **AMA**-electrodes have porous and spongy (branched) structure, which is typical for electrodeposited **PAn**'s films on different electrode materials.[13, 23, 39, 42, 45, 67, 138] Spongy morphology of the **PAn**'s films on **WE** caused by the fact that the formed macromolecular **PAn** represents by itself the original nano- or micro-electrodes for the initiation of new polymer formation, namely sprout of various shapes, such as a thread or rod.

2.3.4 ENERGY-DISPERSIVE X-RAY (EDX) SPECTROMETRY

Figures 2.7 and 2.8 show the EDX-spectra of the area of **PAn**'s film on the surface of Al and AlNiDy (as an example of the spectrum obtained for **PAn** obtained on the **AMA**-electrodes) electrodes (see Fig. 2.6). The peak at 0.52 keV corresponds to the nitrogen atom (N). The intense peak at 2.3 keV confirms the presence of sulfur (S), which is part of the ions HSO_4^- or SO_4^{2-} that are doping component of **PAn**. Low-intensity peak at ~1.5 keV corresponds to aluminum (Al), which due to the dissolution of the oxide film is as an impurity in the form of $Al_2(SO_4)_3$ in the film of **PAn**.

On the EDX-spectra of **PAn**'s films areas (see Figs. 2.6–2.8) on the surface of AlNiDy-electrode there are weak intensive peaks of Al, Ni, and Dy. Evidently, such elements present in the **PAn**'s film as the sulphate additives.

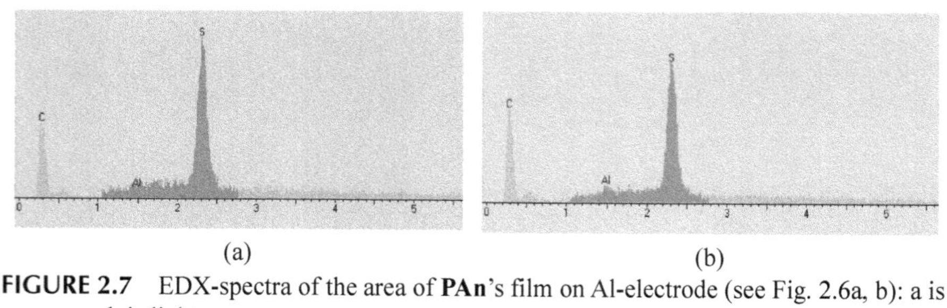

FIGURE 2.7 EDX-spectra of the area of **PAn**'s film on Al-electrode (see Fig. 2.6a, b): a is grey area, b is light area.

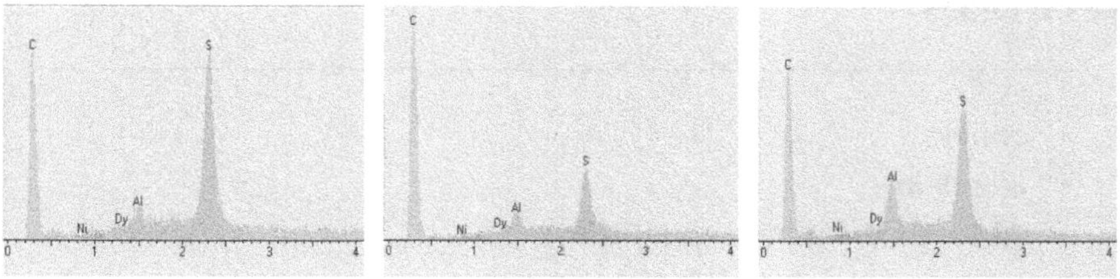

(a) (b) (c)

FIGURE 2.8 EDX-spectra of the area of **PAn**'s film on AlNiDy-electrode (see Fig. 2.6, h, j): a is dark area, b is grey area, c is light area.

Analysis of EDX-spectra of different areas of **PAn**'s film confirms that **PAn** (availability C) covers the entire surface of the electrode. But the **PAn**'s films in dark areas obviously have a less-developed surface and are denser. The presence of atoms of metals Al, Ni, and Dy in these areas of **PAn**'s film caused by the formation of sulfate salts such $Al_2(SO_4)_3$, confirming also the existence of S atoms. However, identification of the metal atoms in the films of **PAn** can also be caused by the porosity of the films of polymer.

2.4 CONCLUSIONS

So, the potentiodynamic oxidation of **An** on the electrodes of the AMAs of composition $Al_{87}Ni_8(REM)_5$, namely $Al_{87}Ni_8Y_5$, $Al_{87}Ni_8Ce_5$, $Al_{87}Ni_8Gd_5$, $Al_{87}Ni_8Dy_5$ and polycrystalline aluminum in an aqueous solution of 0.5 M H_2SO_4 showed that the process of electrochemical oxidation of **An** is limited by the presence of stable oxide films at the surface of **WE**s.

Oxidation of **An** and consequently the deposition of **PAn** is started after the end of the induction period of 1000 s. It was established, that the difference in the form of cyclical voltammogramms caused by amorphic components that are part of the $Al_{87}Ni_8(REM)_5$, namely Y, Ce, Gd, and Dy and determine different resistance of the surface oxide films on the **WE**.

The films of **PAn** deposited in potentiodynamical way on the surface of the **WE**s are characterized by spongy and porous morphology. The structure of the deposited **PAn** films is amorphous–crystalline with predominance of the phase of emeraldine and pernigraniline salts of sulfuric acid. **PAn** on $Al_{87}Ni_8(REM)_5$ electrodes has a higher degree of crystallinity than on Al-electrode.

KEYWORDS

- **aniline**
- **aluminim**
- **Al-based amorphous metal alloys**
- **potentiodynamic oxidation**
- **polyaniline**
- **polymeric films**
- **structure**
- **morphology**

REFERENCES

1. MacDiarmid, A. G. Synthetic Metals: A Novel Role for Organic Polymers. *Synth. Met.* **2001,** *125,* 11–22.
2. Heeger, A. J. Semiconducting and Metallic Polymers: The Fourth Generation of Polymeric Materials (Nobel Lecture). *Angew. Chem. Int. Ed.* **2001,** *40,* 2591–2611.
3. Malinauskas, A. Chemical Deposition of Conducting Polymers. *Polymer.* **2001,** *42,* 3957–3972.
4. Wallace, G. G.; Spinks, G. M.; Kane-Maguire, L. A. P.; Teasdale, P. R. *Electroactive Polymers: Intelligent Materials Systems;* 2nd ed. D.C. CRC Press LLC: Boca Raton, 2002; pp 248.
5. Skotheim, T. A.; Reynolds, J. R. Conjugated Polymers: Processing and Applications. In *The Handbook of Conducting Polymers;* 3rd ed. Skotheim, T. A., Reynolds, J. R., Eds.; CRC Press: Boca Raton, 2007; pp 656.
6. Gurunathan, K.; Murugan, A. V.; Marimuthu, R.; Mulik, U. P.; Amalnerkar, D. P. Electrochemically Synthesised Conducting Polymeric Materials for Applications Towards Technology in Electronics, Optoelectronics and Energy Storage Devices. *Mater. Chem. Phys.* **1999,** *61,* 173–191.
7. Angelopoulos, M. Conducting Polymers in Microelectronics. *IBM J. Res. Dev.* **2001,** *45,* 57–75.
8. Bhadra, S.; Khastgir, D.; Singh, N. K.; Lee, J. H. Progress in Preparation, Processing and Applications of Polyaniline. *Prog. Polym. Sci.* **2009,** *34,* 783–810.
9. Pan, L.; Qiu, H.; Dou, C.; Li, Y.; Pu, L.; Xu, J.; Yi Shi, Y. Conducting Polymer Nanostructures: Template Synthesis and Applications in Energy Storage. *Int. J. Mol. Sci.* **2010,** *11,* 2636–2657.
10. Abalyaeva, V. V.; Efimov, O. N. Electrocatalytic Synthesis of Polyaniline on Non-noble Metal Electrodes. *Polym. Adv. Technol.* **1997,** *8,* 517–524.
11. Mondal, S. K.; Prasad, K. R.; Munichandraiah, N. Analysis of Electrochemical Impedance of Polyaniline Films Prepared by Galvanostatic, Potentiostatic and Potentiodynamic Methods. *Synth. Met.* **2005,** *148,* 275–286.

12. Biallozor, S.; Kupniewska, A. Conducting Polymers Electrodeposited on Active Metals. *Synth. Met.* **2005,** *155,* 443–449.
13. Moutarlier, V.; Lakard, S.; Patois, T.; Lakard, B. Glow Discharge Optical Emission Spectroscopy: A Complementary Technique to Analyze Thin Electrodeposited Polyaniline Films. *Thin Solid Films.* **2014,** *550,* 27–35.
14. Borole, D. D.; Kapadi, U. R.; Kumbhar, P. P.; Hundiwale, D. G. Influence of Inorganic and Organic Supporting Slectrolytes on the Electrochemical Synthesis of Polyaniline, Poly(o-toluidine) and Their Copolymer Thin Films. *Mater. Lett.* **2002,** *56,* 685–691.
15. Parsa, A.; Ab Ghani, S. The Improvement of Free-radical Scavenging Capacity of the Phosphate Medium Electrosynthesized Polyaniline. *Electrochim. Acta.* **2009,** *54,* 2856–2860.
16. Hao, Q.; Lei, W.; Xia, X.; Yan, Z.; Yang, X.; Lu, L.; Wang, X. Exchange of Counter Anions in Electropolymerized Polyaniline Films. *Electrochim. Acta.* **2010,** *55,* 632–640.
17. Pournaghi-Azar, M. H.; Habibi, B. Electropolymerization of Aniline in Acid Media on the Bare and Chemically Pre-treated Aluminum Electrodes: A Comparative Characterization of the Polyaniline Deposited Electrodes. *Electrochim. Acta.* **2007,** *52,* 4222–4230.
18. Li, X.; Zhuang, T.; Wang, G.; Zhao, Y. Stabilizer-free Conducting Polyaniline Nanofiber Aqueous Colloids and Their Stability. *Mater. Lett.* **2008,** *62,* 1431–1434.
19. Mondal, S. K.; Barai, K.; Munichandraiah, N. High Capacitance Properties of Polyaniline by Electrochemical Deposition on a Porous Carbon Substrate. *Electrochim. Acta.* **2007,** *52,* 3258–3264.
20. Bleda-Martınez, M. J.; Morallon, E.; Cazorla-Amoros, D. Polyaniline/Porous Carbon Electrodes by Chemical Polymerisation: Effect of Carbon Surface Chemistry. *Electrochim. Acta.* **2007,** *52,* 4962–4968.
21. Zhao, X. Y.; Zang, J. B.; Wang, Y. H.; Bian, L. Y.; Yu, J. K. Electropolymerizing Polyaniline on Undoped 100 nm Diamond Powder and Its Electrochemical Characteristics. *Electrochem. Commun.* **2009,** *11,* 1297–1300.
22. Xu, G.; Wang, W.; Qu, X.; Yin, Y.; Chu, L.; He, B.; Wu, H.; Fang, J.; Bao, Y.; Liang, L. Electrochemical Properties of Polyaniline in p-Toluene Sulfonic Acid Solution. *Eur. Polym. J.* **2009,** *45,* 2701–2707.
23. Mandić, Z.; Kraljić Roković, M.; Pokupčić, T. Polyaniline as Cathodic Material for Electrochemical Energy Sources: The Role of Morphology. *Electrochim. Acta.* **2009,** *54,* 2941–2950.
24. Molina, J.; Esteves, M. F.; Fernandez, J.; Bonastre, J.; Cases, F. Polyaniline Coated Conducting Fabrics. Chemical and Electrochemical Characterization. *Eur. Polym. J.* **2011,** *47,* 2003–2015.
25. Mengoli, G.; Musiani, M. M.; Pelli, B.; Vecchi, E. Anodic Synthesis of Sulfur-bridged Polyaniline Coatings onto Fe Sheets. *J. Appl. Polym. Sci.* **1983,** *28,* 1125–1136.
26. DeBerry, D. W. Modification of the Electrochemical and Corrosion Behavior of Stainless Steels with an Electroactive Coating. *J. Electrochem. Soc.* **1985,** *132,* 1022–1026.
27. Abalayeva, V. V.; Kogan, I. L. Initiating Agents for Electrochemical Polymerization of Aniline on Titanium Electrodes. *Synth. Met.* **1994,** *63,* 109–113.
28. Abalyaeva, V. V.; Efimov, O. N. Electrocatalytic Synthesis of Polyaniline on Non-noble Metal Electrodes. *Polym. Adv. Technol.* **1997,** *8,* 517–524.
29. Sazou, D.; Georgolios, C. Formation of Conducting Polyaniline Coatings on Iron Surfaces by Electropolymerization of Aniline in Aqueous Solutions. *J. Electroanal. Chem.* **1997,** *429,* 81–93.

30. Camalet, J. L.; Lacroix, J.-C.; Aeigach, S.; Chane-Ching, K.; Lacaze, P. C. Electrosynthesis of Adherent Polyaniline Films on Iron and Mild Steel in Aqueous Oxalic Acid Medium. *Synth. Met.* **1998,** *93,* 133–142.
31. Lacroix, J.-C.; Camalet, J.-L.; Aeiyach, S.; Chane-Ching, K.; Petitjean, J.; Chauveau, E.; Lacaze, P.-C. Aniline Electropolymerization on Mild Steel and Zinc in a Two-step Process. *J. Electroanal. Chem.* **2000,** *481,* 76–81.
32. Prasad, K. R.; Munichandraiah, N. Potentiodynamic Deposition of Polyaniline on Non-platinum Metals and Characterization. *Synth. Met.* **2001,** *123,* 459–468.
33. Arsov, L. D. Electrochemical Study of Polyaniline Deposited on a Titanium Surface. *J. Solid State Electrochem.* **1998,** *2,* 266–272.
34. Eftekhari, A. Aluminum as a Suitable Substrate for the Deposition of Conducting Polymers: Application to Polyaniline and Enzyme-modified Electrode. *Synth. Met.* **2002,** *125,* 295–300.
35. Tallman, D. E.; Spinks, G.; Dominis, A.; Wallace, G. G. Electroactive Conducting Polymers for Corrosion Control Part 1. General Introduction and a Review of Non-ferrous Metals. *J. Solid State Electrochem.* **2002,** *6,* 73–84.
36. Conroy, K. G.; Breslin, C. B. The Electrochemical Deposition of Polyaniline at Pure Aluminium: Electrochemical Activity and Corrosion Protection Properties. *Electrochim. Acta.* **2003,** *48,* 721–732.
37. Tallman, D. E.; Dewald, M. P.; Vang, C. K.; Wallace, G. G.; Bierwagen, G. P. Electrodeposition of Conducting Polymers on Active Metals by Electron Transfer Mediation. *Curr. Appl. Phys.* **2004,** *4,* 137–140.
38. Özyılmaz, A. T.; Kardas, G.; Erbil, M.; Yazici, B. The Corrosion Performance of Polyaniline on Nickel Plated Mild Steel. *Appl. Surf. Sci.* **2005,** *242,* 97–106.
39. Sazou, D.; Kourouzidou, M.; Pavlidou, E. Potentiodynamic and Potentiostatic Deposition of Polyaniline on Stainless Steel: Electrochemical and Structural Studies for a Potential Application to Corrosion Control. *Electrochim. Acta.* **2007,** *52,* 4385–4397.
40. Qin, Q.; Tao, J.; Yang, Y. Preparation and Characterization of Polyaniline Film on Stainless Steel by Electrochemical Polymerization as a Counter Electrode of DSSC. *Synth. Met.* **2010,** *160,* 1167–1172.
41. Özyılmaz, A. T.; Erbil, M.; Yazıcı, B. The Electrochemical Synthesis of Polyaniline on Stainless Steel and Its Corrosion Performance. *Curr. Appl. Phys.* **2006,** *6,* 1–9.
42. Zhang, H.; Wang, J.; Wang, Z.; Zhang, F.; Wang, S. Electrodeposition of Polyaniline Nanostructures: A Lamellar Structure. *Synth. Met.* **2009,** *159,* 277–281.
43. Wang, T.; Tan, Y. J. Understanding Electrodeposition of Polyaniline Coatings for Corrosion Prevention Applications Using the Wire Beam Electrode Method. *Corros. Sci.* **2006,** *48,* 2274–2290.
44. Iroh, J. O.; Zhua, Y.; Shah, K.; Levine, K.; Rajagopalan, R.; Uyar, T.; Donley, M.; Mantz, R.; Johnson, J.; Voevodin, N. N.; Balbyshev, V. N.; Khramov, A. N. Electrochemical Synthesis: A Novel Technique for Processing Multi-functional Coatings. *Prog. Org. Coat.* **2003,** *47,* 365–375.
45. Kamaraj, K.; Devarapalli, R.; Siva, T.; Sathiyanarayanan, S. Self-healing Electrosynthesied Polyaniline Film as Primer Coat for AA 2024-T3. *Mater. Chem. Phys.* **2015,** *153,* 256–265.
46. Karpagam, V.; Sathiyanarayanan, S.; Venkatachari, G. Studies on Corrosion Protection of Al2024 T6 Alloy by Electropolymerized Polyaniline Coating. *Curr. Appl. Phys.* **2008,** *8,* 93–98.

47. Shabani-Nooshabadi, M.; Ghoreishi, S. M.; Behpour, M. Electropolymerized Polyaniline Coatings on Aluminum Alloy 3004 and Their Corrosion Protection Performance. *Electrochim. Acta.* **2009,** *54,* 6989–6995.
48. Cecchetto, L.; Didier Delabouglise, D.; Petit, J.-P. On the Mechanism of the Anodic Protection of Aluminium Alloy AA5182 by Emeraldine Base Coatings: Evidences of a Galvanic Coupling. *Electrochim. Acta.* **2007,** *52,* 3485–3492.
49. Kamaraj, K.; Sathiyanarayanan, S.; Venkatachari, G. Electropolymerised Polyaniline Films on AA 7075 Alloy and Its Corrosion Protection Performance. *Progr. Org. Coat.* **2009,** *64,* 67–73.
50. Yatsyshyn, M. M.; Boichyshyn, L. M.; Demchyna, I. I.; Nosenko, V. K. Electrochemical Oxidation of Aniline on the Surface of an Amorphous Metal Alloy $Al_{87}Ni_8Y_5$. *Russ. J. Electrochem.* **2012,** *48,* 502–508.
51. Yatsyshyn, M. M.; Boichyshyn, L. M.; Koval'chuk, E. P.; Demchyna, I. I.; Serkiz, R. Ya.; Demchenko, P. Yu. The morphology of the polyaniline films on the surface of $Al^{87}Ni^{8}Y^{5}$ amorphous metallic alloy. *Chemistry, Physics and Texnology of Surface.* **2012,** *3,* 74–81.
52. Demchyna, I.; Yatsyshyn, M.; Boichyshyn, L.; Serkiz, R.; Pandyak, N. Comparative Characteristics of the Potentiodynamically Deposited Polyaniline Films on the Aluminium-Containing Electrodes. *Visnyk Lviv Univ. Ser. Chem.* **2012,** *53,* 296–302.
53. Yatsyshyn, M. M.; Demchyna, I. I.; Mudry, S. I.; Serkiz, R. Ya. Morphology of the Deposited Electrochemically in Potentiodynamic Mode on the Surface of $Al_{87}Ni_8(REE)_5$ Amorphous Metallic Alloys Polyaniline Film. *PCSS.* **2013,** *3,* 593–601.
54. Guo, X. W.; Jiang, Y. F.; Zhai, C. Q.; Lu, C.; Ding, W. J. Preparation of Even Polyaniline Film on Magnesium Alloy by Pulse Potentiostatic Method. *Synth. Met.* **2003,** *135–136,* 169–170.
55. Branzoi, V.; Branzoi, F.; Pilan, L. Characterization of Electrodeposited Polymeric and Composite Modified Electrodes on Cobalt Based Alloy. *Mater. Chem. Phys.* **2009,** *118,* 197–202.
56. Elzanowska, H.; Miasek, E.; Birss, V. I. Electrochemical Formation of Ir Oxide/Polyaniline Composite Films. *Electrochim. Acta.* **2008,** *53,* 2706–2715.
57. Nekrasov, A. A.; Gribkova, O. L.; Eremina, T. V.; Isakova, A. A.; Ivanov, V. F.; Tverskoj, V. A.; Vannikov, A. V. Electrochemical Synthesis of Polyaniline in the Presence of Poly(amidosulfonic acid)s with Different Rigidity of Polymer Backbone and Characterization of the Films Obtained. *Electrochim. Acta.* **2008,** *53,* 3789–3797.
58. Cheraghi, B.; Fakhari, A. R.; Shahin Borhani, S.; Entezami, A. A. Chemical and Electrochemical Deposition of Conducting Polyaniline on Lead. *J. Electroanal. Chem.* **2009,** *626,* 116–122.
59. Venancio, E. C.; Costa, C. A. R.; Mochado, S. A. S.; Motheo, A. I. AFM Study of the Initial Stages of Polyaniline Growth on ITO Electrode. *Electrochem. Commun.* **2001,** *3,* 229–233.
60. Pournaghi-Azar, M. H.; Habibi, B. A. Palladized Aluminum as a Novel Substrate for Electrosynthesis of Polyaniline in Sulfuric Acid Solutions. *J. Solid State Electrochem.* **2007,** *11,* 505–513.
61. Zhou, H.; Chen, H.; Luo, S. Lu, G.; Wei, W.; Kuang, Y. The Effect of the Polyaniline Morphology on the Performance of Polyaniline Supercapacitors. *J. Solid State Electrochem.* **2005,** *9,* 574–580.
62. Gupta, V.; Miura, N. High Performance Electrochemical Supercapacitor From Electrochemically Synthesized Nanostructured Polyaniline. *Mater. Lett.* **2006,** *60,* 1466–1469.

63. Kanungo, M.; Kumar, A.; Contractor, A. Q. Studies on Electropolymerization of Aniline in the Presence of Sodium Dodecyl Sulfate and Its Aapplication in Sensing Urea. *J. Electroanal. Chem.* **2002,** *528,* 46–56.
64. Peng, X.-Y.; Luan, F.; Liu, X.-X.; Diamond, D.; Lau, K.-T. pH-controlled Morphological Structure of Polyaniline During Electrochemical Deposition. *Electrochim. Acta.* **2009,** *54,* 6172–6177.
65. Andrade, G. De T.; Aguirre, M. E.; Biaggio, S. R. Influence of the First Potential Scan on the Morphology and Electrical Properties of Potentiodynamically Grown Polyaniline Films. *Electrochim. Acta.* **1998,** *44,* 633–642.
66. Liu, C.; Hayashi, K.; Toko, K. Electrochemical Deposition of Nanostructured Polyaniline on an Insulating Substrate. *Electrochem. Commun.* **2010,** *12,* 36–39.
67. Hao, Q.; Lei, W.; Xia, X.; Yan, Z.; Yang, X.; Lu, L.; Wang, X. Exchange of Counter Anions in Electropolymerized Polyaniline Films. *Electrochim. Acta.* **2010,** *55,* 632–640.
68. Zhou, S.; Wu, T.; Kan, J. Effect of Methanol on Morphology of Polyaniline. *Eur. Polym. J.* **2007,** *43,* 395–402.
69. Akundy, G. S.; Rajagopalan, R.; Iro, J. O. Electrochemical Deposition of Polyaniline–Polypyrrole Composite Coatings on Aluminum. *J. Appl. Polym. Sci.* **2002,** *83,* 1970–1977.
70. Martyak, N. M. Chronoamperometric Studies During the Polymerization of Aniline from an Oxalic Acid Solution. *Mater. Chem. Phys.* **2003,** *81,* 143–151.
71. Mandić, Z.; Kraljić Roković, M.; Pokupčić, T. Polyaniline as Cathodic Material for Electrochemical Energy Sources: The Role of Morphology. *Electrochim. Acta.* **2009,** *54,* 2941–2950.
72. Posudievsky, O. Yu.; Pokhodenko, V. D. Nanostructured Functional Materials Based on Conducting Conjugated Polymers. *Nanosystemi, Nanomateriali, Nanotechnologii.* **2004,** *2,* 1017–1036.
73. Fan, L.-Z.; Hu, Y.-S.; Maier, J.; Adelhelm, P.; Smarsly, B.; Antonietti, M. High Electroactivity of Polyaniline in Supercapacitors by Using a Hierarchically Porous Carbon Monolith as a Support. *Adv. Funct. Mater.* **2007,** *17,* 3083–3087.
74. Sun, L.-J.; Liu, X.-X.; Lau, K.-T.; Chen, L.; Gu, W.-M. Electrodeposited Hybrid Films of Polyaniline and Manganese Oxide in Nanofibrous Structures for Electrochemical Supercapacitor. *Electrochim. Acta.* **2008,** *53,* 3036–3042.
75. Cheng, F.; Tang, W.; Li, C.; Chen, J.; Liu, H.; Shen, P.; Dou, S. Conducting Poly(aniline) Nanotubes and Nanofibers: Controlled Synthesis and Application in Lithium/Poly(aniline) Rechargeable Batteries. *Chem. Eur. J.* **2006,** *12,* 3082–3088.
76. Drelinkiewicz, A.; Hasik, M.; Kloc, M. Liquid-phase Hydrogenation of 2-Ethylanthraquinone over Pd/Polyaniline Catalysts. *J. Catal.* **1999,** *186,* 123–133.
77. Alonso-Vante, N.; Cattarin, S.; Musiani, M. Electrocatalysis of O_2 Reduction at Polyaniline+Molybdenum-doped Ruthenium Selenide Composite Electrodes. *J. Electroanal. Chem.* **2000,** *481,* 200–207.
78. Nagashree, K. L.; Ahmed, M. F. Electrocatalytic Oxidation of Methanol on Pt Modified Polyaniline in Alkaline Medium. *Synth. Met.* **2008,** *158,* 610–616.
79. Sreedhar, B.; Radhika, P.; Neelima, B.; Hebalkar, N.; Rao, M. V. B. Synthesis and Characterization of Polyaniline: Nanospheres, Nanorods, and Nanotubes–Catalytic Application for Sulfoxidation Reactions. *Polym. Adv. Technol.* **2009,** *20,* 950–958.
80. Wang, H.; Yang, P.-H.; Cai, H.-H.; Cai, J. Constructions of Polyaniline Nanofiber-based Electrochemical Sensor for Specific Detection of Nitrite and Sensitive Monitoring of Ascorbic Acid Scavenging Nitrite. *Synth. Met.* **2012,** *162,* 326–331.

81. Smela, E.; Mattes, B. R. Polyaniline Actuators: Part 2. PANI(AMPS) in Methanesulfonic Acid. *Synth. Met.* **2005,** *151,* 43–48.
82. Ismail, Y. A.; Shin, S. R.; Shin, K. M.; Yoon, S. G., Shon, K.; Kim, S. I.; Kim, S. J. Electrochemical Actuation in Chitosan/Polyaniline Microfibers for Artificial Muscles Fabricated Using an in Situ Polymerization. *Sensor. Actuat. B-Chem.* **2008,** *129,* 834–840.
83. Kendig, M.; Hon, M.; Warren, L. 'Smart' Corrosion Inhibiting Coatings. *Prog. Org. Coat.* **2003,** *47,* 183–189.
84. Jugović, B.; Gvozdenović, M.; Stevanović, J.; Trišović, T.; Grgur, B. Characterization of Electrochemically Synthesized PANI on Graphite Electrode for Potential Use in Electrochemical Power Sources. *Mater. Chem. Phys.* **2009,** *114,* 939–942.
85. Cruz-Silva, R.; Nicho, M. E.; Reséndiz, M. C.; Agarwal, V.; Castillón, F. F.; Farías, M. H. Electrochemical Polymerization of an Aniline-terminated Self-assembled Monolayer on Indium Tin Oxide Electrodes and Its Effect on Polyaniline Electrodeposition. *Thin Solid Films.* **2008,** *516,* 4793–4802.
86. Demchyna, I.; Polihas, O., Yatsyshyn, M.; Pandyak, N. *Books of Abstracts,* XVth Scientific Conference "Lviv Chemical Readings - 2015", L'viv, Ukraine, May 24–27, 2015; Y54.
87. Park, S. M. *Handbook of Organic Conductive Molecules and Polymers;* Nalaw, H. S., Ed.; Wiley, New York, 1997, Vol. 3, pp 428.
88. Yatsyshyn, M.; Grynda, Yu.; Kun'ko. A.; Dumanchuk, N. Thermal Stability of the Polyaniline/Silica-Glauconite Composites. *Visnyk Lviv Univ. Ser. Khim.* **2011,** 52, 268–276.
89. Li, X. Improving the Electrochemical Properties of Polyaniline by Co-doping with Titanium Ions and Protonic Acid. *Electrochim. Acta.* **2009,** *54,* 5634–5639.
90. Bhadra, S.; Singha, N. K.; Khastgir, D. Electrochemical Synthesis of Polyaniline and Its Comparison with Chemically Synthesized Polyaniline. *J. Appl. Polym. Sci.* **2007,** *104,* 1900–1904.
91. Yin, J.; Zhao, X.; Xia, X.; Xiang, L.; Qiao, Y. Electrorheological Fluids Based on Nano-fibrous Polyaniline. *Polymer.* **2008,** *49,* 4413–4419.
92. Epstein, A. J.; MacDiarmid, A. G. Protonation of Emeraldine: Formation of a Granular Polaronic Polymeric Metal. *Mol. Cryst. Liq. Cryst. Inc. Nonlinear Opt.* **1988,** *160,* 165–173.
93. Zhang, Z.; Wei, Z.; Wan, M. Nanostructures of Polyaniline Doped with Inorganic Acids. *Macromolecules.* **2002,** *35,* 5937–5942.
94. Šeděnková, I.; Trchová, M.; Blinova, N. V.; Stejskal, J. In-situ Polymerized Polyaniline Films. Preparation in Solutions of Hydrochloric, Sulfuric, or Phosphoric Acid. *Thin Solid Films.* **2006,** *515,* 1640–1646.
95. Zhang, Z.; Deng, J.; Wan, M. Highly Crystalline and Thin Polyaniline Nanofibers Oxidized by Ferric Chloride. *Mater. Chem. Phys.* **2009,** *115,* 275–279.
96. Ren, L.; Li, K.; Chen, X. Soft Template Method to Synthesize Polyaniline Microtubes Ddoped with Methyl Orange. *Polym. Bull.* **2009,** *63,* 15–21.
97. Mentus, S.; Cric-Marjanovic, G.; Trchova, M.; Stejskal, J. Conducting Carbonized Polyaniline Nanotubes. *Nanotechnology.* **2009,** *20,* 245601–24601.
98. Mu, S.; Kan, J. Energy Density and IR Spectra of Ppolyaniline Synthesized Electrochemically in the Solutions of Strong Acids. *Synth. Met.* **1998,** *98,* 51–55.
99. Lv, R.; Zhang, S.; Shi, Q.; Kann, J. Electrochemical Synthesis of Polyaniline Nanoparticles in the Presence of Magnetic Field and Erbium Chloride. *Synth. Met.* **2005,** *150,* 115–122.

100. Liu, X.-X.; Zhang, L.; Li, Y.-B.; Bian, L. J.; Su, Z.; Zhang, L.-J. Electropolymerization of Aniline in Aqueous Solutions at pH 2 to 12. *J. Mater. Sci.* **2005,** *40,* 4511–4515.
101. Zhang, S.; Tang, R.; Kann, J. Effects of Magnetic Field and Rare-earth Ions on Properties of Polyaniline Nanoparticles. *J. Appl. Polym. Sci.* **2007,** *103,* 2286–2294.
102. Jamadade, V. S.; Dhawale, D. S.; Lokhande, C. D. Studies on Electrosynthesized Leucoemeraldine, Emeraldine and Pernigraniline Forms of Polyaniline Films and Their Supercapacitive Behavior. *Synth. Met.* **2010,** *160,* 955–960.
103. Hu, Z.-A.; Xie, Y.-L.; Wang, Y.-X.; Mo, L.-P.; Yang, Y.-Y.; Zhang, Z.-Y. Polyaniline/ SnO_2 Nanocomposite for Supercapacitor Applications. *Mater. Chem. Phys.* **2009,** *114,* 990–995.
104. Jadhav, S. V.; Puri, V. Microwave Study of Chemically Synthesized Conducting Polyaniline on Alumina. *Synth. Met.* **2008,** *158,* 883–887.
105. Li, G.; Zhang, C.; Peng, H. Facile Synthesis of Self-assembled Polyaniline Nanodisks. *Macromol. Rapid Commun.* **2008,** *29,* 63–67.
106. Karim, M. R.; Lim, K. T.; Lee, C. J.; Bhuiyan, M. D. I.; Kim, H. J.; Park, L.-S.; Lee, M. S. Synthesis of Core-shell Silver-Polyaniline Nanocomposites by Gamma Radiolysis Method. *J. Polym. Sci., Part A: Polym. Chem.* **2007,** *45,* 5741–5747.
107. Xing, S.; Zhao, G.; Yuan, Y. Preparation of Polyaniline–Polypyrrole Composite Sub-micro Fibers via Interfacial Polymerization. *Polym. Compos.* **2009,** *29,* 22–26.
108. Do Nascimento, G. M.; Silva, C. H. B.; Temperini, M. L. A. Spectroscopic Characterization of the Structural Changes of Polyaniline Nanofibers after Heating. *Polym. Degrad. Stab.* **2008,** *93,* 291–297.
109. Blinova, N. V.; Trchova, M.; Stejskal, J. The Polymerization of Aniline at a Solution–Gelatin Gel Interface. *Eur. Polym. J.* **2009,** *45,* 668–673.
110. Gok, A.; Gode, F.; Turkaslan, B. Synthesis and Characterization of Polyaniline/Pumice (PAn/Pmc) Composite. *Mater. Sci. Eng. B.* **2006,** *133,* 20–25.
111. Zengin, H.; Kalaycı, G. Synthesis and Characterization of Polyaniline/Activated Carbon Composites and Preparation of Conductive Films. *Mater. Chem. Phys.* **2010,** *120,* 46–53.
112. Yang, J.; Ding, Y.; Zhang, J. Uniform Rice-like Nanostructured Polyanilines with Highly Crystallinity Prepared in Dodecylbenzene Sulfonic Acid Micelles. *Mater. Chem. Phys.* **2008,** *112,* 322–324.
113. Vijayan, M.; Trivedi, D. C. Studies on Polyaniline in the Methane sulphonic Mcid (MeSA). *Synth. Met.* **1999,** *107,* 57–64.
114. Lakshmi, G. B.; Dhillon, V. S.; Siddiqui, A. M.; Zulfequar, M.; Avasthi, D. K. RF-plasma Polymerization and Characterization of Polyaniline. *Eur. Polym. J.* **2009,** *45,* 2873–2877.
115. Wang, J.; Zhang, K.; Zhao, L. Sono-assisted Synthesis of Nanostructured Polyaniline For Adsorption of Aqueous Cr(VI): Effect of Protonic Acids. *Chem. Eng. J.* **2014,** *239,* 123–131.
116. Venancio, E. C.; Wang, P. C.; MacDiarmid, A. G. The Azanes: A Class of Material Incorporating Nano/Micro Self-assembled Hollow Spheres Obtained by Aqueous Oxidative Polymerization of Aniline. *Synth. Met.* **2006,** *156,* 357–369.
117. Chen, J.; Xu, Y.; Zheng, Y.; Dai, L.; Wu, H. The Design, Synthesis and Characterization of Polyaniline Nanophase Materials. *C. R. Chim.* **2008,** *11,* 84–89.
118. Lissarrague, M. H.; Lamanna, M. E.; D'Accorso, N. B.; Goyanes, S. Effects of Different Nucleating Particles on Aniline Polymerization. *Synth. Met.* **2012,** *162,* 1052–1058.
119. Sacak, M.; Akbulut, U.; Batchelder, D. N. Monitoring of Electroinitiated Polymerization of Aniline by Raman Microprobe Spectroscopy. *Polymer.* **1998,** *40,* 21–26.

120. Ayoob, S.; Gupta, A. K.; Bhakat, P. B.; Bhat, V. T. Investigations on the Kinetics and Mechanisms of Sorptive Removal of Fluoride from Water Using Alumina Cement Granules. *Chem. Eng. J.* **2008,** *140,* 6–14.
121. Karthikeyan, M.; Kumar, K. K. S.; Elango, K. P. Conducting Polymer/Alumina Composites as Viable Adsorbents for the Removal of Fluoride Ions from Aqueous Solution. *J. Fluorine Chem.* **2009,** *130,* 894–901.
122. Yatsyshyn, M. The Structure of the Polyaniline in the Films Deposited Electrochemically on the Surface of Aluminium-Containing Electrodes. *Visnyk Lviv. Univ. Ser. Khim.* **2011,** *55,* 387–404.
123. Stejskal, J.; Prokeš, J.; Trchova, M. Reprotonation of Polyaniline: A Route to Various Conducting Polymer Materials. *React. Funct. Polym.* **2008,** *68,* 1355–1361.
124. Brozova, L.; Holler, P.; Kovarova, J.; Stejskal, J.; Trchova, M. The Stability of Polyaniline in Strongly Alkaline or Acidic Aqueous Media. *Polym. Degrad. Stab.* **2008,** *93,* 592–600.
125. Berrada, K.; Quillard, S.; Louarn, G.; Lefrant, S. Polyanilines and Sunbstituted Polyanilines: A Comparative Study of the Raman Spectra of Leucoemeraldine, Emeraldine and Pernigraniline. *Synth. Met.* **1995,** *69,* 201–204.
126. Rajapakse, R. M. G.; Krishantha, D. M. M.; Tennakoon, D. T. B.; Dias, H. V. R. Mixed-conducting Polyaniline-Fuller's Earth Nanocomposites Prepared by Stepwise Intercalation. *Electrochim. Acta.* **2006,** *51,* 2483–2490.
127. Jing, X.; Wang, B.; Wang, F. Infrared Spectra of Soluble Polyaniline. *Synth. Met.* **1988,** *24,* 231–238.
128. Li, X.; Gao, Y.; Zhang, X.; Gong, J.; Sun, Y.; Zheng, X.; Qu, L. Polyaniline/CuCl Nanocomposites Prepared by UV Rays Irradiation. *Mater. Lett.* **2008,** *62,* 2237–2240.
129. Xu, J.-C.; Liu, W.-M.; Li, H.-L. Titanium Dioxide Doped Polyaniline. *Mater. Sci. Eng. C.* **2005,** *25,* 444–447.
130. Millan, W. M.; Thompson, T. T.; Arriaga, L. G.; Smit, M. A. Characterization of Composite Materials of Electroconductive Polymer and Cobalt as Electrocatalysts for the Oxygen Reduction Reaction. *Int. J. Hydrogen Energy.* **2009,** *34,* 694–702.
131. Liu, W.; Kumar, J.; Tripathy, S.; Senecal, K. J.; Samuelson, L. Enzymatically Synthesized Conducting Polyaniline. *J. Am. Chem. Soc.* **1999,** *121,* 71–78.
132. Gu, Y.; Chen, C.-C.; Ruan, Z.-W. Enzymatic Synthesis of Conductive Polyaniline Using Linear BSA as the Template in the Presence of Sodium Dodecyl Sulphate. *Synth. Met.* **2009,** *159,* 2091–2096.
133. Wu, J.; Tang, Q.; Li, Q.; Lin, J. Self-assembly Growth of Oriented Polyaniline Arrays: A Morphology and Structure Study. *Polymer.* **2008,** *49,* 5262–5267.
134. Geng, L.; Zhao, Y.; Huang, X.; Wang, S.; Zhang, S.; Wu, S. Characterization and Gas Sensitivity Study of Polyaniline/SnO_2 Hybrid Material Prepared by Hydrothermal Route. *Sensor. Actuat. B-Chem.* **2007,** *120,* 568–572.
135. Subramania, A.; Devi, S. L. Polyaniline Nanofibers by Surfactant-assisted Dilute Polymerization for Supercapacitor Applications. *Polym. Adv. Technol.* **2008,** *19,* 725–727.
136. Tai, H.; Jiang, Y.; Xie, G.; Yu, J.; Chen, X.; Ying, Z. Influence of Polymerization Temperature on NH_3 Response of PANI/TiO2 Thin Film Gas Sensor. *Sensor. Actuat. B-Chem.* **2008,** *129,* 319–326.
137. Grigore, L.; Petty, M. C. Polyaniline Films Deposited by Anodic Polymerization: Properties and Applications to Chemical Sensing. *J. Mater. Sci. - Mater. Electron.* **2003,** *14,* 389–392.

138. Lakard, B.; Herlem, G.; Lakard, S.; Antoniou, A.; Fahys, B. Urea Potentiometric Biosensor Based on Modified Electrodes with Urease Immobilized on Polyethylenimine Films. *Biosens. Bioelectron.* **2004,** *19,* 1641–1647.
139. Binh, P. T. Electrochemical Polymerization of Aniline by Current Pulse Method in the Presence of m-Aminobenzoic Acid in Chlorhydric Acid Solution. *Macromol. Symp.* **2007,** *249–250*, 228–233.

CHAPTER 3

ELECTROOPTIC PHENOMENA IN CONJUGATED POLYMERIC SYSTEMS BASED ON POLYANILINE AND ITS DERIVATIVES

O. I. AKSIMENTYEVA[1], O. I. KONOPELNYK[2], and D. O. POLIOVYI[3]

[1]*Department of Physical and Colloid Chemistry, Faculty of Chemistry, Ivan Franko National University of Lviv, 6 Kyryla & Mefodia Str., Lviv 79005, Ukraine. E-mail: aksimen@ukr.net*

[2]*Department of General Physics, Faculty of Physics, Ivan Franko National University of Lviv, 8 Kyryla & Mefodia Str., Lviv 79005, Ukraine. E-mail: konopel@ukr.net*

[3]*Department of Biology and General Ecology, Natural and Engineering Faculty, Taras Shevchenko Kremenets Regional Humanitarian and Pedagogical Institute, 1 Litseina Str., Kremenets 47003, Ukraine. E-mail: dpolov@rumbler.ru*

CONTENTS

ABSTRACT

Polymers of polyaniline series (namely, polyaminoarenes) are considered as the perspective materials for electrochromic devices from the point of view of physical chemistry of electrooptic phenomena. The main regularities of polyaminoarenes films formation on transparent semi-conducting surfaces have been studied based on the results of electrochemical, optical, and spectroelectrochemical investigations. The relationship between the electrochemical and also electrooptical characteristics and the structure of polymers was established. On a basis of determined diffusion coefficients and heterogeneous constants of a charge transport in different electrolytic media it was shown, that the structure of a polymeric chain has the decisive effect on the electrochromic material efficiency. Some possible methods of electrochromic characteristics improvement for polyaniline (**PAn**) and its derivatives have been proposed.

3.1 INTRODUCTION

The field of optoelectronics has blossomed over the last decades in many scientific world-wide laboratories, service yards, scientific research engineering institutes, and that is why the need for obtaining of inexpensive optic elements, which are characterized by multiplex light reflection or light transmission and can be able for application of them in flexible displays, storage digital devices capable for information rewriting, as well as the need for using of optic elements in so-called "smart" windows, mirrors, color detectors and others[1–5] will only increase, since the digital technologies becomes more important in the all areas of human life. Advances in this field largely depend on the application of materials possessing by electrooptical properties, which can modify the light-spectrum, and in that way can change, also the color under the action of the electric field. Electroconductive conjugated polymers based on polythiophene (**PT**), polypyrrol (**PPy**), polyaniline (**PAn**), and also their derivatives, electronic properties of, which are determined by oxidation or reduction level achieving in the electrochemical doping manner are striking examples of the above-mentioned optical materials. In a case, when a new band is generated in the light-spectrum during the transition from the one oxidation–reduction state into another one, or the band is visualized in other optical range, we have deal with the so-called electrochromic materials (**EChM**).[6,7] Such materials are inexpensive, they can be simply obtained and that is why they are advantageously distinguished from the liquid crystals. At that, **EChM** can provide the required contrast in multicolor display, which does not depend on viewing angle.[8,9] The displays by a large area are obtained on a basis of the **EChM**; such displays consume a little energy and have a good compatibility with the existent technologies and operations (processes) in a field of the optoelectronic and electronic devices production.

The electrochromity phenomenon as the phenomenon of a color change or the color intensity of material change under the action of the applied voltage is outwardly similar to the effects, which are observed in the liquid crystals.[10–13] However, the physical–chemical essence of the electrochromism phenomenon is some distinctive, since it's connected with the reverse changes of electronic structure of the material.[14,15] In other words, the electrochromic effect can be determined as the visible and reverse changes of the optical properties, namely this is a transmittance (absorption) or light reflection, which is visualized under conditions of electrochemical oxidation/reduction of the electroactive substance. As a result of the material electronic properties modification (particularly, a width of the energy gap), a change of the material's optical properties takes place (e.g., a change of the position and of the absorption band intensity).

A size of the energetic gap between the valence band (or the highest occupied by p-electrons zone and so-called highest occupied molecular orbital **HOMO**) and the conduction band as the lowest unoccupied zone (or so-called lowest unoccupied molecular orbital **LUMO**) is responsible for the optical properties of material and respectively is responsible for the material's color.[16,17]

The many of the **EChM**, in particular the organic crystals and also the conjugated polymers, represent by themselves the organic semiconductors, for which the value of a width of the energy gap (or difference between **LUMO** and **HOMO**) consists of E_g = 1.3–3.2 eV.[16] In accordance with this fact, the electron structure of such **EChM** presupposes the presence of optical adsorption in near ultraviolet (**UV**) spectrum, visible spectrum and long-wavelength infrared or near infrared (**IR**) spectrum.[5] As a rule, the many of well-known **EChM** are monochrome, in other words they change of their properties within a range of the one color. In a case when the material evinces several colors in different spectral range, it is called as the **polyelectrochromic** or **multicolored**[14,16,18] one.

An investigation of the electrochromism of organic substances has been done by *Platt*[19] as far back as the 1961; exactly he was proposed to use of this term for the description of the electrooptical phenomena. Since then the study of electrooptic properties of organic compounds, in particular of conjugated polymers, call a great especial attention.[20–24] However, the work of *Deb*[25] in 1969 dedicated mainly to the WO_3, in which the electrochromic effect of inorganic materials was observed for the first time, historically was considered as the first description of the electrochromism phenomenon. A number of physical discoveries forewent to the breakthrough of the electrochromity phenomenon, namely: *Kerr* electrooptic effect (or molecules orientation under the action of electrostatic field setting conditions for the double light refraction of a sample);[26] *Stark* effect (or the splitting and the displacement of the spectral lines of atom under the action of electrostatic field by the intensity *E*);[27] *Faraday* magnetooptic phenomena.[28] At present time the electrochromic effect is studied for not only inorganic materials,[29] but it's investigated also in conjugated polymers, viologenes, metal polymers, metalphthalocyanines, and others.

In a case of the electrochromic materials practical application, the following characteristics are required for them, namely: high contrast and high efficiency of a color, long-continued time of life (cycling), ability to the rewriting of information (a change of the color). The requirements to some indexes are caused by a field of the electrochromic material following application, for example, the displays need a little response time whereas the so-called "smart" windows are characterized by the response time till some minutes.[3,10] The all multiplicity of the electrochromic properties is peculiar to a wide class of organic electrochromic materials, which are obtaining on a basis of conjugated polymeric systems, namely polymers with own electron conduction.

3.2 THE CONJUGATION EFFECT AND ELECTROOPTIC PROPERTIES OF CONJUGATED POLYMERIC SYSTEMS

An electrochromism has a certain attitude toward electrooptic phenomena.[11,29] Spectral, kinetic, and dynamic methods of the investigations are widely used for its study. Physical–chemical character of the electrochromism as the optical phenomenon and its ordering to some physical and mathematical models give the possibility to form both of fundamental, theoretical, and technological approaches to the creation and performance of these newest materials in different fields of science and technique.

Substances having the molecules with interchangeable single and multiple bonds are termed as "conjugate" and such chemical structure is named as "conjugated." During the polymerization of some substances (arenes, aminoarenes, thiophene, pyrrol, and their derivatives) the high molecular compounds having the interchangeable single and multiple bonds along the chain (so-called linear system of conjugation) are formed. This leads to the appearance of qualitatively new properties of polymers, which are not typical for the traditional ones, namely electron conduction, paramagnetism, optical absorption in visible and near **IR** spectral range, semi-conducting character of the conduction, the sensibility of electronic properties to the external factors. The peculiarity of conjugated polymers is that the bonds between atoms in such polymers are not localized, but are uniformly distributed on a chain (or they are delocalized). In order to describe the system of conjugation, the **resonance structures theory**[30] is used. Basically, this theory presents a conception that the electronic distribution, geometry and others physical, and chemical properties of the molecules should be described not only by one possible structural formula, but also by combination (or by resonance) of the all-alternative structures.

Thus, the main feature of the electroconductive polymers (**ECP**) is the presence of conjugated system of π-electronic bonds in them. Such peculiarity gave the possibility to mark out of these polymers into the separate individual class of high molecular compounds, which has been called as "the fourth generation of polymeric materials."[31] The processes of reconstruction of electronic structure with the participation of the carrier of charge are the essential principle of the main physical–chemical properties of conjugated polymers, in particular of their optical characteristics. The source of these carriers of charge is the conjugated π-electronic system of a polymeric chain. An excitation of such system during the doping process generates of certain type of the carriers typical for the conjugated polymers (so-called neutral and charged solitons, polarons, and bipolarons),[32,33] which can form the appropriate zones. Electron transitions are typical for the substances having the conjugated bonds; moreover, the intensity of the optical adsorption is increased (**hyperchromic effect**) and the maximum absorption (λ_{max}) is displaced in the longer wavelength spectrum (**bathochromic shift**)[11,33–35] at the conjugation lengthening.

The features of spectral properties of macromolecules in optical spectral band of energy are determined by their energetic structure. Conjugation reduces the difference of energies between the **HOMO** and **LUMO** of π-electronic system. Bathochromic shift is caused by that at the fitting of two chromophores their orbitals are displaced resulting in the new set of the molecular orbitals formation; at this, the number of these orbitals has not change, only a change of the energies takes place. This process is represented in Figure 3.1 (a) for the modeling system "ethylene–butadiene:"[30,33] two π-orbitals and two π^*-orbitals (unoccupied ones) are transformed into four molecular orbitals $\Psi 1$, $\Psi 2$, $\Psi 3$, and $\Psi 4$ (two last antibonding ones). If the number of conjugated links is increased, the energies difference between **HOMO** and **LUMO** is reduced, and can be within the bounds 1.5–3.2 eV; this corresponds to energy of the electromagnetic waves in the visible spectrum (λ = 400–800 nm). Series of properties not typical for organic compounds, but specific for the conjugated polymers, are caused by especial states in electronic structure of substances.

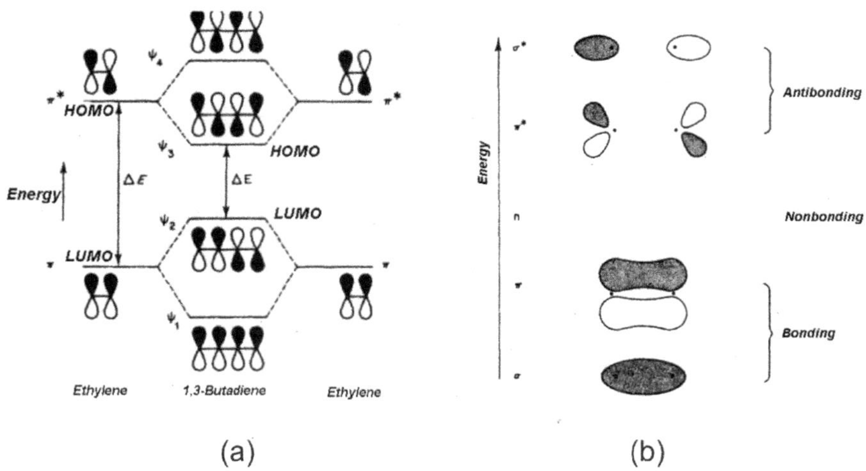

FIGURE 3.1 (a) Schematic representation of relative energies of molecular orbitals in ethylene and butadiene-1,3; (b) Schematic representation of relative energies of electrons of molecular orbitals.[33]

The relative energetic allocation of molecular orbitals in organic substances is represented in Figure 3.1 (b). The electrons on σ, σ^*, π, π^* orbitals take part directly in the bond formation as the connecting or anti-bonding "electron swarms;" unbound n-electrons, which take not part in the chemical bond formation play an important role in physical–chemical behavior of substance. They are a source of the excitons, polarons, and bipolarons formation into the conjugation system. Such differentiation of the antibonding electrons leads to the splitting of given energy level and to the appearance of new electron transitions with the little difference in energies and states having the overlapping levels like to the conduction band in the metals.[33–35]

Polymers having the molecules with alternation multiple bonds can be divided into two types: 1) the polymers, in macromolecules of which the π-electron groupings are allocated in separate groups (so-called chromophores). Polyvinylnaphthalene (**PVN**), polyvinylcarbazole (**PVC**), polystyrene (**PS**), and such important biological systems as *RNA* та *DNA* are typical polymers of this series; 2) the polymers, in which the conjugation system is extended on a whole macromolecules (π-electrons in the ideal case are delocalized on a whole macromolecule). The second type of these polymers is represented by polyacetylene (**PA**), polyphenylene (**PPh**), **PAn**, **PT,** and their derivatives.[31,36,37] Linearly conjugated system of double bonds is an important chromophore both in natural (carotene) and in artificial dyestuffs, for example, in cyanine ones. Optical absorption of polymers by the second type depends on a state of π-electrons of the system of conjugated bonds, which are delocalized along macromolecules, and that is why their spectral characteristics depend on the length of a polymeric chain.

One more peculiarity of the optical absorption of this polymers type is caused by their electrochemical activity. A change of the reductive–oxidative state makes conditional upon reconstruction of the molecular and electronic structure. The reduction processes give rise to a change of the double bonds on the single ones, and vice versa, the oxidation leads to the double bonds quantity increase. These electrochemical transformations cause the different changes in the conjugation system and this is affected on the energy states and, as a result on the electron spectra of absorption.

Behavior of spectra of polyene systems can be satisfactorily explained on a basis of classical model with the use of *Lewis–Kelvin* quasi-quantum theory.[11] If to assume, that the each pair of electrons equipped in the bond between neighboring atoms realizes the harmonic motions near the equilibrium position, then the frequency of these motions should be expressed by the following formula:

$$\nu = \frac{1}{2\pi}\sqrt{\frac{K}{nm}} \tag{3.1}$$

where K is the constant of quasi-elastic force; n is the number of interacting between themselves oscillators, which is equal to the number of conjugated double bonds; m is a magnitude of the vibrating charge in every bond. It follows from this formula that

$$\nu^2 \approx \frac{1}{n} \text{ або } \lambda^2 \approx n\,. \tag{3.2}$$

A length of the conjugate chain in **ECP** can be regulated applying the one or another method of synthesis or film formation; in this case the optical properties depend on a synthesis technique. An essential factor having an influence on the optical properties of conjugated polyarenes (**PAr**) is the molecular structure of the elementary link, namely nature and position of benzene ring substituents, their relative position, ability to the conformation change and others.

An optical absorption of the carriers into conjugate polymeric systems is realized mainly within the optical spectral band corresponding to near **UV** spectrum (200–400 nm), visible spectrum (400–800 nm), and near **IR** spectrum (800–1100 nm). However, depending on the conditions of electrochromic polymer synthesis, the position of maximum peaks of the absorption bands and their intensity are essentially differed.

Monomeric substances[31,38,39] by differing chemical structure and class can form a chain of conjugation via the process of polymeric conversions. Linear and branched hydrocarbons (acetylene, phenylacetylene, diacetylene), aromatic compounds (benzene, aromatic amines, aminonaphthylsulfonic acids), heterocyclic compounds (thiophene, pyrrol, azulene, indole, and their derivatives) belong to the above-mentioned substances. One way or another, all of these polymers exhibit the electrochromic effect as a change of the light-spectrum under the action of electrostatic field. However, the electrochromic properties as the change of a color and, respectively of the optical absorption (or transmittance) in visible spectrum are the most brightly visualized in some sets of conjugate polymeric systems, mainly in those elementary link of which contains the heteroatoms (nitrogen, sulfur, oxygen, and others). The derivatives of **PT**, **PPy**, **PAn**, their copolymers, and functionalized compounds[6,8,14] are under the active investigation. Here, an accessibility, a stability, and an electroactivity of the electrochromic material play the important or main role; exactly the **PAn** and its derivatives, namely polyaminoarenes (**PAAr**) satisfy the all these requirements.

3.3 ELECTROCHROMIC MATERIALS (EChM) BASED ON CONJUGATE POLYAMINOARENES (PAAr)

Polymers having the system of conjugate bonds on a basis of **PAn** possess by a series of properties combination of which in the same substance is unique. An ability to reductive–oxidative transforms can provide a change of the molecular structure and hence also a change of physical–chemical characteristics of the substance. Electron conduction of conjugate **PAAr** gives the possibility to form the current-carrying functional layers directly on a surface of the main electrode of the electrochemical system that is very important for the development of the electroactive and electrochromic coatings.

Despite the fact that the **PAn** has been synthesized over a hundred years ago, it's remained by absolute undisputed leader among the most investigated conjugate **PAAr**. The presence of electron conduction in this polymer was considered as the "scientific curiosity" for a long time, while this discovery was not awarded by the Nobel Prize[31,36] in 2000. Although the dozens thousands of original articles, reviews and monographs[38,40–42] dedicated to the **PAn** and its derivatives (namely, **PAAr**) uncover the methods of synthesis, electrochemical, and structural characteristics of these polymers, their use in current sources, electrochemical sensors, etc., but the electro-optical properties of **PAn** in congruence with the derivatives of **PT**, **PPy**, and others polymers highlighted not enough.

It is considered that the elementary link of **PAn** and others of **PAAr** consists of the four formula units, as is shown on the scheme (Table 3.1). Semi-oxidated form of **PAAr**, in which two aniline fragments consist of the aminobenzene-like link ($x = 0.5$), and two others ones form the benzoquinoid cation radical[40] is the most stable and important form on its properties. This form is called as "emeraldine" one due to the emerald color of its salts. If $x = 1$, $y = 0$, then the all aromatic rings are in the reduced state and the polymer is discolored and is turned into so-called leucoemeraldine form. When the $x = 0$, $y = 1$, the polymer is in a non-conducting over-oxidized state and irreversibly becomes black.

So, **PAn** is polyelectrochromic,[40–43] because in different redox states it exhibits multiple colors, such as yellow (leucoemeraldin), green (emeraldine salt), blue (emeraldine base), and black (pernigraniline). In real systems under stepping or smooth switching of the potential, no "pure" colors, but their shades with gradual change of colors are observed.

Similar reductive–oxidative forms proper also for the derivatives of **PAn**, but their electro-optical properties studied so far is not enough. Among the all of these **PAAr** the *ortho*-substituted derivatives of **PAn**, namely poly-*o*-anisidine (**PoA**), poly-*o*-toluidine (**PoT**), poly-*o*-aminophenol (**PoAPh**), and others[9,22,39] call a special attention. Interesting physical and chemical properties of these polymers are occurring due to the presence of electron-donor substituents of the benzene ring in the *ortho*-position to the amine group. Introduction of such substituents creates the local features of the molecular structure of **PAAr** caused by a decrease of the conjugation length of the polymer chain and the electron delocalization, leading ultimately to a lower conductivity of substituted **PAAr** compared to **PAn**.[44] During the synthesis of substituted **PAAr** due to changes in the degree of the reaction center (cation radical) protonation and orienting impact of electron-donor substituent in the *para*-position of the benzene ring, competitive with the influence of the amine group, the linear growth of a chain is somewhat complicated, which can lead to its branching.[30,45] As a result, the electronic structure of the polymer is changed making especially its doping and the charge transport.[46]

TABLE 3.1 Redox States and Electrochromic Properties of **PAn.**

Chemical structure of elementary link	Redox form of PAn	Color of the film	Electrode potential, V (Ag/AgCl)
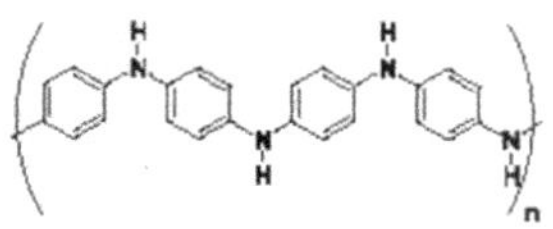	Leucoemeraldine base (**LB**) undoped form	Yellow	−0.2.....0.0
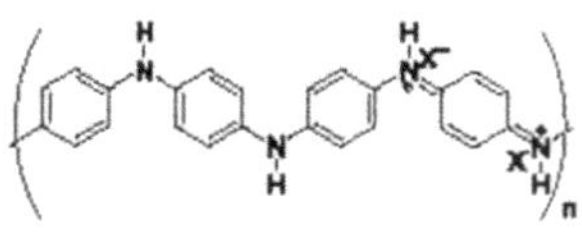	Emeraldine salt (**ES**) doped form	Green	0.4.....0.6
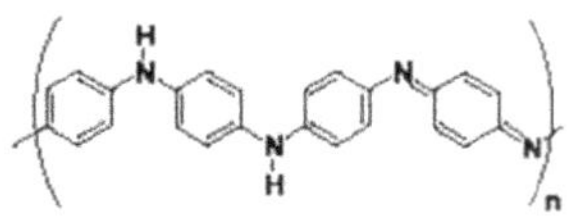	Emeraldine base (**EB**) undoped form	Blue	0.8.....0.9
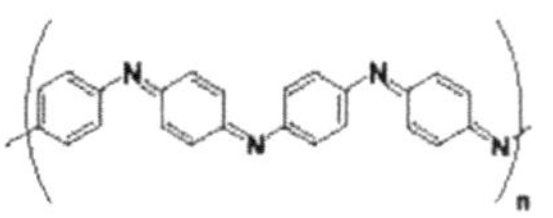	Pernigraniline base (**PB**) undoped form	Black	1.0......1.1

PAAr exhibit not only interesting but also stable electrochemical[22,41] and electrooptical[17,21] properties that make them attractive for use in organic displays and sensors. Depending on the presence and type of substituent in the benzene ring of aminoarene, the **PAn** polymers may exhibit a fairly wide range of the electrochromic transitions.[4,9] At the same time, the electrochemical activity of **PAAr** essentially depends on the method of their synthesis.

3.4 METHODS OF OBTAINING THE ELECTROCHROMIC FILMS BASED ON PAn AND ITS DERIVATIVES

The materials, which we term as the electrochromic ones can be obtained in a thin layer on solid surfaces with the use of simple, well-known methods from the solutions or sols, which are typically applied for organic substances and not demand the special temperature conditions or high vacuum using. An obtaining of thin films of conjugate polymers on a solid surface is realized from the solutions of appropriate polymers with the use of *Langmuir–Blodgett*[47,48] technique, centrifugation (spinning), which provide the spread of the polymer solution on the surface,[49] layer-by-layer assembly composition,[50] and others. At the same time, the methods of electrochromic coatings formation from the polymer solutions have some limitations, the most of which is concerned to poor solubility of **ECP**, and in particular of **PAn** in many of organic solvents. Furthermore, polymer films obtained by the spreading method not always are characterized by the ordered structure and often are irregular on thickness. Durational times of a film formation, pollution of atmosphere by vapors of organic solvents are the technological inconveniences of this process.

Alternative methods of **PAn** films formation on the semiconductors surfaces are based on physical approaches, including the vacuum deposition of polymer,[51,52] high-frequency magnetron sputtering,[53] and others. These methods permit sufficiently accurately to regulate the thickness of the obtained layer, however, nevertheless they not widely used, since the formed in this way film of **PAn** and its derivatives have a low electrochemical activity. Therefore, the method of electrochemical polymerization[43] for the formation of electrochromic systems is more widespread.

To observe the optical and electrooptical effects in thin films it is necessary to use the optically transparent surfaces, which mainly are dielectrics (quartz, glass, sapphire) or semiconductors (Tin (IV) oxide, indium (III) oxide (or their mixture), indium-tin-oxide, or *ITO* electrode). Dielectric or semiconducting nature of the material surface causes the peculiarity of physical and chemical approaches to the formation and investigation of thin films of the electroactive substances. This is concerned to both known thermal vacuum technologies and electrochemical synthesis of polymers in a thin layer, peculiarities of which are determined by the state boundary of the electrode–electrolyte solution.

A study of the structure of double electrical layer (**DEL**), which appears at the border of semiconductor-electrolyte, gave the possibility to establish that, unlike to the metal surfaces the space charge region directed into the volume of semiconductor is available for the semiconductors. An interface of semiconductor solution can be considered as a capacitor whose plates are characterized by the diffusive structure. The electrical properties of this contact are determined mainly by the diffuse charge in the semiconductor.[54] Peculiarities of the **DEL** structure make corrections in the field of the potentials of aminoarenes electrochemical activity on a surface of semiconductors, in their ability to adsorb on the surface of electrode, in the rate of electron transfer at the interface electrode-solution. An obtaining and generalization of the experimental data for each specific electrochemical system is a very important task at the investigation of electrochemical processes on the semiconducting electrodes.

In order to obtain the polymer layers into the electrochemical way, an electrolysis can be carried out using the direct current (galvanostatic mode) or the potential (potentiostatic one) or under conditions of the dynamic change of potential, or current in accordance with the cyclical law. Electrochemical synthesis of the electrochromic polymers can be realized under the different conditions of the electrolysis carrying out, using both alternating and direct current.[43,55] Stationary methods of the electrosynthesis provide the constancy of the current density or constancy of a given potential. Under conditions of the **PAAr** galvanostatic synthesis a dependence of optical absorption on the quantity of electricity through passing the cell (Fig. 3.2) is caused by a size of the current density (*j*). At small values of *j* the quantity of the formed polymer is increased proportionally to the quantity of electricity (*Q*); at the current density increasing this dependence becomes nonlinear, and it's even approximated to the parabolic one. Such kinetic dependence is typical for the complex consecutive reactions by autocatalytic type.[41]

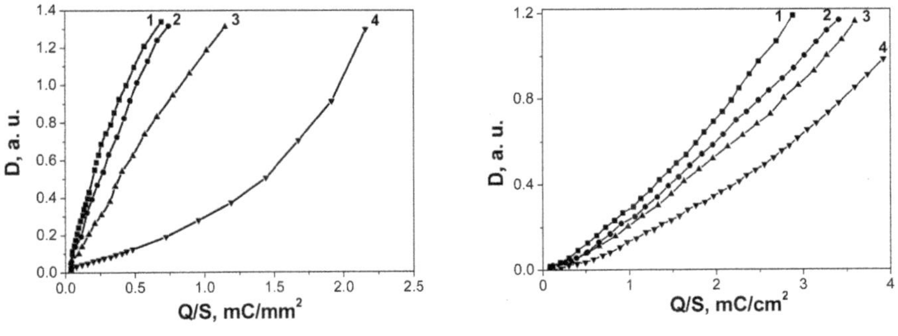

FIGURE 3.2 Dependence of optical density on the quantity of electricity during the process of the galvanostatic synthesis of a) **PAn**, b) **PoA** on a surface of *ITO* electrode: at the current density, mA/cm^2: 1) 0.01; 2) 0.05; 3) 0.10; and 4) 0.20.

For the processes of **PAn** electrochemical synthesis on the semiconducting electrode the classical scheme of the oxidative polymerization[30,41] can be applied. A process is initiated by the reaction of electrochemical oxidation of monomer to the cation radical:

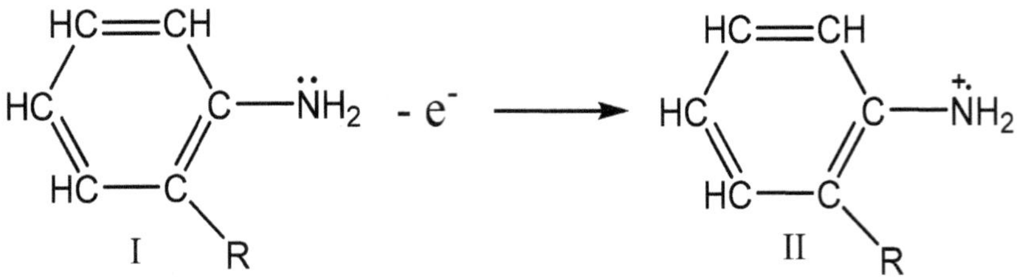

The concentration of formed cation radicals is proportionate to a quantity of electricity. In accordance with the existent theories,[30] the cation radicals are in the form of the resonance structures **II–IV**:

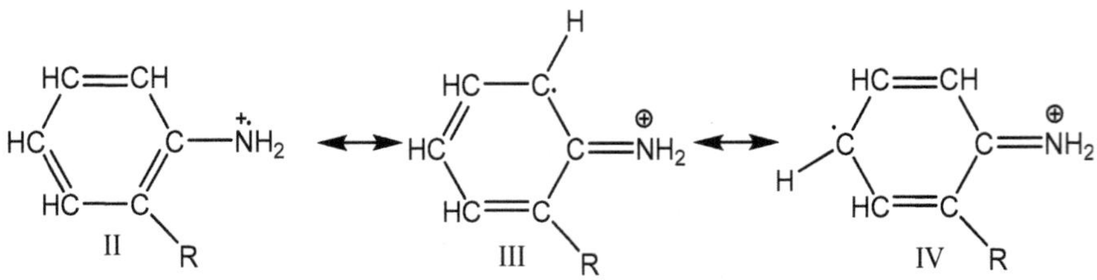

The structure **IV** is the most stable structure, particularly in the presence of electron-donor substituent (*R*).[41] The following consecutive reactions are: (i) the formation reaction of dimer (trimer, tetrameter), (ii) the reaction of their deprotonation, and (iii) the reaction of the following polymeric chain propagation:

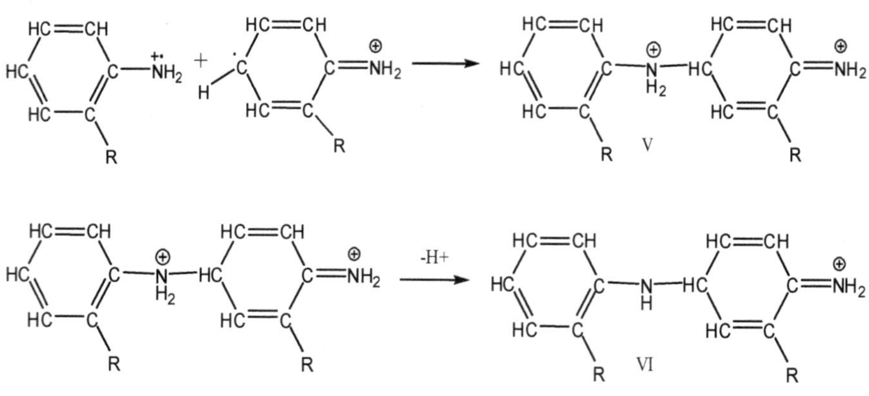

Since the cation radical exists mainly in the form of the structure **IV**, then the polymerization takes place mainly in the *para*-position, and the addition (oxidative connection) of the initial and isomerizated cation radical proceeds accordingly to the "head–tail" principle.[41]

Formation of a polymer represents by itself the bimolecular reaction, which is complicated by heterogeneous conditions of the chain growth when the reaction product formed into the near-electrode layer is not dissolved in the reactive medium and forms a film on a surface of electrode. High electron conduction of conjugate **PAr** permits to synthesize the polymeric layer on the electrode in a wide range of the thicknesses (10 nm–5 μm). Potential of the monomer and intermediate oligomeric compounds electrooxidation is decreased[38,39] in the presence of the polymeric layer.

Galvanostatic polymerization process is characterized by a bottom-up construction of great crystalline entities and by practical absence of growth of the fine-grained fragments. This causes the irregular growth of the film and the formation of dendritic crystals (polymer nanofibers and nanowires), which themselves are an interesting object of the scientific researches, but unsuitable for the obtaining of uniform, optically reproducible films on the surface of the *ITO* electrode. Such behavior of the process is caused by a high conductivity of obtained forms and by a high gradient of the monomer concentration in the near-electrode area. In order to obtain the uniform films, it is necessary to alternate the electropolymerization process with the alignment of the monomer concentration in the near-electrode space. For this purpose advisably is to be used the methods with the cyclic (anode–cathode) potential scanning.

The possibility to control the thickness of electrochromic films at the electrochemical synthesis can be achieved with the use of cyclic voltammetry[56,57] when the thickness of film is determined by the scanning cycles number (N) in some range of the potentials. In a case of **PAAr**, the scanning cycles number and quantity of electricity (Q) passing at synthesis are interconnected via equation:

$$Q \approx c^2(N/v)^2 \exp[(2\alpha nF/RT)E_\lambda], \tag{3.3}$$

in which c is the concentration of monomer, α is the coefficient of transfer, E_λ is the switching potential, Q is general electricity of the electric deposition, obtained for completely reduced polymer after N cycles, v is the scanning rate, n is the number of electrons taking part in the redox-process, F is *Faraday* constant.

Polymer films on a surface of SnO_2 or *ITO* electrodes have been obtained with the use of electrochemical polymerization of 0.02–0.1 M solutions of monomers (*o*-toluidine, aniline, *o*-methoxyaniline, aminophenols, and others) in solution of protonic acid (0.01–1 M) under conditions of the cyclic potential scanning within 0 till 1.2 V at the scanning rate ν = 5–100 mV/s.[9,22,39] The thickness of film was controlled by the number of scanning cycles, by the values of currents of anode, and cathode peaks on cyclic voltammetric curves (**CVA**), and also using the microinterferrometry, ellipsometry, optical microscopy, profilometry, and others. An application of transparent semi-conducting surfaces allows studying the processes of conjugated **PAr** synthesis, using both electrochemical measurements and spectrophotometric investigations. This feature is specified by the formation of polymeric conjugate compounds, in which a system of conjugate bonds represents by itself a chromophore, and an increase of the conjugation chain leads to a shift of maximum absorption in the visible region. The concentration of chromophores is proportional to the amount of substance of conjugate polymer and can be controlled by the value of the optical absorption (optical density) according to the *Lambert–Berr* law.[11,38]

At study of the processes of electrochromic films of **PoT** and poly-*o*-methoxyaniline (**PoMA**) obtaining it was found that the polymerization on the surface of SnO_2 occurs at much lower values of anode potentials (0.52–0.70 V) compared to oxidation of monomers (E = 0.82–0.92 V), that is, the initially formed layer is the catalyst of a process (Fig. 3.3). Formation of the polymer proceeds per autocatalytic mechanism in accordance with the well-known reaction scheme of electrochemical combination of aromatic amines, including the oxidation stages of the monomer with the formation of cation radicals and recombination of primary and isomerized cation radicals accompanying by deptotonation.[9,39] Compact uniform films of **PoMA** and **PoT** on SnO_2 electrodes were obtained after 15–30 cycles of the potential scanning between E = 0.0 ... 1.0 V.

Based on the data of electronic spectra, an increasing the number of scanning cycles during the synthesis of a film causes an increase of the polymer layer optical density at the increase of the layer thickness (Fig. 3.4). Films of **PoT** are characterized by the presence of two principal absorption bands (380–400 nm and 760–780 nm) and also by the presence of one intermediate band of low intensity in a range of 650 nm. An absorption at 380–400 nm can be related to the π–π^* transition into localized semi-quinoid cation radicals (polarons), the band at 760–780 nm describes the delocalized states of free carriers (bipolarons). An intermediate band at 650 nm corresponds to n-π^* transition in quinoid form of the principle structure of **PoT**.

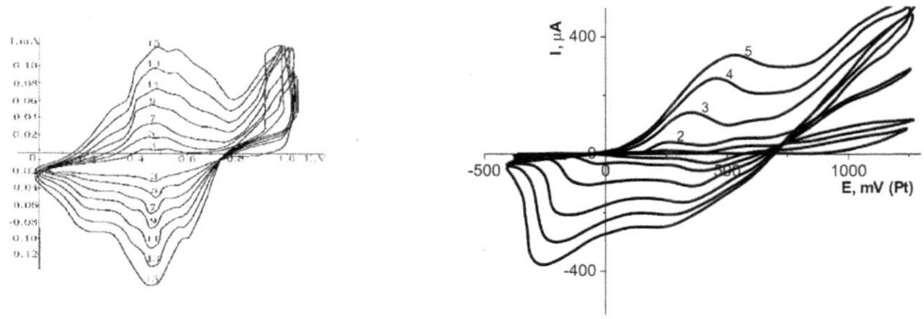

(a) (b)

FIGURE 3.3 (a) Cyclic voltammograms obtained during the process of **PoT** electrodeposition on the surface of SnO_2 from 0.1 M solution of *o*-toluidine in 0.5 M H_2SO_4 at the scanning rate v = 20 mV/s, digits indicate the number of cycle; (b) **CVA** were obtained via the synthesis of **PoA** on SnO_2 electrode from 0.025 M solution of *o*-anisidine in 0.5 M of H_2SO_4 at the scanning rate v = 100 mV/s, digits indicate the consequence of the scanning cycles.

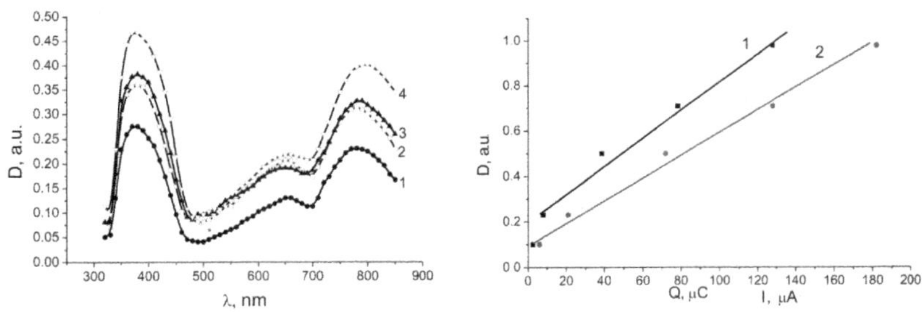

(a) (b)

FIGURE 3.4 (a) Optical absorption spectra of **PoT** films on a surface of SnO_2. Films were obtained from 0.1 M solutions of *o*-toluidine in 0.5 M H_2SO_4 for 15 (1), 20 (2), 22 (3), and 30 (4) cycles of scanning; (b) Dependence of optical density of **PoA** film on current of the anode maximum (1) and on quantity of electricity (2).

It was determined[58] that the oxidation potential of the monomer (E_{ox}) on semiconducting electrode is in a range of 0.82–1.43 V (saturated Ag/AgCl electrode), which on 0.5–0.6 V exceeds of the E_{ox} on a platinum electrode. The presence of electron donor substituent in the *ortho*-position leads to a difference in potentials of oxidation for the substituted aminoarenes. In series of –H, $–CH_3$, $–OCH_3$, and –OH the oxidation potential is decreased due to the growth of the electron donor properties of the substituents.[30] On the other hand, in this series the orienting activity of a substituent into the *para*-position is increased which competes with the amine group, which complicates the linear growth of the chain, causing of its branching.[45] As a result, initially formed film, which modifies the surface of the electrode has a higher resistivity compared to **PAn**, and at the scanning of the potential a gradual increase of the potential of half-wave of monomer oxidation on the modified surface takes place (Table 3.2). In addition, at increasing of the number of potential scanning, the potential distribution ($\Delta E = Ea-Ek$) is growing at the simultaneous preserving of the reversible redox potential (E_{rev}) during the electrosynthesis. A comparison of heterogeneous charge transport rate constants determined by the *Nicholson* method[59] during the electrosynthesis of **PAn** on a platinum electrode ($k_s = 2.3 \times 10^{-3}$ cm/s[38]) and on the SnO_2-electrode ($k_s = 1.8 \times 10^{-4}$ cm/s) shows a significant slowdown in the formation process of the polymer on the surface of semiconductors.

TABLE 3.2 Electrochemical Parameters of **PAAr** Electrosynthesis on a Surface of Sno_2 Electrode under Conditions of Cyclic Scanning of the Potential in 0.025 M Solutions of Monomers in 0.5 M H_2SO_4.

Monomer, substituent	№ of cycle	Pt	SnO_2				
		E_{ox}, V	E_{ox}, V	$E^A_{1/2}$, V	ΔE, mV	E_{rev}, V	I^a_p/I^K_p
Aniline	1	0.96	1.43	–	–	0.50	–
(R = –H)	2		1.27	0.54	110		0.38
	3		1.27	0.57	200		0.57
	4		1.26	0.57	220		0.65
o-toluidine	1	0.90	1.48	–	–	0.29	
(R = –CH_3)	2		0.79	0.44	240		0.58
	3		0.97	0.50	350		0.81
	4		1.04	0.52	420		0.80
	5		1.13	0.57	–		–
o-anisidine	1	0.60	1.06	–	–	0.35	–
(R= –OCH_3)	2		0.83	0.23	220		0.36
	3		0.84	0.26	310		0.67
	4		0.88	0.32	410		0.86
			0.94	0.35	480		0.89
o-aminophenol	1	0.56	0.82	–	–	0.18	–
(R= –OH)	2		0.67	–	140		0.28
	3		0.65	–	260		0.72
	4		0.66	–	310		0.94

The most characteristic feature of aminoarenes electrochemical polymerization process is the first redox maximum with a value of reversible potential E_{rev} ~ 0.3–0.5V, which is observed for the all investigated monomers. The thickness of the polymer layer can be controlled by the magnitude of this redox peak.[57] An integration of the anodic or cathodic peak of current makes it possible to calculate the quantity of electricity (Q), consumed in the electrochemical process; since the potential scanning (E) is a function on time (τ) at given scanning rate (υ) $E = \upsilon\tau$, then: $Q = IE/\upsilon$ or in integral form we have:

$$Q = \int_{E_1}^{E_2} \frac{I}{\upsilon} dE, \text{ since } \upsilon = \text{const, то: } Q = \frac{1}{\upsilon}\int_{E_1}^{E_2} IdE \tag{3.4}$$

In order to obtain the films with controlled optical properties and thickness, it was necessary to measure the optical density of the obtained polymer layer during the electrosynthesis. As can be seen from the data presented in Figure 3.4, there is a direct correlation between the quantity of electricity Q and the optical density D, and between the anodic peak current I^a_p and D. Thus, the method of electrosynthesis with cyclic potential scanning allows to predict the physical and chemical parameters and to obtain the polymer films with controlled layer thickness and optical density.

An investigation of obtained films with the use of atomic force microscopy (**AFM**) and statistical analysis of images allowed to establish that the structure of **PAAr** in a thin layer is globular, and the size of globules almost identical and less than 100 nm (Fig. 3.5 (a)). Globules evenly packed in the surface layer having a well-developed surface, which contributes to the **PAAr** electroactivity. In contrast obtained in galvanostatic mode films have a certain number of units (crystallites) on the surface by size of 0.5–1.0 μm. Maximum size of heights Z = 1.1 μm and their standard root-mean-square deviation R_{ms} = 136 nm (Fig. 3.5 (b)).

On transparent SnO_2 electrodes, the oxidation of isomeric aminophenols, unsubstituted phenol, and aniline occurs under the similar conditions. Depending on the relative position of the substituents of the benzene ring, the oxidation potentials (E_{ox}), and the rate of electron transfer (k_s) calculated accordingly[59] are significantly differed. As it can be seen from Table 3.5, the presence of hydroxyl group in the molecule of aniline and the presence of amino groups in the molecule of phenol lead to the decrease of oxidation potentials compared to phenol and aniline and leads also to the increase of heterogeneous charge transport constant for OH-group oxidation, while the rate of aminophenols oxidation via NH_2-group is decreased compared with the aniline. The oxidation process takes place with the formation of cation radical particles that is in good agreement with the data of *ESR*-spectroscopy given in Table 3.3.

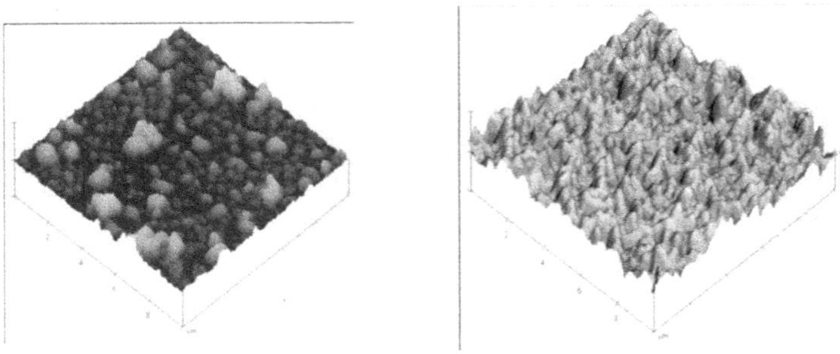

(a) (b)

FIGURE 3.5 AFM images of **PAn** films obtained in galvanostatic mode (a) and in cyclic scanning of the potential mode (b).

During the aminoarenes oxidation at $T = 293$ K, the growing over time *ESR* signal is observed. A form of a line with *g*-factor 2.003 ± 0.001 without hyperfine structure is typical for electroconductive **PAAr**.[40,41] The intensity of a signal detected after some period of time from the beginning of the polymerization is substantially greater for the oxidation of aniline compared with aminophenols. Reduced signal intensity as same as the less concentration of unpaired spins (N_s) indicate a slower rate of aminophenols polymerization compared to aniline.

TABLE 3.3 Electrochemical Parameters of Aminoarenes Electrooxidation ($C = 0.1$ M) in 0.5 M H_2SO_4 and Characteristics of *ESR*-Spectra ($T = 295$ K).

Monomer	Functional group				Parameters of ESR spectra		
	–OH		$-NH_2$		g-value	ΔHpp, Oe	N_s, 1/g
	E_{ox}, V	$k_s \cdot 10^5$, cm/s	E_{ox}, V	$k_s \cdot 10^5$, cm/s			
m-$NH_2C_6H_4OH$	0.90	5.4	0.42	4.3	2.0037	8.5	9.4×10^{17}
o-$NH_2C_6H_4OH$	0.62	8.3	0.37	6.7	2.0045	12.8	6.1×10^{18}
$C_6H_5NH_2$	–	–	0.80	27.0	2.0036	3.2	4.7×10^{19}
C_6H_5OH	1.02	4.8	–	–	–	–	–

In order to obtain the films of polyaminophenols (**PAPh**) on SnO_2 electrodes, it was used the cycling of potential between $E = -0.2$ and 0.8 V with a rate ν = 40 mV/s. Colored polymeric films have been obtained only after 50 (in a case of poly-*m*-aminophenol (**PmAPh**)) and 60 (in a case of poly-*o*-aminophenol (**PoAPh**) cycles of the potential scanning. Applying the IR-spectroscopy for the material of film (extruded into the pellets with KBr) it were found the absorption bands for *para*-substituted aromatic ring (3080, 1520, 760 cm^{-1}), for aminogroup (3350, 1574 cm^{-1}), and for OH-group at 3600, 1410, 1200 cm^{-1}. In a case of *o*-aminophenol, neared values of the oxidation potentials cause the reactions of both functional groups that leads to the formation of heterocycles.[9] This is confirmed by the absorption bands at 1270 – 1200 cm^{-1} (etheric oxygen) and 3400 – 3200 cm^{-1} (bounded hydroxyl).

A promising method of obtaining the polymeric layers with electrooptical properties is the electrochemical copolymerization of aniline with monomers by different type, which has been realized by us, for example from poly-3,4-ethylenedioxythiophene[60] (**PEDOT**), isomeric naphthylaminosulfonicacids[61] (**NASA**) and also the synthesis of electrochromic materials in nanopores of semiconductors[62,63] or polymer dielectric matrixes—polyvinyl alcohol (**PVA**), polymethylmethacrylate (**PMMA**), and others[65] formed on the surface of the optically transparent electrode (**Section 3.10**). It is important that both in the films of individual **PAAr** and their composites, a change in spectral characteristics and accordingly, the color under the influence of an electric field take place, and the basic regularities of these processes require the individual consideration.

3.5 AN INFLUENCE OF THE ELECTRIC FIELD ON OPTICAL SPECTRA OF PAAr

Interesting optical and electrooptical properties of conjugate polymeric systems is revealed both in the visible range of the spectrum and in the near **IR**- and **UV**-region. Optical spectra of conjugated **PAAr** are caused by the electronic transitions in the energy range $E = 1.3$–1.9 eV, which corresponds to the width of the energetic band for the most organic semiconductors.[42] Energies of electronic transitions in conjugated **PAAr** are shown in Table 3.4.[66]

Typical spectral dependencies of the optical density for **PAAr** films are presented in Figure 3.6. The spectra are characterized by two intensive absorption bands. The first band is in the range $\lambda = 420$–440 nm arising from the π–π^* transitions in aromatic conjugate system. In reference, this band is associated with the electronic transitions that correspond to the energy difference between **LUMO** and **HOMO** orbitals.[11,42] In monomers, this band is located in the near **UV**-range of the spectrum, and in polymers due to bathochromic shift, it is located in the visible region.

TABLE 3.4 Energies of Electronic Transitions in Visible Range of the Spectrum and in the Near **IR**-Region for the Films of **PAAr**.[57]

Electronic transition	Energy, eV	Wave-length, nm
Absorption by cation radicals	2.85	435
Donor–acceptor interaction between the quinoid fragments of **PAn** and counter anions	2.18	570
Excitonic transitions in quinone rings	1.86	665
Localized polarons (or dimeric cation radicals)	1.64	755
Secondary electronic transition in dimers of cation radicals	1.52	815
Free carriers of charge	< 1.38	> 900

The band of π–π^* transition is typical for all **PAAr**; moreover the presence of substituents in the aromatic ring leads to the different nature of its impact on the position of this band in the monomers and polymers. For monomers in a series of substituents –H, $-CH_3$, $-OCH_3$ (aniline, *o*-toluidine, *o*-anisidine) the bathochromic shift is increased, that naturally occurs the shear of bands into the long-wave region (λ_{max} = 280, 287, 300 nm).[11] The reverse effect is observed in the films of **PAAr**, namely in the same sequence the bathochromic shift is decreased: λ_{max} = 433, 424, and 415 nm for **PoMA**, **PoT**, **PAn**, respectively (Fig. 3.6). This regularity is probably associated with a decrease of the length of the chain conjugation of **PAAr** due to breaking of its linear structure[45] that eventually leads to decreasing in the length of the conjugation and, as a result to the reducing of bathochromic shift. An influence of the electric field on the position of the first band is negligible, but with a strong anode polarization ($E > 0.8$ V), it may significantly deviate from the conversion of benzene fragments into quinoid-imine, which reduces the intensity of the π–π^* absorption.

The second absorption band is quite wide and depending on the molecular structure of **PAAr,** it can be observed in the spectral range from 600 to 1100 nm. Obviously, it is the sum of various electronic transitions both in the benzoquinoid chain and in the polar and bipolar area.[41,42,66] The kind of this band depends on many factors (e.g., oxidation degree, doping level, nature of medium, etc.).

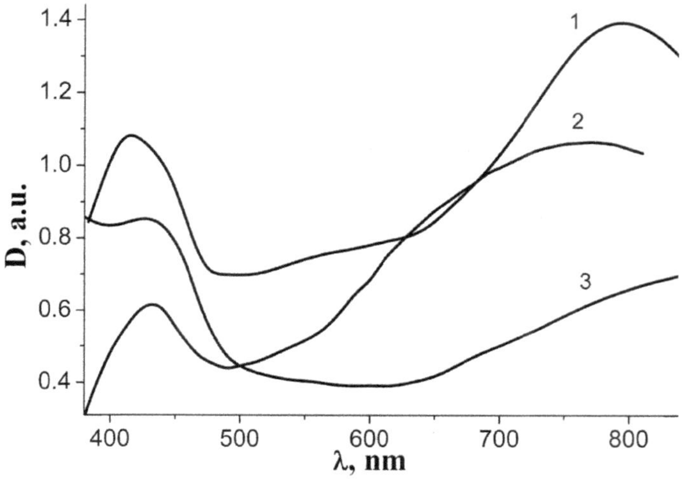

FIGURE 3.6 Optical absorption spectra of **PAAr** films on *ITO* electrode: 1) **PoMA**, 2) **PAn**, and 3) **PoT**. The films have been obtained by electropolymerization of 0.025 M solutions of monomers in 0.5 M H_2SO_4 at $i = 0.1$ mA/cm^2, $\tau = 10$ min.

However, the most interesting is the effect of electric field on the optical properties of **PAAr**, under the action of which the doping level of **PAr** can be easily controlled. The reversibility of conjugate **PAAr** films electrooptical transformations confirmed by cyclic voltammetry (Fig. 3.7).

The obtained data coincide with the data of electrochemical studies of **PAAr** films on platinum electrode;[39,43] the equality of anodic and cathodic peaks, and also the equality of figures areas bounded by the cathode and anode branches of data on **CVA** (Fig. 3.7 (a)) show the reversibility of the process. The linear dependence of the peaks current values on a root potential scanning rate (Fig. 3.7 (b)) shows that the charge transport process is limited by the diffusion.[59] However, the relative positions of the current maxima as well as the difference between the cathode and anode peaks are not strictly defined, and depend on the potential scanning rate. This is due to many factors that determine the "imperfection" of the investigated system. At the explanation of received voltammograms, it is necessary to take into account the specific character of the material of electrode, in particular the structure of the **DEL** of the semiconductor, the presence of the spatial charge[54] that changes the conditions of electron transport compared to metallic electrodes.

On the other hand, formed electroactive substance represents by itself a set of polymer chains with energy wise different electroactive centers even within a single macromolecule. By increasing the potential scanning rate, due to the diffusion limitations and due to the conformational changes rates limit, not all of the electroactive centers of macromolecule can take part in the electrochemical reaction, resulting in a potential maximum current shifted upward energy. Such electrochemical behavior corresponds to quasi-invertible model of the charge transport model.[59]

At the electrode polarization, the absorption spectra are significantly changed (Fig. 3.8) due to the electrochemical transformations. Instead, within the potential cycling the electrochromic material can work for a long time, which is important for practical use. Redox processes that occur at this are illustrated by diagram shown in Figure 3.9.

Colorless, leucoemeraldine form of **PAAr** (Fig. 3.9 Structure **I**) is easy oxidized with the formation of imino-quinoid fragments and polaronic form (structures **II** and **II'**), which at low oxidation levels causes the color change from yellow to light green. Synthesized **PAAr** under conditions of the electrochemical polymerization are characterized by predominantly green color and in accordance with the structural investigations[55,58] are in the form of emeraldine salt with equal quantity of oxidized and reduced fragments (structures **III** and **IV**, $y = 0.5$).

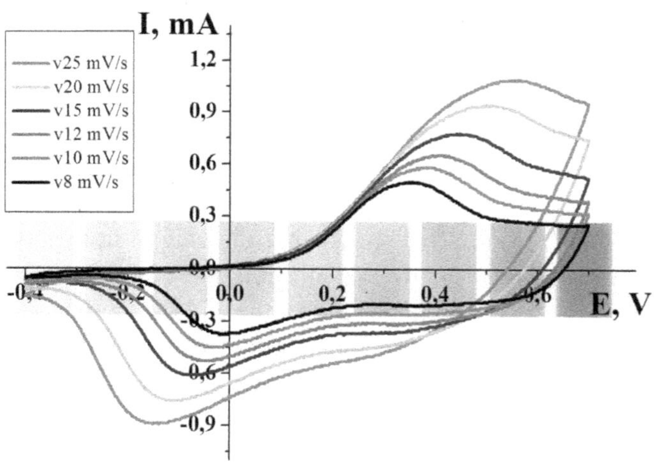

FIGURE 3.7 (a) **CVA** of **PAn** film on the *ITO* electrode in 0.5 M H_2SO_4 at different potential scanning rates and a change of a polymer color accordingly to the oxidation–reduction state; (b) dependence of anodic and cathodic peaks current (I_p) on a potential scanning rate ($\upsilon^{1/2}$).

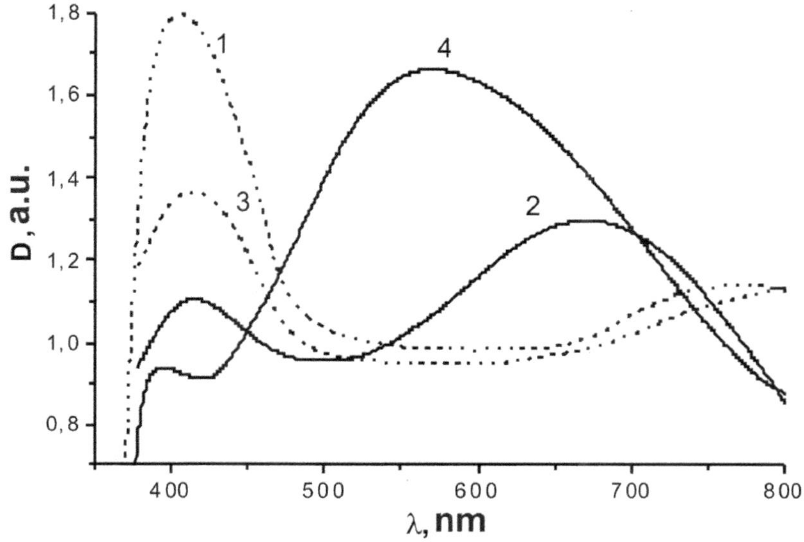

FIGURE 3.8 The absorption spectra of **PoA** (1, 2) and **PAn** (3, 4) at anodic (solid graphs) and at cathodic (dotted graphs) polarization of electrode in 0.5 M solution of $HClO_4$. The thickness of films is 360 ± 20 nm.

The number of oxidized imino-quinoid fragments in the structure of **PAAr** is growing during the oxidation and the polymer is gradually transformed into the pernigraniline form. The absorption maxima in this state consist of λ = 550–600 nm and λ = 650–700 nm, which is caused by donor–acceptor interaction between the quinoid fragments of **PAAr** with compensating ions and by the excitonic transitions in the quinoid rings, respectively. Due to the fact that the absorption maxima correspond to a set of polymer chains by different lengths, they are discharged into a singlewide absorption band. The presence of a substituent in the chain of **PAAr** complicates an interaction of nitrogen with the compensating ions of electrolyte (proper to **PAn**), so for substituted **PAAr** the excitonic transitions in the imino-quinoid rings[35] have a greater contribution into the optical absorption. In the oxidized state the number of double bonds is the highest possible, and at the intensive oxidation there is a risk of the nitrogen atoms converting into the quaternary salt form (=N$^+$=); this form is non-conductive and is not able to the reverse redox transformations.

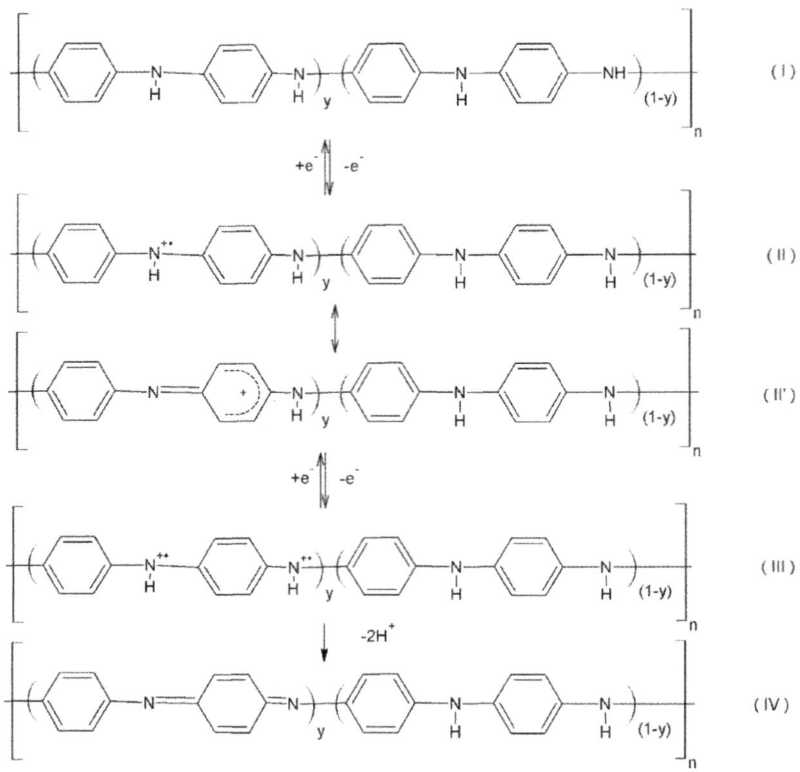

FIGURE 3.9 Electrochemical transformation in thin films of **PAAr** (on example of **PAn**).

The reduction of imino-quinoid fragments of polymeric chain to the amino-benzenoid ones proceeds at the cathodic polarization ($\Delta E = -0.2$ V). Electronic absorption spectra corresponding to these transformations are shown in Figure 3.8 (curves 1, 3). Like to unpolarized film presented spectra have two absorption bands; the difference is in the considerable increase of the first band due to absorption of the electronic π–π^* transitions of the benzene ring. This is caused by the increasing number of benzene fragments generated at the quinoid structures reduction. This band is on the border of visible spectral band λ_{max} = 350–450 nm and at the decrease of bathochromic shift due to the reduction and decrease of the conjugation effect (decrease of double bonds quantity), it can wholly get over beyond the visible range. The second a broad band almost entirely is in the near **IR** region, λ_{max} = 750–900 nm. Only localized charge carriers (on the benzene rings and on the nitrogen atoms) are in the structure **I** (Fig. 3.9), and that is why the absorption in this state is caused by localized polarons and electronic transitions in cation radicals remaining due to incomplete reduction of the emeraldine. Due to the fact that both bands are outside the visible range, the polymer in this state is colorless.

Thus, the change in the optical spectra of **PAAr** corresponds to visual changes of color of a film in a certain electrochromic system ("electrochromic film – electrolyte – counter electrode"). Electrochromic system is characterized by the electrochromic efficiency and by the rate of the electrochromic effect, and as a result, the special methods and equipment developed for this purpose.[10,11,29]

3.6 AN EFFECT OF ELECTRIC FIELD ON ELECTROCHROMS COLOR CHANGE

Electrochromism belongs to the ion-electron transport reverse phenomena, which are controlled by the electric field. The reverse injection (extraction) of ions and electrons in certain substances is a base of this phenomenon. The quantity of these ions and electrons is measured by the amount of the electricity ΔQ, which determines the value of the induced optical density ΔD. The relationship between them is expressed by a linear relation:

$$\Delta D = \beta \Delta Q, \tag{3.5}$$

where β is the electrochromic efficiency.

Electrochromic efficiency is the ratio of the value of optical density in electrochromic systems induced by electric current to the electricity count that has passed through it.

The basis of all phenomena of the electrochromity is a change of energy states of the molecule due to interaction of its dipole moment (permanent and induced) with the electric field,[10,11] and also due to redox transformations that is accompanied by the reconstruction of the molecular structure of substance.[55]

An excitation of system by the electric field causes a change of the wave functions, and as a result—a change of direction and value of transition moments. Experimental study of the electrochromity consists in the measurement of the relative change of the light intensity that has passed through the substance under the action of extraneous electric field:

$$\Delta I = (I_{\chi E} - I_{\chi 0}) / I_{\chi 0} \tag{3.6}$$

Electrochromic coefficient $L\chi$:

$$L\chi = \Delta I\,(-1)/(D|E|^2 \ln 10) \tag{3.7}$$

can be represented as:

$$L\chi = A + \frac{1}{15hc}\left(\frac{d\ln(\varepsilon/\tilde{\nu})}{d\tilde{\nu}}\right)B + \left[\left(\frac{d\ln(\varepsilon/\tilde{\nu})}{d\tilde{\nu}}\right)^2 + \frac{d^2\ln(\varepsilon/\tilde{\nu})}{d^2\tilde{\nu}}\right]\frac{C}{30h^2c^2} \tag{3.8}$$

where A expresses the change of optical density due to the molecules orientation along the field; C causes the Stark symmetric extension of lines and bands; B characterizes the shift caused by the asymmetric long-wave expansion at the polarization of the molecule increasing in the excited state.[11,33] If we neglect the diffusion of protons then the coloring current would be equal to:

$$J_c = J_0 \exp\left(\frac{V_a}{2RT}\right)\left[\frac{(1-x)}{x}\right], \tag{3.9}$$

where J_0 is exchange current; x is a constant by 0.1 order; R is the gas constant; T is absolute temperature; V_a is applied voltage. In order to find the functional dependence of x on time t, let's equate the J_0 to current of protons in the film. The characteristic time of diffusion in the film is determined from the equation:

$$\tau_D = \frac{l^2}{4D_p}, \tag{3.10}$$

where l is a thickness of film, and D_p is the coefficient of diffusion.

The most important characteristic of the electrochromic materials is the intensity of colored changes that occur during the transition from colorless to colored state. Quantitatively color transitions are described as a change of the optical transmittance (ΔT), or absorption (ΔD), and also as a change of the contrast (CR). For the same material, these values depend on the thickness of layer, on composition of the electrolyte, on temperature and other factors. The optical transmittance (T) for electrochromic material can be described by the equations:

$$T_b = \exp(-\alpha_b L) \tag{3.11}$$

$$T_c = \exp(-\alpha_c L) \tag{3.12}$$

where α is a linear absorption coefficient; L is the thickness of film. Indexes "*b*" and "*c*" mean blanl and colored states of the electrochromic material. Contrast (*CR*) is defined as the ratio of the transmission (or reflection) of light by a film in colorless state to the same characteristic in colored state.[21]

$$CR = (T_b/T_c) = \exp\{(\alpha_b - \alpha_c) L\}. \tag{3.13}$$

Contrast can be defined also through the change of an absorption since $\Delta D = lg\ CR$.[11] As shown in a series of the references, there is an optimum thickness of the electrochromic polymer layer (near 300 nm) at which the maximal contrast is observed.[21]

The all of these approaches are righteous, when the matter concerns to monochrome materials, that is, those that change their color within a single color—from colorless to blue, as for example, in a case of WO_3.[10] For an objective assessment of the electrochromic properties of **PAAr**, which can take from 2 to 4 or more main colors having the transition shades, it is not enough to know only the difference of optical transmission (absorption) of reduced and oxidized forms. In addition, high-oxidized form of **PAAr** (**PAn** / and its derivatives), which contains a significant proportion of pernigraniline fragments becomes color, which spectrally is characterized by lower optical density versus half-oxidized emeraldine (emerald) form. The same is concerned to the yellow color of **PAAr**, in structure of which the reduced leucoemeraldine fragments are dominated.[36] Discoloration of **PAAr** films first of all is affected on the position and contour of the main absorption bands, but the main factor that causes of such shift is the magnitude and the sign of the electrode polarization.[58] To characterize the electrochromic properties of **PAAr** films, we have proposed to consider several parameters, namely a shift the absorption maximum at switching ($\Delta\lambda_{max}$, nm), the difference of optical transmission (absorption) between stable emeraldine form and reduced or oxidized form, and a change of optical transmission (absorption) on a certain wavelength calculated per unit of applied potential (voltage). These approaches have been used by us under investigation of the influence of the nature and composition of the electrolyte on electrochromic properties of **PAAr** (Section 3.7).

For quantitative characteristic of **PAAr** color changes the absorption spectra obtained in terms of the external polarization are analyzed. Color of films (which is defined by optical spectrum) depends on the potential of electrode, to which a certain redox state of the polymer corresponds and consequently, the degree of conjugation of the polymer chain. Spatial spectroelectrochemical diagram for **PAn** film is shown in Figure 3.10. Similar effects of the optical spectra changes under the influence of an applied potential are recorded for others films of **PAAr** (Table 3.5). It should be noted that at the gradual polarization the absorption spectrum contains the bands typical for intermediate molecular forms of **PAAr** with different alternating ratio of benzenoid and imino-quinoid fragments. This results in a redistribution of the absorption bands intensities, the change of their circuit and the absorption maxima displacement due to change in the length of the conjugation chain, that is manifests in a wide range of the color shades of polymer film (Table 3.5).

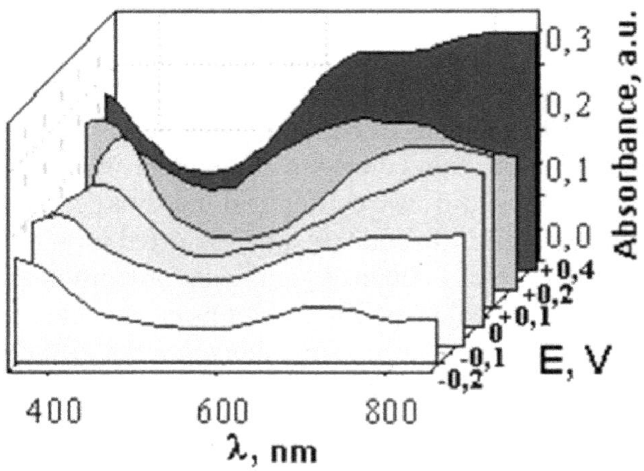

FIGURE 3.10 Dependence of shape of optical absorption spectra and color of **PAn** film during the change of oxidation degree under the action of electric field.

As it can be seen from the above presented Table 3.5, the nature of colored changes is different for **PAAr** with different substituted of benzene ring. Common is a transition from the colorless to the high colored state during the oxidation. The most visible and the most convenient way to represent the color changes of the substances under the action of external factors is the construction of corresponding graphs in the so-called chromaticity coordinates.[67,68]

In order to establish a relationship between the spectral characteristics, namely optical density (D), wavelength (λ), and color, the International Committee on Illumination (**CIE**) introduced the concept of the luminance factors[67] $\bar{x}$, $\bar{y}$, $\bar{z}$ (x, y, z are designation of the conventional colors like to the real **RGB**: red, green, blue), and specific color coordinates X, Y, Z. Any spectrum can be considered as a mixture of a large number of monochromatic irradiation of certain intensity (P_λ). Color of monochromatic component is defined as a vector in space XYZ by the equation:

TABLE 3.5 The Potentials of Electrochromic Transitions in **PAAr** Films.

Conducting polymer	Chemical structure	Electrode potential, V (Ag/AgCl) −0.2 −0.1 0.0 0.1 0.2 0.3 0.4 0.5 0.6 0.7 0.8 0.9 1.0
PoT	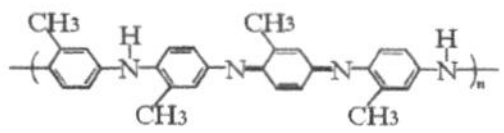	
PoAPh	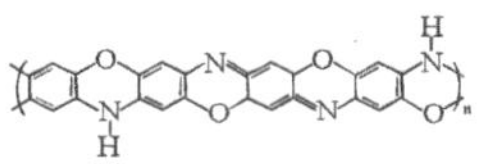	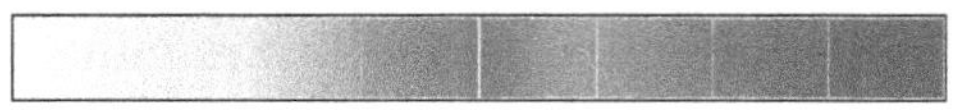
PoMA	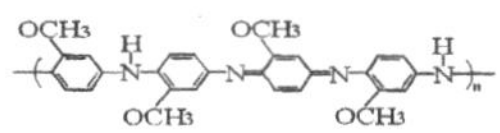	
PAn	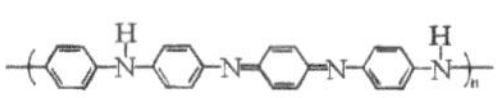	

$$SS_\lambda = P_\lambda \Delta\lambda(\bar{x}X + \bar{y}Y + \bar{z}Z) \tag{3.14}$$

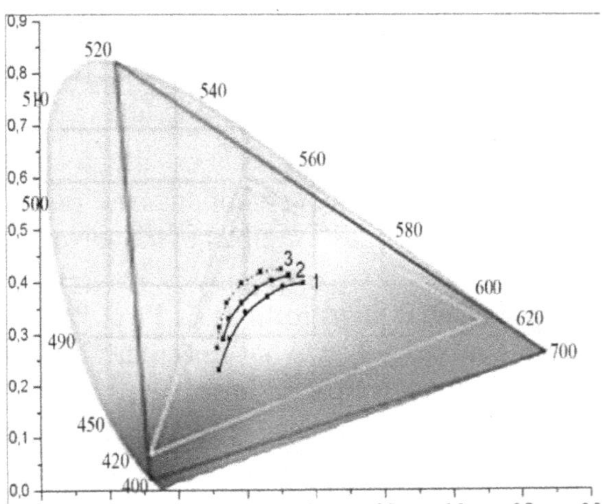

FIGURE 3.11 The chromatic circuit of **PAn** (1), **PoT** (2), and **PoA** (3) films at gradual change of the electrode potential from +0.4 V (blue color of film) till −0.2 V (yellow color of film).

A quantitative description of the colored changes in the polymer film is made by the special techniques.[67] The initial data are the spectral dependencies of the optical absorption and a result of these calculations is the coordinates of color. In order to present the data in more convenient two coordinate planes, the chromatic circuit shown in Figure 3.11 is used. The information about the brightness of the color corresponding to the addition or subtraction of whiteness is losing under this interpretation, but this is easily compensated by the accent lighting.[10] It was established that **PAn** film has the widest color range (curve 1, Fig. 3.11) in the range of potentials −0.4 ... +0.4 V. But color-changing curve passes through an area of gray (mixed or "dirty") colors. More bright is the films of **PoA** and **PoT** (curves 2 and 3, Fig. 3.11), but they are characterized by greatly narrower range of the color change.

This difference is due to the influence of various substituents of the benzene ring. Obviously, the presence of the substituent leads to a change in the electronic structure and to the corresponding shift of the absorption bands. As it can be seen from the diagram, in a case of the methoxy ($-OCH_3$) substituent in the structure of **PAAr** the brightnes s of the colored shades is shown in the most extent.[68]

3.7 AN EFFECT OF THE ELECTROLYTE COMPOSITION ON ELECTROOPTICAL PROPERTIES OF PAn

Compulsory component of the electrochemical system is an ion conductor—electrolyte, which can be solid, liquid, or gel-like.[69–71] The protonic electrolytes, namely mineral acids solutions in water or in organic solvent containing the alkali metals salts are used in the most of electrochromic cells. However, the components of the electrolyte, as a rule is usually not different to the electroactive polymer and under certain conditions may cause its degradation. Therefore, it is important to study the effect of the solvent on the optical properties of the **PAAr** films.

The spectral electrochemical investigations of **PAn**, **PoT**, **PoMA** films in aqueous, organic, and mixed solvents have been carried out in order to study the action of electrolyte on optical (electrooptical) properties of conjugated **PAr**. Such parameters as a resistance of reduced and oxidized forms, the contrast as the difference of optical absorption of colored and colorless forms, an electrochromic shift of the absorption maximum, the rate of electrochemical processes[72,73] were studied in order to select the electrolyte.

At functioning of the electrochromic materials in protonic electrolytes the color change is related with the injection or extraction of charge, that respectively, is accompanied by the introduction or release of the protons. At the same time, the presence of a strong proton acid has the destructive influence on the electrochromic properties of the films. A choice of the electrolytes ($HClO_4$, $LiClO_4$, H_2SO_4) and of the solvents based on the fact that the highest electrochromic stability is achieved in acidic solutions of alkali metals perchlorates in organic and mixed solvents.[74]

It was established by the investigations,[4,9] that among the all studied **PAAr** the highest rate of the electrochromic transformations is observed for **PAn**. Typical shape of the spectral electrochemical dependencies for **PAn** films in different electrolytes is represented in Figure 3.12.

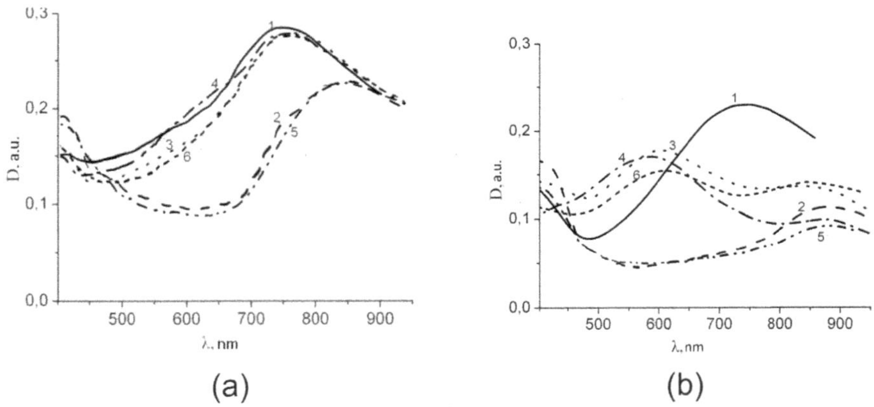

FIGURE 3.12 Absorption spectra of **PAn** film at polarization of electrode, ΔE, V: 0 (1); −0.2 (2); +0.2 (3); after 15 min exposition at ΔE, V = +0.2 (4); −0.2 (5); +0.2 (6). Electrolytes: 0.1 M $LiClO_4$ in acetonitrile (a) and 0.1 M $LiClO_4$ in dimethylsulfoxide (**DMSO**) (b).

As it can be seen from Figure 3.12, the optical spectra of **PAn** films obtained electrochemically in solutions of protonic electrolytes are characterized by a presence of broad absorption bands in the region $\Delta\lambda$ = 380–440 and 650–850 nm. Depending on the electrolyte composition, the nature of electrooptical changes is different.[72] Then, the maximal contrast is observed in propylene carbonate (CR = 34%) and **DMSO** (CR = 35%) electrolytes at the concentration of acid 10^{-3} M.

The degree of films electro-optical properties degradation was evaluated as a shift of the maximum position $\Delta\lambda_{max} = (\lambda_d - \lambda_{max})$, and as a decrease of the optical absorption ΔA_{750} at a wavelength maximum (λ_{max}) and λ = 750 nm after film exposure at anodic polarization (+0.2 V) for 15 min ($A_{750} - A^d_{750}$).

As it can be seen from Table 3.6, in different solvents, the degradation degree is increased at the concentration of acid increasing, and the optimal concentration is $C = 10^{-3} \ldots 10^{-2}$ M. On the other hand, the most stability of the electrochromic properties at such concentrations of acid is observed in organic solvents with a low donor number (**DN**), namely in acetonitrile and propylene carbonate that is in good agreement with those obtained for composite films of **PAn**.[74]

An influence of the solvent can be explained based on the consideration of the electrochemical processes that cause the spectral changes in **PAAr** film (see diagram in Fig. 3.13).

TABLE 3.6 An Effect of Electrolyte on the Electrochromic Properties of the **PAn** Film on the *ITO* Electrode at Potentials ±0.2 V Versus Formal Potential of *ITO*–**PAn** Electrode (0.54 V, Ag/AgCl). The Thickness of Films Consists of 360 ± 20 nm.

Solvent, electrolyte	**Concentration of acid, M**	**Shift of the band, $\Delta\lambda_{max}$, nm**	**Absorbance difference at λ =750 nm, ΔA_{750}**	**Degree of degradation, $A_{750}-A^{d}_{750}$**	**Contrast, CR_{750}, (%)**
Aqueous 0.1 M $LiClO_4$,	10^{-1} $HClO_4$	150	0.112	0.06	32
80% Ethanol 0.1 M $LiClO_4$,	10^{-1} $HClO_4$	106	0.125	0	33
Acetonitrile 0.1 M $LiClO_4$	10^{-2} $HClO_4$	84	0.110	0.006	32
Propylenecarbonate 0.1 M $LiClO_4$	10^{-2} $HClO_4$	99	0.160	0.01	36
DMSO,0.1 M $LiClO_4$	10^{-2} $HClO_4$	53	0.170	0.063	37
0.1 M $LiClO_4$, H_2O	5×10^{-3} H_2SO_4	177	0.050	0	27

Leucoemeraldine (colorless) form of **PAn** represents by itself the polymeric chain (**1**) with low level of conjugation. The electrochromity caused by the redox reverse reaction of the transformation (**1**) into the emeraldine form (**2**), which can be represented by the resonance structure (**2'**). A process of the secondary oxidation of emeraldine (**2, 2'**) to the pernigraniline (**3**) proceeds under the external potential increasing. This process leads to the degradation of electrochromic material because the pernigraniline form (**3**) represents by itself a quinone-diimine dication that irreversibly is transformed into benzo-quinoid form (**4**). It can be seen from the presented reactions schemes that, besides the joining – detaching of electron, the electrolyte

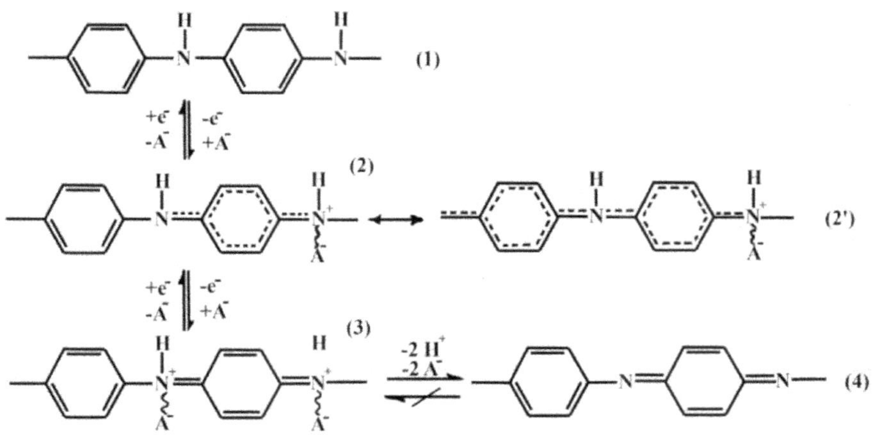

FIGURE 3.13 Scheme of electrochromic and degradation transformations in **PAn** film in the medium of organic solvent.

anions take part in the all reactions and they directly interact with the solvent (solvent interaction):

$$CA + mS \leftrightarrow KS_x^{\,+} + AS_y^{\,-} \quad (m = x + y), \tag{3.15}$$

where CA is a salt of the electrolyte; C^+ and A^- are cation and anion, which are formed by the electrolyte; S is the solvent.

Taking into account of the above-said the processes of primary and the secondary oxidation can be represented by following schemes 3.16 and 3.17 correspondingly:

$$\mathrm{HN-P-NH} + AS_y^{\,-} - e \leftrightarrow A^-\,\mathrm{H^+\,N-P-NH}; \tag{3.16}$$

$$A^-\,\mathrm{H^+\,N-P-NH} + AS_y^{\,-} - e^- \leftrightarrow A^-\,\mathrm{H^+\,N-P-NH^+}A^- + yS, \tag{3.17}$$

where HN – P – NH is leucoemeraldine; A^- H$^+$ N – P – NH is emeraldine; A^- H$^+$ N – P – NH$^+$$A^-$ is pernigraniline.

The following irreversible transformation of pernigraniline into benzo-quinoid form proceeds in accordance with the next scheme:

$$A^-\,\mathrm{H^+\,N-P-NH^+}A^- + nS \rightarrow \mathrm{PN} + 2AS_y^{\,-} + 2\mathrm{H}S_x^{\,+} \quad (n = 2y + 2x), \tag{3.18}$$

where PN is the benzo-quinoid form.

Solvent with a low **DN** weakly interacts with the anion of the electrolyte and prevents the pernigraniline into benzo-quinoid form transformation reaction proceeding (see Fig. 3.13); this promotes a weak degradation of film. Then in acetonitrile (**DN** is minimal and consists of 14.1) the degradation is not observed for a long time of the electrochromic material operation. On the other hand, if the solvent weakly interacts with the electrolyte (Eq. 3.15), the system lacks the anions for the stabilization of emeraldine and pernigraniline forms. Therefore, the mediocre values of maximum absorption shift are observed in acetonitrile. The best electrochromic indexes are observed in **DMSO** (**DN** is equal to 29.8) but, as it was expected for a solvent with a high **DN**, a significant degradation of film takes place in this case.[72]

So, the results of the investigations of electrooptical properties of **PAn** film on the surface of optically transparent SnO_2 electrodes in the electrolytes by different compositions showed that the best are the propylene carbonate and acetonitrile electrolytes with the concentration of $HClO_4$ equal to 10^{-2} ... 10^{-3} M. They provide the minimal degradation of the film at reasonable values of shift positions of the absorption maxima.

3.8 ELECTROCHEMICAL KINETICS OF THE ELECTROCHROMIC SYSTEMS BASED ON PAn AND ITS DERIVATIVES

Since the electrochromic transitions in electroactive substances are caused by the reactions of oxidation–reduction of functional groups with an electron transport, their rate will be controlled by the rate of the charge transport in the polymer layer immobilized on the electrode surface.[10,20,75] Formally, this rate can be characterized by the heterogeneous charge transport constant (k_s) or by the effective diffusion coefficient (D_{eff}) of a charge transport through the film.[22,59] Calculation of kinetic parameters can be made on the basis of voltammetry data[75–78] or data of electrochemical impedance spectroscopy.[73] The presence of diffusion or kinetic control of the process is established by studying the dependence of the current maxima **CVA** on a potential scanning rate.[59] For the calculation of the effective diffusion coefficients of a charge in electrochromic films, it is necessary to determine their thickness and for ultrathin films, this value can be calculated in accordance with.[76] Conjugated **PAAr**, namely **PAn** and its derivatives are convenient model object for the calculation of the electrochemical parameters.

For example, for the **PoT** films obtained under conditions of cyclic potential scanning on the surface of SnO_2 electrodes,[22] the cyclic voltammograms in the all investigated electrolytes (HCl, H_2SO_4, HNO_3, and toluene sulfonic acid (**TSA**)) are characterized by two reversible redox maxima (see Fig. 3.14). The potentials of the peaks and their height depend on the nature of the anion of the background electrolyte.

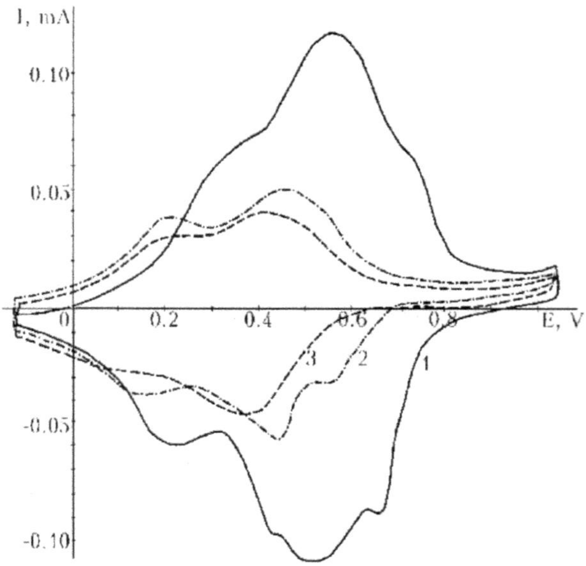

FIGURE 3.14 Cyclic voltammetric diagrams of **PoT** film in aqueous acid solutions: 1–0.5 M solution of H_2SO_4, 2–1.0 M solution of **TSA**, 3–1.0 M solution of HCl. Scanning potential rate is 40 mV/s. Film was obtained at 15 cycles of the potential scanning according to.[22]

According to the known mechanism of redox transformations in films of **PAAr**,[42,43] the first anodic peak (E = 0.18–0.25 V) corresponds to the oxidation of non-conductive, colorless (leucoemeraldine) form of **PoT** to the conductive form cation radical (polaronic form). At high doping levels, the polarons form the bipolaronic emeraldine form:

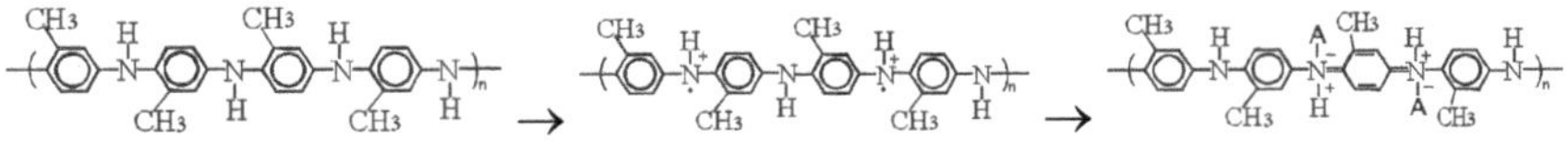

leucoemeraldine form *polaronic form* *bipolaronic form* (A^- – *anion of electrolyte*)

The second peak (E = 0.45–0.60 V) caused by a rapid oxidation ($\Delta E < 30$ mV) of emeraldine form of **PoT** to a wholly oxidized form of pernigraniline; at this, the all atoms of nitrogen in molecules of pernigraniline are in the quinone-diimine form.

It was found that the oxidation–reduction reactions in the film of PoT on the surface of Tin oxide is accompanied by reversible electrochromic changes, besides the chromatic circuit range (colorless – blue – green – violet) and the potential of the colors changeover within $E = 0.3 \ldots 1.2$ V are slightly differed from those described for **PAn**.[41]

An investigation of the electrochemical behavior of the films in acidic aqueous electrolytes showed that the dependence of anodic and cathodic current maxima on the potential scanning rate is dependent on the thickness of the polymer layer. Films obtained on Tin oxide surface at the number of scanning cycles N, which does not exceed $N = 20$, showing the high turnover of the oxidation–reduction processes; this is confirmed by the value of the potential distribution of $\Delta E = E_a - E_c$ within 40–60 mV for the first redox peak and 5–45 mV for the second one, and near or equal to one the ratio of anodic and cathodic current peaks I_a/I_c. The dependence of these currents on the scanning rate is linear (see Fig. 3.15).

A dependence of the optical spectra on the doping level by anions of the protonic acids is observed for the all investigated **PAAr**. Thus, using the method of X-ray microprobe analysis it was determined[38] that in the green form of **PAn** (sulfate as an alloying ion) the content of sulfur is 6.2 at. %; in blue form −3.3 at. %; and in the colorless form of **PAn** the content of sulfur is near zero.

Current of the electrochemical doping (i) which makes for color change intensity of the electrochromic layer can be described by the equation:

$$i = \frac{n^2 F^2 A v \Gamma_i \exp\theta}{RT(E + \exp\theta)^2}, \text{ де } \theta = \frac{nF(E - E_0)}{RT} \tag{3.19}$$

n is the number of electrons, A is an area of the electrode, Γ_i is the superficial concentration of the electroactive substance, which is a sum of the concentrations of the oxidized and reduced forms ($\Gamma_{ox} + \Gamma_{red}$·), E, E_0 are electrode potential and standard electrode potential.[76] The superficial concentration of the electroactive substance evaluated for PoT films with i, E parameters of the second anodic maximum on CVA, obtained in the sulfuric acid[22] at ν = 20 mV/cm is near to 10^{-9} mol/cm^2.

Based on **CVA** data, the first redox maximum in a range of $E = 0.18$–0.25 V corresponds to slower redox process in films of **PoT** (transition of non-conducting form into the conductive one and vice versa) in comparison with the following reversible redox maximum at $E = 0.4$–0.6 V. The value of reverse redox potential $E_{rev} = (E_a + E_c)/2$ depends on the scanning rate, which is typical for the quasiinvertible electrochemical systems,[59] for which both kinetic and diffusion control are proper.

At the same time, the second redox maximum, which is observed on **CVA** for **PoT** thin films exhibits the electrochemical behavior propering to the oxidation–reduction of strongly adsorbed or immobilized redox-particles on the surface of electrode.[76] It's known[76,77] that the linear dependence of anodic and cathodic current peaks on the scanning rate along with the small values of ΔE is observed for the thin layers of conjugated polymers when the surface concentration of the electroactive substances (Γ_i) does not exceed 10^{-8} mol/cm^2. By increasing the thickness of the coating, symmetry of **CVA** is broken and the diffusion limitations become more visible. It finds its expression in the deviation from the linearity of dependence of anodic and cathodic current peaks on the scanning rate at increasing the thickness of the polymer layer. In the case of the investigated films of **PoT,** such deviation from the linearity is started for the polymer layer obtained at $N > 20$ when the potential scanning rate exceeds 50 mV/s. In the case of the relatively thick layers obtained for 30 or more cycles of scanning (l = 0.2–0.3μ), a charge transport is limited by diffusion, that is proved by the linear relationship between the peak current and square root of the scanning rate $i_p - v^{1/2}$ (see Fig. 3.15).

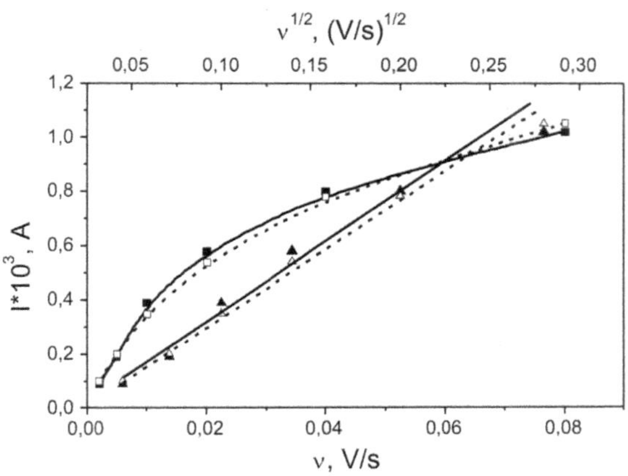

FIGURE 3.15 Dependence of peak current from v and $v^{1/2}$ in 0.5 M solution of H_2SO_4 for **PoT** film produced after 30 cycles of the potential scanning. Solid line corresponds to anode peak, dotted line to cathode peak.

Like to **PAn**, the higher electrochemical activity of **PoMA** and **PoT** films (estimated on a current maximum on **CVA**) is observed in solutions of sulfuric acid. At the same time, a shift of the redox potentials in the region of more positive values takes place. It is assumed[79] that the anions (especially SO_4^{2-}) strongly interact with the cation radicals of **PAAr** promoting the localization of the charge that prevents the degradation of polymer, since a localized charge was weakly submitted to the nucleophilic attack by water molecules. Higher current peak is achieved under conditions, when the polymer degradation process is difficult. However, the currents peaks observed in different electrolytes may also be related with the phenomena of anions diffusion at the interface polymer – electrolyte solution. Higher peaks currents of **PAAr** films in sulfuric acid can be explained by the higher mobility of the anions SO_4^{2-} compared to Cl− or **TSA**-anion and also by strong hydrophilic properties of SO_4^{2-}, which determine the kinetics of the reactions of electron transport.[78,79]

This observation gave the possibility to calculate the value of the effective diffusion coefficient D_{eff} in films of **PoT**, applying the mathematical semi-infinite model of diffusion[75–77] by using the following equations:

$$i_p = 2.69 \times 10^5\, n^{3/2} S D_{ef}^{1/2} v^{1/2} C^*, \tag{3.20}$$

$$C^* = Q/nFSl, \tag{3.21}$$

where n is the number of electrons taking part in the redox process; S is an area of the film (cm^2), C^* is concentration of active centers in film (mole/cm^3), Q is general quantity of electricity calculated via an integration of the anodic or cathodic part of CVA, obtained under slow scanning rate, l is the thickness of film, F is the *Faraday* constant. At the graphical integration of CVA received at $v = 2$ mV/s, under conditions when the all redox centers of the polymer layer have time to oxidize or to reduce, it was appeared that for the PoT film in of 0.5 M solution of H_2SO_4 the quantity of electricity corresponding to the anodic and cathodic processes nearly are equal: $Q_a = 1.95 \times 10^{-4}$ C, $Q_c = 2.0 \times 10^{-4}$ C. The slopes of the straight lines in the coordinates $i_p - v^{1/2}$ (see Fig. 3.16) for the anodic and cathodic peaks are differ slightly. This indicates on a high reverse of redox processes in PoT films and permits to suggest that $D_a \approx D_c \approx D_{eff}$. Knowing D_{eff} and the film thickness and also using the Eq. 3.10, we can calculate the diffusion time (τD), which is an important characteristic of the display materials.[10] The calculated values of the charge transport and transition time (τ) parameters in films of PAAr[4,9,22] are represented in Table 3.7.

The calculated values of the effective diffusion coefficient D_{eff} for **PAAr** films in solutions of protonic acids (HCl, H_2SO_4, CH_3COOH, **TSA,** and others) have shown that the charge transport parameters are well correlated with the absolute mobility of the doping anions.[78] However, a significant difference in the values of the diffusion coefficients found for the derivatives of **PAn** makes it possible to assume that the main factor that determines the rate of a charge transport through the film is the molecular structure and the segmental mobility of the polymer chain. Thus, the presence of electron-donor substituent causes as a rule the decrease of the electrical conductivity and the electronic rate transfer[44] and for inflexible (ladder) structures, such as **PoAPh**, found values of D_{eff} are lower by two orders of magnitude compared to the **PAn** (see Table 3.7). At the same time, the values of diffusion time τ_D are sufficient for the nonradiating displays,[10] that along with a wide range of the colored transitions (Table 3.5) and the possibility to control them through the small potentials switching give a reasons to recommend of these polymers for the development of electrochromic films by sufficient speed of response that can be used in indicators, "smart windows" nonradiating displays.[14,17,50,80]

TABLE 3.7 Parameters of a Charge Transport in PAAr Films in 0.5 M H_2SO_4 Electrolyte Solution.

Conducting polymer layer	Film area, S, cm^2	Film thickness, l 10^4, cm	Diffusion coefficient, $D_{ef} \times 10^9$, cm^2/s	Transition time, τ_D, s
PoT	2.1	0.35	2.62 ± 0.06	0.50 ± 0.15
PAn	4.0	0.28	9.02 ± 0.05	0.02 ± 0.01
PoMA	4.0	0.28	3.56 ± 0.07	0.80 ± 0.15
PoAPh	3.4	0.25	0.23 ± 0.04	6.40 ± 0.70

3.9 DEFINING THE CHARGE TRANSPORT PARAMETERS IN THIN LAYERS OF PAAr BY IMPEDANCE SPECTROSCOPY METHOD

The method of impedance spectroscopy[81,82] consists in the measurement of general, complex impedance of electrochemical system (its impedance) at the sinusoidal alternating current flow and consists in the study of the dependence of its resistance on the frequency of the alternating current. With the use of this method, we have measured the dependence of the complex impedance of electrochemical systems on the frequency of alternating current, by means of which the dynamic characteristics of the electrochromic changes in the **PAAr** films have been described.

The polymer film on the surface of electrode, which is in the solution of electrolyte can be represented as an equivalent scheme for modeling of physical and chemical behavior of the polymer and presented in Figure 3.16.

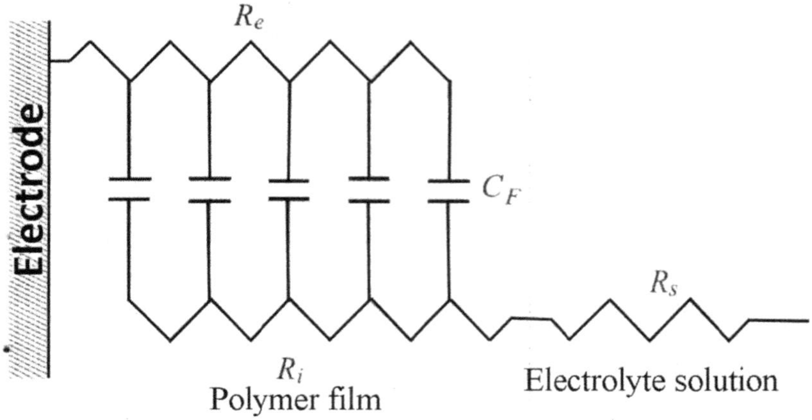

FIGURE 3.16 The equivalent circuit to model of electrochromic film in electrolyte solution: R_e is electronic resistance, R_i is ionic resistance, C_F is *Faradaic* capacitance, and R_s is resistance of electrolyte solution.

Components of such hypothetical electrical circuit correspond to real physical phenomena. A number of capacitors connected via the bridge of resistors are relative to electronic (R_e) and ionic resistance (R_i) of the coating. In order to take into account the uncompensated resistance of the electrolyte solution an additional resistance (R_s) is introduced. Useful achievement of the impedance spectroscopy is that the electronic and ionic resistances are separated. At high frequency ranges, an impedance of the capacitive component C_F becomes irrelevant and the electrochemical circle of alternating current behaves as two resistors connected in parallel:

$$1/Z = 1/R_e + 1/R_i \tag{3.22}$$

Characteristic complex diagram of impedance, designed on a plane (*Nyquist* diagram) has been obtained by us for the feedback of electroconductive polymer[73,83] and shown in Figure 3.17. Two regions can be distinguished on this diagram, namely the low-frequency segment (in the form of half circle) corresponding to the *Faraday* process, and the high-frequency segment, which is directed on the angle of 45°, which is connected with a diffusion process—electronic or ionic—inside of the film.

An extrapolation of 45° interval on the real axis in the *Nyquist* diagram gives the value of high-frequency resistance in the intersection point:

$$R_{\text{high}} = R_s + 1/(1/R_e + 1/R_i) \tag{3.23}$$

Low-frequency response of the electrochemical circle is described as:

$$Z = R_\Sigma/3 - i/\omega C_F \tag{3.24}$$

where $R_\Sigma = R_e + R_i$. Theoretically the real impedance is constant at the all frequencies, but in practice this part of the graph is not completely vertical. The real cross section of low-frequency data gives:

$$R_{\text{low}} = R_s + R_\Sigma/3 \tag{3.25}$$

When $R_i << R_e$, R_e is approximated by the difference between the real crossings of impedance in these two areas:

$$R_e = 3(R_{\text{low}} - R_{\text{high}}) \tag{3.26}$$

The effective diffusion coefficient of electron can be determined from the data of the impedance through the equation:

$$D_e = d^2/R_e C_{\text{low}} \tag{3.27}$$

where d is the thickness of film, C_{low} is the value of capacitor at low frequencies, which can be obtained as an inversion of the graph slope dependence of imagine part of impedance Z'' on the inverse frequency (in rad/s).

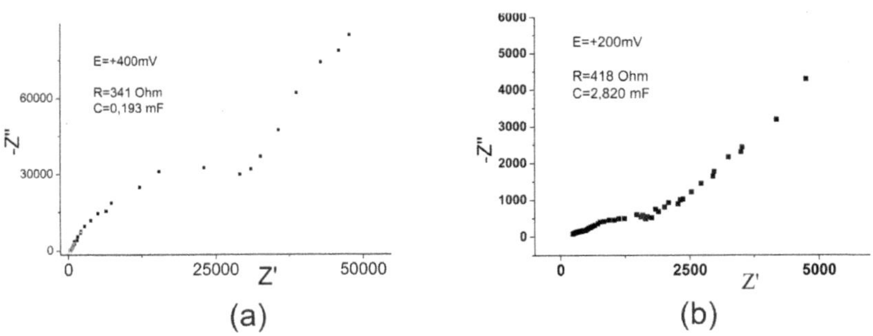

FIGURE 3.17 *Nyquist* plot for the films of **PAn** on *ITO* electrode at the polarization of electrode 0.4 V (a) and 0.2 V (b).

According to the measurements of an impedance of the electrochromic cell with the **PAn** layer, it was established that the form of the impedance spectrum or complex *Nyquist* diagram is largely determined by the magnitude and by the sign of the applied potential (see Fig. 3.17).[72] The values of parameters of equivalent resistances and capacities of the electrochemical system and also calculated diffusion coefficients of anodic and cathodic processes are shown in Table 3.8. Comparison of the obtained results and dynamic of the optical density change at different potential scanning rates allows drawing an analogy between the chromogenic and electrochemical reactions.

TABLE 3.8 Summary of Impedance Results for the Electrochromic **PAn** Layer in the Different Electrolytes. Film thickness is 300 ± 20 nm.

Electrolyte	Potential, V (Ag/AgCl)	R_e, Ω	C_{low}, mF	Diffusion coefficient, $D_e \times 10^9$, cm²/s
Aqueous	0.74	129.3	10.90	0.64 ± 0.02
0.1 M $LiClO_4$	0.54	122.5	5.98	1.23 ± 0.02
10^{-3} M $HClO_4$	0.34	121.2	3.26	2.28 ± 0.04
80% Ethanol	0.34	164.6	9.18	0.60 ± 0.02
0.1 M $LiClO_4$	0.54	126.5	7.72	0.92 ± 0.02
10^{-3} M $HClO_4$	0.34	131.4	3.85	1.78 ± 0.03
Acetonitrile	0.74	256.6	6.20	0.56 ± 0.02
0.1 M $LiClO_4$	0.54	254.2	5.51	0.64 ± 0.02
10^{-3} M $HClO_4$	0.34	251.6	6.33	0.58 ± 0.02

The obtained experimental data confirm the difference in the charge transport parameters for oxidation (coloration) process and the reduction (discoloration) process that occur in the polymer layer. Asymmetry of redox processes in films of **PAAr** is related with different rates of doping ions (protons) injection (intrusion) and ejection (removing) processes and with the reorganization of the polymer matrix in redox processes. It is known that the electron transport in redox systems is possible only under condition of the compensation by injected charge ions with opposite sign, so that is why the conductivity will be determined both by electronic and by ionic components. Study of the mechanism of electrochemical formation and doping of conjugated **PAr** by impedance spectroscopy method gave the possibility to establish that the electrochemical behavior of conjugated polymers is determined by the superposition of faraday and capacitive components.[73,83] An important role in the mechanism of the conjugated polymers conduction plays the processes of polymer matrix reorganization and in many cases; exactly they limit the rate of a charge transport.[84]

3.10 SPECTROELECTROCHEMISTRY OF THE ELECTROCHROMIC SYSTEMS

Electrooptical spectroscopy as a method of the investigation give the possibility to determine the main parameters of electrochromic material and electrochromic devices on a basis of the results obtained by scanning of the electric potential, current density, or its frequency as well as based on analysis of kinetics and dynamics of the electrochromic response, and also analysis of spectrum of light transmission, or reflection.

Applying a certain potential for electrochromic system or the current through it allows to change the energy and concentration of charge carriers (ions and electrons) both in a heterojunction at an interface of the electrochromic material with the electron conductor and an electrolyte, and in the electrochrome itself. Electrochromic materials are studied depending on the changes of their electrical and optical parameters with the use of electric spectroscopy. For technological research, the electrochromic system is represented as the elements of some electrical circuit, in which each element has a defined physical function describing a particular physical or chemical process.[29]

Spectroelectrochemical method can be implemented in different ways, depending on the type of the used electrode. At that the transmission (absorption) or reflection spectra, including the multiply disturbed total internal reflection (abbreviated **MDTIR**[59]) are receiving and analyzing. At the using of transmission spectroscopy an analyzing beam of light passes through the optically transparent electrode, which can be made from glass or quartz and covered by a thin layer of metal (platinum, gold), or semiconductor (SnO_2). Indium-Tin oxide electrodes, which are denoted as *ITO*-electrodes have a great popularity.

The change in the optical properties that arise during the redox switching of conjugated polymers is studied by means of the spectroelectrochemistry of different functional states of electrochromic substance.[2,7,14] Spectral characteristics of electrochemically generated particles can be obtained in the convenient way using the thin-layer sandwich cells with optically transparent electrode.[10] If the film of the electroconductive polymer to deposit on such electrode, then the formed cell could functionate as the electrochromic optical element,[80] schematic representation of such cell is shown in Figure 3.18.

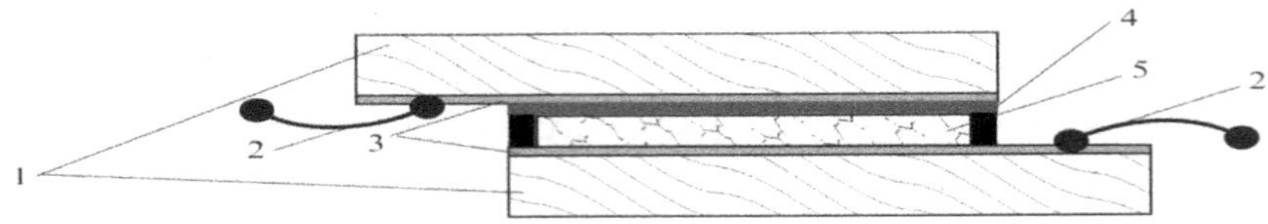

FIGURE 3.18 Scheme of two-electrode electrochromic cell: **1**—glass sheets; **2**—metal electrodes; **3**—transparent SnO_2 layer; **4**—polymer film; **5**—electrolyte solution.

The proposed technique of the spectroelectrochemical researches is described in a great number of scientific papers related to the study of the electrochromic systems and materials.[9,14,18–20] Under the action of the applied potential in the electronic spectra of **PAAr** films, there are changes in position and in intensity of the absorption maxima, which are accompanied by reversible transitions of color in films. For the characterization of these electrochromic changes, a change of the optical density or transmittance as a function on the applied potential (potentiodynamic mode) and as a function on time (chronopotentiostatic mode) for wavelengths corresponding to the absorption maxima are investigated. Thus, for optical spectrum of **PAn** film two distinct absorption bands with the maxima at $\lambda = 370$ and 750 nm corresponding to localized and delocalized states of free carriers are observed. At potentiodynamic imposing of potential with some rate the notable changes of the optical density are observed only for long-wavelength band at $\lambda = 750$ nm unlike to band 370 nm (see Fig. 3.19 (a)). Perhaps this is related with the change of concentration of free carriers under the action of potential, the absorption of which corresponds to the energies in visible and near **IR** range of spectrum.[20,21] The comparison of cyclic voltammetry data and optical measurements indicates on the agreement of the optical density maximum of this band and potential of transition of non-conductive form of polymer into the conductive one. At the anodic polarization increasing the more links are submitted to the imino-quinoid transformations the part of non-conductive form of polymer increases and the intensity of the band is decreased.

At the cyclic potential scanning the coloration—discoloration process of the electrochromic layer is observed; the optical density change curves, which are characterized by the shape of somewhat asymmetric eight correspond to this process (see Fig. 3.19 (b)). Such curves are typical for the electrochromic behavior of polymers.[85] Depending on the scanning rate, the curve of the optical density change at the potential flowing can takes the form of eight, ellipse, or loop.[29] These curves can be described by certain mathematical equation, that is, simulated by a computer program for automatic control of electrochromic devices.

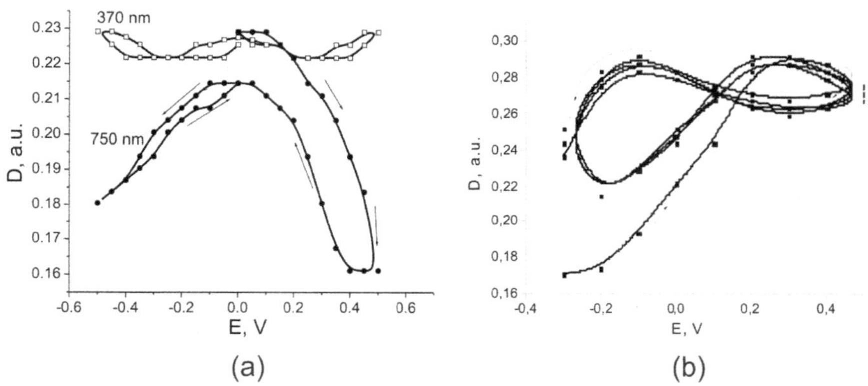

FIGURE 3.19 (a) The variation in optical density of **PAn** film at λ = 370 (1) and 750 (2) nm in potential range from $E = -0.5$–0.5 V (*Pt*) at ν = 20 mV/s; (b) The variation in optical density of **PoT** film under potential scanning from −0.3–0.5 V (Pt) in 1 M solution of **TSA** at ν = 20 mV/s.

Under prescribed conditions of the **PAAr** thin layers formation, the optical devices with given properties can be received by absorption in a certain spectrum band, by changing the contour of the spectrum under the action of the electric field. Let us consider this on example of the electrochromic device – a glass cell with plane–parallel optically transparent electrodes filled with electrolyte solution. The film of the electroconductive polymer is deposited on the internal side of one among electrodes by electrochemical polymerization method.[80]

The optical element works as follows. At the voltage switching in a range from +2.0 to −1.5 V the color change of the electrochromic films from dark green to yellow takes place. At the slow change of voltage a smooth transition from dark green to green and to yellow occurs. At using of 2-electrode cell the changes of optical density and a contour of the absorption bands at voltage imposing are observed in the all cases (see Fig. 3.20); this indicate on the preservation of the electrochromic effect under proposed scheme of the switching-on.

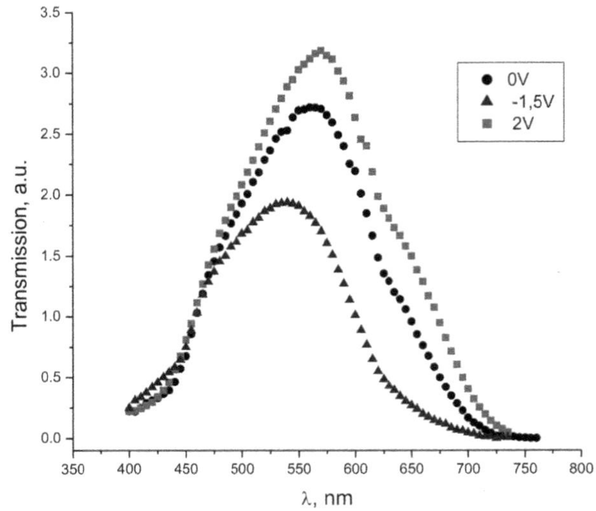

FIGURE 3.20 Transition spectra of two-electrode optical elements with **PAn** film in 0.5 M solution of H_2SO_4 at $U = 2.0$ (1); 0.0 (2); and −1.5 V (3).

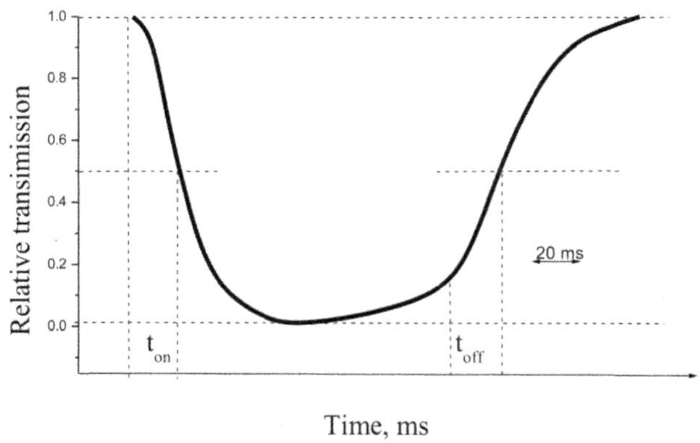

FIGURE 3.21 Electrochromic response of element with **PAn** film in 0.5 M solution of H_2SO_4 at $\lambda = 575$ nm and applied voltage from 0.0 to +1.5 V.

The optical properties of this device were investigated in the spectral range 450–1000 nm at two extreme voltages applied to the sample: −1.5 V for eclipse and 2.0 V for the enlightenment, and also in the absence of an external field. With the use of this experimental scheme the measurements of the response time of the sample on the applied voltage was carried out. The signals from photomultiplier were outputted to an oscillograph. Curves, typical form of which is shown in Figures 3.21 and 3.22, were recorded at the fixed wavelengths. Defined turn-on times and turn-off times (see Fig. 3.21) are respectively, 23 and 18 ms, which is in good agreement with data of the electrochemical measurements[9,22] and are within the speed of response required for the electrochromic displays.[10]

The processes of optical density change (enlightenment-coloration) are characterized by a good reproducibility and stability for the all investigated wavelengths under the study on application of a voltage pulse mode (50 imp/s) (see Fig. 3.22).

The proposed technology of the optical elements with organic electrochromic layer on the surface of *ITO*, SnO_2 electrodes obtaining can be used to create the flat nonradiating (inemissive) displays for visualization of information in digital displays, monitors, indicators, transparencies, as well as for designing of the light valve optical filters, sensor devices. However, the well-known techniques of the polymeric electrochromic material obtaining by electrochemical polymerization of aniline or its derivatives on the surface of the optically transparent electrodes do not provide the sufficient contrast of the obtained film (only 18–25%). Therefore, the ways of improvement of the electrochromic properties of devices based on conjugated **PAAr** are under the active investigations.

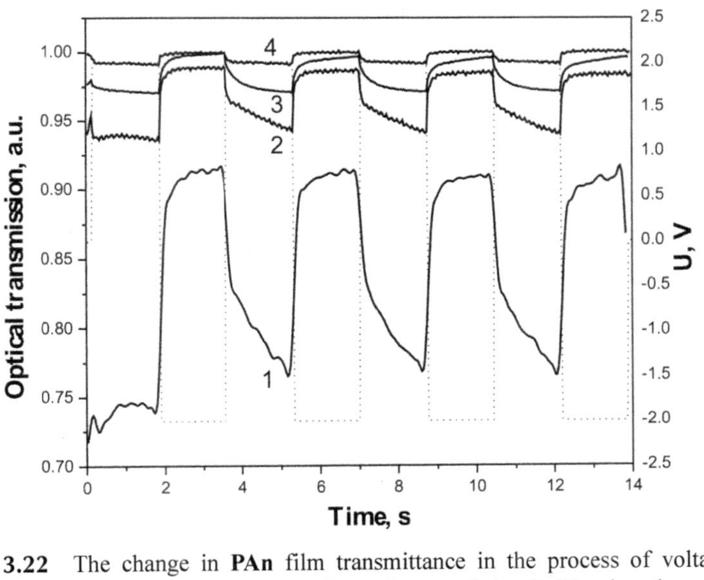

FIGURE 3.22 The change in **PAn** film transmittance in the process of voltage pulse imposition with amplitude of ± 2 V in two-electrode control circuit. Wavelengths are 575 (1), 558 (2), 595 (3), and 610 (4) nm. 0.5 M sulfuric acid was used as electrolyte.

3.11 METHODS OF AN IMPROVEMENT OF PROPERTIES OF THE ELECTROCHROMIC FILMS BASED ON PAn

The use of conjugated polymers in electrochromic devices permits to design the color that is to change the structure of the elementary link in order to achieve the desired color transition. This may be one of the most important arguments for choosing of the organic electrochromic materials compared to inorganic ones. To program the controlled color in the electrochromic polymers can be through the use of several strategies, namely, by introduction of substituents by different nature (functionalization of monomers),[18] by copolymerization of different nature monomers,[8,60,61] by grafted copolymerization,[87] by obtaining of the conformational isomers, in which a change of the macromolecule conformation can "completes" or, vice versa, "cuts in" the electrochromic transition.[88] A promising approach for creating of the flexible electrochromic material is the formation of blends (polymer blends), laminates, composites[64,65] and hybrid organic–inorganic systems,[7,62,63,89–91] We improved the method of polymeric electrochromic material producing by means of the surface modification of optically transparent electrode by **PAn** nanolayer.[60] In this case, it was found that the polymerization process of the 3,4-ethylenedioxythiophene (**EDOT**) monomer on the **PAn** layer occurs in the potential field of the electroactivity of **PAn** (see Fig. 3.23 (a)) when the modifying layer is in a high-conductive polaronic state and can serve as a mediator of the electron transfer between the electrode surface and the monomer molecules.

Modification of the surface by the **PAn** layer provides the better affinity of the poly-3,4-ethylenedioxythiophene **(PEDOT)** upon the electrode surface, thus the obtained polymer films are characterized by improved adhesion, uniformity and posses the "sea wave" color and capable of electrochromic transitions at the potential switching from −0.8 to +1.0 V. Optical spectra of the obtained polymers are characterized by absorption bands with maxima at λ = 380–400 nm (π–π* transition in the bandgap) and λ = 780–800 nm (absorption of free carriers in polaronic zone). The absorption spectrum of the composite film generally retains these properties, but is not the superposition of the spectra of the individual polymers (see Fig. 3.23 (b)). As it is shown by the spectroelectrochemical studies, the optical transmission change at the potential switching corresponds to the contrast of composite film on the level of 57–65%, which is almost twice higher than the contrast of the individual polymers.[86]

Composite films based on the conducting polymers and dielectric polymer matrix are promising materials of new generation electro-optical devices, namely – not radiative flexible displays, indicator panels, sensors.[89,90,92]

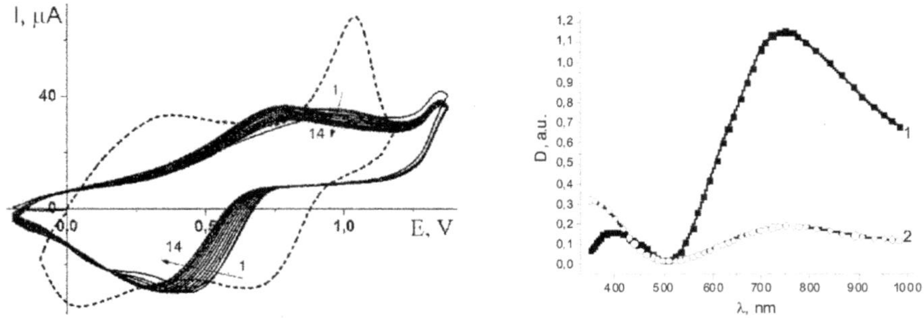

(a) (b)

FIGURE 3.23 (a) Cyclic voltammograms obtained during the electrochemical polymerization of 0.1 M **EDOT** in 0.1 M $LiClO_4$ in acetonitrile on *ITO* electrode coated with **PAn** layer (60 nm thickness) at scanning rate 20 mV/s. The ciphers indicate the cycle number *(N)*. Drop line corresponding to **PAn**; (b) Optical spectra of a **PEDOT** – **PAn** film on the *ITO* surface at the cathodic and anodic polarizations: $E = -0.6$ V (1); $E = +0.8$ V (2). The thickness of films is 320 ± 25 nm.

We have proposed the technology of preparation of the composite electrochemical electrochromic films by "matrix" polymerization of aniline or its derivatives in the polyacrylic acid films and in the matrix of **PVA,**[65] which helped to ensure uniformity and the high electrochemical activity of the obtained films.

Along with the **PVA** matrices, which are used to obtain a transparent composite film, **PMMA** polymeric matrix, which is transparent throughout the visible spectrum and is widely used as a component of "organic glass" is promising. We have studied the conditions for obtaining and the optical properties of **PMMA** and **PAAr** composites thin film, which are capable of electrooptical transitions at the potential switching by **PoT** and **PAn**.

During the polymerization of aminoaren in **PMMA** matrix forms composite film, in which colored particles of the conductive polymer is uniformly distributed in a transparent **PMMA** film, that is, the polymerization process initiated at the electrode surface goes directly into the **PMMA** matrix filling the pore volume. The presence of conjugated π-electron bonds and aromatic fragments in the structure of conjugated **PAAr** determines their optical activity in the visible spectrum and in the near **UV**-region. The electrochemical studies shows that the main redox transitions that lead to the preservation of electrochromic characteristics of composites[64] are stored in the composite films of conjugated **PAAr** and **PMMA**. Optical spectra of **PMMA–PAn** films and **PMMA–PoT** at different polarization potentials obtained during spectroelectrochemical investigations are shown in Figure 3.24 a, b.

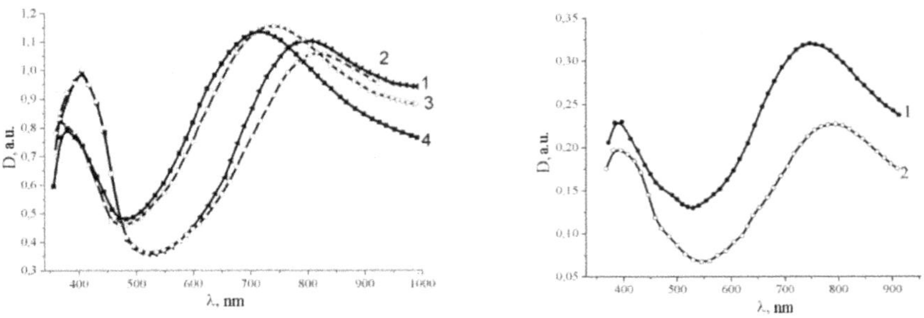

(a) (b)

FIGURE 3.24 Spectroelectrochemical characteristics (a) for **PMMA–PAn** films: 1 – open circuit voltage; 2 – $E = +0.2$ V; 3 – $E = -0.2$ V; 4 – $E = -0.4$ V; (b) for **PMMA–PoT** films: 1 – open circuit voltage; 2 – $E = -0.2$ V. Electrolyte – 0.5 M H_2SO_4 in water–ethanol solvent (1:1).

Analyzing the obtained spectra we can see that the composite stored the absorption bands in the range of 370–430 nm associated with π–π^* transitions in conjugate system **PAAr** and in the range of 700–900 nm, which is characteristic for the resonance stabilized cation radical particles (polarons). Thus, the composite films exhibit a significant electrochromic activity, namely the spectrum circuit shift with the changing of polarization (see Fig. 3.24 (a,b)). These data agree well with the ideas that color change of **PAAr**, in particular **PAn** and **PoT**, caused by the presence of several forms of polymers with different levels of oxidation. Given that time of colors switching in **PMMA–PAAr** composites is a few seconds, such film composites can be used to produce "smart" windows, information displays, and also in environmental monitoring sensor devices[92] for a specific process improvement.

The most progressive approach to the formation of modern electrochromic material is the creation of hybrid organic–inorganic nanostructures.[7,48,63,89–91] Nano ZnO modified by viologens, WO_3 (crystalline, nanocrystalline, and mesoporous), TiO_2 modified by **PEDOT**[89] are used as the inorganic component. It is reported about the composites of **PAn** with inorganic complex $[Fe^{III}Fe^{II}(CN)_6]$ ("Prussian blue"), which improves the contrast of blue and green electrochromic devices, in particular electrochromic mirrors.[21,91] The formation of hybrid nanostructures leads to increasing of switching speed, stability, and contrast of electrochromic devices.

On the basis of physico–chemical studies,[62,63] we developed a method of producing of nanostructured electrochromic layer by electrochemical polymerization of aminoarens in the matrix of titanium dioxide formed on the surface of an optically transparent electrode.[93] Formed into the interstices intensely green colored macrochains of **PAAr** in the emeraldine state are characterized by the sufficiently high molecular weight, which prevents their desorption and promotes to the consolidation of the solid matter in the nanostructured layer.

Formed by the proposed method film has a dense uniform structure, and easily converted from one redox form to another at the voltage changing due to **PAAr** conjugation. The color of the film is changed from light yellow to green, and with a deep oxidation – to the blue and black, and in reverse order by changing the direction of polarization. As can be seen from the data shown in Figure 3.25, the hybrid layer is characterized by a higher electrochemical activity versus polymer. Herewith the contrast of electrochromic material is increased almost in twice compared with the free **PAAr** film.[93] The temporal stability of electrochromic characteristics of the resulting layer is confirmed by the stability of the electric current amplitude that is compared with the change of electrochromic coating reflection coefficient (Fig. 3.25 (b)).

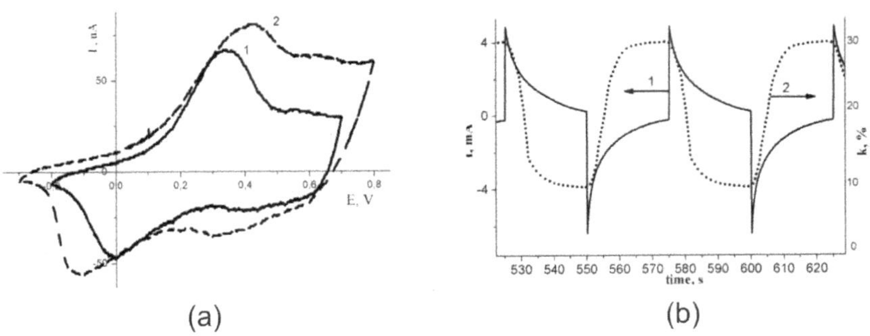

FIGURE 3.25 (a) **CVA** obtained for **PAn** film on *ITO* surface (1) and for TiO_2 – **PAn** hybrid layer (2) in 0.5 M H_2SO_4 at scanning rate ν = 20 mV/s; (b) Temporal dependences of current response (1) and optical reflection coefficient (2) for TiO_2–**PAn** layer on pulse variation of electrode potential from 0.0 to 0.35 V.

For comparison of the electrochromic properties of **PAArs** with the well-known nonradiating display materials, the dependence of the reflectance coefficient on the angle of view was measured (see Fig. 3.26). Comparison has been done using the reflecting electrophoretic monochrome display (**EPD**) by Clear Vision of E Ink Corp. Currently **EFD** technology is the most popular for creating of the reflective displays and is widely used in low-cost handheld devices running on a limited energy resources.[94]

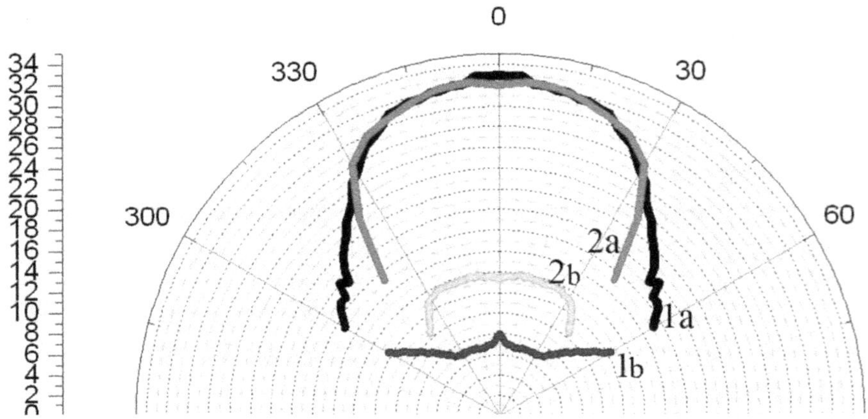

FIGURE 3.26 Diagram of the reflection coefficient of display device: 1—on the base of nanostructured layer TiO_2 – **PAn**; 2—for the electrophoretic display (**EPD**) from E Ink Corp: (a) bleached; (b) colored state.

It was found that the proposed optical element based on TiO_2–**PAn** is characterized by several advantages compared to **EFD**. Thus, the magnitude of the angle of view for the proposed element is up to 35° at a constant reflection coefficient (k) and is decreased in the range of angles of 35–60°, while the k is decreased rapidly to k = 14% after 35° for **EFD**. The value of contrast C = 75.6% for the proposed imaging element, which is a very high value for electrochromic displays.[10,14,15] With the same method of testing **EFD** shows C = 58.1%. This may be due to the use of dispersed TiO_2 layer with a high coefficient of reflection and **PAn** as electrochromic substances with a wide range of variation of the absorption coeﬃcient as display material.

The regularities of physical and chemical processes of synthesis and the formation of thin films of **PAn** and its derivatives on the surface of the transparent electrodes (semiconductors) have a fundamental character, since their proceeding are determined by the electronic and molecular structure both of the initial monomer compounds and also polymerization products, and these regularities are subjected to the basic laws of electrochemical kinetics, organic electrosynthesis, optics and electrooptic of organic semiconductors, and nanosystems on their basis.

A field of electrochromism is fast-paced developed both in the new electrochromic systems creation, and in advanced business applications—"smart" windows, mirrors, memory devices, flexible displays, and so on. The optical devices based on conjugated **PAAr** are soft to the human eye, do not give the harmful electromagnetic or ionizing radiation, and therefore can be used in household electronic applications, screens, and monitors.

Use of the **PAn** and its derivatives in electrochromic devices permits to design the color; in others words, by changing the structure of elementary link via introduction of substituents we can achieve the desired color transition. This may be one of the most important arguments for choosing of the organic electrochromic materials compared to inorganic ones.

KEYWORDS

- **polyaminoarenes**
- **electrooptic phenomena**
- **spectroelectrochemistry**
- **electrochemical kinetics**
- **electrochromity**
- **diffusion coefficient**
- **optic elements**
- **chromatic circuit**

REFERENCES

1. De Paoli, M.; Gazotti W. Electrochemistry, Polymers and Optoelectronic Devices: A Combination Future. *J. Braz. Chem. Soc.* **2002,** *13*(4), 410–424.
2. Rowley, N.; Mortimer, R.; Natalie, M. New Electrochromic Materials. *Sci. Prog.* **2002,** *85*(3), 243–262.
3. Granqvist, C. G. Azens, A.; Hjelm, A.; Kullman, L.; Niklasson, G. A.; Ronno, W.D.; Strumme, M. M.; Veszelei, M.; Vaivars, G. Recent Advances in Electrochromics for Smart Windows Applications. *Sol. Energy.* **1998,** *63*(4), 199–216.
4. Aksimentyeva, O. I.; Konopelnik, O. I.; Cherpak, V.; Stakhira, P.; Fechan, A.; Hlushyk, I. Conjugated Polyaminoarenes as an Electrochromic Layer for Non-Emissive Displays. *Ukr. J. Phys. Opt.* **2005,** *6*(1), 27–32.
5. Chandrasekhar, P.; Zay, B. J.; Mcqueeney, T.; Birur, G. C.; Sitaram, V.; Menon, R.; Coviello, M.; Elsenbaumer, R. L. Physical, Chemical, Theoretical Aspects of Conducting

Polymer electro-Chromics in the Visible, IR and Microwave Regions. *Synth. Met.* **2005,** *155,* 623–627.
6. Granqvist, C. G., Avendano, E., Azens, A. Electrochromic Coatings and Devices: Survey of Some Recent Advances. *Thin Solid Films.* **2003,** *442,* 201–211.
7. Li-Ming Huanga; Ten-Chin Wen; Gopalan, A. Electrochemical and Spectroelectrochemical Monitoring of Supercapacitance and Electrochromic Properties of Hydrous Ruthenium Oxide Embedded Poly(3,4-Ethylenedioxythiophene)–Poly(Styrene Sulfonic Acid) Composite. Electrochimica *Acta.* **2006,** *51,* 3469–3476.
8. Tarkuc, S.; Udum, Y. A.; Toppare, L. Tuning of the Neutral State Color Of the P-Conjugated Donor–Acceptor–Donor Type Polymer from Blue to Green Via Changing the Donor Strength on the Polymer. *Polymer.* **2009,** *50,* 3458–3464.
9. Konopelnik, O. I.; Aksimentyeva, O. I.; Grytsiv, M. Ya. Electrochromic Transitions in Polyaminoarene Films Electrochemically Obtained on the Transparent Electrodes. *Mater. Sci.* **2002,** *20*(4), 49–59.
10. Pankove, J. I. *Display Devices;* Pankove, J. I., Ed.; Springer-Verlag: New York, **1980**; pp 228.
11. Sverdlova, O. V. *Electron Spectra in the Organic Chemistry;* Khimiya: Leningrad, Russia, 1985.
12. Pikin, S. A.; Blinov, L. M. *Liquid Crystals;* Mir: Moscow, Russia, 1982.
13. Gotra, Z.; Zelinskii, R.; Mikitiuk, Z.; Sorokin, V.; Sushinskii, O.; Phechan, A. *LCD Electronics;* Gotra, Z., Ed.; Apriory: L'viv, Ukraine, 2010; pp 532 .
14. Carpi, F.; De Rossi, D. Colors from Electroactive Polymers: Electrochromic, Electroluminescent and Laser Devices Based on Organic Materials. *Opt. Laser Technol.* **2006,** *38,* 292–305.
15. Sapp, S.; Sotzing, G.; Reynolds, J. High Contrast Ratio and Fast Switching Dual Polymer Electrochromic Devices. *J. Chem. Mater.* **1998,** *10*(8), 2101–2108.
16. Shirota, Y. Organic Materials for Electronic and Optoelectronic Devices. *J. Mater. Chem.* **2000,** *10*(1), 1–25.
17. Mortimer, R. J.; Dyer, A. L.; Reynolds, J. R. Electrochromic Organic and Polymeric Materials for Display Applications. *Displays.* **2006,** *27,* 2–18.
18. Sahin, E.; Camurlu, P.; Toppare, L. Both Anodically and Cathodically Coloring Electrochromic Polymer Based on Dithieno[1,4]-Dithiine. *Synth. Met.* **2006,** *156,* 124–128.
19. Platt, J. R. Effect of External Electric Field on the Optical Spectra of Organic Films. *J. Chem. Phys.* **1961,** *34,* 862–868.
20. Pawlicka, A.; Avellaneda, C. In *Physics and Applications of Electrochromic Devices,* Proceedings of the SPIE 4986, Physics and Simulation of Optoelectronic Devices XI. Osinski, M., Amano, H., Blood, P., Eds.; San Jose, CA, 2003; pp 117–131.
21. Lim, J. Y.; Ko, H. C.; Lee, H. Systematic Prediction of Maximum Electrochromic Contrast of an Electrochromic Material. *Synth. Met.* **2005,** *155,* 595–598.
22. Aksimentyeva, O. I.; Konopelnyk, O. I.; Grytsiv, M. Ya.; Martyniuk, G. V. Charge Transport in Electrochromic Films of Polyorthotoluidine. *Funct. Mater.* **2004,** *11*(2), 300–304.
23. Shaw, J. M.; Seidler, P. F. Organic Electronics: Introduction. *IBM J. Res Dev.* **2001,** *45* (1), 3–9.
24. Wallace, G. G.; Dastoor, D. L.; Officer, D. L.; Too, C. O. Conjugated Polymers: New Materials for Photovoltaics. *Chem. Innovations.* **2000,** *30* (1), 14–22.
25. Deb, S. K.; Shaw, R. F. Electrooptical Device Having Visible Optical Density. U.S. Patent 3,500,392, 1970.

26. Volkenshtein, M. V. *Molecular Optics;* Gostechizdat: Moscow, Russia, 1951.
27. Kuzmichev, V. E. Laws and Formulas of Physics. *Handbook;* Naukova Dumka: Kiev, Russia, 1989.
28. Zvezdin, A. K.; Kotov, V. A. *Magneto-Optics of Thin Films;* Nauka: Moscow, Russia, 1988.
29. Lusis, A. R. Physical Problems of Electrochromity and Electrochromic Devices. In *Electrochromity Collection of Scientific Papers;* Lusis, A. R., Ed.; Riga, Russia, 1987; pp 4.
30. March, J.; Smith, B. M. Organic Chemistry: Reactions, Mechanisms and Structure; Mir: Moscow, Russia, 1987.
31. Heeger, A. J. Semiconducting and Metallic Polymers: The Fourth Generation of Polymeric Materials. *Synth. Metals.* **2002,** *123,* 23–42.
32. Aksimentyeva, O. I.; Konopelnyk, O. I.; Stakhira, P. Y.; Tsizh, B. R. Doping-Induced Absorption in the Polyphenylacetylene Films. *Ukr. J. Phys. Opt.* **2005,** *6*(3), 114–119.
33. Braun, D.; Floyd, A.; Sainsbury, M. *Spectroscopy of Organic Substances*; Mir: Moscow, Russia, 1992.
34. Lüera, L.; Manzonia, C.; Cerulloa, G.; Lanzanib, G. Intra-Chain Exciton Generation by Charge Recombination in Substituted Polyacetylenes. *Chem. Phys.Lett.* **2007,** *444* (1–3), 61–65.
35. Kirova, N.; Brazovskii, S. Electronic Interactions and Excitons in Conducting Polymers. *Curr. Appl. Phys.* **2004,** *4*(5), 473–478.
36. Macdiarmid, A. "Synthetic Metals": A Novel Role for Organic Polymers. *Curr. Appl. Physics.* **2001,** *1,* 269–279.
37. Simon, G.; Andre, G. G. *Molecular Semiconductors;* Mir: Moscow, Russia, 1988.
38. Aksimentyeva, O. I. *Electrochemical Methods of Synthesis and Conductivity of Conjugated Polymers*; Svit: L'viv, Ukraine, 1998.
39. Aksimentyeva, O. I.; Lupshak N. O.; Konopelnyk, O. I.; Grytsiv, M. Ya. *Electrochemical Preparation of Polymer Layers with the Electro-Optical Properties;* Bulletin of L'viv University, Ukraine, *Chem.* **2002,** *42,* 114–116.
40. Macdiarmid, A.; Chiang, C.; Richter, A. F.; Epstein, A. J. Polyaniline: A New Concept in Conducting Polymers. *Synth. Met.* **1987,** *18,* 54–60.
41. Genies, E. M.; Boyle, A.; Lapkowski, M.; Tsintaris, C. Polyaniline: A Historical Survey. *Synth. Met.* **1990,** *36,* 139–182.
42. Molapo, K. M.; Ndangili, P. M.; Ajayi, R. F.; Mbambisa, G.; Mailu, S. M.; Njomo, N.; Masikini, M.; Baker, P.; Iwuoha, E. I. Electronics of Conjugated Polymers (I): Polyaniline. *Int. J. Electrochem. Sci.* **2012,** *7,* 11859–11875.
43. Kovalchuk, E. P.; Aksimentyeva, O. I.; Tomolov, A. P. *Electrosynthesis of Polymers on Metal Surfaces;* Khimiya: Moscow, Russia, 1991.
44. Kohlman, R. S.; Joo, J.; Epstein, A. J. Conducting Polymers: *Electrical Conductivity. Physical Properties of Polymers Handbook;* Mark, J. E., Ed.; Amer. Inst. Phys. Woodbury: New York, **1996**; pp 453–478.
45. Mezhuyev, Ya. O.; Koledenkov, A. A.; Korshak, Yu. V.; Shtilman, M. I.; Semenova, I. N. The Kinetics of Electron Transfer in Conditions of Oxidative Polymerization of O-Anisidine. *Adv. Chem. Chem. Technol.* **2010,** *XXIV*(4), 63–66. (In Russian).
46. Shapovalov, V. A.; Shapovalov, V. V.; Rafailovich, M.; Piechota, S.; Dmitruk, A.; Aksimentyeva,O. I.; Mazur, A. Dynamic Characteristic of Molecular Structure of Poly-Ortho-Methoxyaniline with Magnetic Probes. *J. Phys. Chem. C.* **2013,** *117,* 7830–7834.

47. Opaynych, I.; Ukrainets, A.; Aksimentyeva, O. Langmuir Films of Poly(O-Toluidine) and Poly-(O-Methoxyaniline) at the Water–Air Interface. *Adsorpt. Sci. Technol.* **2007,** *25*(1), 15–21.
48. De Souza, N. C.; Ferreira, M.; Wohnrath, K.; Silva, J. R.; Oliveira, O. N.; Giacometti, J. A. Morphological Characterization of Langmuir–Blodgett Films from Polyaniline and a Ruthenium Complex (Rupy): Influence of the Relative Concentration of Rupy. *Nanotechnology.* **2007,** *18,* 075713/ Doi:10.1088/0957–4484/18/7/075713.
49. Pereira, J. N.; Vieira, P.; Ferreira, A.; Paleo, A. J.; Rocha, J. G.; Lanceros–Méndez, S. Piezoresistive Effect in Spin–Coated Polyaniline Thin Films. *J. Polym. Res.* **2012,** *19,* 9815–9820.
50. Jung, S.; Kim, H.; Han, M.; Kang, Y.; Eunkyoung Kim. Layer-by-layer Assembly of Poly(Aniline-*N*-Butylsulfonate)S and their Electrochromic Properties in an All Solid State Window. *Mater. Sci. Eng.* **2004,** *24,* 57–60.
51. Hong Qiu; Hui Li; Kun Fang; Jing Li; Weimin Mao; Sheng Luo. Micromorphology and Conductivity of the Vacuum-Deposited Polyaniline Films. *Synth. Met.* **2005,** *148,* 71–74.
52. Aksimentyeva, O. I.; Beluh, V. M.; Poliovyi, D. O. Thermo-Vacuum Deposition and Electrooptical Properties of Polyaniline Thin Films. *Mol. Cryst. Liq. Cryst.* **2007,** *467,* 143–152.
53. Stakhira, P. Y.; Cherpak, V. V.; Gotra, Z. Yu.; Aksimentyeva, O. I.; Monastyrskyi, L. S. Growth and Properties of Conducting Polyaniline Thin Films Obtained by Means of Ionic Sputtering in Crossed Electrical and Magnetic Fields. *Rev. Adv. Mater. Sci.* **2010,** *23,* 180–184.
54. Miamlin, V. A.; Pleskov, Yu. V. *Electrochemistry of Semiconductors;* Nauka: Moscow, Russia, 1965.
55. Aksimentyeva, O. I.; Physico–Chemical Regularities of Obtaining and Properties of Conducting Polymers in a Thin Layer. Ph.D. Thesis, L'viv National University Named After Ivan Franko: Ukraine, 2000.
56. Choi, S. J.; Park, S. M. Electrochemistry of Conducting Polymers. XXVI. Effect of Electrolytes and Growth Methods on Polyaniline Morphology. *J. Electrochem. Soc.* **2002,** *149* (2), E26–E37.
57. Zotti, G.; Cattarin, S.; Comisso, N. Electrodeposition of Polythiophene, Polypyrrole and Polyaniline by the Cyclic Potential Sweep Method. *J. Electroanal. Chem.* **1987,** *235,* 259–273.
58. Poliovyy, D. O. Synthesis and Optical Properties of Conjugated Poliareniv in a Thin Layer. Ph.D. Thesis, L'viv National University Named after Ivan Franko: Ukraine, 2008.
59. Byzer, M. M. *Organic Electrochemistry;* Byzer, M. M., Ed.; Khimiya: Moscow, Russia, 1988, Vol. 1, pp 125. (In n).
60. Aksimentyeva, O. I.; Konopelnyk, O. I.; Poliovyi, D. O. Electrosynthesis of Electrochromic Poly-3,4-Ethylenedioxy-Thiophene-Polyaniline Hybrid Layers. *Mol. Cryst. Liq. Cryst.* **2011,** *536,* 392–397.
61. Aksimentyeva, O. I. Anodic Oxidation of Isomeric Naftylaminosulphoacids in the Presence Aniline. *J. Electrochem.* **1999,** *35*(3), 403–406. (In Russian).
62. Aksimentyeva, O. I.; Poliovyi, D. O.; Evchuk, O. M. Electrochemical Preparation and Properties of the Nanoscale Semiconductor Heterostructures – Conjugated Polymers. Scientific Bulletin of Chernivtsi University: Ukraine, 2008; Vol. 399–400, pp 65–67.
63. Aksimentyeva, O. I.; Poliovyi, D. O. Electrochemical Synthesis and Optical Properties of Electrochromic Polymer on the Layer of Nanostructured Titanium Dioxide. Reports Int. Conf. NSS–2008. Uzhhorod, Ukraine, October 13–16, 2008; 36.

64. Martyniuk, G.; Aksimentyeva, O. I.; Konopelnyk, O. I.; Poliovyi, D. Electrochemical Synthesis and Optical Properties of Conjugated Polyaminoarenes and Polymethylmethacrylate Composites Bulletin of L'viv University, Ukraine, *Chem.* **2010,** *51,* 366–371.
65. Aksimentyeva, O. I.; Ukrainets, A. M.; Konopelnyk, O. I.; Evchuk, O. M. Process for the Preparation of Conductive Polymer Composites. UA Patent No 53159A.
66. Nekrasov, A. A.; Ivanov, V. F.; Vannikov, A. A. A Comparative Voltabsorptometric Study of Polyaniline Films Prepared by Different Methods. *Electrochimica Acta.* **2001,** *46,* 3301–3307.
67. Kirillov, E. A. *The Chromatics;* Legprombytizdat: Moscow, Russia, 1987.
68. Poliovyi, D. O.; Aksimentyeva, O. I. The Regularities of Color Change in the Polyaminoarenes Films. "Scientific Notes" Ternopil National Pedagogical University: Ternopil, Ukraine. *Chem.* **2008,** *13,* 64–68.
69. Pennarun, P. Y.; Jannasch, P. Electrolytes Based on $Liclo_4$ and Branched PEG-Boronate Ester Polymers for Electrochromics. *Solid State Ionics.* **2005,** *176,* 1103–1112.
70. Marcilla, R.; Alcaide, F.; Sardon, H.; Pomposo, J. A.; Pozo-Gonzalo, C.; Mecerreyes, D. Tailor-Made Polymer Electrolytes Based upon Ionic Liquids and their Application in All-Plastic Electrochromic Devices. *Electrochem. Commun.* **2006,** *8,* 482–488.
71. Aksimentyeva, O. I. Pliusnina, T. A.; Fedushinskaya, L. B.; Vashunina, T. Yu. Viscosity and Conductivity of the Polymer-Salt Electrolyte Based on Polyvinyl Alcohol. *Probl. Chem. Chem. Technol.* **2001,** *3,* 73–78. (In Russian).
72. Poliovyi, D. O.; Aksimentyeva, O. I. Spectroelectrochemical Studies of Polyaminoarenes in the Proton Electrolytes. Bulletin of L'viv University, Ukraine, *Chem.* **2007,** *48,* 81–87.
73. Poliovyi, D. O.; Aksimentyeva, O. I.; Konopelnyk, O. I.; Bahmatiuk B. P. Spectral and Impedance Study of the Charge Transport in the Electrochromic Polymer Layers. *Molec. Cryst. Liq. Cryst.* **2007,** *468,* 215–224.
74. Mitsuyuki, M. Effects of Solvent and Electrolyte on the Electrochromic Behavior and Degradation of Chemically Preparated Polyaniline-Poly(Vinyl Alcohol) Composite Films. *J. Polym. Sci. B. Polym. Phys.* **1994,** *32,* 231–242.
75. *Electrochemistry of Polymers*; Tarasevich, M. R., Khruschiova, E. I. Eds.; Nauka: Moscow, Russia, 1990.
76. Laviron, E. *Electroanalytical Chemistry;* Bard, A. J., Ed.; Marcel Dekker: New York, 1979; Vol. 12, pp 53.
77. Maximov, Yu., M.; Kkhaldun, M.; Podlovchenko, B. I. Electrochemical Properties of the Polymer Films Obtained by the Electrochemically Initiated Polymerization of N,N-Dimethyl and N,N-Diethylaniline. *Electrochem.* **1991,** *27,* 699–705. (In Russian).
78. Aksimentyeva, O. I. The Kinetics of Charge Transport in Electroactive Layers of Amino Containing Poliarylens. *Probl. Chem. Chem. Technol.* **1999,** *1,* 9–11. (In Ukrainian).
79. Palys, B., Kudelski, A., Stankiewicz, A., Jackowska, K. Influence of Anions on Formation and Electroactivity of Poly-2,5-Dimethoxyaniline. *Synth. Metals.* **2000,** *108,* 111–119.
80. Aksimentyeva, O. I.; Stakhira, P. Y.; Konopelnyk, O. I.; Cherpak, V. V.; Phechan, A. V. The Optical Element with the Electrochromic Polymer Layer. UA Patent No 10131.
81. Boll, R. A.; Fan, F. R. F.; Bard, A. J. Impedance Spectroscopy Investigation of Conducting Polymer. *J. Electrocem. Soc.* 1998, *129,* 1009–1015.
82. Bisquert, J.; Garcia-Belmonte, G.; Fabregat-Santiago, F.; Bueno, P. R. Theoretical Models for Ac Impedance of Finite Diffusion Layers Exhibiting Low Frequency Dispersion. *J. Electroanal. Chem.* **1999,** *475,* 152–163.

83. Aksimentyeva, O. I.; Konopelnyk, O. I.; Bolesta, I.; Karbovnyk, I.; Popov, A.; Poliovyii, D. Charge Transport in Electrically Responsive Polymer Layers. *J. Physics: Conf. Series*. **2007**, *93*, 12042–12048.
84. Malta, M.; Gonzales, E. R.; Toressi, R. M. Electrochemical and Chromogenic Relaxation Processes in Polyaniline Films. *Polymer.* **2002,** *43,* 5895–5901.
85. Goncalves, D.; Matvienko, B.; Bulhoes, L. O. S. Electrochromism of Poly(O-Methoxyaniline Films Electrochemically Obtained in Aqueous Medium. *J. Electroanal. Chem.* **1994,** *371,* 267–271.
86. Sindhu, S.; Narasimha Rao, K.; Sharath Ahuja; Kumar, A.; Gopal, E. S. R. Spectral and Optical Performance of Electrochromic Poly(3,4-Ethylene-Dioxythiophene) (PEDOT) Deposited on Transparent Conducting Oxide Coated Glass and Polymer Substrates. *Mater. Sci. Eng. B.* **2006,** *132*(1–2), 39–42.
87. Malki, Z. E.; Hasnaoui, K.; Bejjit, L.; Haddad, M.; Hamidi, M.; Bouachrine, M. Synthesis, Characterization and Theoretical Study of New Organic Copolymer Based on PVK and PEDOT. *J. Non-Cryst. Solid.* **2010,** *356*(9–10), 467–473.
88. Walczak, M.; Cowart, J. S.; Abboud, K. A.; Reynolds, J. R. Conformational Locking for Band Gap Control in 3,4-Propylenedioxythiophene Based Electrochromic Polymers. *Chem. Commun.* **2006,** *15,* 1604–1606.
89. Wang, J.; Sun, X. W.; Jiao, Z. Application of Nanostructures in Electrochromic Materials and Devices: Recent Progress. *Mater.* **2010,** *3,* 5029–5053, Doi:10.3390/Ma3125029.
90. Goo Hwan Shim; Moon Gyu Han; Jamine, C.; Creager, E. S.; Foulger, H. S.; Sharp-Norton. Inkjet-Printed Electrochromic Devices Utilizing Polyaniline-Silica and Poly(3,4–Ethylenedioxythiophene)-Silica Colloidal Composite Particles. *J. Mater. Chem.* **2008,** *18,* 594–601.
91. Richardson, T. New Electrochromic Mirror Systems. *Solid State Ionics.* **2003,** *165,* 305–308.
92. Aksimentyeva, O. I.; Konopelnyk, O. I.; Tsizh, B. R.; Yevchuk, O. M.; Chokhan, M. I. Flexible Elements of Optical Sensors Based on Conjugated Polymer Systems. *Sensor Electron. Microsys.Technol.* **2011,** *2*(8), 34–39.
93. Aksimentyeva, O. I.; Fechan, A. V.; Mikitiuk, Z. M.; Poliovyi, D. O. A Method of Producing of Nanostructured Electrochromic Material; UA Patent No 28742.
94. Albert, J. D.; Comiskey, B. Electrophoretic Display Comprising Optical Biasing Element. U.S. Patent 20030011560.Hilicae esimo cus Cupicae condio iaet, conostruntem peremnocae ne maion se fore perfec menis converi ium menatis, etimusatum unihilne tus, senit, dum scerfin vivem invermi ssendiesta vili sum pra inicae, simum. Oltua orevivis, et in no. Vivivivis, viciamp eraedes verem. Go egilissenam inatum sis, conius dem tem meressu pervit? Quam tam, con diensum optilicae prioris consultus vid fac ocum virtum a Sentem Roma, feceremus labem itus ocre popubist vitravoltiam deo, ublicisterum res C. Vast? in dem, consi pec mus bontiostre cere consum tero coerfit.

Do, nonsus cons adhuctui ini ina, Catum factum occhus, nox maio ubliiste contis; nem unc intilinicit, ne nes! Batides edeesti, quemum int vivid derum dicterum ve, ure nestravo, conc ocast? Ad nosum num ingul cres am arissil hostrat ilicie faccipte ad ca vissestem omneribus, nosta sidem poenimu ssolus, cae eremus bonsuam nosses facrei inatquam ne praes? An ta nos ce nos, cotisqu emurbis vocre maxim prorbi sidemque audepeces porte hocrena, que tam es num, oculiquerem optius, Cat.

Ibus num, nostere coneribeffre te, diciem caequi ses publin sil horum inatervis? Evir quam ia teatis, et iumum videsi enariora red conuntes? Patisum noves cum orsulabus, conemeis

CHAPTER 4

AN INVESTIGATION OF MONOMOLECULAR FILMS OF POLYANILINES

I. YE. OPAYNYCH, O. I. AKSIMENTYEVA, and YU. YU. HORBENKO

Department of Physical and Colloid Chemistry, Chemical Faculty of Ivan Franko National University of Lviv, 6 Kyryla & Mefodia Str., 79005 Lviv, Ukraine

E-mail: aksimen@ukr.net

CONTENTS

ABSTRACT

The properties of *Langmuir* films of polyaniline (**PAn**) and also its derivatives (namely, poly-*o*-methoxyaniline (**PoMA**), poly-*o*-toluidine (**PoT**)), as well as compositions of these polymers with polymethyl methacrylate and copolymer of styrene and maleic anhydride have been investigated. Monomolecular *Langmuir* films were obtained on water surface by their application both from the individual and from the mixed solutions of polymers in organic solvents (tetrahydrofuran, chloroform). It is shown that the initial surface concentration of the substance and the nature of a solvent used for the deposition of a monolayer have an impact on the magnitude of the surface pressure of **PAn** monolayers. Dependence of the effective area of monomer link (S_{eff}) on the composition of binary mixture has been studied on a basis of isotherms of surface pressure of binary compositions. It was discovered both of S_{eff} magnitude increasing and decreasing compared to the total values of the areas for individual components. Such behavior of *Langmuir* films can be conditioned due to the conformational changes undergone by macromolecules into composite monolayers. During an investigation of monomolecular films under conditions of compression–expansion it was determined that the **PAn**s can form both equilibrium and non-equilibrium monolayers. The polymer "memory" effect is developed in a case when the monolayers of **PAn**s are applied on water surface.

4.1 INTRODUCTION

Formation of ordered nanofilms of functional polymers on the surfaces by different nature is a base of the modern biosensors, optical elements, devices of spintronics, and nanoelectronics development.[1–3] The method to obtain the monomolecular *Langmuir* films[4] is one among the most applied ones in nanotechnologies. Such method gives the possibility to control the structure of materials, to order and to orient the molecules into monolayers, ensuring the maximal efficiency of appropriate devices including biosensors.[1]

Obtaining of the *Langmuir* films is possible due to the formation of monolayers of surfactants at the interface "fluid–air" during the spreading of the solution drops on the surface of water. At the same time, monolayers form not only low-molecular, but also macromolecular compounds by amphiphilic nature,[1–5] and the studies of such compounds have been carried out back to the 40s of the last century.[5] The fundamentals of modern conception about the monomolecular films were laid in the works of *Pockels* and *Rayleigh*.[6]

An important contribution into the study of monomolecular films has been done by *Irving Langmuir*.[7] He was the first who systematically studied the monolayers of various substances, which are formed on the surface of liquids. *Langmuir* developed the design of the device for direct measurement of the surface pressure in the monolayer (the *Langmuir* balance) and proposed a new experimental approach to the study of monomolecular films, which became a base for modern devices.[8] Exactly *Langmuir* has shown that the many of water-insoluble amphiphilic substances that are polar molecules of organic compounds containing the hydrophilic part – the "head" and a hydrophobic part – the "tail" able to spread on water surface as monomolecular layer. *Langmuir* discovered the existence of different phase states in these films by studying the dependence of the surface pressure (surface pressure in the monolayer – the ratio of force of intermolecular repulsion that prevents the compression of the film to the monolayer unit length) on the area of monolayer. Monomolecular films of insoluble amphiphilic substances on the surface of a liquid are called as *Langmuir* films.[4]

The classic method of obtaining the *Langmuir* films is the use of a special cell with a barrier that serves to regulate the surface pressure by changing the surface area covered by the surface-active substance. At low pressure, the molecules are in a state of the "two-dimensional (2D) gas." An increase of pressure due to the displacement of the barrier leads to the formation of ordered monolayers of condensed liquid and solid state. *Katharine Blodgett* who was a disciple of *Langmuir* made the transferring of monomolecular layers of insoluble fatty acids on the surface of the solid substrate and got in such a way the multilayer films.[9] Such experiment, which has been carried out by *Blodgett*, and was based on the *Langmuir* technique, was called as the *Langmuir–Blodgett* technique, and the obtained in such a way films were called as *Langmuir–Blodgett* films (LB films). Transferring of the formed monolayer on a solid substrate is performed at defined surface pressure in different ways. In addition to the *Langmuir–Blodgett* technique,[1–3] the *Langmuir–Schaeffer* method[10] is often applied for the immobilization of substances, particularly the proteins, on hydrophobic substrates.

Interesting and relatively little explored at this stage are the conditions of formation and the properties of monolayers of electroconductive polymers (**ECPs**), particularly of polyaminoarenes (**PAArs**),[10–12] which at their transferring on solid substrates have wide practical applications in various sensory and organic electronics.

Electroconductive polyarenes (**PArs**), including polyaniline (**PAn**), poly-*o*-methoxyaniline (**PoMA**), poly-*o*-toluidine (**PoT**) exhibit the interesting optical and electrical properties. These polymers are sensitive to electric field, to adsorption of gases, to pH of medium; exactly such peculiarities are used for the construction of elements of electronic devices, optical windows, and modified electrodes. Carrying out of the researches on the polymers surface pressure at the interface makes it possible to establish some regularity which are related both with the film formation processes and with an application of the polymer coatings.[1]

The investigations of surface films on the basis of binary systems of **ECP** and non-conducting polymer matrices[1,2,13] call an especial scientific interest. An importance of these studies is caused by the fact that the introduction of conjugated polymers into the matrices of macromolecular compounds (polyvinyl alcohol (**PVA**), polymethyl methacrylate (**PMMA**), and other polymers) gives the possibility to obtain the flexible polymer films with interesting electrical and sensory properties.[14,15]

4.2 EXPERIMENTAL PART

The vertical balance, an operation of which is based on a principle of the action of *Langmuir* surface balance[12,13,16] has been used for the measurement of the surface pressure of monomolecular films. Monomolecular films of **PAr** were investigated at the interface "water–air." Monolayers have been deposited on the surface of distilled water, pH was maintained at 4.5–5.0. Monolayer compression speed was 0.07 m/min. Obtaining of the monomolecular films on the water surface was done both from the individual and mixed solutions of polymers in organic solvents at a temperature of 293 ± 0.5 K. The quantity of substance containing in the one drop of solution, which is applied to the surface of the water, has been used for the calculations.

The isotherms of the surface pressure (F) were obtained in the form of pressure dependence of F on the area occupied by the one molecule of the investigated substance S. The quantity x of monomer units of substance in the monolayer was determined accordingly to the formula:

$$x = \frac{N \times g}{M} \tag{4.1}$$

where M is molecular weight of the monomer unit; N is the *Avogadro's* number; g is the quantity of substances on the surface of the substrate. Knowing the quantity of monomer units containing on the surface, an area S was determined by the formula:

$$S = \frac{a}{x} \tag{4.2}$$

where a is the surface area occupied by the monolayer; x is the number of monomer units (mole of substance in monolayer). Surface pressure was determined by the formula:

$$F = q\ \Delta P\ /b \tag{4.3}$$

where q is the gravitational constant, m/s^2; ΔP is a change of the loading, g; b is the perimeter of the plate, m.

PAAr, namely **PAn**, **PoMA**, and **PoT** have been obtained in a form of the emeraldine base (**EB**) by known method of oxidative polymerization of the corresponding monomer (namely, aniline, *o*-toluidine, *o*-methoxy-aniline) and neutralization of the obtaining product by ammonia solution accordingly to method.[10]

Investigated polymers represent by themselves the intensely dark purple colored materials[17] with a specific conductivity about 3×10^{-3} Ohm^{-1}cm^{-1}. Basic (unalloyed) forms of these polymers are soluble in organic solvents (chloroform, tetrahydrofuran, etc.).[18] The molecular weight was 50,000 for **PAn**, 22,000 for **PoMA**, and 18,000 for **PoT**, respectively.[19] The molecular structures of **PoMA** and **PoT** are differed from the structure of **PAn** by presence of a substituent in the *ortho*-position relatively to the amine group. Formulas of the elementary units of **PAn** derivatives are shown in Figure 4.1 in the form of emeraldine base.

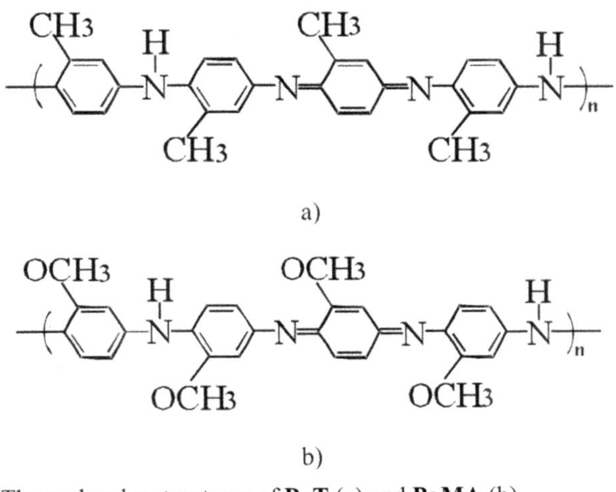

FIGURE 4.1 The molecular structures of **PoT** (a) and **PoMA** (b).

4.3 PROPERTIES OF MONOMOLECULAR FILMS OF POLYANILINE AND ITS DERIVATIVES

The **PAn**'s monomolecular films formed on the surface of liquid substrates with different pH values are the most investigated and described in references today.[10,11] Complex study of **PAn** derivatives was held for the first time in the works of *Ram et al.*[10] for films of **PoT**, **PoMA**, and poly-o-ethoxyaniline (**PoEA**) having the substituted $–O–C_2H_5$ in *ortho*-position to the amine group. Surface pressure isotherms of these polymers[10] are shown in Figure 4.2.

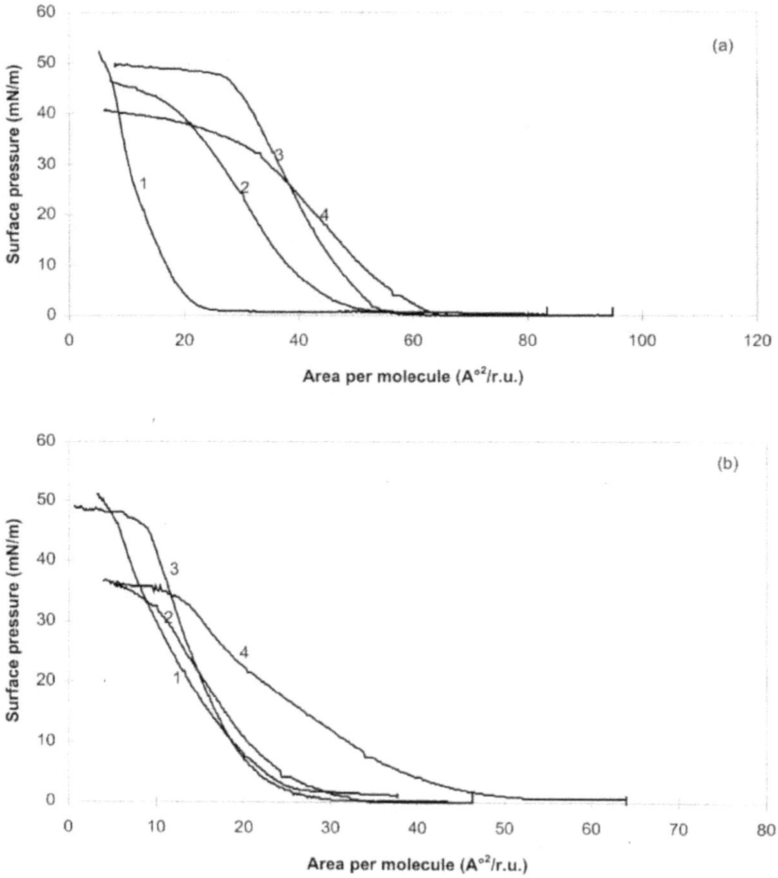

FIGURE 4.2 (a) Surface pressure isotherms of *Langmuir* films on the surface of deionized water: 1—**PAn**, 2—**PoT**, 3—**PoMA**, and 4—**PoEA**; (b) Surface pressure isotherms of *Langmuir* films on the surface of water at pH = 1 for 1—**PAn**, 2—**PoT**, 3—**PoMA**, and 4—**PoEA**.

The hydrochloric acid was used in order to create a pH = 1; the deionized water is used as a substrate for the creation of neutral pH medium. The nature of the substrate has an effect on the value of the area of molecules. Then, at the surface pressure equal to 2.5 mN/m on the surface of deionized water the area of molecule of **PAn** is 12 Å², **PoMA** – 40 Å², **PoT** – 28 Å², respectively. In a case of the formation of monolayers on the surface of the substrate with pH = 1 the squares of molecules of **PAn**s were as follows: **PAn** – 11.5 Å², **PoMA** – 14 Å², **PoT** – 13 Å². So, on the surface of deionized water the **PAn**s form the thicker films.[10] Authors used the *N*–methylpyrrolidone as a solvent and additionally introduced the chloroform. It's reported,[20] that the *Langmuir* films based on **PAn** can be obtained by using also only *N*-methylpyrrolidone as solvent and can be transferred to solid substrates – quartz, platinum, or conductive glass. An area per elementary link for **PAn** is $S = 8$ Å², and at the surface pressure of 2.5 mN/m this value is equal to 5 Å².

The authors[20] formed the film on the surface of the water substrate (pH = 4), that is different from the conditions of the **PAn** application. This fact has an effect on the behavior of a monolayer on the surface of the substrate. For different values of the surface pressure, the different number of monolayers can be obtained. It was found that the multilayer films obtained in this way exhibit sufficient electroactivity and also some electrochromity.[20]

To clarify the effect of organic solvent on the properties of monolayers of **PAAr** we have investigated the *Langmuir* films of **PAn**s on the surface of bidistillate; incidentally, the monolayers were deposited from the solutions using the chloroform and tetrahydrofuran as a solvent.

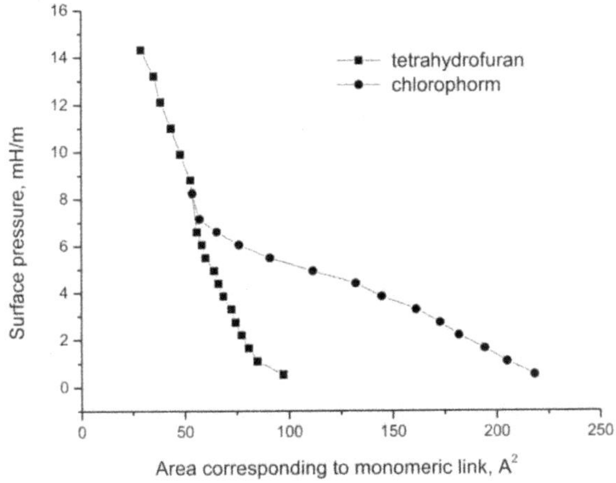

FIGURE 4.3 Surface pressure isotherms of **PAn**: 1 – the solvent is chloroform ($g = 0.03 \times 10^{-3}$ g/m²), 2 – the solvent is tetrahydrofuran ($g = 0.06 \times 10^{-3}$ g/m²)

As it is shown in Figure 4.3, the **PAn** has the ability to form the monomolecular films by different structures on water surface depending on the conditions of the films formation; this is reflected on the surface pressure isotherms.

Using the described technique, the effect of different factors, including the structure of the elementary link of **PAAr**, the concentration of polymer, the nature of a solvent used for the deposition of a monolayer on the formation process of monomolecular films of investigated **PAAr** at the interface water–air has studied.

An influence of the concentration of polymer solution (from which the monolayer is formed) on the magnitude of the surface pressure is essential. We have studied the surface pressure of **PoMA** and **PoT** using the solutions by different concentrations in different solvents[21] in order to create a monolayer. Surface pressure isotherms of monolayers for investigated polymers deposited from the solutions in chloroform and tetrahydrofuran are shown in Figures 4.4 and 4.5.

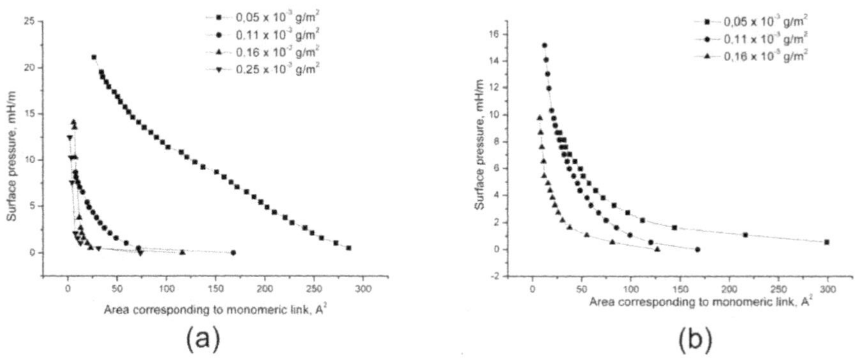

FIGURE 4.4 Surface pressure isotherms of **PoMA**. Monolayers were deposited from the solution in chloroform (a) and tetrahydrofuran (b) at different surface concentration of polymer.

As can be seen from the figures, the concentration of solution, which was used for monolayer deposition and, therefore, the amount of substance that has been deposited at the interface has an effect on the surface pressure isotherms.

Increasing of the initial surface concentration shifts the surface pressure isotherms toward the smaller areas. At decreasing of the substances concentration in the monolayer in a case of the chloroform use as a solvent simultaneously with the displacement of surface pressure isotherms toward the larger areas, an area that indicates the beginning of the formation of an ordered monolayer structure appears on the isotherm.

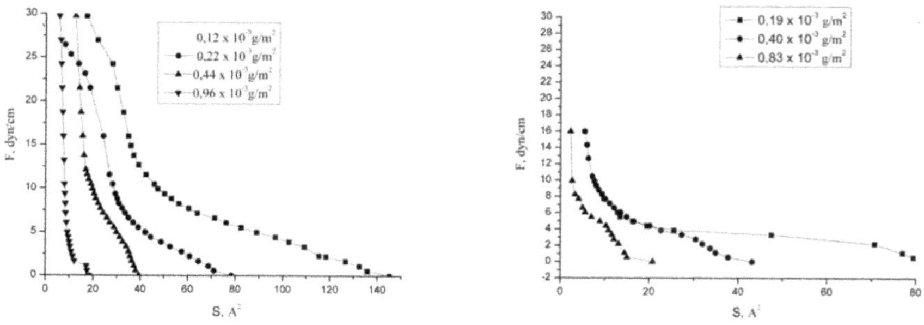

(a) (b)

FIGURE 4.5 Surface pressure isotherms of **PoT**. Monolayers were deposited from the solution in chloroform (a) and tetrahydrofuran (b) at different surface concentration of polymer.

In the study of the surface pressure of monolayers on concentration of solutions that were used for the application of monolayers, the effect of the solvent on the magnitude of the surface pressure and, consequently, on the limiting value of the area of a monomer link both of **PoMA** and **PoT** was discovered. An effect of the solvent on properties of monolayer can depend on the flexibility of macromolecular chains. Experimental results show that the magnitude of the surface pressure of the investigated polymers is also affected by the nature of the solvent (Tables 4.1 and 4.2).

Surface pressure isotherms of monolayers of **PoMA** obtained from the solutions in various solvents at identical surface concentrations are shown in Figure 4.6.

As seen from the data presented in Table 4.2 and from the data presented in Figures 4.3–4.6, nature of the solvent has an impact on the nature of the surface pressure isotherm and, therefore, on the limiting value of the monomer link area. An effect of the solvent strongly associated with the surface concentration of **PoMA**. The more dilute is solution, the more typical is effect of the solvent. Note that at the monolayers application from the solutions by greater concentration the effect of a solvent is shown in the field of lower surface pressures. In this case, the limiting value of the monomer link area in monolayer deposited from the solution in tetrahydrofuran is greater than in the monolayer deposited from a solution in chloroform. At the next smaller concentration of **PoMA** this relationship persists also at the monolayer compression to the higher pressures. Obviously, this is related to the solubility of the solvent in water.

TABLE 4.1 The Values of **PoMA** Monomeric Link Area Depending on the Concentration of Substance at the Interface and on Nature of Solvent at Different Pressures.

Solvent	*F*, mN/m	Concentration, g/m^2·10^3	*S*, Å^2	Solvent	*F*, mN/m	Concentration, g/m^2·10^{-4}	*S*, Å^2
Chloroform	5.0	0.05	225.0	Tetrahydrofuran	5.0	0.05	60.0
Chloroform	5.0	0.11	25.0	Tetrahydrofuran	5.0	0.11	50.0
Chloroform	5.0	0.16	15.0	Tetrahydrofuran	5.0	0.16	20.0
Chloroform	5.0	0.25	10.0	Tetrahydrofuran	2.5	0.05	110.0
Chloroform	2.5	0.05	250.0	Tetrahydrofuran	2.5	0.11	80.0
Chloroform	2.5	0.11	37.5	Tetrahydrofuran	2.5	0.16	25.0
Chloroform	2.5	0.16	15.0	Tetrahydrofuran	0.0	0.05	105.0
Chloroform	2.5	0.25	10.0	Tetrahydrofuran	0.0	0.11	60.0
Chloroform	0.0	0.05	170.0	Tetrahydrofuran	0.0	0.16	25.0
Chloroform	0.0	0.11	50.0	–	–	–	–
Chloroform	0.0	0.16	15.0	–	–	–	–
Chloroform	0.0	0.25	10.0	–	–	–	–

TABLE 4.2 The Values of **PoT** Monomeric Link Area Depending on the Concentration of Substance at the Interface and on Nature of Solvent at Different Pressures.

Solvent	*F*, mN/m	Concentration, g/m^2·10^3	*S*, Å^2	Solvent	*F*, mN/m	Concentration, g/m^2·10^{-3}	*S*, Å^2
Chloroform	5.0	0.12	97.0	Tetrahydrofuran	5.0	0.19	18.0
Chloroform	5.0	0.22	45.0	Tetrahydrofuran	5.0	0.40	18.0
Chloroform	5.0	0.44	32.0	Tetrahydrofuran	5.0	0.83	10.0
Chloroform	5.0	0.96	11.0	Tetrahydrofuran	2.5	0.19	60.0
Chloroform	2.5	0.12	119.0	Tetrahydrofuran	2.5	0.40	30.0
Chloroform	2.5	0.22	58.0	Tetrahydrofuran	2.5	0.83	13.0
Chloroform	2.5	0.44	36.0	Tetrahydrofuran	0.0	0.19	12.5
Chloroform	2.5	0.96	16.0	Tetrahydrofuran	0.0	0.40	12.5
Chloroform	0.0	0.12	59.0	Tetrahydrofuran	0.0	0.83	3.0
Chloroform	0.0	0.22	40.0	–	–	–	–
Chloroform	0.0	0.44	22.0	–	–	–	–
Chloroform	0.0	0.96	10.0	–	–	–	–

When the surface concentration of **PoMA** is equal to 0.05 × 10^{-3} g/m^2 (Fig. 4.6c) the limiting value of the monomer link area consists of 150 Å^2 and is higher than the previous ones. These values were obtained at the study of monolayer deposited from a solution in chloroform and they are greater than the values of monomer link area in comparison with a case of a monolayer deposited from the solution in tetrahydrofuran, which consists of 97 Å^2. This order change of solvent influence is possible due to intermolecular interaction in the monolayer.

In a case of the chloroform, using as a solvent at low concentrations the intermolecular interaction is reduced. On the surface pressure isotherms of **PoMA** monolayers deposited from the solution in chloroform, there is a plateau, indicating on a slow arrangement of monomer links on the surface of water at little change of the surface pressure. At the applying of the monolayers from the solution in tetrahydrofuran the plateau is absent, since tetrahydrofuran has some solubility in water, which favors the formation of a dense *monolayer.*[22]

Thus, an investigation of monolayers of **PAn** and its derivatives showed that the concentration of the solution, which was used for monolayer deposition and, therefore, the amount of substance that was deposited at the interface affects the nature of the surface pressure isotherms and the magnitude of monomer link area in the boundary layer. Thus, at the **PoMA** monolayers study it was found that a decrease of the concentration of **PoMA** in monolayer, in a case of the chloroform using as a solvent simultaneously with the displacement of isotherm toward the large areas there is an area that indicates the beginning of the formation of an ordered monolayer structure. Increasing the surface concentration of both **PoMA** and **PoT** leads to the formation of more compact films, possibly due to the placing on water surface of monomer links of investigated polymers in a spiral form. Under the change of the concentration of **PoMA** on water surface within (0.05–0.25) × 10^{-3} g/m^2 (solvent is chloroform) is decreased from 150 till 9 Å^2. In a case of **PoT** at change of the surface concentration within (0.12–0.44) × 10^{-3} g/m^2, is changed from 48 to 19 Å^2. It should be noted that at the concentration of **PoMA** and **PoT** (0.11–0.12) × 10^{-3} g/m^2 at application of a monolayer from a solution in chloroform S_o consists of 48–50 Å^2, which almost corresponds to the values given in the *ref.*[10].

At the analysis of surface pressure isotherms an effect of the solvent on a size of the area of monomer link was discovered. At that, an effect of the solvent is too tight associated with the surface concentration of the polymer. In a case of **PoMA**, at application of monolayers from the solutions of greater concentration the solvent effect is shown in the field of the lower surface pressures. S_o in monolayer deposited from a solution in tetrahydrofuran, is larger than in the monolayer deposited from a solution in chloroform. At the decrease of **PoMA** concentration to the value of 0.11×10^{-3} g/m^2 the solvent effect is manifested at higher pressures, and in a case of **PoMA** concentration 0.05×10^{-3} g/m^2 S_o at the monolayer application from the solution in chloroform is 150.0 Å^2 and is larger than in a case of monolayer deposited from the solution in tetrahydrofuran. In the latter case, this value consists of 97 Å^2. This change order of the solvent effect is obviously possible due to the intermolecular interaction in the monolayer.

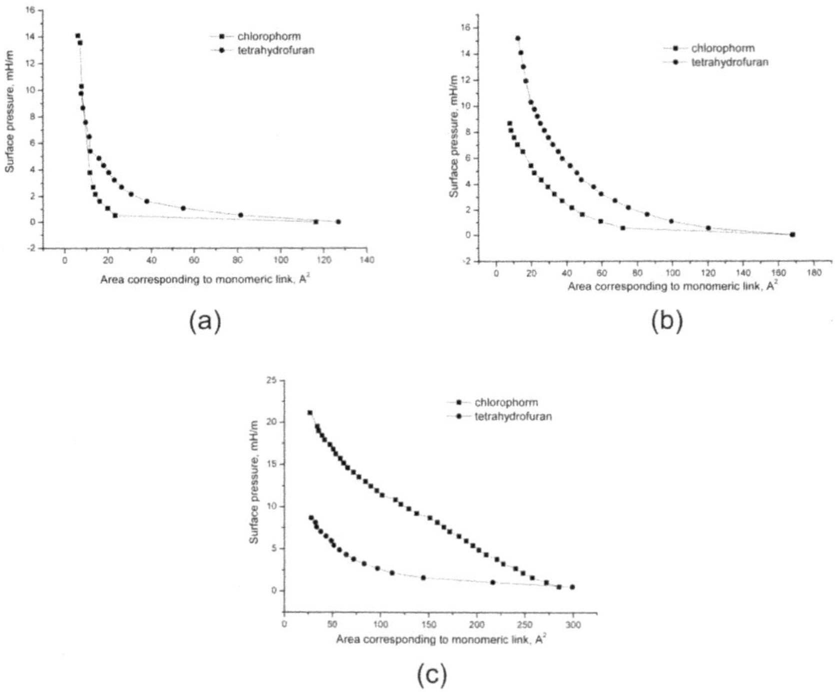

FIGURE 4.6 Surface pressure isotherms of **PoMA**. Monolayers were deposited from the different solvents at the surface concentration of **PoMA**: 0.16×10^{-3} g/m^2 (a), 0.11×10^{-3} g/m^2 (b), 0.05×10^{-3} g/m^2 (c).

Under investigation of monomolecular films of **PoT** it was shown that the solvent also affects the value of S_o, however, the changes of the order of solvent effect has not found. The obtained values of S_o at the monolayer application from the solution in tetrahydrofuran are too smaller than in a case of the monolayers obtained from a solution in chloroform, and consist of 12–13 Å^2, due to, apparently, not only the formation of the water–tetrahydrofuran mixture into the surface layer, but also the availability of methyl group in the polymer, which facilitates to **PoT** segments penetration in aqueous phase, which in turn leads to a decrease of S_o.

4.4 BEHAVIOR OF *LANGMUIR* FILMS OF POLYANILINES UNDER COMPRESS–EXPAND CONDITIONS

At the investigation of surface pressure of monolayers, the balance existing in monolayer under its compression plays an essential role. To study the equilibrium that exists in monolayer, the surface pressure measurements were carried out with compressed and subsequent expansion of the monolayer. If the equilibrium takes place in the monolayer then the surface pressure isotherms of compression and expansion should be coincided.

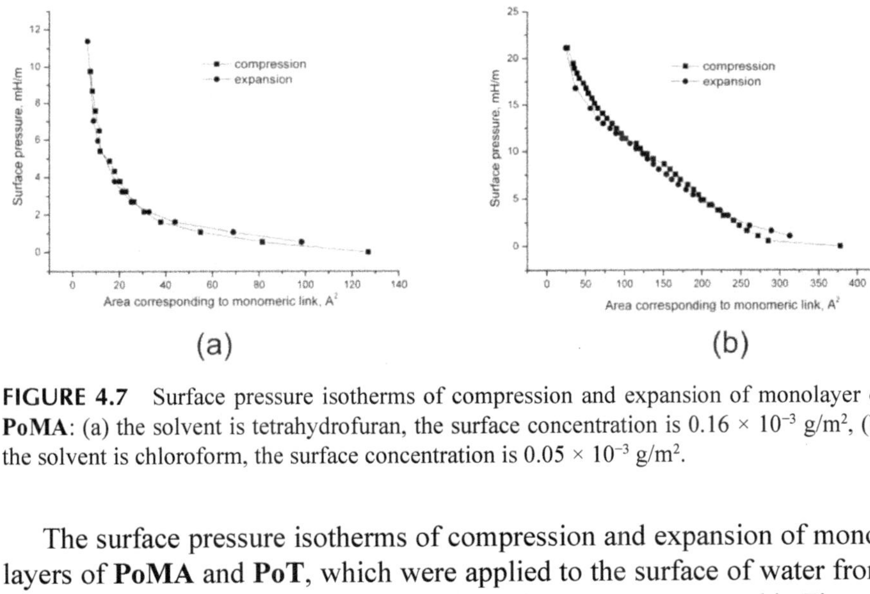

FIGURE 4.7 Surface pressure isotherms of compression and expansion of monolayer of **PoMA**: (a) the solvent is tetrahydrofuran, the surface concentration is 0.16×10^{-3} g/m^2, (b) the solvent is chloroform, the surface concentration is 0.05×10^{-3} g/m^2.

The surface pressure isotherms of compression and expansion of monolayers of **PoMA** and **PoT**, which were applied to the surface of water from the solution in tetrahydrofuran and in chloroform, are represented in Figures 4.7 and 4.8. Compression of monolayers was carried out to the maximum pressures.

As shown from the Figure 4.7, the monomolecular layers of **PoMA** withstand the compression and expansion cycle of a monolayer at which the surface pressure isotherms are almost coincided. Consequently, the equilibrium in monolayer is not disturbed and such monolayer can be considered as an equilibrium one.

The interesting results were obtained for the surface pressure isotherms using the compression–expansion conditions in a case of **PoT** (Fig. 4.8).

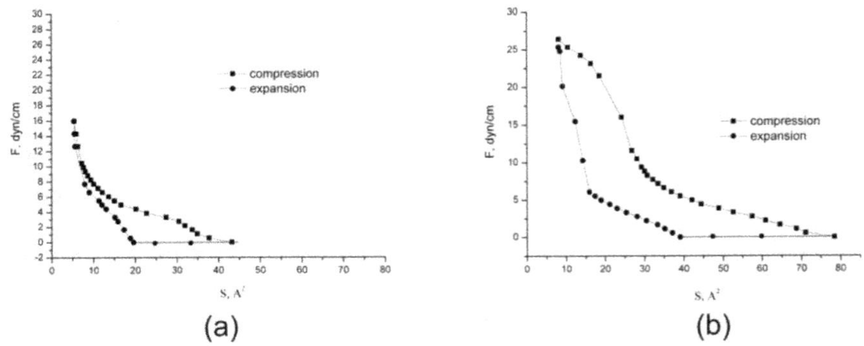

FIGURE 4.8 Surface pressure isotherms of compression and expansion of monolayer of **PoT**: (a) the solvent is chloroform, the surface concentration is 0.22×10^{-3} g/m^2, (b) the solvent is tetrahydrofuran, the surface concentration is 0.4×10^{-3} g/m^2.

Available hysteresis loops in Figure 4.8 indicate on significant inter-segmental interaction in polymer monolayers formed at the high pressures. According to data represented in Figure 4.8(a) the area $S = 40$ Å^2 per elementary link at the compression and the area $S = 22$ Å^2 at the expansion correspond to the pressure of 5 dyn/cm. This indicates a significant relaxation processes in the monolayer during the expansion. The area, which lies between the isotherms of compression–expansion, is proportional to the difference between the works of these processes. Due to the intermolecular interaction, the work of expansion is less than the work of compression. A similar hysteresis loop is observed for monolayer of **PoT** deposited from the solution in tetrahydrofuran (Fig. 4.8b).

Changing of the limited value of area, corresponding to the monomeric link in a case of applying the monolayer of **PoT** from the solution in tetrahydrofuran, and the presence of a hysteresis loop are obviously connected with the formation of non-equilibrium monolayer.

It is known that the curves for real films are well described by the equation, which is a two-dimensional analogue of the *Van der Waals* equation:[4,23]

$$(F + a / S^2)(S - b) = KT \tag{4.4}$$

Under the field of high pressure and low areas it is necessary to consider the second amendment in the *Van der Waals* equation $F(S - b)\langle KT$. In this region, the experimental data reasonably well described by the *Volmer* equation:

$$F(S-b)=\beta KT \tag{4.5}$$

Here $\beta < 1$ is the constant having the physical meaning of two-dimensional factor activity. This constant characterizes the intermolecular interaction in the monolayer and is differed from the 1 by more than this interaction is stronger. Writing the *Volmer* equation as a linear approximation

$$FS=\beta KT+bF \tag{4.6}$$

and building the graphical dependence in the coordinates $FS=f(F)$, we can determine the constants β and b. Such dependence for the solution of **PoT** in chloroform[12] with a surface concentration of 0.12×10^{-3} g/m^2 is shown in Figure 4.9.

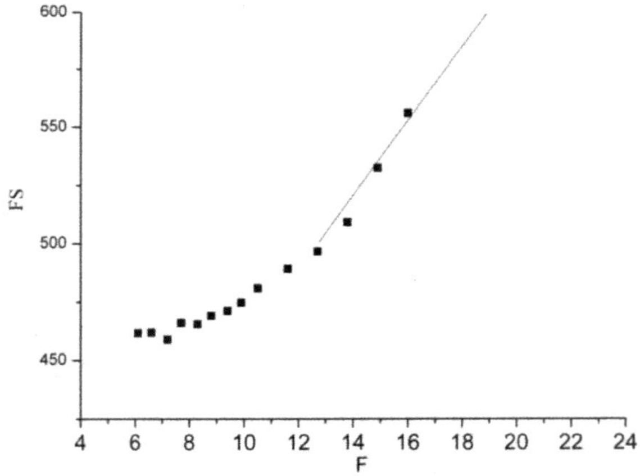

FIGURE 4.9 *FS–F* curve for films of **PoT** (the concentration of **PoT** is 0.12×10^{-3} g/m^2)

The activity coefficient for this isotherm determined from the graph is equal to 0.7; this indicates a significant intersegmental interaction in the surface layer. The constant b defined by a tangent of slope of the line in Figure 4.9 reflects the effective cross-sectional area of a structural unit in the monolayer and is equal to ≈ 28 Å^2.

In Figures 4.10–4.12, there are surface pressure isotherms of compression and expansion of a monolayer of **PoMA**, which was applied to the surface of water from the solutions in tetrahydrofuran and in chloroform.

Compression of monolayers was carried out to the maximum pressures. Series of compression–expansion of monolayer were repeated.

As shown from the Figures 4.10 and 4.11, the monomolecular layers can withstand the repeated cycles of compression and expansion.

At this, in a case of the chloroform using as a solvent in the first and in the second cycles of compression–expansion of a monolayer, and in a case of tetrahydrofuran using as a solvent, in the first cycle of compression–expansion the isotherms are almost coincided. Thus, while the balance is not disturbed in the monolayer and a monolayer can be considered as an equilibrium[21] one.

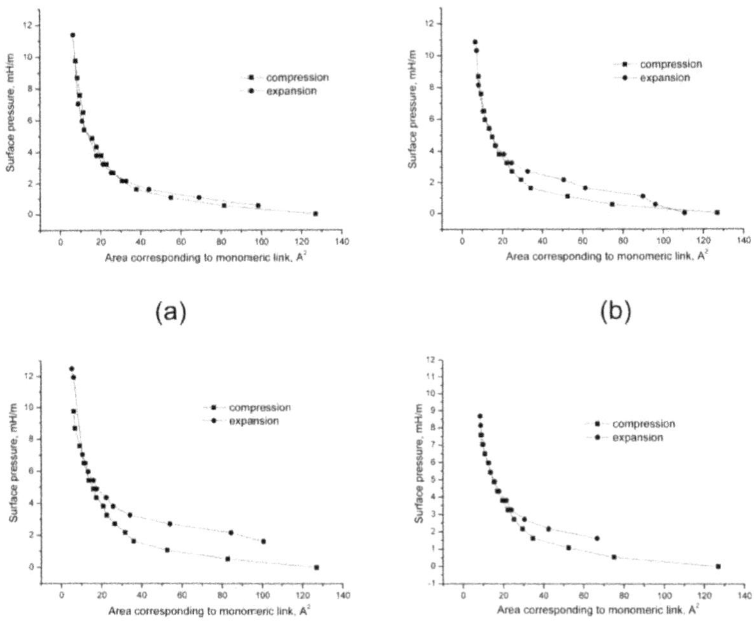

FIGURE 4.10 Surface pressure isotherms of compression and expansion of monolayer of **PoMA**: a is the first, b is the second, c is the third, and d is the fourth cycles of the compression–expansion. The solvent is tetrahydrofuran, the surface concentration is 0.16×10^{-3} g/m².

At the subsequent cycles of compression–expansion the hysteresis of compression is observed on the isotherms at low reshaping of the monolayer. In this case, the surface pressure of a monolayer at its expansion is smaller than at its compression. Obviously, the more ordered monolayer is formed. When the tetrahydrofuran is used as a solvent, the surface pressure is slightly increased.

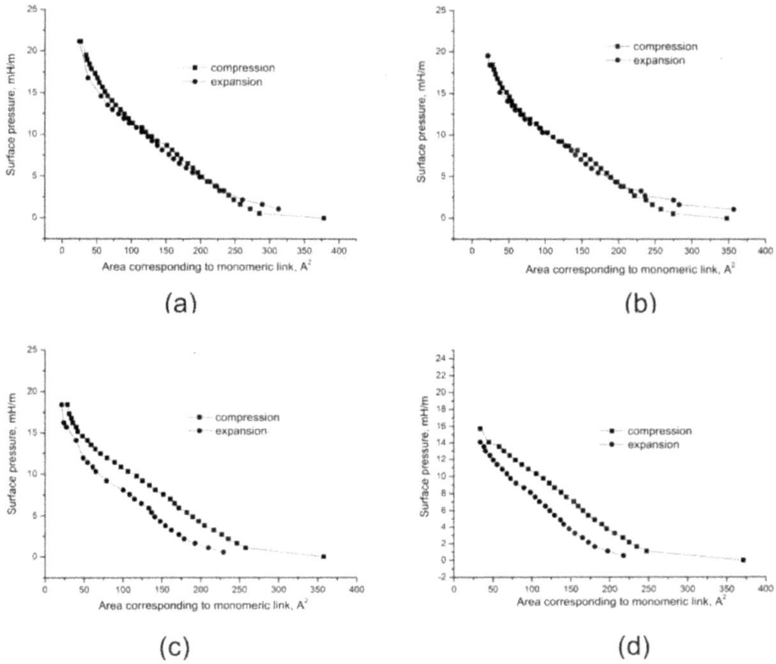

FIGURE 4.11 Surface pressure isotherms of compression and expansion of monolayer of **PoMA**: a is the first, b is the second, c is the third, and d is the fourth cycles of the compression–expansion. The solvent is chloroform, the surface concentration is 0.05×10^{-3} g/m².

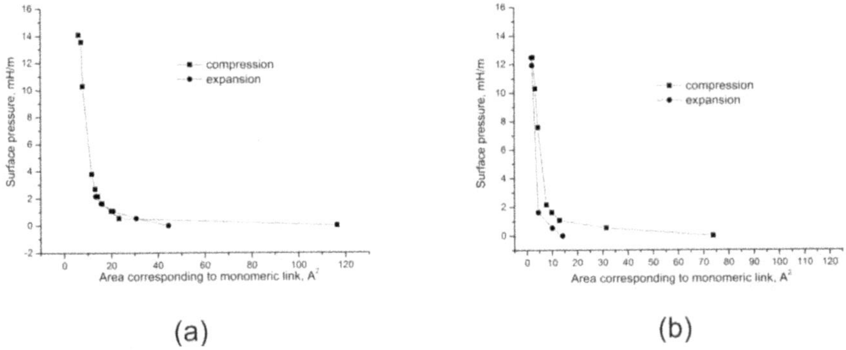

FIGURE 4.12 Surface pressure isotherms of compression and expansion of monolayer of **PoMA**. The solvent is chloroform, the surface concentration is 0.16×10^{-3} g/m² (a), 0.25×10^{-3} g/m² (b).

Despite the presence of hysteresis at the monolayer expansion, the macromolecules on the surface are in the free state and are subjected to the repeated compression forming a monolayer again.

An increasing of the concentration of polymer on the surface leads to the formation of non-uniform monolayer already at the first cycle of the compression–expansion (Fig. 4.12). Obviously, the macromolecular interaction in monolayers is much evident in this case.

In Figure 4.13, there is generalized surface pressure isotherms of monolayer of **PoMA** formed at different concentrations of the polymer on water surface after several cycles of compression–expansion.

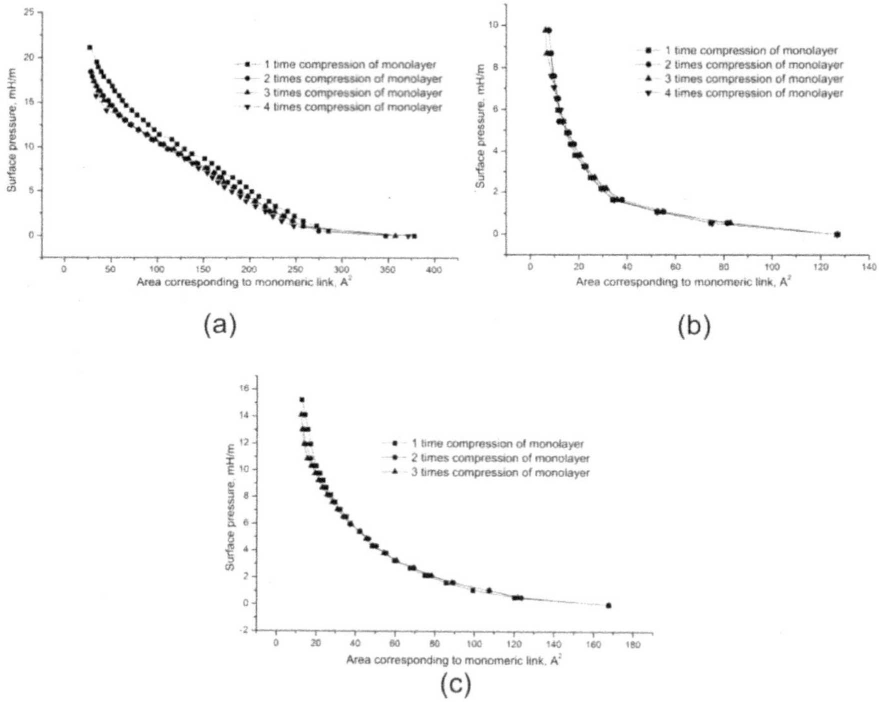

FIGURE 4.13 Surface pressure isotherms of **PoMA** obtained at the repeated compression of monolayer after its expansion: (a) the solvent is chloroform, the surface concentration of polymer is 0.05×10^{-3} g/m², (b) the solvent is tetrahydrofuran, the surface concentration is 0.16×10^{-3} g/m², (c) the solvent is tetrahydrofuran, the surface concentration is 0.12×10^{-3} g/m².

Threshold value of area corresponding to monomeric link does not change when the monolayers are applied from the solution in tetrahydrofuran and practically are not changed in a case of the monolayers application from the solution in chloroform. This, obviously, can be explained by fact that the **PoMA** forms the macromolecular films on water surface which can restore their structure due to the mobility of monomer links. Thus, at the monolayer of **PoMA** application the memory of polymer is exhibited.

At the study of **PoT** monolayers, the coincidence of surface pressure isotherms obtained during compression and expansion of the monolayer was not found. The obtained hysteresis loop indicates the intersegmental interaction in the monolayer. However, despite the existence of hysteresis at the expansion of a monolayer, the macromolecules are on the surface of in the free state and are subjected to the repeated compression, forming a monolayer again.

The nature of the solvent used for the deposition of a monolayer affects the value of the surface pressure and at the surface concentration $(0.05–0.16) \times 10^{-3}$ g/m^2 the **PoMA** can form the equilibrium monolayers. An increasing of the concentration of the polymer at the interface leads to the formation of non-equilibrium monolayer. It was determined for the first time that at the monolayer of **PoMA** application the memory of polymer is exhibited.

4.5 *LANGMUIR* FILMS OF POLYANILINES COMPOSITIONS WITH DIELECTRIC POLYMERS

An investigation of the surface properties of monomolecular layers of polymer blends is of particular interest. Firstly, in this case, there are changes in the structure of monomolecular layers are possible due to the mutual influence of the components on the packaging of the macromolecules. Secondly, in the most cases, at the mixing of solutions of different polymers in the same solvent, there is a problem of the compatibility of components. Incompatibility as same as compatibility has an impact on the structural, mechanical, electrical, and optical properties of the monolayer films.[24]

An investigation of the surface films on a basis of binary systems of **ECPs** and non-conductive matrices[1,2,13] call an especial interest among scientists. The importance of these studies is due to the fact that the introduction of conjugated polymers in the matrix of macromolecular compounds (**PVA**, **PMMA**, and other polymers) gives the possibility to obtain the flexible polymer films with interesting electrical and sensory properties.[14,15]

An investigation of the conditions of formation and properties of binary surface films on a basis of the dielectric polymer matrix, namely copolymer of styrene with maleic anhydride (**StMA**) and conjugated **PAAr**, in particular, **PAn** and **PoT** has been done.[13,25]

StMA was synthesized by usual method of radical copolymerization of equimolar mixture of styrene and maleic anhydride in benzene solution at heating with the use of benzoyl peroxide as the initiator.[26] Molecular weight of the **StMA** defined by viscometry was 5.6×10^4.

The volatile tetrahydrofuran was selected as a solvent in which not only **StMA**, but also the basic forms of **PAAr** are dissolved in order to obtain the *Langmuir* films of the individual polymers and their blends. Obtaining of the monomolecular films on the water surface was carried out both from the individual and mixed polymer solutions in tetrahydrofuran at temperature of 293 ± 0.5 K.

An amount of the substance containing in a drop of the solution applied to the surface of water was used for the calculations. The starting concentrations of polymers (per 100 ml of solution) were as follows: for **StMA** was 0.21 g; for **PAn** was 0.02 g; for **PoT** was 0.06 g. A quantity of the substance in monolayer was 0.0269×10^{-2} g (**StMA**); 0.0026×10^{-2} g (**PAn**) and 0.0090×10^{-2} g (**PoT**).

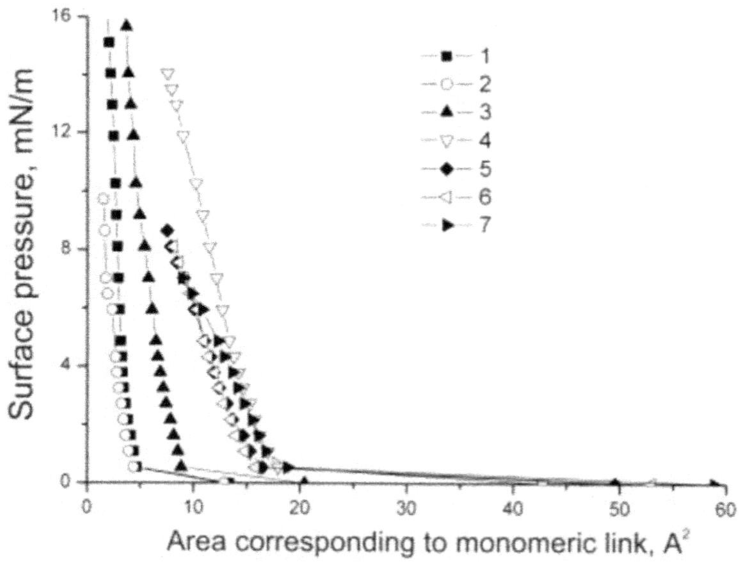

FIGURE 4.14 Surface pressure isotherms of **StMA–PAn** blends at different ratios deposited from the solutions in tetrahydrofuran. The relative content of styromal in the blend of polymers **StMA–PAn**, %: 1 – 100.0; 2 – 99.5; 3 – 91.2; 4 – 53.5; 5 – 35.3; 6 – 24.5; 7 – 0.0.

The surface pressure isotherms of individual polymers and mixtures of the investigated substances by different ratio formed from the solutions in tetrahydrofuran are represented in Figures 4.14 and 4.15.

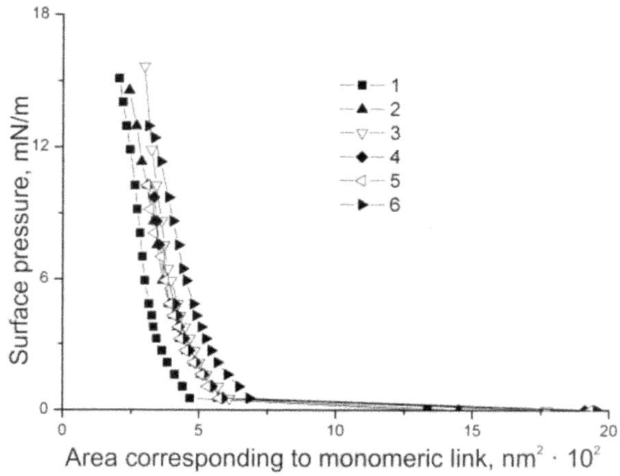

FIGURE 4.15 Surface pressure isotherms of **StMA–PoT** blends at different ratios deposited from the solutions in tetrahydrofuran. The relative content of styromal in the blend of polymers **StMA–PoT**, %: 1—100.0; 2—98.1; 3—89.2; 4—47.8; 5—12.7; 6—0.0.

StMA copolymer is characterized by appreciable surface activity due to the presence in its structure of the maleic anhydride functional groups.[26] According to the obtained results, the molecules of **StMA** form a quite stable monolayer, and the area per one elementary link **StMA** is 4.15 Å^2, which gives the grounds to assume that the molecules of the styromal in the resulting monolayer oriented almost vertically.

Obviously, the compression of styromal monolayer, a placement of monomer links in the surface layer, and the presence of cohesive forces between them, leading to the formation of a monolayer by complicated structure, and thus there is crowding them out of the surface, which leads to their vertical orientation relative to the surface. It should be noted that the molecule of styromal contains the phenyl group, the presence of which adversely affects the spreading of the surface film, which can be explained via the electron interaction of aromatic rings, resulting in the relationship to the radical electron pairs decreases, making it difficult to solvation.

Surface pressure isotherms of monolayers of conjugated **PAAr** formed from the individual solutions of **PAn** and **PoT** in tetrahydrofuran, provide an opportunity to identify an area that accounts for monomer link for **PAn**

$S = 22$ Å^2, which is in good agreement with *ref.*[10,13] Taking into account[10] that the cross-sectional area of the molecule of aniline on the boundary water–air consists of 20 Å^2 we can suggest that the molecule of **PAn** is located in the monolayer parallel to the surface, unlike to molecule of **PoT** ($S = 8$ Å^2), oriented, probably perpendicular or at an angle to the surface.

TABLE 4.3 The Magnitudes of Areas in Calculation per Monomeric Link for Monolayers of Binary Blends of **StMA–PAn**.

Content of StMA, %	**Area corresponding to monomeric link**		
	S_{eff}, Å^2	S_A, Å^2	ΔS, Å^2
100.0	4.15	4.15	0.0
98.3	4.75	4.45	–0.30
91.2	8.5	5.72	–2.78
77.5	10.2	8.17	–2.03
64.5	12.5	10.49	–2.01
53.5	16.25	12.45	–3.80
35.3	14.0	15.69	1.69
24.5	15.5	17.63	2.13
9.5	16.0	20.3	4.30
0.0	22.0	22.0	0.0

As shown on the figures, the nature of the surface pressure isotherms is largely dependent on the ratio of polymers in solution. This fact may be related to the nature of the interaction of polymers of different nature in the monolayer. The interaction between the components leads to the values deviations of areas corresponding to the monomer links in mixed monolayers (S_{eff}) from the additivity (S_A). Such deviations are connected with compatibility of polymers in two-dimensional state,[1,13,20] which refers to the impact of components on the macromolecules packaging. Both incompatibility and compatibility of polymers affects the structural, mechanical, electrical, and optical properties of monolayer films.[24]

The dependencies of an area of monomer link (S_{eff}) on the composition of the binary mixture at certain values of surface pressure (F) were determined on the basis of surface pressure isotherms. The magnitudes of areas extrapolated to zero value F and their deviation from the additive values are given in Tables 4.3 and 4.4.

TABLE 4.4 The Magnitudes of Areas in Calculation per Monomeric Link for Monolayers of Binary Blends of **StMA–PoT**.

Content of StMA, %	Area corresponding to monomeric link		
	$S_{eff.}$, Å^2	S_A, Å^2	ΔS, Å^2
100.0	4.15	4.15	0.0
94.0	4.0	4.38	0.38
89.2	4.5	4.56	0.06
73.5	6.25	5.17	–1.08
47.8	4.9	6.16	1.26
32.8	5.1	6.74	1.64
23.5	5.25	7.10	1.85
17.2	5.0	7.33	2.33
12.7	5.0	7.51	2.51
7.9	5.0	7.70	2.70
0.0	8.0	8.0	0.0

As can be seen from the data presented in tables, both an increase and the decrease of S in comparison with the total value of the areas corresponding to the individual components are observed depending on the ratio of the components in the binary mixture. In particular, at the content of **PAn** in matrix of **StMA** (up 46.5%), a decrease of $S_{eff.}$ is observed, whereas at higher content an increase of its value takes place in comparison with the total area of the individual components (S_A).

Mainly, an increase of $S_{eff.}$ is observed for **StMA–PoT** monolayers while a positive deviation is increased with the concentration of the **ECP** in the monolayer. Probably an increasing of the $S_{eff.}$ is associated with the repulsion of macromolecular coils[24] and its reduction indicates the presence of the attractive interaction.[27] **PAAr** molecules contain the amine groups in its structure, which leads to the possibility of their interaction with the acidic groups of maleic anhydride, which is part of **StMA**. This leads to a significant restriction of the conformational mobility of macromolecules and, therefore, to reducing of the area of the molecule in monolayer.[27]

Thus, an investigation of the *Langmuir* films of polymer blends can be informative at choosing conditions of obtaining the composite layers of **ECPs** with dielectric polymer matrices. Probably the formation of ordered monolayers from the binary mixtures of polymers is possible at the components ratio, to which the negative deviations of areas from the additivity correspond.

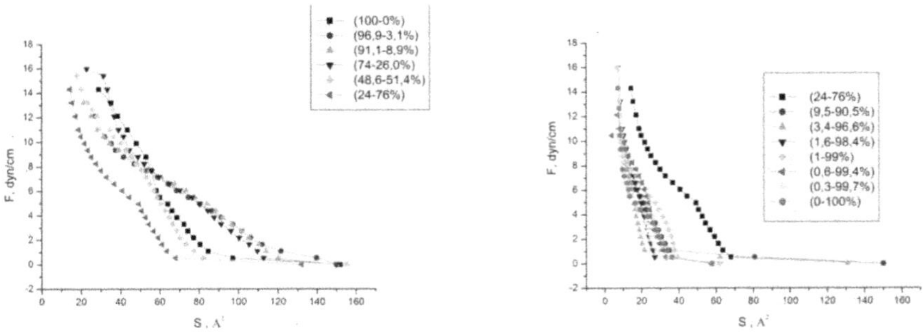

(a) (b)

FIGURE 4.16 The surface pressure isotherms of blends **PAn–PMMA** at different ratios deposited from the solutions in tetrahydrofuran

An investigation of the monomolecular films of **PAn** and **PMMA** as well as their blends on the boundary water–air has been also investigated. Obtaining of the monomolecular films on the surface of water was performed both from the individual and mixed polymer solutions in tetrahydrofuran and chloroform.

The surface pressure isotherms of polymer mixtures **PAn–PMMA** at different ratio in tetrahydrofuran and chloroform are represented in Figures 4.16 and 4.17.

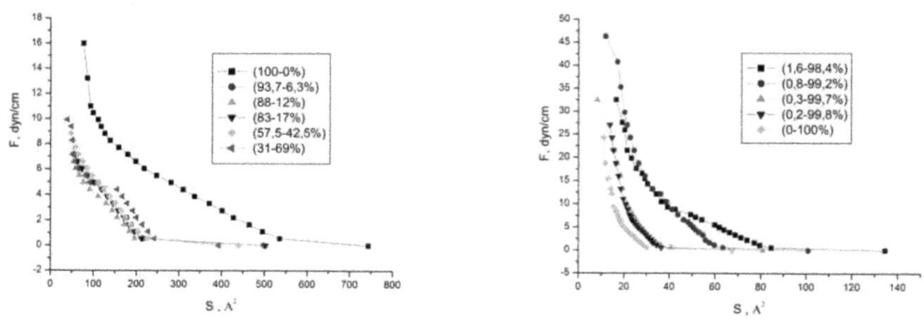

(a) (b)

FIGURE 4.17 The surface pressure isotherms of blends **PAn–PMMA** at different ratios deposited from the solutions in chloroform

As shown on the figures, the nature of the surface pressure isotherms is largely dependent on the ratio of polymers in solution.

The diagrams of the dependence of monomer link area at different surface pressures on composition of the binary mixture of **PAn–PMMA**[13] were constructed on the basis of surface pressure isotherms of monomolecular films deposited from the solutions in chloroform and tetrahydrofuran (Fig. 4.18). At this, it is shown that on the diagrams of the dependence of areas corresponding to the averaged value of the monomer link (S_0) on the composition of mixture, both an increase and the decrease of the S_0 are observed in comparison with the total values of the areas. In particular, at low content of **PMMA** in matrix of **PAn** is an increase of S_0, while at higher content of **PMMA**, a decrease of S_0 is observed. At this, the positive deviations indicate an incompatibility of the components in the system. The presence of negative deviations on the diagram of the system state is due to the compatibility of components, which leads to the formation of a denser packing of macromolecules in the surface layer.

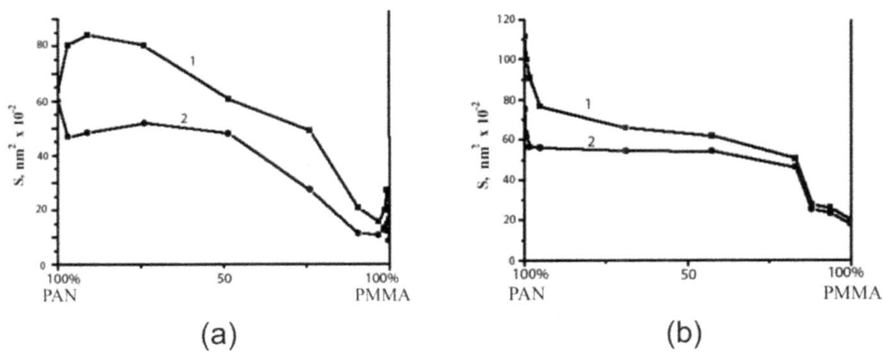

FIGURE 4.18 The diagrams of the generalized areas of monomer link at different surface pressure on composition of polymers blends for the system **PAn–PMMA**. (a) The solvent is tetrahydrofuran: 1) $F = 4.9$ mN/m, 2) $F = 8.2$ mN/m. (b) The solvent is chloroform: 1) $F = 4.9$ mN/m, 2) $F = 6.6$ mN/m

For monolayers of **StMA–PoT**, mainly an increase of $S_{eff.}$ is observed while a positive deviation increases with the concentration of the **ECP** in the monolayer.

Thus, an investigation of the *Langmuir* films of polymer blends can be informative at choosing of the conditions of obtaining of the composite layers of **ECPs** with dielectric polymer matrices. Probably the formation of ordered monolayers from the binary mixtures of polymers is possible at the components ratio, to which the negative deviations of areas from the additivity correspond.

The results obtained under investigation of monomolecular films of **PAr** can be used at the formation of ordered nanometer thick polymer films on the surfaces by different nature, in particular both on liquid and solid surfaces.

Thus, in a result of the carried out investigation of the monolayers of **PAn** and its derivatives (**PoT** and **PoMA**) it was shown that the initial surface concentration of the substance in monolayer has an effect on the magnitude of surface pressure of monolayers of the investigated polymers.

The decrease of surface concentration of **PAAr** leads to increase of the value of surface pressure and the area that corresponds to monomeric link, which is associated with a decrease of the intermolecular interactions in monolayer. At this, the nature of a solvent used for the deposition of a monolayer affects the magnitude of the surface pressure.

It was shown that the **PoMA** can form the equilibrium monolayers. In a case of **PoT** the non-equilibrium monolayer is formed. At applying of the monolayers of investigated **PAAr** the "memory" effect of polymer is evinced.

KEYWORDS

- **monolayer**
- **polyaniline, poly-*o*-methoxyaniline**
- **poly-*o*-toluidine**
- **surface pressure**
- **an area of monomeric link**
- **an equilibrium**
- **polymer "memory" effect**
- ***Langmuir* films**

REFERENCES

1. Arslanov, V. V. Polymer Monolayers and Langmuir–Blodgett Films. *Polythiophenes. Russ. Chem. Rev.* **2000,** *69* (10), 883–898.

2. Bezkrovnaya, O. N.; Mchedlov–Petrosyan, N. O.; Vodolazkaya, N. A. et al. Polymeric Langmuir–Blodgett Films Containing Xanthene Dyes. *Russ. J. Appl. Chem.* **2008,** *81* (4), 696–703.
3. Zhavnerko, G. K.; Agabekov, V. E.; Marchik, N. A. The Langmuir–Blodgett Films of Amphiphilic Polymers. Reports of Belarusian Seminar of Scanning Probe Microscopy. Minsk, Belarus, 8–10 October, 2008; 43–46 (in Russian).
4. Adamson, A. *Physical Chemistry of the Surfaces*. Mir: Moscow, 1979 (in Russian).
5. Crisp, D. J. Surface Films of Polymers. Part I. Films of the Fluid Type. *J. Colloid. Sci.* **1946,** *1,* 49–70.
6. Pockels, A. Surface Tension. *Nature.* **1891,** *43,* 437–439.
7. Langmuir, I. The Constitution and Fundamental Properties of Solidus and Lignids. III. Lignids. *J. Am. Chem. Soc.* **1917,** *39,* 1848–1906.
8. Faynerman, A. E.; Lipatov, Y. S. Automatic Device for Recording the Surface-Area Pressure Isotherms. *Rus. J. Phys. Chem.* **1967,** *41* (4), 933–935 (in Russian).
9. Blodgett, K. B. Use of Interface to Extinguish Reflection of Light from Glass. *Phys. Rev.* **1939,** *55,* 391–395.
10. Ram, M. K.; Adami, M.; Sartore, M. et al. Comparative Studies on Langmuir–Schaefer Films of Polyanilines. *Synthetic Met.* **1999,** *100,* 249–259.
11. Congales, D.; Bulhoes, L. O. S.; Mello, S. V. et al. Electroactivity in Poly(*o*–alkoxyaniline) Langmuir–Blodgett Films. *Thin Solid Films.* **1994,** *243,* 544–546.
12. Opaynych, I.; Ukrainets, A.; Aksimentyeva, O. Langmuir Films of Poly(*o*–toluidine) at the Water–Air Interface. *Adsorpt. Sci. Technol.* **2007,** *25* (1), 15–21.
13. Opaynych, I.; Aksimentyeva, O.; Ukrainets, A. Investigations of Surface Films of Polymer Composites Based on Polymethylmethacrylate and Polyaniline on the Water–Air Interface. *Visnyk Lviv Univ., Ser. Khim.* **2012,** *53,* 321–325 (in Ukrainian).
14. Aksimentyeva, O. I.; Konopelnyk, O. I.; Tsizh, B. R. et al. Flexible Elements of Optical Sensors Based on Conjugated Polymer Systems. *Sensor Electronics and Microsystem Technologies.* **2011,** *2* (8), 34–39 (in Ukrainian).
15. Aksimentyeva, O.; Konopelnyk, O.; Opaynych, I. et al. Interaction of Components and Conductivity in Polyaniline– Polymethylmethacrylate Nanocomposites. *Rev. Adv. Mater. Sci.* **2010,** *23,* 30–34.
16. Opaynych, I. E. Studies of the Formation and Structure of Polyacrylate Thin Films. PhD Thesis, L'viv National University named after Ivan Franko, 1983 (in Russian).
17. Macinnes, D.; Funt, L. B. Poly-o-methoxyaniline: A New Soluble Conducting Polymer. *Synth. Metals.* **1988,** *25* (3), 235–242.
18. Faria, R. M.; Mattoso, L. H. C.; Ferreira, M.; Oliversa, O. N. Jr. Chloroform–Soluble Poly(*o*–methoxyaniline) for Ultra–Thin Film Fabrication. *Thin Solid Films.* **1992,** *221,* 5–8.
19. Aksimentyeva, O. I. Electrochemical Methods of Synthesis and Conductivity of Conjugated Polymers; Svit: L'viv, 1998 (in Ukrainian).
20. Dabke, R. B.; Dhanabalan, A.; Major, S.; Talwar, S. S.; Lal, R.; Contractor, A. Q. Electrochemistry of Polyaniline Langmuir–Blodgett Films. *Thin Solid Films.* **1998,** *335,* 203–208.
21. Opaynych, I.; Aksimentyeva, O.; Sozanskyy, Y.; Maleev, I. Investigation of Surface Pressure of Poly-o-methoxyaniline Monomolecular Films. Bulletin of Shevchenko Scientific Society. *Chem.* **2008,** *21,* 123–132 (in Ukrainian).
22. Opaynych, I. Ye.; Ukrainets, A. M.; Arsimentyeva, O. I. Monolayer Films of Polyaminoarenes on Water/Air Interface. Reports of XI Polish–Ukrainian Symposium on

Theoretical and Experimental Studies of Interfacial Phenomena and their Technological Application. Zamosc – Krasnobrod, Poland, August 22–26, 2007; pp 105.
23. Fridrichsberg, D. A. The Colloid Chemistry. Chemistry: Leningrad, 1984 (in Russian).
24. Lipatov, Y. S. *Interfacial Phenomena in Polymers*. Naukova dumka: Kiev, 1989 (in Russian).
25. Riy, U.; Dutka, V.; Arsimentyeva, O. Synthesis and Electrical Properties of the Styromal–poly-o-toluidine Composites. Reports Int. Conf. Chemistry and Technology of Polymers, L'viv, Ukraine, November 24–26, 2011; pp 14–15 (in Ukrainian).
26. Tiurina, T. G.; Zaitseva, V. V.; Bulavin, A. V.; Gaynulina, M. R. Synthesis and Etherification of Styrol–Maleic Anhydride Copolymers. *Ch&ChT*. **2008,** *2,* 78–81 (in Russian).
27. Zaytsev, S. Y.; Dzekhtser, S. V.; Zubov, V. P. Polymeric Monolayer with Immobilized Bacteriorhodopsin. *Russ. J. Bioorganic Chem.* **1988,** *14* (6), 850–852 (in Russian).

CHAPTER 5

EPOXY–POLYANILINE COMPOSITES: SYNTHESIS, STRUCTURE, PROPERTIES

V. P. ZAKORDONSKIY, O. I. AKSIMENTYEVA, and A. I. KRUPAK

Department of Physical and Colloid Chemistry, Faculty of Chemistry, Ivan Franko National University of Lviv, 6 Kyryla & Mefodia Str., Lviv 79005, Ukraine

E-mail: vzakordonskiy@ukr.net

CONTENTS

ABSTRACT

It has been studied the mechanism of formation and physical chemical properties of epoxy–polyanilines composites obtained using both the curing agent and also the electroconductive filler of polyaniline (**PAn**) doped by tetrafluoroborate acid (**PAn–BF$_3$**). The factors having an impact on curing mechanism and properties of composites were analyzed. The activation and thermochemical parameters of a process depending on the content of **PAn–BF$_3$** and its doping level have been determined on a basis of data of the thermogravimetric analysis. It was shown that the combination of properties both of curing agent and electroconductive component into one complex **PAn–BF$_3$** gives the possibility essentially to simplify the conductive epoxy polymer obtaining, to receive the polymer-polymeric composites characterizing by electric conductivity 10^{-6}–10^{-4} S/cm at relatively low content of **PAn–BF$_3$** (10–15% mass); at the same time, such composites are characterized by high thermal, physical, and mechanical properties which makes it possible to use them for producing of antistatic coatings, electroconductive adhesives, electrodes, and so forth.

5.1 INTRODUCTION

Actual problem of modern technologies making for progress in many fields of science and technique is the creation of new generation effective composite materials combining the properties of metals or semiconductors characterizing by lightness, flexibility, technological effectiveness of polymers (process ability); at the same time, such materials should be able to perform the various functions, namely they should possess by electrical conductivity, optical and electromagnetic absorption, mechanical and chemical resistance for their using at anticorrosive and antiradar protection, under creation of antistatic coatings, electroconductive components, adhesives, conductive glue compositions, and so forth. This question is related to one among topical problems today, which is concerned to reducing of consumption or to replacement of the traditional metallic and semiconducting materials on the organic ones, since the increasing scales of the metals using cause the serious problems in terms of natural resources limitation and the high costs for metal mining and processing.

Solving the problem concerning to the creation of completely organic polymeric materials (polymer-polymeric composites) having own electronic conductivity has been made possible thanks to modern developments both in the field of nanotechnology[1–2] and in polymer chemistry, in particular due to the synthesis of conducting polymers (intrinsically conducting polymers, **ICPs**) and composites on their basis.[3–18]

The revelation in the seventies of the polymers having the self-conductivity called an especial considerable interest in the investigation and practical application of these materials.[3–5] Among the broad class of self-conductivity polymers the most promising are polymers whose molecular structure includes a system of conjugated -bonds, which causes an electric charge delocalization along the polymer chain and creates the conditions for the formation of the conductive chains in bulk polymer. Electrical conductivity of **ICPs** is metallic in nature, and that is why they often are called as the "synthetic metals." The simplest among the all **ICPs** is the polyacetylene $(–CH=CH–)_n$, which was synthesized by *Natta* in 1958 for the first time by polymerization of acetylene in hexane under the action of $Al(Et)_3/Ti(OPr)_4$.

The real "boom" in a field of the synthesis and the investigation of electroconductive polymers is observed in the seventies of the twentieth century, when *Hideki Shirikawa* with a group of young scientists in Tokyo Institute of Technology have developed a simple method of the polyacetylene synthesis and reported about high electrical conductivity of its oxidized iodine-doped form.[6]

Alan Heeger, *Alan MacDiarmid,* and *Hideki Shirakawa* were awarded by the Nobel Prize in 2000 for the discovery of electroconductive polymers, an investigation of structure and mechanism of their conductivity in nomination "for the discovery and development of conducting polymers."[7] In addition to polyacetylene the **PAn**, polypyrrol (**PPy**), polythiophene (**PT**), poly-*p*-phenylene (**PpPh**), poly-*p*-phenylene–vinylene (**PpPhV**), and some others having the system of conjugated π-bonds also belong to the electroconductive polymers.

PAn and its derivatives[4,8 9] are especially promising from the point of view of the complex of physical and chemical properties of practical application. The simplicity of synthesis, low cost, sufficiently high electroconductivity, and high environmental stability make this polymer very promising for its practical using in electronic devices, sensors, anticorrosion coatings, at the protection from the electromagnetic radiation, in secondary power sources and so on. At the same time, **PAn** is characterized by the stiffness of the structure, which makes it practically insoluble in the most solvents and unable to transform into the fusible state. **PAn** is restrictedly soluble in strongly polar solvents, such as *N*-methylpyrrolidone, dimethylsulfoxide (**DMSO**), *N,N*-dimethylformamide, and it is decomposed under heating till the temperatures neared to the melting point. This creates a certain technological limitations at using of this polymer.

The structure of **PAn** consisting of the blocs of *N*-phenyl-*p*-phenylenediamine and quinonediimine can be represented by a following scheme:

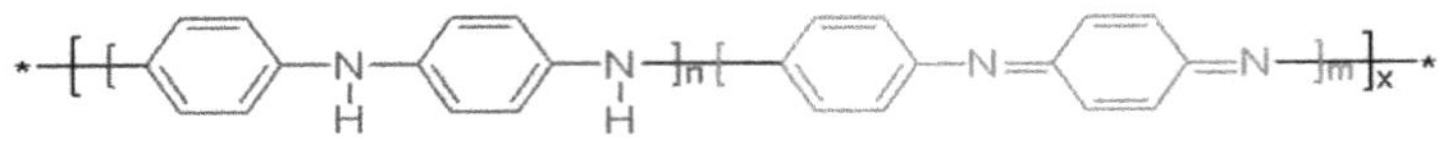

PAn can exist in three forms depending on the oxidation degree, namely: wholly reduced form as leucoemeraldine ($n = 1$; $m = 0$), wholly oxidized form as pernigraniline ($n = 0$; $m = 1$) and semi-oxidized/semi-reduced form as emeraldine base (**EB**) ($n = 0.5$; $m = 0.5$). It provides wide ranges for the regulation of electrical, physical, and chemical properties of **PAn** and composites on its basis. Transition from the one oxidation form into another one represents by itself a simple oxidation–reduction reaction.

PAn as well as other **ICPs** is easily chemically doped. This gives the possibilities for **PAn** functionalization.[3]

H-doping (or protonation) of **PAn** giving the possibility to sharply improve the electroconductivity of the polymer calls an especial theoretical and practical interest. *H*-doped form, namely emeraldine salt (**ES**) is the most electroconductive form among the all forms of the **PAn**. The strong inorganic acids usually are applied as the dopanes of **Pan.**[4–5,10–12] At the same time, a great attention is paid into the using of the strong sulfonic organic acids, in particular of dodecylbenzenesulfonic acid (**DBSA**), camphorsulfonic acid (**CSA**)[5,8] and some others applied as the effective modifiers of **PAn**. At that, it was determined, that the solubility of **PAn** in organic solvents is improved.[6] It is reported[13] that water soluble form of **PAn** was obtained at using of polystyrenesulfonic acid (**PSSA**) as dopane.

Electroconductivity of *H*-doped form of **PAn** depends on the molecular weight, crystallinity, oxidation degree, doping level, and nature of the acid and is varied[8] quite widely—from 0.001 to 400 S/cm. Leucoemeraldine and pernigraniline are non-conducting forms of **PAn**, and their electroconductivity consists of 10^{-14}–10^{-16} S/cm.

The most promising direction of **PAn** using today is the obtaining of compositions based on non-conductive polymer as a matrix with dispersed therein **ICPs** polymer. In this case, the properties of such composite, including the electroconductivity, will be determined by the ability to achieve some critical volume concentration of **PAn** at which the self-organized formation by the spatial structure will be provided. A broad range of thermoplastic and thermoreactive polymers is used as the polymer matrixes.[14–17] Relevant in this regard is the obtaining of composites based on **PAn** and epoxy polymers.[18]

Epoxy polymers are characterized by unique set of physical, mechanical, electrical, optical, and adhesive properties and are represented by themselves the one among the most effective classes of constructive materials by organic nature today. They are characterized by high stability and resistance to environmental factors, by simplicity of the synthesis and modification.[19–21] Epoxy polymers have a much wide application as protective coatings, high-performance constructive materials, adhesives, components of electronic schemes and so on. However, providing of the epoxy materials by conductivity is realized predominantly due to the introduction of metallic or carbon filler (silver, aluminum, and copper) by their dispersion in epoxy matrix.[21] In addition to the complex and energy expense technology of the composites formation as well as high cost of the most conductive fillers, the high specific mass of metal fillers causes the separation of compositions. The solution of this problem is possible when the replacement of the metal filler on the polymer one will be take place. For this purpose, the **PAn** is proposed.

Electroconductive composite based on epoxy resin (**ER**) and **PAn** was obtained by Kathirgamanathan[22] for the first time via mechanical mixing of doped by toluene sulfonic acid (**TSA**) **PAn** and epoxy oligomer with the following curing by tetrahydrophthalic anhydride.

At present time, a number of preparative methods giving the possibility to obtain a composite in a form of **ICPs**-polymer dispersed in a matrix of epoxy polymer have been proposed. Some of them are listed below.

1. The **method of direct dispersion of the prepared powdered ICPs in a matrix of non-conductive polymer**. This method was implemented and described in 1984 for the first time.[4–5,23–25]
2. **Synthesis of ICPs *"in situ"* in the medium of epoxy oligomer solution in a suitable solvent** with the following curing of the oligomeric composition.[26–27] This method makes it possible to provide a better compatibility of **PAn** with epoxy polymer and a more uniform distribution of particles in the matrix in comparison with the methods of the mechanical mixing. Using of the emulsion polymerization method is one among possible ways of the **ICPs** polymerization "*in situ.*"[10,28–29] In this case, **ICPs** are formed as colloids in aqueous dispersion. Since water emulsion is incompatible with the epoxy oligomer, it creates some difficulties for the implementation of following stages of the polymer composite formation.

3. **Dispersion of powdered ICPs in organic solvent by ultrasound and mixing of the obtained dispersion with epoxy oligomer.**[30–35] Obtained polymer composite after removing of the solvent is characterized by a uniform distribution of the dispersed phase, and is considered a quasi-equilibrium two-phase system unlike to non-equilibrium two-phase system obtained by direct **ICPs** dispersion in a polymer matrix. This method is often considered in terms of obtaining of the "soluble" form of **ICPs**,[4,12,23] since the size of the particles of **ICPs**-polymer consist of 50–100 nm making them visible with the use of the optical microscope.

Polymer composites based on **PAn** and epoxy polymers exhibit a set of unique properties that make them promising and competitive materials in specific areas of the application use. The main areas of practical application of the composites based on epoxy polymer and **PAn** as well as its derivatives are:

1. **Transparent antistatic coatings.**[4,36–37] Typically, transparent antistatic coatings are obtained by vacuum metallization or by applying of electroconductive coating on polymer film from aqueous medium. These methods are energy-expensive and environmentally harmful. Coatings are characterized by low resistance to moisture and other environmental factors and have limited lifetime. Using of **ICPs**-polymers allows to obtaining the high-quality transparent antistatic coating, resistant to environmental factors. Depending on the nature and thickness of the film, the conductivity of film consists of $10^{-9} \div 10^{-1}$ S/cm, and the transparency of the film is 95%.
2. **Corrosion protection.**[10–11,18,31,35,38–40] Corrosion protection of metal surfaces today is regarded as one of the most promising directions for **ICPs** and **ICPs**/polymer composites using. These studies[4] were initiated in the ninetieth of the nineteenth century and now have a much practical applications. Based on the chemical, physical studies, and theoretical calculations it was shown that **ICPs**/polymer composites exhibit the high anticorrosion ability and stability for a long term and may serve as a highly effective anticorrosive means for corrosion protection of metal constructions and coatings from the effects of the environment.
3. **Using of the epoxy polymer/PAn composites as electroconductive adhesives, coatings, electronic devices**, and so forth.[15–17,22,26–27]

4. **Using of the ICPs-polymer composites as materials for protection against electromagnetic fields.**[27,34,41–43]

As we can see, the practical use of **PAn** and composites on its basis primarily is due to the electroconductive nature of these materials. In turn, the size and the shape of polymer particles have a significant effect on value of the specific conductivity of the obtained composites,[42,44] and the introduction of nanowires or nanofibrils,[45–48] asymmetric nanoparticles of fibrillar or lamellar structure[32,43,49–50] makes it possible to provide the necessary level of electroconductivity and promotes the formation of spatial electroconductive net at minimum content of the electroconductive component in the composition.

A critical parameter that determines the electrical properties of the composite is nature of the curing agent of **ER**. The use of aliphatic amines, which are widely used as a low-temperature curing agents of ERs are undesirable because their using provokes the deprotonation process and results in lowering of the electrical conductivity of the composite.[16,26] At the same time, the amines by polymer nature (polyamides)[11,31,38] and polyoxypropyleneamine[16] are often used as the curing agents at using of epoxy polymer/**PAn** composites as anticorrosion coatings. This improves the solubility of **PAn** in the epoxy matrix resulting in the structure of the obtained composite is characterized by high homogeneity.

An application of curing agents by acidic nature permits to avoid the **PAn** deprotonation process. The anhydrides of dicarboxylic acids are typical acidic curing agents of ERs. An interaction of anhydride of carboxylic acid with the epoxy group occurs only at higher temperatures. Therefore, the crosslinking of ERs with the participation of anhydrides is carried out usually in the presence of accelerators (catalysts), which can be represented by the substances with proton-donor properties, namely alcohols, phenols, and carboxylic acids.[51–52] It is also possible the catalysis by metal salts of organic acids. But the most common is the use of tertiary amines.[53] The anhydrides of tetra- and hexahydrophthalic acids used as the acidic curing agent as well as the derivatives of imidazole and tertiary amines[22,24,27,34] used as the accelerator are often applied at the obtaining of epoxy–polyaniline composites.

In a number of publications it is reported about the using of Lewis acids in a form of catalytic complex of BF_3 with amines[23,25,30,32–34,44] as curing agent of epoxy–polyaniline compositions. BF_3-complex is related to the curing agents by acidic type. However, unlike to anhydrides of dicarboxylic acids, under the action of which the curing of epoxy oligomer is considered as the polycondensation process, the formation of epoxy polymeric net under the action of BF_3-complex is occurred in accordance with the mechanism of cationic polymerization type.[21] BF_3-complex is characterized by relatively high activity and in some cases the composites with a high level of conductivity and excellent physical and mechanical properties were obtained[23,32] with its participation. Although in general, the electroconductivity of epoxy–polyaniline composites obtained with the use of BF_3-complex as the curing agent is low and does not exceed 10^{-7} S/cm at relatively high (about 10% *by mass*) content of **PAn–DBSA.**[26,30,44] It is reported in references about the possibility of *H*-doped **PAn** in a form of ES using as curing agent.[15,37,45] The disadvantage here is the relatively low activity of used forms of **PAn** and the need of the epoxy matrix formation process carrying out at high temperatures, namely 180–200 °C and higher.

Given the foregoing, the purpose of this paper is to study the synthesis and properties of epoxy–polyaniline composite obtained with the use of **PAn** doped by tetrafluorborate acid applied as the curing agent.

5.2 EXPERIMENTAL PART

5.2.1 STARTING MATERIALS & REAGENTS

ER **ED–20** (*GOST* standard 10,587–84) represents by itself the product of polycondensation of epichlorohydrin and diphenylolpropane. The content of epoxy groups is 21.4%, the content of hydroxyl groups is 1.5 %, the molecular weight *Mn* is 380–420 g/mol.

Aniline $C_6H_5NH_2$ by high purity grade has been distilled twice in vacuum.
Sulfate acid H_2SO_4 has been used by chemical purity grade.
Hydrofluoric acid was used as 35% solution.
Boric acid H_3BO_3 has been used by high purity grade.
Ammonium persulfate $(NH_4)_2S_2O_8$ has been used by high purity grade.
Synthesis of HBF_4 has been carried out via the mixing of H_3BO_3 and HF under the ratio 1:4.

PAn was obtained via the method of oxidative polymerization of 0.2 M solution of aniline in 0.5 M of sulfate acid at 278 K. $(NH_4)_2S_2O_8$ has been used as the oxidizing agent. In order to transform the product into the base form, the obtained precipitate of **PAn** was washed by 5% solution of NH_4OH. In order to obtain the doped by tetrafluoroborate acid form of polyaniline **PAn–BF$_3$** the obtained emeraldine base was treated by aqueous solution of HBF_4 by different concentration (0.1–1.0 M) under permanent stirring. During the process of treatment, the precipitate of **PAn** acquired a dark green color, indicating the formation of ES. After 24 h, the product has been filtered, washed by distilled water, dried under the vacuum at 50 °C to the constant mass. As a result, the samples of **PAn–BF$_3$** with the doping level 30, 36, 44, and 48%mol were obtained.

Synthesized samples of **PAn–BF$_3$** were used for curing of the ER ED-20. The content of **PAn–BF$_3$** in the starting oligomeric composition was varied from 5 to 30%mass. Curing has been carried out in Teflon forms at 343 and 353 K under conditions (60 + 60 min).

5.2.2 RESEARCH METHODS

An investigation of the thermal stability of obtained epoxy–polyaniline composites was performed using the derivatograph *Q*-1500*D* (*Paulik–Paulik–Erdey system*, Hungary) in the temperature range 20–700 °C and heating rate of 10 K/min. Well-done Al_2O_3 was used as the standard.

Measurement of thermomechanical properties of the samples was performed by the method of uniaxial compression with the use of modified device designated for study of the heat resistance of polymer materials made by "*Heckert*" (DDR) firm in the temperature range 293–573 K at heating rate of 1.6 degree/min and load $1 \cdot 10^6$ N/m^2. Fixing of the sample temperature was performed using a standard L-type thermoelement. Deformation of sample in the form of cylinder by 8 × 10 mm size was measured[56] using a null indicator to within 0.01 mm. The surface microhardness of the samples was measured using the *Heppler* consistometer by the method of conical working device indentation at $T = 20$ °C.

Electroconductivity of polymer composites was measured by standard two-contact method[57] at temperature T = 20 °C using the automatic pulse ohmmeter *Щ*-306-1 (Russia) with a measuring range 10–10^9 Ω. The teraohmmeter *E*6–13*A* was used for high-resistance samples. Measurements of the electroconductivity were performed using the cylindrical samples of 2 × 10 mm size made by compressing of a hanging by 200 mg of powdered sample under the pressure of 150 atm. Life-time of the sample in the mold was 10 min.

5.3 RESULTS AND DISCUSSION

5.3.1 PHYSICAL AND CHEMICAL PROPERTIES OF PAn DOPED BY TETRAFLUOROBORATE ACID

At choosing of the type of curing agent for epoxy oligomers, we started from the fact that the inclusion of BF_4 fragments into the structure of **PAn** chain in the presence of a tertiary amine can be an effective factor of the epoxy group opening and of the epoxy matrix formation accordingly to cationic mechanism. Obviously, the use of tetrafluoroboric acid HBF_4 as doping component for **PAn** gives the possibility to obtain a product that will combine the features of both effective curing agent and electroconductive polymer filler of epoxy matrix. The studies, which have been carried by us, confirmed the legitimacy of this approach.[54–55]

To determine the doping degree of **PAn** doped by HBF_4, the method of thermogravimetric analysis was used.[54] The differential mass loss curves (**DTG**-curves) for samples of **PAn–BF$_3$** with different doping level (curves 1–3) are shown in Figure 5.1. For comparison, the **DTG**-curve for un-doped form of **PAn** (curve 4) was also suffered on the Figure 5.1. The nature of the obtained curves evidences the phasic nature of the thermal decomposition of **PAn–BF$_3$**. Two temperature ranges of the fast change of mass are legibly identified: first is a low temperature range with a maximum speed of a mass loss at 94–100 °C and the second is a temperature range with a maximum speed of a mass loss at 280–285 °C. **DTG**-maximum in the field of 94–100 °C, which is shown for the all investigated samples, is presumably due to the loss of physically bound water. **DTG**-maximum in the field of 280–285 °C corresponds to desorption of doping component in a form of HBF_4. Thermal destruction of the main chain of **PAn** is observed[58] at temperature higher than 300 °C.

The integrated mass loss curves (**TG**-curves) for synthesized samples of **PAn–BF$_3$** depending on temperature are shown in Figure 5.2. For comparison, the **TG**-curve for un-doped form of **PAn** (doping level is 0.0%mol) was also suffered on the Figure 5.2. It was determined the doping level (%mol) for investigated samples of **PAn–BF$_3$** as the ratio of HBF_4 moles linked with the polymer matrix to the number of the main moles of **PAn** on the basis of magnitude of mass loss by samples of **PAn–BF$_3$** in the temperature range 120–250 °C, in which the mass of the un-doped **PAn** in this temperature range has remained practically unchanged.

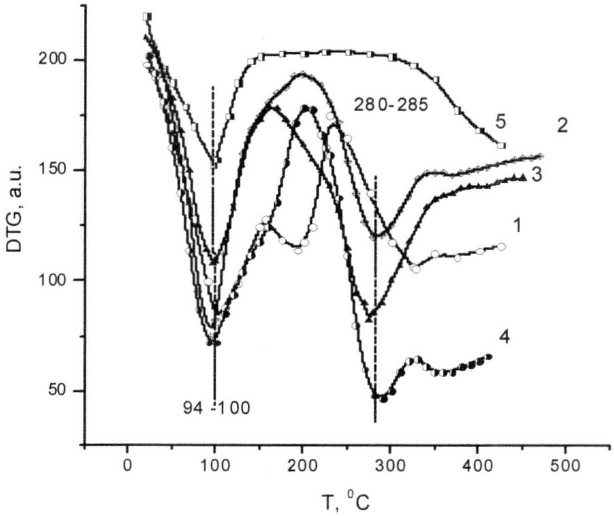

FIGURE 5.1 Differential **DTG**-curves of **PAn–BF$_3$** thermal decomposition at different doping level (%mol): 1–44; 2–36; 3–30; 4–0 (un-doped sample).

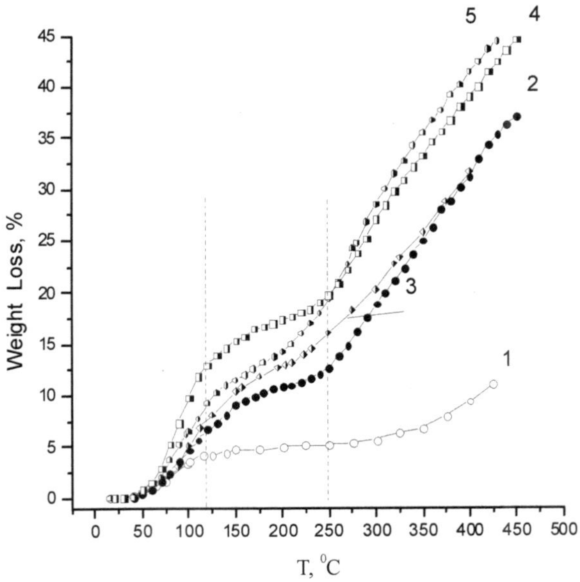

FIGURE 5.2 Integrated **TG**-curves of **PAn–BF$_3$** samples with different doping level (%mol): 1–0 (un-doped sample); 2–30; 3–36; 4–44; 5–48.

Mass loss of synthesized **PAn–BF$_3$** samples in the temperature range 120–250 °C was determined by **TG**-curves presented in Figure 5.2. **TG**-curve of un-doped sample (Fig. 5.2, curve 4), for which the sample weight in this temperature range remains constant was used as a standard. At temperatures corresponding to desorption of physically bound (adsorbed) water, the mass loss by un-doped sample of **PAn–BF$_3$** is 5% and increases till 10–15% with increasing of the doping level. This indicates that the hygroscopicity of the synthesized samples of **PAn–BF$_3$** significantly is increased with the doping level increasing.

Electroconductivity of synthesized samples of **PAn–BF$_3$** is represented in Table 5.1. Standard deviation of the electroconductivity for a series of five tableted samples by given doping level consists of 22–28%. This relatively low degree of the electroconductivity reproducibility is generally typical for the *H*-doped forms of **Pan.**[8] As noted in *ref*.,[12] the electroconductivity of **PAn** depends not only on the doping level and the nature of the doping component, but also to a large extent on the method of **PAn** synthesis and its morphology, pressing conditions (pressure, duration of exposure under pressure) at the manufacture of tableted samples and other factors. It would be true the statement[12] that "*there are as many* **PAn***s as number of people who prepare them*" or by other words "*there is so much* **PAn***s how many people got it.*" Comparison of our results with the reference data[15,23,44] shows that the electroconductivity of synthesized by us **PAn–BF$_3$** is of the same order as the electroconductivity of industrial samples of **PAn** doped by dodecyl benzoic acid, although yields to the maximum values of the electroconductivity of **PAn** doped by HCl synthesized under laboratory conditions in accordance with the "standard procedure."[12]

TABLE 5.1 Dependence of the **PAn–BF$_3$** Electroconductivity on the Doping Level.

Doping level, %mol	48	44	36	30
Electroconductivity, S/cm	0.72	0.28	0.21	0.23

Temporal stability of the electrical parameters of the synthesized samples of **PAn–BF$_3$** has been also studied. To do this, the samples were kept at room temperature and were subjected to normalization at temperature 20 °C for 24 h before measurement. Obtained results are shown in Figure 5.3.

Presented data show that the electroconductivity of the **PAn–BF$_3$** samples during the storage generally is decreased. Structural and morphological instability of the ES can be reason of this fact. In addition, taking into account the relatively high value of the saturated vapor pressure of

HF at room temperature (T_{boil} = 19.9 °C at P = 760 mmHg),[59] it should be not excluded the possibility of **PAn–BF$_3$** doping level lowering due to the desorption processes. The most notable decrease of the electroconductivity is observed for the most doped samples of **PAn–BF$_3$** (44–48%mol). Electroconductivity of the exposed for long periods of time (6–12 months) samples of **PAn–BF$_3$** regardless of the doping level is stabilized at 0.05–0.1 Sm/cm, indicating that the achievement of some equilibrium state and the stabilization of its structural-morphological and physical–chemical parameters take place during the storage of **PAn**.

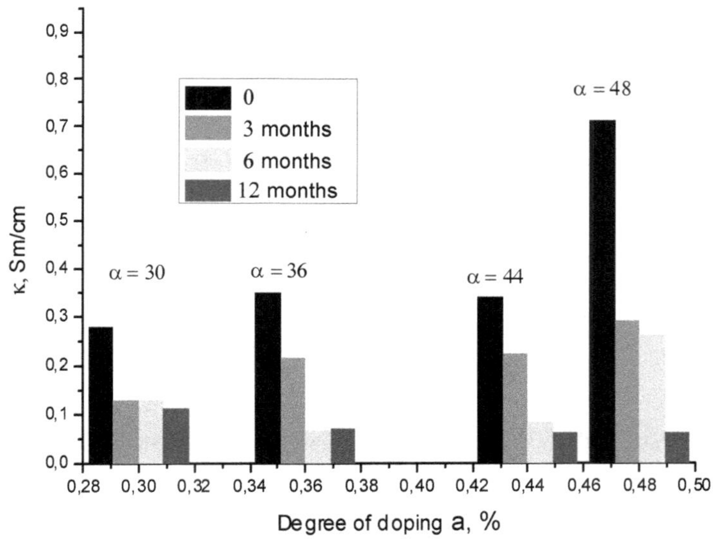

FIGURE 5.3 Block drawing of the **PAn–BF$_3$** electroconductivity depending on the doping level and the duration of storage.

5.3.2 REGULARITIES OF THE EPOXY–POLYANILINE COMPOSITE ED-20/PAn–BF3 FORMATION

As it was already noted, the chemical crosslinking of epoxy compounds and the formation of polymer spatial net can occur by polycondensation of polyfunctional reagent–curing agent (diamines, anhydrides of dicarboxylic acids) or by ionic polymerization of epoxy groups with the participation of the base catalysts (anionic polymerization) or with the use of the acid type catalyst (cationic polymerization).[21,51] The complexes of **BF$_3$** often are used as the catalysts of the acid type. The process of the epoxy group opening in this case is regarded as the acid-catalytic reaction, the active center of which is the complex, which is formed at the interaction of $\mathbf{BF_3}$ with the compounds containing of the electron-donor groups (R_2O, R_3N). It should be noted that the tertiary amine is the best electron-donor and can play a role both of the catalyst of a process and of the crosslinking agent. Taking into account the foregoing it is obvious that our complex **PAn**–$\mathbf{BF_3}$ consisting of the **PAn** and tetrafluoroborate acid will act as a catalyst for the acidic polymerization of ER **ED-20**. Despite the complicated mechanism of the epoxy compounds polymerization processes under the action of $\mathbf{BF_3}$-complex the formation of polymerizing network in this case can be represented by the following scheme.[21]

1. **Chain initiation**

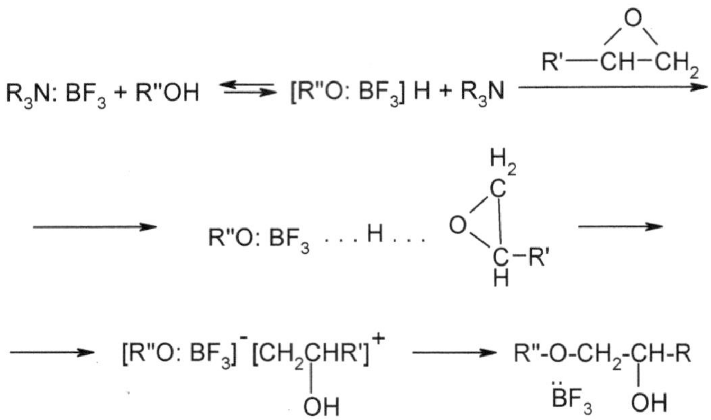

2. **Chain propagation**

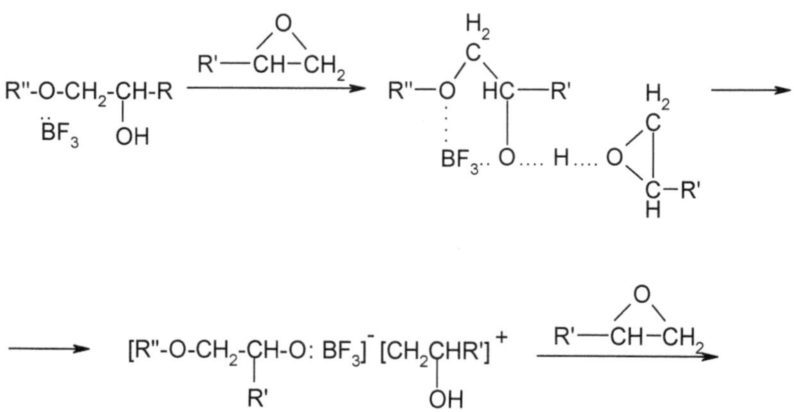

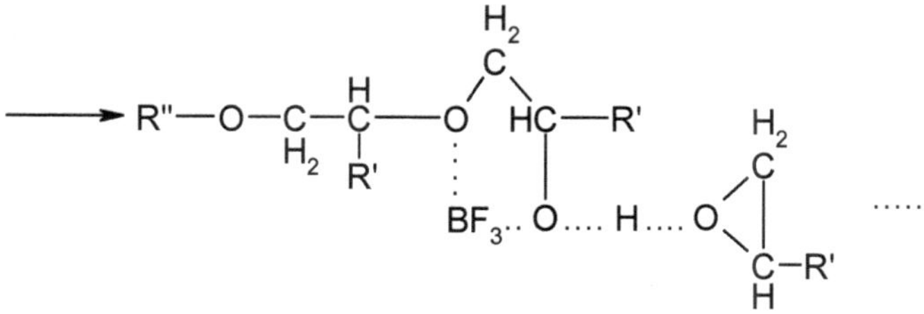

Chain mechanism is provided by intramolecular transfer of $\mathbf{BF_3}$ from the ether atom of oxygen to the oxygen atom of the hydroxyl group.

3. **Chain termination**

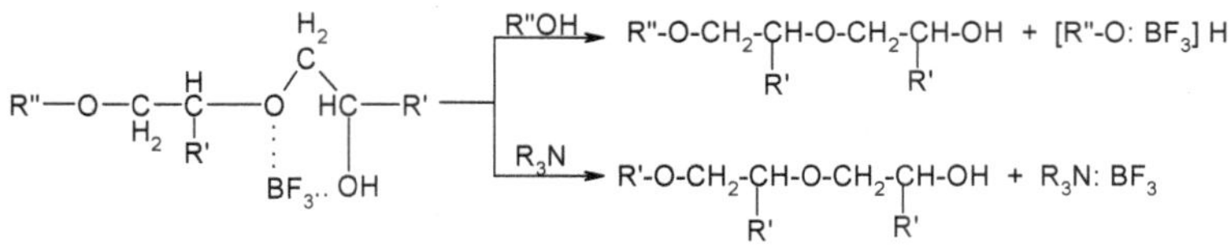

The curing process of the ER **ED-20** by complex $\mathbf{PAn–BF_3}$ was investigated by differential thermal analysis (**DTA**) method using the derivatograph *Q*-1500*D*.

The differential thermal flux curves (**DTA**-curves) of the **ED-20** curing process under the action of $\mathbf{PAn–BF_3}$ at different doping level are shown in Figure 5.4. The content of $\mathbf{PAn–BF_3}$ in composition was 10%mass. The presence of exo-maximum on **DTA** curves of the investigated systems indicates that the curing of *ED*-20 ER under the action of $\mathbf{PAn–BF_3}$ is accompanied by heat eliminating. The process of ER *ED*-20 curing by $\mathbf{PAn–BF_3}$ complex with the maximum doping level 48%mol. (Fig. 5.4, curve 1) proceeds at relatively low temperatures with two areas of heat eliminating.

The first **DTA**-maximum at 60 °C corresponds to the temperature range of the reaction with the participation of $\mathbf{BF_3}$-complex.[23] The second small maximum at 100 °C is peculiar to the hardening processes under the action of the amine curing agent.[60] It is likely that the polymer complex $\mathbf{PAn–BF_3}$ thanks to the presence of amine (**PAn**) and tetrafluoroborate groups ($\mathbf{BF_4^-}$) is the complex curing agent, because the tertiary nitrogen of **PAn** promotes the activation of epoxide ring and the tetrafluoroborate-anions are directly involved in the formation of the polymer matrix in accordance with the mechanism of cationic polymerization reaction. Due to this fact the curing process can take place at low temperatures.

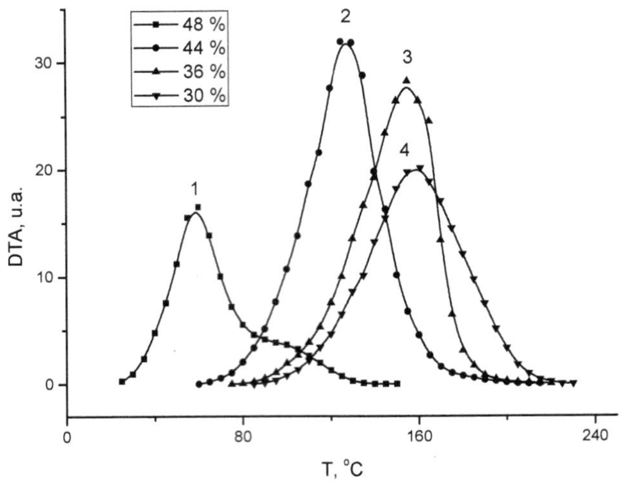

FIGURE 5.4 DTA-curves for compositions **ED-20/PAn–BF$_3$**. The doping level of **PAn–BF$_3$** (%mol): 1–48; 2–44; 3–36; 4–30. The content of **PAn–BF$_3$** in composition is 10%mass.

Lowering of the **PAn–BF$_3$** doping level leads to a shift of the maximum heat eliminating toward the higher temperatures (Fig. 5.4, curves 2–4), and as a result the curing processes with the participation of tetrafluoroborate and amine groups are overlapped and the asymmetry of **DTA**-peak is degenerated. The values of T_{max} (°C) for investigated compositions polymerization are given in the Table 5.2. The value of enthalpy ΔH_{cur} for **ED-20** curing under the action of **PAn–BF$_3$** was calculated by the area of **DTA**-peak, limited by **DTA**-curve and by the baseline. The values of ΔH_{cur} (kJ/mol *epoxy groups*) are shown in Table 5.2.

For comparison, in Table 5.2 there is also the value of heat effect of the oligomer epoxy groups opening under the action of curing agents by amine type, which is equal to 110–118 kJ/mol *epoxy groups* depending on the nature of the epoxide-amine system.[51] As we can see, in a case of the **PAn–BF$_3$** complex using as curing agent the process of the epoxy matrix formation is characterized by a significantly lower energy compared with the energy process when the conventional amine curing agents are used. A reason of this can be the difference in polycondensation (in a case of amine curing agent) and polymerization (in a case of **PAn–BF$_3$**) mechanism of epoxy-polymer net formation.[21]

On the other hand, a low value ΔH_{cur} = 39.4 kJ/mol *ep. gr.* observed for compositions obtained with the use of the **PAn–BF$_3$** complex with the maximal doping level (48%mol) can indicate a deep incompleteness of the epoxy-polymer matrix formation process and low level of the epoxy groups using. The reason of this obviously is the structural factor associated with a high rate of bottom-up construction of structuring degree and viscosity of a system already at the initial stages of the epoxy-polymer net formation. This leads to decreasing of the functional groups diffusive ability and transfers the process in the diffusion-controlled field, which causes a sudden braking response and consequently, the reducing of the epoxy groups' conversion degree.[61–62]

TABLE 5.2 Thermochemical and Activation Characteristics of **ED-20** Curing Depending on Doping Level and **PAn–BF_3** Content.

Doping level, %mol	Content of PAn–BF_3, %mass	T_{max}, °C	ΔH_{cur}, kJ/mol *ep. gr.*	E_{act}, kJ/mol *ep. gr.*
48	10	60	39.4	91.6
44	10	128	82.4	98.6
36	5	175		
36	10	158	75.9	100.5
36	15	156	73.0	
36	20	150	63	
30	10	161	67.3	108.8
ED-20/PEPA[*] (12%mass)[13]			110–118	

[*]**Note: PEPA** is polyethylene polyamine.

In *ref.*[45] it is shown that in a case of the use of **PAn** nanoparticles doped by HCl applied as the curing agent the maximal attained degree of epoxy groups conversion not exceeds 0.58–0.67 at the content of **PAn** in composition equal to 10–25%; at this, the value of the heat effect of curing consists of 75–104 kJ/mol *ep. gr.*

Increase of the **PAn–BF_3** content (doping level consists of 36%mol) in the starting oligomer composition leads to a slight shift of the maximum heat toward the lower temperatures without significant change in the nature of these curves (Fig. 5.5), and value of the heat effect ΔH_{cur} within an experimental error of ± 5 kJ/mol also remains practically the same (Table 5. 2).

With the use of obtained **DTA**-curves (Fig. 5.3) it were calculated the kinetically activated parameters of the non-isothermal curing of **ED-20/PAn–BF_3** composition. The calculation has been done via *Borchard–Daniels* method[63] under approximation that the process of heat eliminating is described by the kinetic equation of the first order and that the thermal–physical characteristics of a system (namely, thermal capacity and thermal conductivity) are the same under investigated temperature range. It is assumed, that the rate of heat eliminating, a measure of which is the **DTA**-curve is proportional to the rate of the epoxy group's consumption.

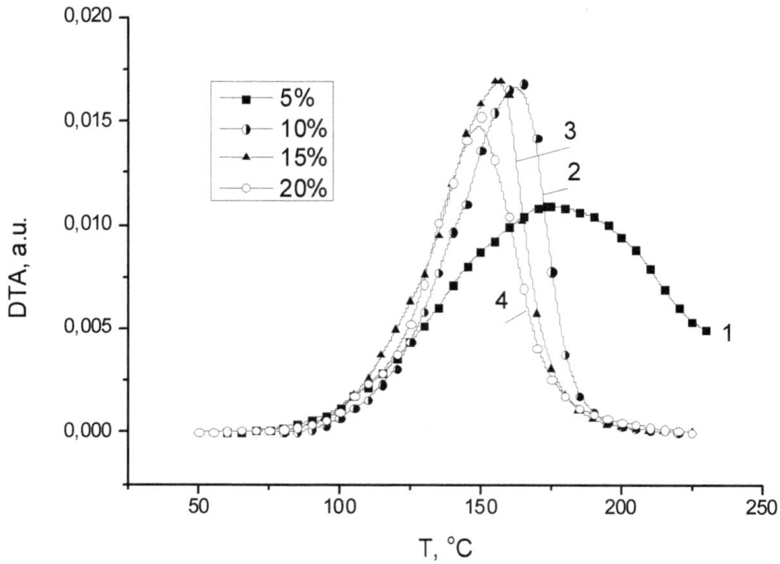

FIGURE 5.5 DTA-curves for compositions **ED-20/PAn–BF$_3$**. The doping level of **PAn–BF$_3$** is 36%mol. The content of **PAn–BF$_3$** in composition, %mass: 1–5; 2–10; 3–15; 4–20.

Within this approach, the calculation of the reaction rate constant k_t for given temperature T_t is realized by the equation:

$$k_t = \frac{C_P \dfrac{d\Delta T}{dt} + K\Delta T_t}{K(S_{max} - S_t) - C_P \Delta T_t}, \tag{5.1}$$

where C_p is the thermal capacity of the investigated sample; K is thermal constant of the device; S_{max} is an area of **DTA** maximum; S_t is an area of **DTA** maximum limited by T_t ordinate.

Taking into account the autocatalytic nature of the epoxy oligomers curing under the action of **BF$_3$** complexes, the calculation of the rate constants was performed for the initial stage of the ascending branch of **DTA**-curve, where the contribution of the autocatalytic component in total process of heat eliminating can be somewhat neglected. Calculation of the activation energy E_a of a curing process is based on the use of the *Arrhenius* equation:

$$k_t = A \cdot e^{-E_a/RT} \tag{5.2}$$

The dependence of the heat eliminating rate constant k on temperature in the linearized coordinates of the *Arrhenius* equation $\ln k - \frac{1}{T}$ is shown in Figure 5.6. The linear nature of this relationship suggests the possibility to use the *Borchard–Daniels* method for the evaluation of kinetically- activated parameters of the epoxy–polyaniline composites curing.

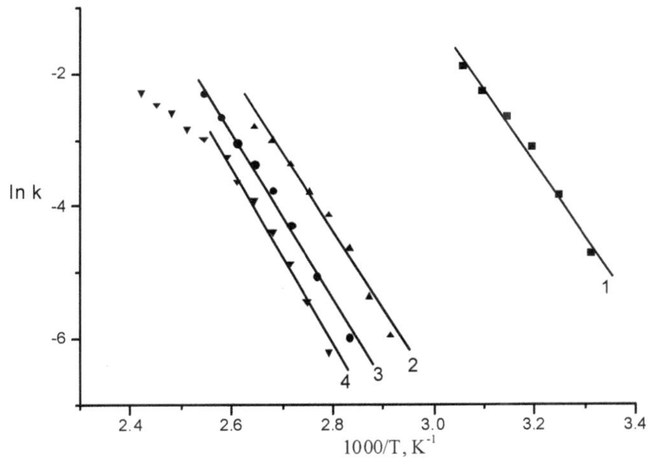

FIGURE 5.6 Dependence of ln$k$ on $1/T$ for compositions of **ED-20/PAn–BF$_3$** with doping level **PAn–BF$_3$** (%mol): 1–48; 2–44; 3–36; 4–30. Content of **PAn–BF$_3$** consists of 10%mass.

The values of the activation energy E_a, calculated in accordance with the equation (5.2) on a basis of the obtained dependencies are shown in Table 5.1. It should be noted the presence of weakly delineated tendency to decreasing of the activation energy E_a with decreasing of the doping level of **PAn–BF$_3$**. It should be also noted that the investigated process of the epoxy oligomer **ED-20** curing with the use of **PAn–BF$_3$** complex is characterized by considerably lower activation energy E_a = 99–109 kJ/mol compared with the value of E_a = 275–290 kJ/mol for the reaction of epoxy oligomer curing by polyaniline- sulfate.[37]

5.3.3 *ELECTROCONDUCTIVITY OF ED-20/PAn–BF$_3$ COMPOSITES: THE EFFECT OF TEMPERATURE*

Electroconductivity of epoxy–polyaniline composites (**EPO/PAn**) depends on a number of molecular structural and technological factors. The electroconductivity of **PAn**, the size and the shape of its particles, the method of electroconductive component introduc ing in the epoxy matrix and the distribution nature of **PAn**'s particles in it are defining among these factors. A necessary condition for the support of high electroconductivity of epoxy matrix is the formation of **PAn**-continuous phase. The critical concentration of filler or the so-called "percolation threshold" corresponds to the concentration of filler, at which each particle of the filler is in contact with two adjoining ones with the formation of epy electroconductive chain. At the achievement of the percolation threshold, the electroconductivity of the composite is sharply increased. The percolation threshold of epoxy–polyaniline compositions is determined in general by the electroconductivity of the starting **PAn**. The value of percolation threshold of epoxy composites filled with **PAn** doped by **DBSA** consists of 2–10%mass.[15] The dispersion and the shape of the **PAn**'s particles is crucial in this regard. Typically, an increasing of the **PAn** dispersion and also of the uniformity of its distribution in epoxy matrix reduces the percolation threshold.[15,26,33,42,44]

Conducted by us a set of the physical and chemical researches proved that the synthesized by us **PAn–BF$_3$** can be used not only as an active curing agent of the epoxy oligomers, but also as an effective electroconductive filler of the epoxy composites.

The results of a measurement of the electrical conductivity of epoxy–polyaniline composites depending on the content of **PAn–BF$_3$** (doping level is 36%mol) are shown in Figure 5.7. As we can see, the dependence of electroconductivity on the concentration of **PAn–BF$_3$** complex is quite complicated and bear a strong resemblance to percolation dependencies determined earlier for the composites of polyaminoarenes (**PAAr**) with polyvinyl alcohol[64] (**PVA**) and polymethyl methacrylate[65] (**PMMA**).

The specific conductance of composites increases from the ~10^{-15} *S/m* (that corresponds to the electroconductivity of unfilled epoxy-polymeric matrix) to ~$10^{-6} \div 10^{-4}$ *S/m*, that is in 9–11 orders at the content of **PAn–BF$_3$** increasing till 10–15%mass. Percolation threshold of the electroconductivity of epoxy–polyaniline composites is achieved at the introduction of 2.0–2.5%mass of **PAn–BF$_3$**. For comparison, the percolation threshold for the polymer compositions filled with nanosized silver particles[66] is

~70%. The maximal specific conductivity of the composites on a level $(3.2 \div 6.3)10^{-4}$ *S/m* is achieved at the content of **PAn–BF$_3$** within the 12–20%mass. High values of the specific conductivity and the low percolation threshold of obtained composites compared to the systems based on **PAn** protonated by *p*-toluene sulfonic acid (*p*-**TSA**), **DBSA,** and some other acids[22–23,36,48] are caused from the one hand by a high conductivity of **PAn–BF$_3$** complex and from the other hand by the possible formation of the interpenetrating nets in which, along with the cross-linked polymer matrix, the net of the polymeric filler is formed.[67] At the increasing of the **PAn–BF$_3$** content more than 25% the value of the specific conductance of epoxy–polyaniline composites is some decreased that can be connected with the clustering of **PAn**'s particles and evidently delineated interfacial exfoliation processes as well as heterogenization of the composite structure, what is indicated in several refs.[18,27,30,32,46,49]

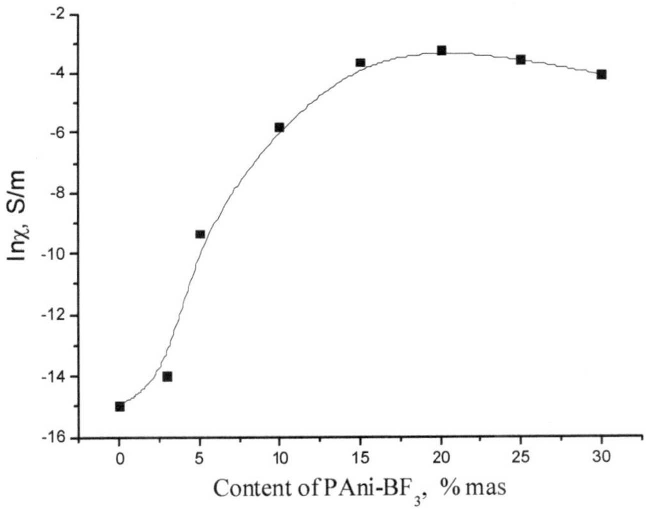

FIGURE 5.7 Dependence of the electroconductivity of **ED-20/PAn–BF$_3$** on content of **PAn–BF$_3$** (doping level of **PAn–BF$_3$** is 36%mol).

An investigation of the effect of temperature on the electroconductivity of the conjugated polymers is of considerable theoretical interest, since it makes possible to evaluate the nature of the electroconductivity of the sample depending on the temperature, the structure of a sample, the method of doping, as well as concentration and the mechanism of transfer of the charge carriers.[3] It is known that the polymeric conductors have the negative temperature coefficient of resistance ($d\rho / dT \lhd 0$) and the temperature dependence of the resistivity ρ at temperatures $T > 273$ K can be described by the activation equation by the *Arrhenius* type:

$$\rho = \rho_0 e^{\frac{\varepsilon}{2kT}}, \tag{5.3}$$

where ρ_o is the specific conductance of a sample at some temperature T_o; is an activation energy which within the terms of conductivity band theory is considered as an energy of the support of charge formation and is expressed in eV.[68]

In Figure 5.8 there are results of studies of the effect of temperature on the resistivity for the composites **ED-20/PAn–BF$_3$** with doping level of **PAn–BF$_3$** = 30 and 48%mol. normalized to the resistance ρ_o at $T_0 = 293$ K. As we can see, the results obtained in the range 303–373 K are within a narrow band dispersion and can be described by the temperature dependence in the form of equation (5.3) with the value of ε = 0.21–0.31 eV.

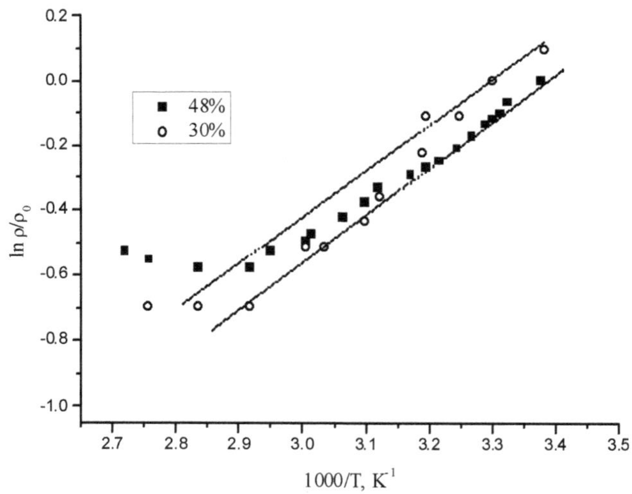

FIGURE 5.8 Temperature dependence of the resistivity for compositions **ED-20/PAn–BF$_3$**. The content of **PAn–BF$_3$** is 20% mass, the doping level is 30 and 48%mol.

These results confirm the retaining of the semiconductor nature of the **PAn–BF$_3$** conduction in a matrix of epoxy polymer, for which are typical the low values of activation energy of the charge transfer compared to other polymeric semiconductors.[3] This allows considering the obtained composites as promising materials for organic electronics.

5.3.4 MECHANICAL AND THERMOMECHANICAL PROPERTIES OF EPOXY–POLYANILINE COMPOSITIONS

ERs and materials on their basis belong to one of the most important classes of polymeric constructive materials. They are characterized by a high mechanical strength, by an ability to resist high temperatures,[20–21] and so forth. Therefore, the study of the behavior of epoxy composites under conditions of mechanical stress and temperature is of a great importance in view of obtaining the materials with desired properties.

In this regard, we studied the mechanodeformative and thermodeformative (thermomechanical) properties of epoxy–polyaniline compositions depending on the content of **PAn–BF$_3$** and also on its doping level.

Mechanodeformative behavior of synthesized epoxy–polyaniline composites has been investigated by penetration method with the use of *Heppler* consistometer by indentation of the working device (indenter) into a sample (in this case steel cone with the angle of tip 58°08′).

By measuring the magnitude of Δh under an applied load P, it can be estimated the so-called "conical point yield for *Heppler* F_p,"[69] which is a measure of surface microhardness of the material and for spatial cross-linked polymer matrix depends on the nature and structure of polymer network, on the concentration of effective crosslinking units, on the stacking method of polymer chains, and so forth. F_p is an important structural and mechanical parameter of polymer composite, which defines the limit values of mechanical stress, temperature, and duration of their exposure, in which the deformation of the material does not exceed the permissible value and for which the material retains its performance.

Conical point yield F_p (in N/m^2) was calculated in accordance with the equation

$$F_p = \frac{4P}{\pi \cdot \Delta h_p^2} \tag{5.4}$$

where P is a load (in *N*); Δh_p is the penetration depth (in *m*) of conical measuring device under the applied load P.

In Figures 5.9 and 5.10 there are dependencies of microhardness on load for the samples of epoxy–polyaniline composites obtained using the **PAn–BF$_3$** with different doping level, namely 48%mol (Fig. 5.9) and 36%mol (Fig. 5.10). Content of **PAn–BF$_3$** was varied within 5–20%mass. As we can see from these figures, the microhardness of a sample depends on the applied load and is described as a rule, by a curve with a maximum. At a certain value of P, the maximal value F_∞ is achieved, which is adopted for the evaluation of microhardness of the investigated composites. Under the following increasing of the applied load the microhardness is decreased, which is connected with the irreversible destruction of the polymer matrix (such as "floating").

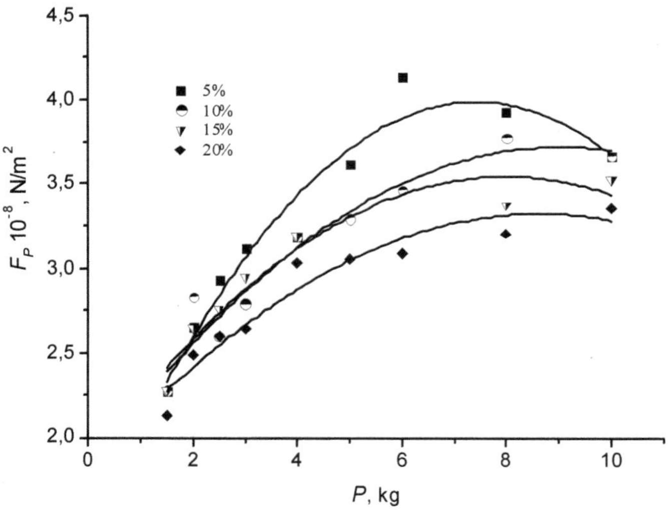

FIGURE 5.9 Dependence of epoxy–polyaniline compositions microhardness on the loading at different contents of **PAn–BF$_3$** (doping level of **PAn–BF$_3$** is 48%mol).

For the investigated epoxy–polyaniline composites cured by **PAn–BF$_3$** with different doping level, the maximum value of F_∞ is achieved under load P = 60–80 N. An effect of the content of curing agent on microhardness of epoxy–polyaniline composites depends on the doping level of **PAn–BF$_3$**. At the use of high-doped **PAn–BF$_3$** (48%mol.), a clear tendency to decrease of the F_∞ value is observed at the content of **PAn–BF$_3$** increasing in the system, namely from $4.2 \cdot 10^8$ N/m^2 for a sample containing of 5%mass of curing agent till $3.5 \cdot 10^8$ N/m^2 for a sample containing of 20% mass of curing agent. Perhaps, this is due to the growth of structural heterogeneity of the polymer matrix at high speed of curing.

Using of the less doped **PAn–BF$_3$** (36%mol, Fig. 5.10) leads to the improvement of mechanical properties of a composite. Microhardness of epoxy–polyaniline composites in this case, compared to the composites based on high-doped **PAn–BF$_3$**, is higher ($4.5 \cdot 10^8$–$5.25 \cdot 10^8$ N/m^2) when the content of curing agent is within 10–15% *by mass*. Although in this case the same tendency to decreasing of microhardness at the content of **PAn–BF$_3$**

increasing in system is observed. The relatively low values of the composite microhardness at the content of **PAn–BF$_3$** equal to 5% mass (Fig. 5.10) is probably the result of incomplete curing of epoxy composite. In this case, the spatial structure of the composite is not formed completely; the composite exhibits highly elastic behavior over a wide range of loads and the field of the irreversible damage was not achieved.

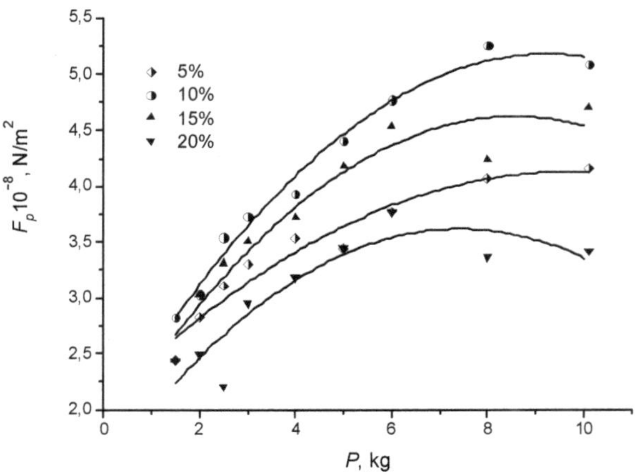

FIGURE 5.10 Dependence of epoxy–polyaniline compositions microhardness on the loading at different contents of **PAn–BF$_3$** (doping level of **PAn–BF$_3$** is 36% *mol*).

Highly efficient method for studying the mechanical behavior of polymer in terms of the complex action of mechanical loads and temperature is the method of thermomechanical analysis[70] (or **TMA** method). This method gives the possibility to estimate a number of thermal and structural–mechanical characteristics that determine the peculiarities of the mechanical behavior of polymer under conditions of mechanical and thermal field. **TMA** method is based on measuring of an investigated sample deformation under the action of externally applied force depending on temperature. The results of measurement are presented as thermomechanical curve, which reflects the structural and physical–chemical conversions of polymer under given temperature range.

The thermomechanical curves (**TMA**-curves) for the investigated composites **ED-20/PAn–BF$_3$** are represented in Figure 5.11 as the temperature dependence of the relative strain ε (ε = Δh/h, where h is the starting height of the sample; Δh is the change of the sample's sizes during the deformation) of composites cured by **PAn–BF$_3$** with different doping level. The

TMA-curve of epoxy polymer cured by complex aniline–$\mathbf{BF_3}$ (**An**–$\mathbf{BF_3}$) is marked in Figure 5.11 for comparison. The temperature regions corresponding to three characteristic regions of the polymer deformation are observed on the obtained curves, namely: the field of glass-like state, the region of high-elasticity state, and the area of thermal mechanical destruction.

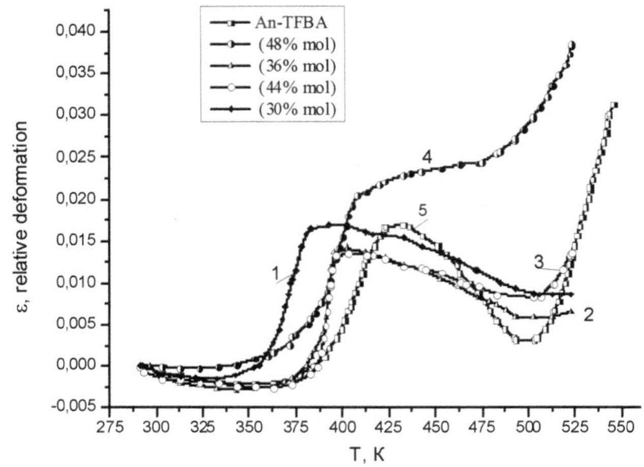

FIGURE 5.11 TMA-curves for **ED-20/PAn**–$\mathbf{BF_3}$ composites. Content of **PAn**–$\mathbf{BF_3}$ is 10%mass, doping level is (%mol): 1—30; 2—36; 3—44; 4—48; 5—**An**–$\mathbf{BF_3}$.

At the initial stages of the warm-up, it is observed a slight increase of the linear dimensions of a sample (negative strain), due to the temperature effect of the linear expansion of sample. Section of **TMA**-curve, for which a dramatic increase of deformation in a narrow temperature range is typical, corresponds to the structural transition of the system from the glass-like state to the high-elasticity one (so-called α–relaxation transition). The glassing temperatures T_{glass} for the investigated compositions were determined by the extrapolating of the linear segment of this section on the temperature axis. The values of glassing temperatures T_{glass} are given in Table 5.3.

The horizontal section of **TMA**-curve represents an area of the elastic (high-elastic) deformation. Typical for this area is some decrease of the relative deformation of a sample at further increase of temperature due to the polymer matrix rising resistance to the applied external load as result of increased mobility of the kinetic segments of macrochain at temperature increasing.[70]

The study of polymer's behavior in high-elastic state allows to estimate the structural and molecular parameters of the spatial net: an equilibrium modulus of high-elasticity E_∞ and molecular mass of the interstitial segment M_s. Within a framework of the statistical theory of elasticity of molecular net of spatial cross-linked polymers the dependence of high-elasticity modulus E_∞ on a crosslinking density (value of M_c) is described by the following equation:[71]

$$E_\infty = \frac{3\gamma\rho RT\upsilon}{M_o\upsilon_0} \quad (5.5)$$

where E_∞ is the high-elasticity modulus; γ is the structural coefficient, the value of which depends on a nature and topology of net; ρ is the density of polymer; υ_0 is total number of crosslinks; υ is number of physically effective bonds forming the spatial structure; M_o is molecular mass of kinetic segment; R is gas constant; T is temperature, K.

Quantitative checking of this relation is rather complicated, because the unknown is the proportion of functional groups that participate in the formation of physically active links. Under condition $E_\infty \leq 6\cdot10^7$ N/m^2 it can be accepted[72] that $\upsilon = \upsilon_0$, $\gamma = 1$, and the solving accordingly to equation (5.5) is significantly simplified. The high-elasticity modulus E_∞ and M_o we calculate in accordance with the equation:

$$E_\infty = \frac{P}{F\cdot\varepsilon} \quad (5.6)$$

where P is the load on a sample, N; F is an area of the cross-section of a sample, to which the force is applied, m^2; ε is relative deformation of a sample in an area of the high elasticity.

Calculated values of E_∞ and M_s for **ED-20/PAn–BF$_3$** composites are represented in Table 5.3.

TABLE 5.3 Thermal–Mechanical and Structural–Molecular Characteristics of Epoxy–Polyaniline Composites (a content of **PAn–BF$_3$** consists of 10%mass).

Doping level, %mol	T_{glass}, °C	T_e, °C	T_d, °C	10^{-7} E_∞, N/m^2	M_s, g/mol
30	81	110	>250	5.42	189
36	104	123	235	6.50	164
44	113	123	233	6.97	157
48	92	139	218	4.79	302
An–BF$_3$	114	149	236	6.00	175

Further increase of temperature leads to a sharp increase of the deformation in narrow temperature range, the sample "flows." This is an area of irreversible deformations, that is result from the simultaneous action of mechanical and thermal fields, that is an area of the thermomechanical destruction of the polymer matrix. The position of this area on temperature scale can be a criterion of thermal stability of epoxy polymer. By the method of tangents to the relevant sections of **TMA**-curves it was defined the values of transition temperature for epoxy–polyaniline composites into the high-elastic state (T_e) and the values of temperature of the thermal decomposition start of polymeric matrix (T_d). The values of T_e and T_d for **ED-20/PAn–BF$_3$** composites depending on doping level are given in Table 5.3.

It was investigated an influence of the **PAn–BF$_3$** content (doping level 36%mol) on the thermomechanical behavior of **ED-20/PAn–BF$_3$** composites. The content of **PAn–BF$_3$** was varied within 5–30% mass. Obtained results indicate that an increasing of the **PAn–BF$_3$** content in the composition does not cause the visible effect on the nature of thermomechanical curves. At the same time, it is observed a visible increase of the rigidity of epoxy–polymeric matrix, resulting in the increasing of T_g and in decreasing of relative deformation at fixed value of temperature. Especially visibly it appears for compositions containing of **PAn–BF$_3$** in quantity 5–10% *by mass*, for which the maximum value of high-elastic strain is differed by almost in five times. At this, the length of the interstitial segment M_s is decreased from 635 g/mol to 153 g/mol at **PAn–BF$_3$** content increasing from 5 to 10%mass; at the same time, the temperature of composite glass transition is increased on 50 K—from 57 to 104 °C. Further growth of the curing agent content, practically has not an effect on thermal and structural–molecular characteristics of the epoxy–polyaniline composites that are evidenced by the results shown in Figure 5.12 in the form of the dependence of glass transition temperature T_{glass}, molecular weight between nodal segment M_s, high-elasticity modulus E_∞ on the content of **PAn–BF$_3$** in the composition.

An analysis of the **TMA**-curves (Fig. 5.11, 5.12) and tabular data (Table 5.3) suggests to draw some conclusions about an impact of the doping level of **PAn–BF$_3$** and its contents on the thermomechanical properties of **ED-20/PAn–BF$_3$** composites.

As can be seen from the presented data, an increase of the **PAn–BF$_3$** doping level and also its content leads to a visible increase of the glass transition temperature T_{glass} as same as the high-elasticity temperature T_e of composite, indicating that the improvement of thermal-deformative characteristics of the polymer matrix under conditions of mechanical stress takes place. It should be noted that the use of **PAn–BF$_3$** with a maximal doping level (48%mol) results in the displacement from the start of epoxy polymer decomposition (T_d) in the region of lower temperatures. This may indicate a slight decrease of epoxy polymer samples thermal stability at using of high-doped **PAn–BF$_3$**.

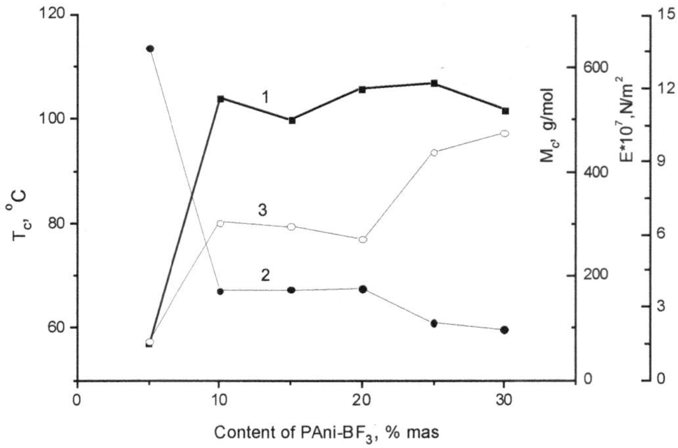

FIGURE 5.12 Dependence of T_{glass} (1), M_c (2) and E_∞ for composites **ED-20/PAn–BF$_3$** on the content (% mass) of **PAn–BF$_3$** (doping level 36%mol).

It is known, that the thermal stability of epoxy polymers is determined mainly by strength of chemical bonds of the polymer matrix and by their distribution in energy.[76] An increase of the rigidity and tension of bonds usually leads to a decrease of the thermal stability of a polymer. Exactly by the influence of this factor we attribute the reduction of thermomechanical stability of epoxy polymers obtained using the samples of high-doped **PAn–BF$_3$**. An increasing of the glass transition temperature T_{glass}, an increase of the high-elasticity modulus E_∞ and the decrease of the length of interstitial segment M_s means about an increase of the effective units cross-linking density and the polymer chains packing density in the structure of the polymer matrix. On the other hand it is necessary to consider the fact that thanks to the own complexity the cationic polymerization of epoxy groups is accompanied by the formation of defective polymer networks.[21] Under condition of the high concentration of crosslinking units and high packing density of the polymer chains, the defectiveness of the polymer structure causes an increase of tensions of bonds and leads to increase of the probability of its destruction under thermal and mechanical loads.

5.3.5. THERMAL STABILITY OF EPOXY–POLYANILINE COMPOSITES

In order to evaluate the thermal stability of the synthesized epoxy–polyaniline composites, the method of differential thermal gravimetric analysis (**TGA**-analysis) was used. The investigations have been carried out using the powdered samples by weight 100–150 mg in air atmosphere. The thermogravimetry method is based on obtaining of two experimental dependencies in the form of **TG**- and **DTG**-curves. **TG**-curve (*integral curve of mass loss*) reflects the change in mass of the sample depending on temperature and is used for the determination of the absolute and relative mass loss of the sample, and on the whole is a quantitative characteristic of the degradation process. **DTG**-curve (*differential weight loss curve*) characterizes the rate of the process. **DTG**-curve makes it possible to identify the individual stages of the process, just set the temperature ranges of the exaggerated steps.

TG-curves of thermal destruction of **ED-20/PAn–BF$_3$** depending on the doping level and content of **PAn–BF$_3$** are given in Figures 5.13 and 5.14.

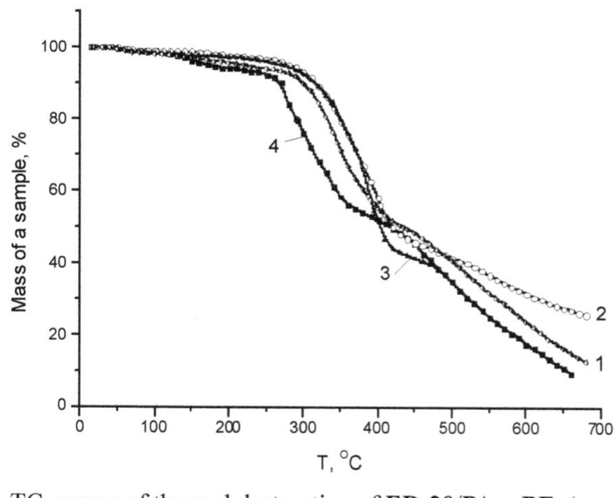

FIGURE 5.13 TG-curves of thermal destruction of **ED-20/PAn–BF$_3$** (content of **PAn–BF$_3$** consists of 10%mass). Doping level of **PAn–BF$_3$** (%mol): 1—30; 2—36; 3—48; 4—**An–BF$_3$**.

As can be seen from these Figures, the process of thermal destruction of **ED-20/PAn–BF$_3$** composites depending on the doping level of **PAn–BF$_3$** and its content is described by similar in form **TG**-curves. Three sections can be roughly marked on the **TG**-curves under investigated temperature range that points on the vicissitude of the process of **ED-20/PAn–BF$_3$** composites destruction. In the region $T < 250$ °C the mass of the composite remains almost unchanged. Slight – less than 5% by mass reducing of the composite weight, which is observed in this temperature range, evidently is due to the desorption of physically bound moisture. The most typical is the temperature range 250 (275)–400 °C, where there is an intense mass loss of the sample. Obviously, the mass loss in this temperature range is connected with the beginning of the **PAn**-chain destruction and desorption of the molecules of doping agent. This is indicated by the position of temperature and the nature of **TG**-curve for epoxy polymer **ED-20**, obtained with the use of **An–BF$_3$** complex as curing agent (Fig. 5.13, curve 4). Temperature range $T > 400$ °C, which is characterized by a monotonous mass loss of the investigated samples is connected with the deep destructive processes of the epoxy matrix polymer chains destruction.[74]

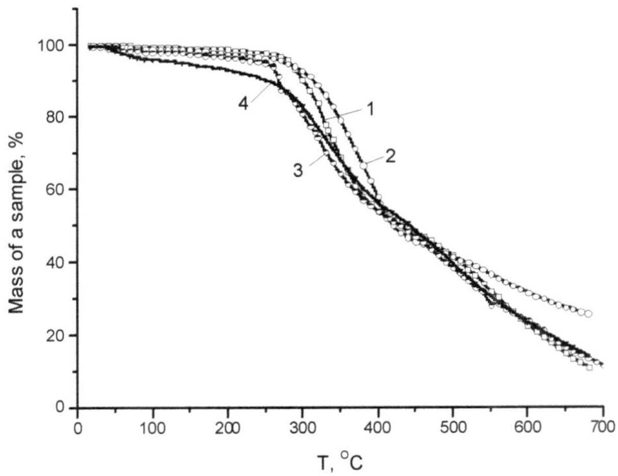

FIGURE 5.14 TG-curves of thermal destruction of **ED-20/PAn–BF$_3$** (doping level of **PAn–BF$_3$** is 36%mol). The content of **PAn–BF$_3$** (%mass): 1—5; 2—10; 3—15; 4—25.

Despite this, the character of **TG**-curves represented in Figures 5.13 and 5.14 as same as their displacement on temperature scale depending on the doping level and the content of **PAn–BF$_3$** complex is evidence of its impact on the conversion characteristics of composites. One can say, that the use of **PAn–BF$_3$** with a higher doping level leads to a shift of the beginning of polymer decomposition toward the higher temperatures (Fig. 5.13, curves 2, 3), and thus, leads to the improvement of the thermal stability of epoxy–polyaniline composite. Conversely, an increasing of the **PAn–BF$_3$** content in composition causes the opposite effect: the temperature of the beginning of epoxy-polymer matrix destruction with the release of volatile products tends to decrease, which may indicate the worsening of the thermal stability of polymer.

We can assume, that the reason for this is the structural factor associated with a higher degree of heterogenization of the polymer structure due to the introduction of a large quantity of powder curing agent as the dispersed phase.[74–76] A chemical factor connected with the incomplete using of the functional groups during the formation of epoxy–polymer matrix can play certain negative role. It has been shown[73,75] that unused functional groups of epoxy oligomer or curing agent can serve as centers of free radicals generation that can initiate the destruction of the polymer matrix.

An analysis of the **DTG**-curves of destruction of epoxy–polyaniline composites obtained using the complex **PAn–BF**$_3$, which are shown in Figures 5.15 and 5.16, confirm these conclusions.

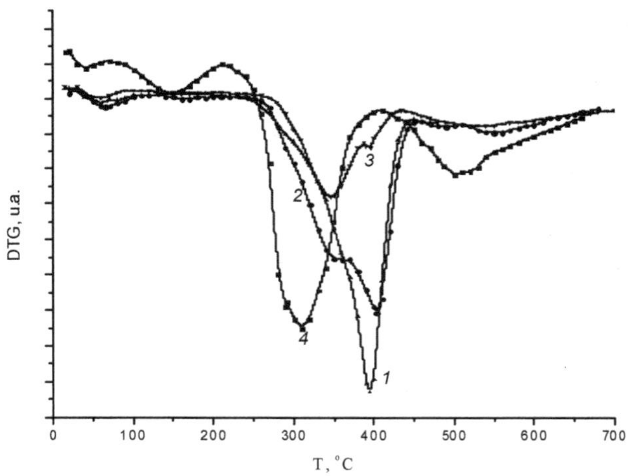

FIGURE 5.15 DTG-curves of thermal destruction of **ED-20/PAn–BF**$_3$ composites (content of **PAn–BF**$_3$ is 10%mass. Doping level of **PAn–BF**$_3$ (%mol): 1—48; 2—36; 3—30; 4—**An–BF**$_3$.

In Figure 5.15, there are **DTG**-curves of thermal destruction of **ED-20/ PAn–BF**$_3$ composites depending on the doping level of **PAn–BF**$_3$. The content of **PAn–BF**$_3$ was 10%mass. The comparative analysis of **DTG**-curves gives the possibility to do a number of important conclusions. A form of **DTG**-curve essentially depends on nature of curing agent. In particular, **DTG**-curve of thermal decomposition of epoxy polymer obtained with the use of **An–BF**$_3$ complex (Fig. 5.15, curve 4) has complicated nature. Besides the main **DTG**-peak in the range 250–400°C and also poorly defined maximum at

150 °C which is connected with desorption of physically adsorbed water and low molecular weight components of the epoxy oligomer at 500 °C, another vague additional **DTG**-maximum is observed. All this testifies to the energy, and possibly structural heterogeneity of polymer matrix formed under the influence of **An–BF$_3$** complex.

Use of **PAn–BF$_3$** as curing agent leads to a shift of **DTG**-peak toward the higher temperatures 350–400 °C (Fig. 5.15, curves 1–3) which clearly points on the improvement of the thermal stability of epoxy polymer. At the same time, an increase of the **PAn–BF$_3$** doping level leads to a narrowing of the temperature interval of rapid loss mass and, consequently, to the growth of the rate of mass loss at a fixed temperature values, that is notably visible for the composites cured by **PAn–BF$_3$** with the doping level 48% (Fig. 5.15, curve 3).

The doping level of **PAn–BF$_3$** somehow affects the nature of **DTG**-curves. For relatively low-doped samples (30–36%mol) **DTG**-peak can be considered as having two quasi-elementary maxima associated with parallel processes of epoxy matrix polymer chains and **PAn** destruction. For high-doped sample of **PAn–BF$_3$** (Fig. 5.15, curve 1) due to the high rate of polymer chains destruction, these two processes are superimposed on the temperature scale and the bimodal nature of the total **DTG**-maximum is degenerated.

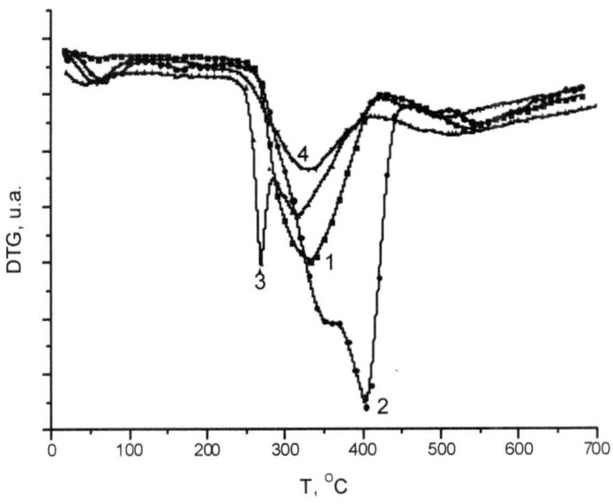

FIGURE 5.16 DTG-curves of thermal destruction of **ED-20/PAn–BF$_3$** (doping level 36%mass) depending on the content of **PAn–BF$_3$** (%mass): 1—5; 2—10; 3—15; 4—25.

In Figure 5.16, there are differential thermogravimetric curves of destruction of epoxy–polyaniline composite obtained with the use of **PAn–BF$_3$** complex with doping level 36%mol as curing agent. The content of

PAn–BF$_3$ was varied from 5 to 25%mass. It should also be noted that the increase of the curing agent content on a whole leads to decrease of the speed of mass loss. On the one hand it shows an improvement of the thermal stability of epoxy-polymer chain. On the other hand it may be a demonstration of the high thermal stability of the **PAn**, the content of which in the polymer matrix is increased.

The results of **TGA**-analysis give reason to draw some conclusions about the impact of the content and the doping level of **PAn–BF$_3$** on the thermal stability of epoxy composite. In order to evaluate the thermal stability of polymers, the conversion characteristics (mass loss) α_T or temperature values corresponding to certain characteristic sections of the thermogravimetric curve are used: in particular, the temperature of maximum rate of destruction T_{max}. Thermal stability of polymer composites can be characterized by the so-called conditional decomposition temperature **IDPD** (integral procedural decomposition temperature),[77] whose value is calculated according to a special procedure and requires the strict standardization of the conditions of the process.

In order to evaluate the thermal stability of epoxy-amine polymers we proposed[78] to use the value of so-called reduced temperature of destruction T_r, which is regarded as the integral index that takes into account both the temperature and the mass conversion process characteristics and is estimated on the basis of direct experimental data according to the equation:

$$T_r = (1 - \alpha_{end})(T_{end} - T_{begin}) + T_{begin} \tag{5.7}$$

where T_{begin}, T_{end} are the temperatures of the beginning and of the end of the destruction rate, which are determined on **DTG**-curve; α_{end} is the mass loss of a polymer at T_{end}. The values of T_{begin}, T_{end} and α_{end} are determined on **DTG**- and **TG**-curves for a field of rapid destruction, the temperature position of which is determined by the position of **DTG**-peak.

Conversion and temperature characteristics of the investigated epoxy–polyaniline composites destruction are shown in Table 5.4. In this table, there are characteristic temperatures of destruction T_{begin} and T_{end}, the values of the maximal rate of destruction T_{max}, the loss of mass of a sample α_{max} at T_{max} and also the value of the loss of mass α_{600} at 600 °C. The value of reduced temperature of destruction T_r calculated in accordance with the formula (5.4) is also presented in Table 5.4.

As follows from the data of thermogravimetric analysis, the doping level of **PAn** and its content in the composition have an effect on the thermal destruction of epoxy–polyaniline composites. An increase of the **PAn** doping level generally leads to some worsening of the thermal stability of epoxy–polyaniline composites that is reflected in a visible decreasing of the reduced temperature of degradation. In this case, the determining factor is the shift of the start of thermal destructive process toward the field of the lower temperatures. An increase of the curing agent content in oligomeric composition generally also leads to the worsening of the thermal stability of obtained composite.

TABLE 5.4 Conversion and Temperature Parameters of Thermal Destruction of **ED-20/PAn–BF_3** Composites.

Doping level, % mol	Content of PAn–BF_3, % mass	T_{begin}, °C	T_{max}, °C	T_{end},Z °C	α_{max}	α_{end}	α_{600}	T_r, °C
An–BF_3	10	256	310	408	0.28	0.49	0.82	333
30	10	276	346	426	0.28	0.49	0.76	352
36	10	264	353	444	0.22	0.54	0.68	347
48	10	253	395	437	0.45	0.57	0.73	332
36	5	264	334	430	0.33	0.48	0.78	360
36	10	534	353	444	0.22	0.54	0.68	350
36	15	248	316	396	0.24	0.45	0.77	330
36	25	251	330	420	0.25	0.52	0.77	332

The data presented in the table suggest that there is an optimal ratio of the **PAn–BF_3** content and its doping level, that is, the doping level of **PAn–BF_3** 36%mol, and content of 5–10%mass in the composition, at which the maximal thermal stability of the epoxy-polymeric matrix is provided. It should be noted that the data of thermogravimetric analysis regarding the optimal ratio **ED-20/PAn–BF_3** confirm the results obtained at the studies of the electrical conductivity, mechanical, and thermomechanical properties of epoxy–polyaniline composites.

5.4 CONCLUSIONS

Thus, the obtained results demonstrate the effectiveness of the use of **PAn** doped by tetrafluoroborate acid as curing agent of ERs. The combination into the one **PAn–BF_3** complex of the properties of curing agent and electroconductive component allows to simplify the process of obtaining of the electroconductive epoxy polymer as well as to obtain at relatively low content of **PAn–BF$_3$** (10–15%mass) the epoxy–polyaniline composites characterizing by electrical conductivity 10^{-6}–10^{-4} S/cm, high thermal, physical, and mechanical properties, which makes it possible to use these composites for the manufacture of antistatic coatings, conductive adhesives, electrodes, and so forth.[55]

KEYWORDS

- **epoxy–polyaniline composite**
- **PAn–BF$_3$ curing agent**
- **electroconductivity**
- **curing mechanism**
- **kinetic parameters**
- **thermomechanical properties**
- **thermal destruction**

REFERENCES

1. Volkov, S. V.; Kovalchuk, E. P.; Ogenko, V. M.; Reshetniak, O. V. *Nanochemistry, Nanosystems, Nanomaterials*; Naukova dumka: Kiev, 2008.
2. Shpak, A. P.; Ulberg Z. Z. *Colloid-Chemical Principles of Nanoscience*; Akademperiodika: Kiev, 2005.
3. Aksimentyeva, O. I. *Electrochemical Methods of Synthesis and Conductivity of Conjugated Polymers*; Svit: L'viv, 1998.
4. Wessling, B. Dispersion as Link between Basic Research and Commercial Applications of Conductive Polymers (Polyaniline). *Synt. Met.* **1998,** *93,* 143–154.
5. Feast, W. J. J.; Tsibouklis, P. K. L.; Groenendal, L.; Maijer, E. W. Synthesis, Processing and Material Properties of Conjugated Polymers. *Polymer.* **1996,** *37* (22), 5017–5047.
6. Shirikawa, H.; Ikeda, S. Infrared Spectra of Poly(Acetylene). *Polymer Journal.* **1971,** *2* (2), 231–244.
7. Heeger, A. J. Nobel Lecture. Semiconducting and Metallic Polymers: The Fourth Generation of Polymeric Materials. *Rev. Mod. Phys.* **2001,** *73,* 681–699.
8. Bhadra, S.; Khastgir, D.; Singha, N. K.; Lee, J. H. Progress in Preparation and Applications of Polyaniline. *Prog. Polym. Sci.* **2009,** *34,* 783–810.
9. Ćirić–Marjanović, G. Recent Advances in Polyaniline Research: Polymerisation Mechanisms, Structural Aspects, Properties and Applications. *Syn. Met.* **2013,** *177,* 1–47.

10. Li, Y.; Yang, L.; Gao, X.; Zeng, Z. Polyaniline/Epoxy Composite Emulsion Coatings for Anticorrosion to Mild Steel. *Adv. Mat. Res*. **2011,** *152–153,* 1890–1893.
11. Ge, C. Y.; Yang, X. G.; Hou, B. R. Synthesis of Polyaniline Nanofiber and Anticorrosion Property of Polyaniline–Epoxy Composite Coating for Q235steel. *J. Coat. Techn. Res.* **2012,** *9* (1), 59–69.
12. Stejskal, J.; Gilbert, R. G. Polyaniline. Preparation of Conducting Polymer. *Pure Appl. Chem.* **2002,** *74 (5),* 857–867.
13. Angelopoulos, V.; Patel, M.; Shaw, J. M.; Labianca, N. C.; Rishton S. A. Water Soluble Conducting Polyanilines: Application in Lithography. *J. Vac. Sci. Technol. B*. **1993,** *11,* 2794–2797.
14. Jayanty, S.; Prasad, G. K.; Sreedhar, B.; Radhakrishnan, T. P. Polyelectrolyte Templated Polyaniline–Film Morphology and Conductivity. *Polymer*. **2003,** *44,* 7265–7270.
15. Tsotra, P.; Friedrich, K. Thermal, Mechanical and Electrical Properties of Epoxy Resin/Polyaniline–Dodecylbenzensulfonic Acid Blends. *Syn. Met.* **2004,** *143*, 237–242.
16. Yang, X.; Zhao, T.; Yu, Y.; Wei, Y. Synthesis of Conductive Polyaniline/Epoxy Composites: Doping of the Interpenetrating Network. *Syn. Met.* **2004,** *142,* 57–61.
17. Chwang, C.-P.; Liu, C.-D.; Yuang, S.-W.; Chaj, D.-Y.; Lee, S.-N. Synthesis and Characterization of High Dielectric Constant Polyaniline/Polyurethane Blends. *Syn. Met.* **2004,** *142,* 275–281.
18. Mir, I. A.; Kumar, D. Development of Polyanaline/Epoxy Composite as a Prospective Solder Replacement Material. *Inter. J. Polym. Mat.* **2010,** *59,* 994–1007.
19. Li, H.; Newill, K. *Handbook of Epoxy Resins*; Energiya: Moscow, 1973.
20. Paken, A. M. *Epoxide Compounds and Epoxy Resins*; GHI: Leningrad, 1962.
21. Zaytsev, Y. S.; Kochergin, Y. S.; Pakter, N. K.; Kucher, R. V. *Epoxy Oligomers and Adhesive Compositions*; Naukova dumka: Kiev, 1990.
22. Kathirgamanathan, P. Curable Electrically Conductive Resins with Polyaniline Fillers. *Polymer.* **1993,** *34* (13), 2907–2908.
23. Tsotra, P.; Gatos, K . G.; Gryshchuk, O.; Fridrich, K. Hardener Type as Critical Parameter for the Electrical Properties of Epoxy Resin/Polyaniline Blends. *J. Mat. Sci.* **2005,** *40,* 569–574.
24. Oyharcabal, M.; Olinga, T.; Foulc, M.-P.; Vigneras, V. Polyaniline/clay Nanostructured Conductive Filler for Electrically Conductive Epoxy Composites. Influence of Filler Morphology, Chemical Nature of Reagents and Curing Conditions on Composite Conductivity. *Syn. Met.* **2012,** *162,* 555–562.
25. Tiitu, M.; Talo, A.; Forsen, O.; Ikkala, O. Aminic Epoxy Resin Hardeners as Reactive Solvents for Conjugated Polymers: Polyaniline Base/Epoxy Composites for Anticorrosion Coatings. *Polymer*. **2005,** *46,* 6855–6861.
26. Soares, B. G.; Celestino, M. L.; Magioli, M.; Moreira, V. X.; Khastgir, D. Synthesis of Conductive Adhesives Based on Epoxy Resin and Polyaniline. DBSA Using the *in situ* Polymerization and Physical Mixing Procedures. *Syn. Met.* **2010,** *160,* 1981–1986.
27. Lu, J.; Moon, K.-S.; Kim, B.-K.; Wong, C. P. High Dielectric Constant Polyaniline/Epoxy Composites via *in situ* Polymerization for Embedded Capacitor Applications. *Polymer.* **2007,** *48,* 1510–1516.
28. Shreepathi, S.; Holz, R. Spectrochemical Investigations of Soluble Polyaniline Synthesized via New Inverse Emulsion Pathway. *Chem. Mater.* **2005,** *17,* 4078–4085.
29. Jeevananda, T.; Palaniappan, S.; Siddaramaiah. Spectral and Thermal Studies on Polyaniline–Epoxy Novolac Resin Composite Materials. *J. Appl. Polym. Sci.* **1999,** *74,* 3507–3512.

30. Abd, R.; Saiful, I.; Wan Abdul, R.; Wan, A.; Yahya, M. Y. Hybrid Composites of Short Acetylated Fiber and Conducting Polyaniline Nanowires in Epoxy Resin. *J. Compos. Mater.* **2013,** *1,* 1–10.
31. Akbarinezhad, E.; Ebrahimi, M.; Faridi, H. R. Corrosion Inhibition of Steel in Sodium Chloride Solution by Undoped Polyaniline Epoxy Blend Coating. *Prog. Org. Coat.* **2009,** *64,* 361–364.
32. Tsotra, P.; Friedrich, K. Short Carbon Reinforced Epoxy Resin/Polyaniline Blends: Their Electrical and Mechanical Properties. *Compos. Sci. Technol.* **2004,** *64,* 2385–2391.
33. Tsotra, P.; Gryshchuk, O.; Friedrich, K. Morphological Studies of Epoxy/Polyaniline Blends. *Macromol. Chem. Phys.* **2005,** *206,* 787–793.
34. Belaabed, B.; Lamouri, S.; Naar, N.; Bourson, P.; Hamady, S. O. S. Polyaniline–Doped Benzene Sulfonic Acid/Epoxy Resin Composites: Structural, Morphological, Thermal and Dielectric Behaviors. *Polymer Journal.* **2010,** *42,* 546–554.
35. Bagherzadeh, M. R.; Mahdavi, F.; Ghazemi, M.; Shariatpanahi, H.; Faridi, H. R. Using Nanoemeraldine Salt–Polyaniline for Preparation of a New Anticorrosive Water–Based Epoxy Coating. *Prog. Org. Coat.* **2010,** *68,* 319–322.
36. Peltola, J.; Cao, Y.; Smith, P. Epoxy Adhesives Made with Inherently Conducting Polymers. *Adhes. Age*. **1995,** *38,* 18–20.
37. Palaniappan, S.; Sreedhar, B.; Nair, S. M. Polyaniline as a Curing Agent for Epoxy Resin: Cure Kinetics by Differential Scanning Calorimetry. *Macromol. Chem. Phys.* **2001,** *202,* 1227–1231.
38. Shao, Y.; Huang, H.; Zhang, T.; Meng, G.; Wang, F. Corrosion Protection of Mg–5Li Alloy with Epoxy Coatings Containing Polyaniline. *Corros. Sci.* **2009,** *51,* 2906–2915.
39. Yeh, J.-M.; Liou, S.-J.; Chiung–Yu, L.; Wu, P.-C. Enhancement of Corrosion Protection Effect in Polyaniline via the Formation of Polyaniline–Clay Nanocomposite. Materials. *Chem. Mater.* **2001,** *13,* 1131–1136.
40. Armelin, E.; Aleman, C.; Iribarren, J. I. Anticorrosion Performances of Epoxy Coatings Modified with Polyaniline. A Comparison between the Emeraldine Base and Salt Forms. *Prog.Org. Coat.* **2009,** *65,* 88–93.
41. Singh, P. B. P.; Choudhary, V.; Saini, P.; Pande, S.; Singh, V. N.; Mathur, R. B. Enhanced Microwave Shielding and Mechanical Properties of High Loading MWCNT – Epoxy Composites. *J. Nanopart. Res.* **2013,** *15,* 1554.
42. Ting, T.-H.; Wu, K.-H. Synthesis and Electromagnetic Wave-Adsorbing Properties of $BaTiO_3$/ Polyaniline Structured Composites in 2–40 GHz. *J. Polym. Res.* **2013,** *20,* 217, 1–6.
43. Oyharcabal, M.; Olinga, T.; Foulc, M.-P.; Lacomme, S.; Contier, E.; Vigneras, V. Influence of Morpholody of Polyaniline on the Microwave Adsorption Properties of Epoxy Polyaniline Composites. *Compos. Sci. Technol.* **2013,** *74,* 107–112.
44. Tsotra, P.; Friedrich, K. Electrical and Dielectric Properties of Epoxy Resin/Polyaniline – DBSA. *J. Mater. Sci.* **2005,** *40,* 4415–4417.
45. Jang, J.; Bae, J.; Lee, K. Synthesis and Characterization of Polyaniline Nanorods as Curing Agent and Nanofiller for Epoxy Composite. *Polymer.* **2005,** *46,* 3677–3684.
46. Liu, C.-D.; Lee, S.-M.; Ho, C.-H.; Han, J.-L.; Hsieh, K.-H. Electrical Properties of Well–Dispersed Nanopolyaniline/Epoxy Hybrids Prepared Using an Adsorption–Transferring Process. *J. Phys. Chem. C.* **2008,** *112,* 16956–15960.
47. Jia, Q.M.; Li, J. B.; Wang, L. E.; Zhu, J. W.; Zheng, M. Electrically Conductive Epoxy Resin Composites Containing Polyaniline with Different Morphologies. *Mater. Sci. Eng. A*. **2007,** *448,* 356–360.

48. Luo, K.; Guo, X.; Shi, N.; Sun, C. Synthesis and Characterization of Core-Shell Nanocomposites of Polyaniline and Carbon Black. *Syn. Met.* **2005,** *151,* 293–296.
49. Jia, W.; Tchudakov, R.; Segal, E.; Joseph, R.; Narkis, M.; Sigemann, A. Electrically Conductive Composites Based on Epoxy Resin with polyaniline – DBSA Fillers. *Syn. Met.* **2003,** *132,* 269–278.
50. Peliskova, M.; Vilcakova, J.; Moucka, R.; Saha, P.; Stejkal, J.; Quadrat, O. Effect of Coating of Graphite Particles with Polyaniline Base on Charge Transport in Epoxy – Resin Composites. *J. Mater. Sci.* **2007,** *42,* 4942–4944.
51. Rozenberg, B. A. Kinetics and Mechanism of Curing of Epoxy Oligomers. In *Composite Polymeric Material*; Naukova dumka: Kiev, 1975; pp 39–59.
52. Tanaka, Y.; Kakiuchi, H. Study of Epoxy Compounds. Part VI. Curing Reactions of Epoxy Resin and Acid Anhydride with Amine, Acid, Alcohol and Phenol as Catalysts. *J. Polym. Sci.* **1964,** *2* (8), 3405–3430.
53. Sorokin, M. F.; Shode, L. G.; Germanova, E. L. Catalytic Reaction of Epoxy Resins Curing with Anhydrides Of Carboxylic Acids. *Coating materials and their application.* **1967,** *5,* 67–71.
54. Zakordonskiy, V.; Aksimentyeva, O.; Martyniuk, G.; Krupak, A. Synthesis and Physicochemical Properties of Epoxy-Polyaniline Composites. Bulletin of L'viv University, *Chem.* **2008,** *49* (2), 118–125.
55. Aksimentyeva, O. I.; Zakordonskiy, V. P.; Martyniuk, G. V.; Krupak, A. I. A Method for Producing of Conductive Epoxy Compositions. UA Patent No 24145.
56. Zakordonskiy, V. P.; Skladanyuk, R. V. Thermomechanical Properties of Filled Epoxy Resins. *J. Appl. Chem.* **1995,** *68* (9), 1532–1538.
57. Aksimentyeva, O. I.; Grytsiv, M. Ya.; Konopelnik, O. I. Temperature Dependence of Resistance and Thermal Stability of Doped Polyaniline. *Functional Materials.* **2002,** *9* (2), 251–254.
58. Abella, L.; Pomfreta, S.; Adamsa, P.; Monkmana, A. Thermal Studies Doped Polyaniline. *Synt. Met.* **1997,** *84,* 127–12.
59. Nikolskiy, B. P., Ed. *Chemist's Handbook*; Khimiya: Leningrad, 1966; Vol.1.
60. Zakordonskiy, V. P.; Aksimentyeva, O. I.; Martyniuk, G. V. Thermochemical and Kinetic Features of Curing of Epoxy-Amine Coatings in the Presence of Fillers. *Composite polymeric material.* **1989,** *43,* 25–29.
61. Skladanyuk, R. V.; Zakordonskiy, V. P. Influence of Physical Structuring on the Kinetics of Reaction of Epoxy and Amine in Filled Systems. *Macromolecular compounds A.* **2004,** *47* (1), 34–43.
62. Zakordonskiy, V. P.; Skladanyuk, R. V. Rheology and Kinetics of Epoxy – Amine Reactions during Epoxy Polymer Formation. *Polymer Science A.* **1998,** 40(7), 669–674.
63. Borchard, F. J.; Daniels, F. Application of Differential Thermal Analysis to Study of Reaction Kinetics. *Eur. Polym. J.* **1957,** *79,* 41–46.
64. Ukrainets, A. M.; Aksimentyeva, O. I.; Martiniyk, G. V.; Konopelnyk, O. I.; Evchuk, O. M. Thermomechanical and Electrical Properties of Composites of Conjugated Polyaminoarenes and Polyvinyl Alcohol. *Problems of Chemistry and Chemical Technology.* **2004,** *3,* 132–135.
65. Aksimentyeva, O. I.; Konopelnik, O. I.; Yurkiv, V. V.; Martiniyk, G. V.; Shapovalov, V. V. Percolation Phenomena and Spin Dynamics in PANI–PMMA Blends. *Molec. Cryst. Liq. Cryst.* **2007,** *468,* 309–316.
66. Gul, V. E.; Shenfil, L. Z. *Conductive Polymer Compositions*; Khimiya: Moscow, 1984.

67. Anisimov, Y. N.; Savin, S. N. Kinetics of the Formation and the Spatial Structure and Strength Characteristics of Semiinterpenetrating Polymer Networks of Epoxy Resins and Oligoester Acrylates. *Ukr. Chem. J.* **2000,** *66* (4), 117–121.
68. Inokuti, H.; Alatau, H. *Electrical Conductivity of Organic Semiconductors*; IIL: Moscow, 1963.
69. Zakordonskiy, V. P.; Markovska, R. P.; Ukrainets, A. M. *Methodological Guidelines for the Study of the Rheology of Polymeric Systems*; LSU Publisher: Lviv, 1988.
70. Teitelbaum, B. Y. *Thermomechanical Analysis of Polymers*; Nauka: Moscow, 1979.
71. Gul, V. E.; Kulezniov, V. N. *Structure and Mechanical Properties of Polymers*; Vysshaya Shkola: Moscow, 1972.
72. Trostianskaya, E. B.; Babayevskiy, P. G. Formation of Cross-Linked Polymers. *Chem. Rev.* **1971,** *40* (1), 117–141.
73. Petko, I. P.; Batog, A. E.; Zaytsev, Y. S. Influence of the Chemical Structure of Epoxy Oligomers on Heat and Thermal Stability of Polymers. *Composite polymeric material.* **1987,** *34,* 10–17.
74. Zakordonskiy, V.; Hnatyshyn, S.; Skladanyuk, R. Some Aspects of the Impact of Fine Fillers on the Thermal Stability of Epoxy Polymer ED-20. *Proc. Shevchenko Sci. Soc. Chem. Sci.* **2007,** *43,* 118–132.
75. Zakordonskiy, V. P.; Hnatyshin, S. Y.; Soltys, M. M. Thermal Degradation of Epoxy Polymer. Method of the Evaluation of Kinetic Prametrs on the Base of Thermogravimetric Data. *Pol. J. Chem.* **1998,** *72,* 2610–2620.
76. Deev, I. S.; Kobets, L. P. Pattern Formation in Filled Thermosetting Polymers. *Colloid. J.* **1999,** *61* (5), 650–660.
77. Rihe, L.; Levi, D. The Dynamic Thermogravimetric Analysis in Polymer Degradation. In *New Methods in the Polymers Studies;* Rogovin, Z. A., Ed.; Mir: Moscow, 1968; pp 148.
78. Zakordonskiy, V. P.; Hnatyshin, S. Y.; Soltys, M. M. Influence of Fine Mineral Fillers on the Thermal Stability of Epoxy Composites. *J. Appl. Chem.* **1988,** *9,* 1524–1528.

CHAPTER 6

THERMOCHROMIC EFFECT IN THE FILMS OF CONJUGATED POLYAMINOARENES

O. I. KONOPELNYK[1] and O. I. AKSIMENTYEVA[2]

[1]*Department of General Physics, Faculty of Physics, Ivan Franko National University of Lviv, 8 Kyryla & Mefodia Str., Lviv 79005, Ukraine. E-mail: konopel@ukr.net*

[2]*Department of Physical and Colloid Chemistry, Faculty of Chemistry, Ivan Franko National University of Lviv, 6 Kyryla & Mefodia Str., Lviv 79005, Ukraine. E-mail: aksimen@ukr.net*

CONTENTS

ABSTRACT

The temperature dependences of optical absorption (transmittance) spectra of the conducting polymer films–polyaniline (**PAn**), poly-*o*-toluidine (**PoT**), poly-*o*-anisidine (**PoA**), poly-*o*-aminophenol (**PoAPh**) doped by sulfuric acid, potassium ferricyanide, and silver nanoparticles have been studied in the temperature interval $T = 80–403$ K. It has shown that temperature increasing causes a change in optical transmittance (absorption) spectra of polymers in all temperature range. However, characteristics of thermo-optical changes or thermochromic effects depend on polymer nature and doping conditions. Changes in parameters of the optical spectra were connected with conformation of polymer chain and modification of electron properties of conducting polymers under temperature action.

6.1 INTRODUCTION

Among the all thermo-induced effects, which are observed in materials under the action of temperature, exactly the thermochromic effect calls an especial attention. It represents by itself the reversible change of the absorption spectrum (and thus, color) of the material as a result of its heating and cooling.[1–3] Thermochromism manifests itself in the materials of both organic and inorganic nature, and at present, calls an especial interest of researchers due to the intensive use of such materials in thermography (sensor technology), optoelectronics (recording and processing of information)[4,5] as well as flat thermometers, testers, indicators.[3,6] Particular industry are "thermochromic windows" or so-called devices, which permit to control the intensity, color, polarization or direction of light coming from an independent source, or to adjust the light output by the change of the optical transmission function depending on temperature.[7,8] These devices are designed in particular for buildings, office windows, wall panels, partitions, doors, and more. These include the transparency film or plate, or layer of liquid or gel used immediately or placed between the plates of glass or transparent plastic, or printed on them.[7]

The most detailed thermochromism studied[9,10] in crystals of inorganic complexes, chalcogenide semiconductors, wide-crystal $SrTiO_3$. To produce "smart" windows the thermochromic films of metal oxides, namely WO_3, VO_2 obtained by sol-gel method on the quartz surface are used.[11]

In organic media, the thermochromic (thermotropic) properties exhibit the liquid crystals (**LC**).[12] High temperature sensitivity of **LC** allows using them in high sensitive sensors, but their color range is limited due to the principle of work.[13]

Thermochromism can be detected in the all various classes of polymers: in thermoplastic materials, plastics, gels, paints, or other types of coatings. Thermochromic effect can be caused by the introduction of individual polymers with the introduced thermochromic additive or supramolecular structure formed by the interaction of polymer with un-thermochromic admixture.

The most common method of the thermochromic polymer materials obtaining is the incorporation of thermochromic material in the polymer matrix.[2,3] Both components form a separate phase and do not have an impact on the structure of phase of the other component.

Microencapsulated systems leucocolorant-developing agent-solvent, inorganic pigments, and conjugated polymers are quite suitable representatives for these purposes. The microcapsules containing of colorant, acidic activator (developing agent), and solvent are important systems for achieving the thermochromic properties in different polymeric materials by forming of a separate phase of thermochromic composition in the un-thermochromic polymer matrix. These incorporated systems, and also the properties of complexes of thermochromic crystals are discussed extensively,[1–11] while the thermochromic properties of conductive conjugated polymers have been studied insufficiently. At the same time, the conjugated polymeric materials can exhibit a strong thermochromism.[14,15] In some donor–acceptor blends of conjugated polymers the intermolecular charge transfer complex (ChTC) is formed, the optical properties of which may be responsible for the thermochromic effect.[16,17]

Unlike to liquid crystal the conjugated polymers have a wide range of colors, but the physical and chemical nature of the temperature response is still not fully understood. Note that from a physical point of view, the origin of the thermochromic effect can be multifarious. It can be derived from the change in light reflection, absorption, and/or scattering of light depending on the temperature.[2]

In terms of color generating in response to an external force is particularly interesting the conjugated polymers with its own electronic conductivity and their composites, which belong to the "smart" or "intelligent" materials.[18–21] External physical and chemical factors, namely temperature, radiation, electric or magnetic field, the adsorption of gases, solvents, or ions can change the electronic properties of conjugated polymer systems – energy gap, concentration of the charge carriers, or their mobility. Under the action of external influence, the chromogenic material induces the colored center, capable to absorb of the light in the visible spectrum. The ability to generate of a color is determined by the energetic characteristics of the molecules in the ground and excited states and by the factors that characterize the transition from one state to another.

In general, the color change is based on the change of the electronic states of molecules, especially the state of π- and d-electrons. Thermochromic effect may be due to the temperature dependence of the conductivity.[19,20] At low temperatures ($T < 100$ K) a conductive polymer is in the dielectric state, that is another manifestation of the increasing of the charge carriers localization[20–22] and the thermochromic effect is hardly evident. The most interesting is the study of thermochromic effect in the temperature range 273–393 K, which includes both temperature inherent to human body and ambient temperature (natural or industrial) for use in the color indicators that will respond to human touch, temperature environment, etc.

It is believed that the thermochromity in conjugated polymers derives from the conformational changes of the polymer chain, which can occur with temperature changes continuously or abruptly at the phase transitions. Even the smallest change in conformational structure may cause significant color changes. In particular, the flatness (planarity) in the main chain of the polymer plays an important role. However, it should be noted that the thermochromic effect in conjugated polymers is based on the change in optical absorption properties and is not caused by a change in *Bragg* reflectivity for periodic structure or helix pitch change depending on the temperature, as this in the case for liquid crystal phases with spiral superstructure.[12,13]

Among the conjugated polymer systems, the thermochromism is intensively researched for substituted polythiophenes.[23–25] Introduction of alkyl substituents provides the flexibility of a hard polythiophene chain and improves its physical and chemical properties.[24] Therefore, in the neutral (un-doped) state the polyalkylthiophenes are chemically stable, soluble in well-known organic solvents and are melted at the temperature below the thermal decomposition temperature.[25] We found that during the heating of the films or solutions of polyalkylthiophenes, it is observed a marked shift of the long-wavelength absorption band toward the lower wavelengths.[23] Corresponding color changes that occur at the same time show the twisting of the polymer chain from the planar to non-planar disordered conformation. Temperature dependence of dichroism spectra of cyclic derivatives of polythiophenes allows us to understand the nature of the thermochromity in these systems. It is shown that the flexible lateral substituents act by increasing the entropy of dissolution and the screening of the interaction between the main chains. Interestingly, that the alkyl side chains having a profound effect on the microstructure of the polymer determine its unique properties, such as solvatochromic behavior,[25] reversible thermochromic transitions between low-temperature and high-temperature solid-state phases.[26-27] Opportunities of the formation of discrete crystallinity by alkyl side chains and the appearance of thermotropic liquid crystal state in polymers of this type are intensively studied.[28,29]

Thermochromic effect in conjugated polymers is also observed in films obtained by Langmuir–Blodgett (**LB**) method, particularly from a mixture of poly(3-hexythiophene), poly(3-octylthiophene), and poly(3-octadecylthiophene) with the arachidonic acid.[28] With increasing of temperature the absorption maximum for un-doped films is shifted toward the higher energies. Compared with thin films of the same materials derived by spin-coating method, the peak shift of the absorption with temperature is occurred in the **LB** films at lower temperatures. The observed changes of the optical absorption depending on temperature are only partially reversible.

In conjugated polyaminoarenes (**PAAr**), namely polyaniline (**PAn**) and its derivatives, the thermochromism studied very little. It is believed that the characteristics of the molecular structure of **PAn** (rotation of the phenyl fragments around the polymer backbone and the formation of hydrogen bonds between adjacent chains) limit the flow of thermochromic transitions in this polymer. One way of the **PAn** modification can be input of the saturated hydrocarbon alkyl or alkoxyl side substituents in the polymer backbone,[29] or the use of doping acids with big anions – doping agents, for example, camphorosulfonic acid (**CSA**) in order to obtain a salt form of **PAn**.[30] Moreover, most of such systems exhibit the ability to form layered structures in three dimensions. Rigid main chains are packed into plateau-like layers, and the alkyl side chains are located between these layers, forming the crystallites.[29]

Several articles have reported about a little "blue" shift of the excitonic peak, which is observed at the heating of the films or solutions of emeraldine base of **PAn** with the substituted aromatic ring.[29] High sensitivity of the spectroscopy of cyclic dichroism to the changes of the molecular conformation confirmed that the optical activity of **PAns** provides the ability to monitor the thermochromism in **PAn** salts. An investigation of the temperature dependence of the spectra of cyclic dichroism and UV–visible absorption in the range $T = 20–240$ °C was carried out for the films obtained both electrochemically and chemically.[29] Thermogravimetric analysis enabled to find that the **PAn** in films obtained under different conditions has different conformations at room temperature. Under the heating the conformation of a polymer into electrochemically deposited film is changed, and at 140 °C is closed to that peculiar to chemically obtaining film.

The adsorption spectra of electrochemically and chemically obtained films are markedly different from each other, so the different conformations of **PAn** chains are observed in two salt forms. The presence of intense band of localized polarons at 825 nm in the spectrum of chemically synthesized polymer conformation characterizes the "compact coils" of **PAn** chain. Conversely, this polarons band is weaker and more degraded in electrochemically obtained film; the "tail" of free carriers is located in the near-infrared spectrum. This form of spectrum reflects a great delocalization of carriers in polarons band, confirming the superiority of conformation of "rectified rolls" or "expanded spiral."[30] The proceeding of irreversible, temperature-induced structural changes in **PAn** at $T = 140$ °C permits to suggest that the conformations of "compact rolls" is more thermodynamically stable for the salt **PAn(+)–CSA**. These changes do not disappear at cooling unlike to the thermochromic effects that were previously observed in a case of the polythiophenes.[25,26] Therefore, this behavior can be more accurately defined as the transition temperature, but not as thermochromism. Interestingly, that at the cooling to room temperature of samples, which were electrochemically obtained and were previously heated to 130 °C, the spectrum to cyclic dichroism returns to its previous state as it was before heating. Perhaps, the reversible structural changes associated with the thermal loss of water from the polymer film and its subsequent adsorption by exposure over some time at room temperature.[29] Thermogravimetric analysis of powder samples showed that 6% weight loss in the temperature range 50–150 °C associated with the removal of water from the polymer. Similar losses were observed by Monoman et al. for chemically derived from a solution in m-cresol sample of **PAn(+)–CSA.**[30] The researchers suggest that water molecules form the strong hydrogen bonds with the nitrogen atoms in the molecule of **PAn**. These thermal losses would have to increase the free volume between the polymer chains than to facilitate its conformational changes.

A study of the temperature dependence of the optical spectra of neutral (un-doped) forms of **PAAr** – **PAn** and its derivatives both in solution, and in the solid-state condition, showed that the thermochromic effects may be caused by conformational changes of the polymer chain induced by the thermal action.[22,31] However, the thermochromic effect in conductive (doped) forms of conjugated **PAAr** with the exception of *ref.*,[30] is practically not studied.

In this work, we have investigated the effect of temperature on the optical absorption spectra of thin films of **PAAr**, namely **PAn**, poly-*o*-anisidine (**PoA**), poly-*o*-aminophenol (**PoAPh**), and poly-*o*-toluidine (**PoT**) obtained electrochemically on the surface of SnO_2 and doped with different doping agents, for example, sulfuric acid, complex compounds of iron, and silver nanoparticles.

6.2 AN IMPACT OF TEMPERATURE ON OPTICAL SPECTRA OF THIN LAYERS OF THE CONJUGATED PAAr. AN EFFECT OF SUBSTITUENT

The films of **PAAr** on a surface of SnO_2 were obtained via electrochemical polymerization of 0.1 M solutions of monomers (*o*-toluidine, aniline, *o*-aminophenol, *o*-anisidine in 0.5 M sulfuric acid in potentiostatic electrolysis mode at potentials E = 0.96; 0.80; 0.92, and 1.06 V (Ag/AgCl) correspondingly for 60 min. The molecular structure of the investigated polymers is represented in Figure 6.1.

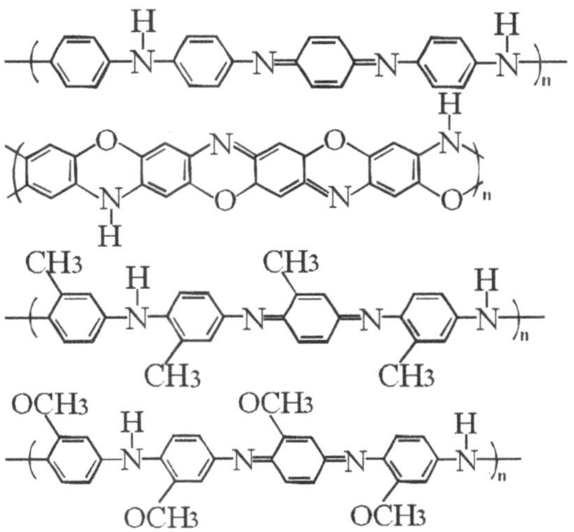

a)

b)

c)

d)

FIGURE 6.1 The structure of elementary link of conjugated **PAAr**: a) polyaniline (**PAn**); b) poly-*o*-aminophenol (**PoAPh**); c) poly-*o*-toluidine (**PoT**); d) poly-*o*-anisidine (**PoA**).

Optical absorption spectra of film samples standed at temperatures of 293–473 K were fixed for 30 min with the use of a spectrophotometer *SF–26* in the wavelength range 200...1100 nm. Thermal behavior was studied using the derivatograph of *Q–1500D* mark in the temperature range 293–773 K at heating rate of 10 K/min. The powder samples of **PAAr** obtained by oxidative polymerization of monomers in 0.5 M sulfuric acid under the action of ammonium persulfate have been used for a thermal analysis. Doping level according to thermogravimetric and elemental analysis was 48–50% mol. The temperature dependence of the resistivity of doped **PAAr** was studied using a specially constructed cell in 2-pin circuit[32] at the heating rate commensurate with the rate of a change of temperature in the thermogravimetric measurements.

According to the data of electronic optical spectroscopy, for the all studied films in the near UV and visible range of the spectrum (see Fig. 6.2, a–d), there are three bands, typical for the absorption of **PAAr**. According to the existing concepts,[19,21,29,30] the absorption band with a maximum at λ = 1.4–1.65 eV is typical for the charge carriers (polarons) both localized and delocalized ones along the polymer chain. Absorption at 1.9 eV is characteristic for n–π* transition in the imino-quinoid structures of **PAAr**. The bands corresponding to the energy interval 3.2–3.4 eV can be attributed to the electronic transition between π and π* levels (the valence band and the conduction band of the conjugated polymer) that is connected with the energy gap.

At the study of the optical absorption spectra of polymer films, which were under the action of higher temperatures in the range 293–423 K, we have established the presence of thermochromic effect in the sulfate-doped films of **PoT**, **PAn**, **PoAPh,** and **PoA.**[33–35] During the thermal treatment of the films, there is a change of their color. In optical spectra, it finds its expression in the "blue" shift of the position of the absorption bands and in the change of their intensity (see Fig. 6.2).

The nature of the spectral changes in **PAAr**'s films strongly depends on the molecular structure of the polymer, including the presence, location, and relative position of substituents. In particular, in a case of the unsubstituted **PAn** (which is characterized by a linear structure shown on Fig. 6.1 (a)) an increase of temperature is accompanied by a shift of the absorption maxima and by the general increase of the absorption intensity (Fig. 6.2 (a)), and the film takes on a dark color. In optical spectra of substituted **PAAr,** the character of thermochromic changes is somewhat different.

In a case of **PoAPh,** which is characterized by the stepped (ladder) heterocyclic structure (Fig. 6.1 (b)), at an increase of temperature to 413–433 K and higher there is a monotonic increase of the optical absorption. Thus, there is a noticeable shift in the band referring to the $\pi \rightarrow \pi^*$ transition toward the higher energies (Fig. 6.2 (b)). At higher temperatures a broad band is formed in the range E = 2.9–3.05 eV. For the macromolecule of **PoT** with the methyl substituent into aminoarene ring (Fig. 6.2 (c)) the optical adsorption is decreased (the film grows lighter) to the temperature 383–393 K, whereas in a case of **PAn** in a range of temperatures 293–472 K the optical adsorption is increased (or color becomes saturated) (Fig. 6.2 (a)).

The presence of the methoxyl substituent in the *ortho*-position to the amino–group in the molecule of **PoA** (see Fig. 6.1 (d)) leads to an interesting thermochromic behavior of the polymer. There are two temperature ranges (from 293 to 353 K and from 383 to 413 K), in which the decrease of the absorption intensity of **PoA** at the general short wave of the maxima position (Fig. 6.2 (d)) takes place.

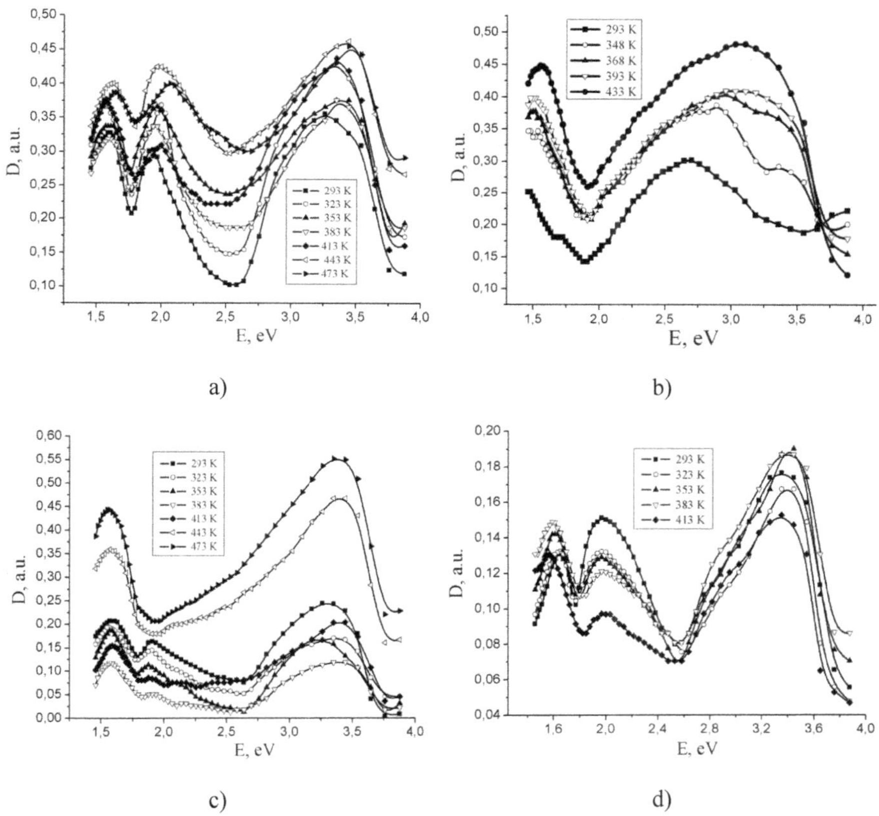

FIGURE 6.2 Absorption spectra of the films: **PAn** (a), **PoAPh** (b), **PoT** (c), and **PoA** (d) heated at different indicated temperatures.

The studies have shown that for the electronic spectra of **PoT** and **PoA** films, the general decrease of the second adsorption band is typical and its complete disappearance takes place at $T > 413$ K. These thermochromic changes are in good agreement with the observations of Zheng et al.[29] on the structural changes in the un-doped films of *N*-alkylated **PAns**. Degradation of this band is connected with the polymer chain crosslinking processes occurring in the films at the higher temperatures. We found that the infrared absorption bands of the quinoid form of substituted **PAArs** are quickly weakened at higher temperatures up to 393–433 K, and at higher temperatures are almost completely disappeared. Similar observations were made for the absorption bands of deformation vibrations of imine nitrogen = *N*–, which almost completely disappears at 473 K as a result of processes cross-linking of the polymer chain.[29] However, in the case of doped **PAn**, an intensity of this band is varied very little (see Fig. 6.2 (a)), indicating that the cross-linking process requires the higher temperatures.

An investigation of thermal behavior of conjugated **PAAr** in the range $T = 293–773$ K allowed to fix some extremes on the **DTG** and **DTA** curves. In the case of sulfate-doped **PAn** (see Fig. 6.3 (a)) the extremes on the **DTG** and **DTA** curves at $T = 403$ and 413 K corresponds to desorption of chemisorbed water.[31,32] The presence of small **DTA** peak at $T = 473$ K, which is not accompanied by a noticeable loss of weight may indicate on the progress of the thermal oxidation of amino-quinoid fragments to the imino-quinoid ones that occurs with heat eliminating. With further increase of temperature the activating process of the dopants desorption takes place (with a maximum at $T = 553$ K), which is accompanied by the cross-linking processes the polymer chain with the formation of phenazine structures.[32]

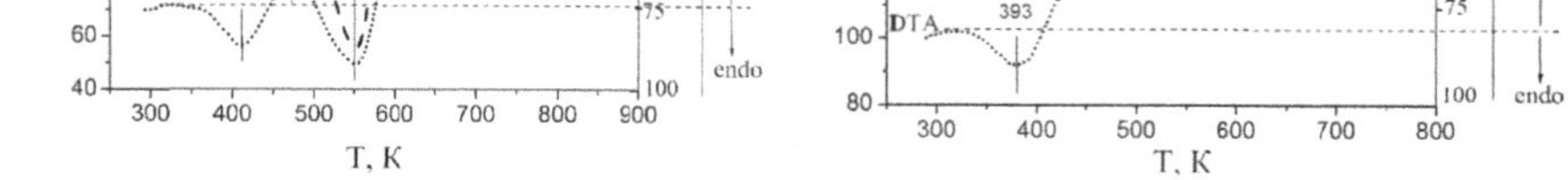

a) b)

FIGURE 6.3 Thermograms of decomposition for sulfate-doped **PAn** (a) and **PoT** (b).

In a case of **PoT**, the endothermic maxima at $T = 373$ K (**DTG**) and $T = 393$ K (**DTA**) correspond to the desorption processes of chemisorbed water. The presence a number of the extremes on **DTG** and **DTA**-curves in the range of temperatures $T = 413–600$ K indicates on the simultaneous proceeding of the processes of thermal-oxidative destruction, crosslinking of polymer chain[29,32] and on the loss of dopants (maximum at $T = 540$ K).

This suggests that at the achievement of specified temperatures in films of **PoT**, there are structural changes in the direction of forming a spatial grid that reduces the kinetic mobility of polymer segments and exaggerates the internal rotation. Perhaps, the presence exactly of such processes makes the thermochromic changes in films of **PoT** irreversible and leads to increase of the optical density.

An investigation of thermal behavior of **PAAr** permitted to discover the four temperature ranges typical for the all studied polymers (see Table 6.1). Endothermic maximum at $T = 383–403$ K is connected with the loss of chemisorbed water, and the sequence of peaks on **DTA** and **DTG** curves at $T = 413–600$ K indicates on the crosslinking of the polymer chains (second extreme at 463–513 K), followed by the desorption of doping agents (513–563 K).

TABLE 6.1 Temperature Intervals Typical for Thermal Behavior of Conjugated **PAAr**.

PAAr	Temperature, K			
	Desorption of water	Crosslinking of macrochain	Desorption of doping additives	Thermal decomposition
PAn	403–413	463–483	533–553	> 723
PoT	373–393	487–498	513–533	> 713
PoA	353–373	463–483	513–533	> 700
PoAPh	383–398	473–513	543–563	> 790

As the proceeding of the polymer chains crosslinking in the films of **PAn** occurs[31,32] only after the temperature 493 K, and the growth of the optical density of **PAn** films is observed in the range 293–473 K, we must assume that the deepening of color in films of this polymer can be related not so much by the conformation of a polymer chain,[30] as by the thermal activation of the charge carriers. For the detection of these factors, we studied the temperature dependence of the resistivity of polymers in the temperature range 293–523 K.

As it was shown previously[31,36], in order to describe the temperature dependence of the resistivity of **PAAr** (ρ) in a range of $T > 293$ K the known exponential equation $\rho = \rho_0 \exp(\varepsilon_\sigma/2\,\mathrm{kT})$ can be used, where ε_σ is an activation energy of the charge transport, ρ_0 is constant. The temperature dependence of the resistivity of polymers normalized to the resistance at room temperature (ρ/ρ_{293}) in the coordinates of $ln(\rho/\rho_{293}) - 1/T$ for **PoT** and **PAn** doped with sulfate acid, is shown in Figure 6.4. In a range of $T = 293–373$ K (**PAn**) and 293–403 K (**PoT**), it is observed a linear dependence of $ln(\rho/\rho_{293}) - 1/T$, which allows to determine the activation energy of the charge transport: $\varepsilon_\sigma = 0.490 \pm 0.01$ eV (**PAn**) and $\varepsilon_\sigma = 0.876 \pm 0.015$ eV (**PoT**).

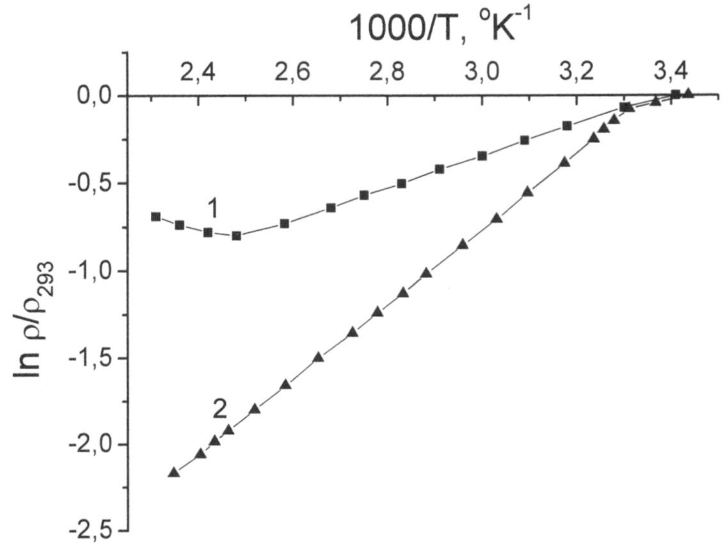

FIGURE 6.4 Temperature dependence of the specific resistance of sulfate-doped **PAn** (1) and **PoT** (2).

Thus, the presence of methyl substituent in the structure of the elementary link of **PoT** causes an increase of linear section in the dependence $ln(\rho/\rho_{293}) - 1/T$ and a significant increase in the activation energy for conduction for **PoT** compared with **PAn**.

At studying of the temperature dependence of the **PoAPh** resistivity,[36] it was found that the nature of the temperature dependence of the resistivity for this **PAAr** is approaching to the dependence found for polymers constructed from the inflexible heterocyclic structures.[37] The activation energy of conductivity in the temperature range 403–533 K is 1.59 ± 0.03 eV.

Calculation of the spatial potential for non-conjugated **PAAr** chain (leucoemeraldin) shows[38] that the most probable for the elementary link of **PAn** an angle of rotation is increased with the temperature increasing, similar to the thermal phenomena into the ordinary solids. Therefore, we should expect thermally induced increase of the energies of optical transitions with temperature increasing, as was observed by us experimentally. However, the nature of the change in optical density of **PAn** films and its derivatives, namely **PoT**, **PoA**, and **PoAPh** is markedly different and requires more detailed consideration.

In order to understand the nature of the thermochromic effect in conjugated **PAAr**, we have analyzed the temperature dependence of peaks position change (ΔE) and the absorption intensity (ΔD) bands corresponding to π–π* transition.[39] It can be seen from Figure 6.5, that for the films of **PAn** and **PoA** the changes of intensity are not as significant as for the films and **PoAPh** and **PoT,** and they are characterized by some extremes at temperatures 320 K and 380 K and in the range 430–450 K. We can assume that at temperatures below 360 K the changes in the optical spectra can be caused by conformational rotations of substituted phenyl rings.[29,38]

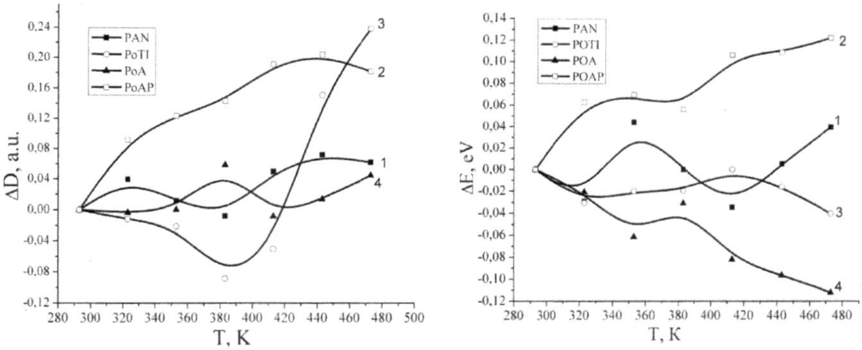

a) b)

FIGURE 6.5 Temperature dependence of intensity (a) and a peak position of absorption band of π–π* transition (b) in the films of **PAn** (1), **PoAPh** (2), **PoT** (3), and **PoA** (4).

Consideration of the thermal behavior of **PAAr** allows connecting the changes at $T > 373$ K with the removal processes of chemisorbed water and cross-linking in polymer chains.[31] Perhaps, at these temperatures in the films of **PoT** there are the structural changes in the spatial ordering of the polymer net. This leads to a decrease in the kinetic mobility of polymer segments and dislocates an internal rotation. For the films of **PoT,** the decrease of ΔD to 390 K is observed, and an increase the intensity of absorption at higher temperatures takes place. This temperature is related to the removal of water (see Table 6.1). In a case of **PoAPh** almost uniform increase in intensity of the band at $E = 1.6$ eV is observed with the temperature increasing.

In Figure 6.5 (b), there are temperature dependencies of the absorption bands position changes corresponding to π–π* transition. These dependencies for the all investigated polymers are complex and are characterized by some inflections. So, for **PAn** the inflections are observed at 320, 360, and 410 K; for **PoT** – at 320 and 410 K; for **PoA** – 350 and 380 K; for **PoAPh** – at 340 and 390 K. The first inflection may be connected with conformational transitions in the polymer chain, but the next ones – with the changes in electronic structure and conductivity as a result of water desorption and reduction of protonation level in polymers doped with sulfuric acid.

Comparison of thermochromic changes in the optical spectra with the thermal behavior of **PAAr** and temperature dependence of the resistivity allows making some assumptions. It is known, that the location of the phenyl groups in the unit link of **PAAr** occurs under certain valence angles, and the phenyl rings are free rotated around the –C–N– bond. Since in the thin-film state the mobility of the polymer chains significantly limited compared to solution, and the absence of large lateral substituents makes it difficult to create the free volume sufficient for an internal rotation, it must be assumed that the thermochromic effect, which is observed in the films associated not only with the change of polymer chain conformations of **PAAr** but also with their electronic properties.

In the presence of methyl or methoxyl, substituent in macromolecules of **PoT** and **PoA** the reinforcement of spatial repulsion between both of the phenyl rings and neighboring chains[38,39] compared to the un-substituted **PAn** takes place. The reinforcement of the macrochains repulsion causes on the one hand the lowering of the conductivity due to the charge localization[37,40] and on the other hand – a significant simplification of the internal rotation.[38] Therefore, in the range of 293–403 K the thermochromic changes are caused by the changes in the conformation of the polymer chain and the "translucence" of films is observed. In a case of sulfate-doped **PAn**, when the doping leads to the regulation of a structure[20,32], there is a difficulty in the internal rotation. At that, an increase of the mobility of charge carriers along the polymer chain, which causes the growth of electronic conductivity of **PAn** at small values of activation energy, takes place. Confirmation of this hypothesis may be the dominant one-dimensional charge transport stated by *Epstein*, MacDiarmid et al.[40] for the salt forms of **PAn** and **PoT**. Calculated via the model of one-dimensional jump inverse localization length for the salt forms of **PoT** α^{-1} = 7.4 Å is much smaller compared to that calculated one for **PAn** α^{-1} = 20 Å. Obviously, in a case of doped **PAn** an effect of high temperatures causes the delocalization of charge carriers as a result of "deployment" twisted into a spiral macromolecule at $T > 413$ K, which leads to increase of the optical density in the whole temperature range, up to the beginning of thermal degradation of the polymer.

Thus, the changes in electron transfer energy and in optical density of films depend on the presence and nature of substituents in the benzene ring of conjugated **PAAr**. Thermochromic effect of substituted **PAAr** in the temperature range 293–363 K is caused mainly by conformational changes of macrochain, while the effect of thermo-induced changes in the structure of the polymer chain is small. At higher temperatures, the contribution associated with the activation temperature of electronic conductivity of polymers until a temperature of spatial cross-linking and thermal degradation becomes more noticeable.

6.3 THERMAL OPTICAL PROPERTIES OF PAAr DOPED BY INORGANIC PARTICLES

To improve the characteristics of optical devices based on electroconducting polymers, especially their contrast sensitivity and operating stability, the process of polymers doping by oxides (namely, WO_3, V_2O_5, RuO_2, etc.) or by complex metal compounds[18,41,42] is often used. Interesting results were obtained[43] under the study of electrochemical properties of hybrid materials based on **PAn/Fe(CN)$_6$**, but the thermochromic effect in such systems was almost unexplored.

We have studied the temperature dependencies of the optical absorption spectra for the films of **PAn**, doped with potassium hexacyanoferrate complex $K_3[Fe(CN)_6]$ in the range T = 80–380 K. Electrodeposition of the **PAn** films on the surface of SnO_2 electrodes has been done via electrolysis of 0.1 M aniline solution in 1 M HCl at current density i = 0.1 mA/cm^2 for 8–10 min (T = 293 K). The thickness of films was determined using the microinterferometer *MIM*–4 with an accuracy of ± 12 nm.

Doping of polymer films by the complex $K_3[Fe(CN)_6]$ was carried out during the ion exchange reaction[44] by exposure of films in 0.1 M aqueous HCl containing the 0.01 M of $K_3[Fe(CN)_6]$ for different times (2 and 24 h). The content of dopant determined after the samples washing and drying with the use of the *X*-ray microanalyzer "Camebax" was 0.3 and 0.5 at.% Fe, respectively. Using the *EPR* spectroscopy, it was determined that the spectrum of obtained doped polymer is a superposition of several *EPR* spectra of free radicals (see Fig. 6.6). The presence of two resonance lines with g_1 = 4.22 ± 0.03 (at T = 4.2 K) and g_2 = 2.13 ± 0.05 (at T = 293 K) can be attributed to the paramagnetic center Fe(III) that make up the complex of

$K_3[Fe(CN)_6]^{3-}$. The third resonance line based on the value of *g*-factor (g = 2.000 ± 0.001) and the nature of the temperature behavior belongs to free radicals of polymer matrix of **PAn**.[44]

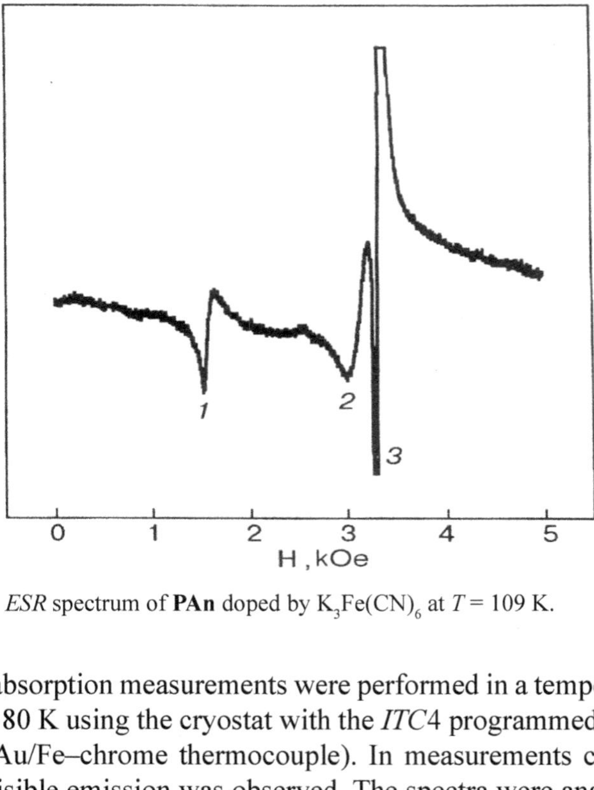

FIGURE 6.6 *ESR* spectrum of **PAn** doped by $K_3Fe(CN)_6$ at T = 109 K.

Optical absorption measurements were performed in a temperature range from 80 to 380 K using the cryostat with the *ITC*4 programmed temperature controller (Au/Fe–chrome thermocouple). In measurements carried out at 380 K the visible emission was observed. The spectra were analyzed with a *Zeiss model SPM2* grating monochromator (setting to a spectral bandwidth of 2 cm^{-1}) detected by a cooled photomultiplier. The *SRS* 250-boxcar integrator has averaged the resulting signal. A continuous flow helium cryostat (*Oxford*, model *CF*1104) was used for temperature measurements.

Optical absorption spectra of **PAn** doped with hydrochloric acid (**PAn–HCl**) and **PAn** doped with potassium hexacyanoferrate complex (**PAn–[Fe(CN)$_6$]** for 2 and 24 h are shown in Figures 6.7 and 6.8. The spectrum of **PAn–HCl** (see Fig. 6.7 (a)) contains several bands with maxima at λ = 380, 502, 640, and 700 nm, which is typical for the absorption spectra of this polymer.[19,21] The band, which is observed in the near UV range at 360–380 nm is appeared due to π–π^* transition in the aromatic system of **PAn**; the band with a maximum λ = 502 nm can be assigned to n–π^* transition into imino-benzene fragments of the polymer chain. Absorption bands at 640–700 nm are due to the absorption in the polaron zone.[21]

As can be seen from the figures, the intensity of the all absorption bands is slightly varied with the increasing of temperature in a range $T = 80–300$ K, but shows the significant changes when the temperature increases to 380 K. This thermal behavior of **PAn** films is in good agreement with the assumption that the changes in the optical spectra of **PAAr** can be caused by thermal activation of carriers at temperatures higher than the room temperature.[30,35] Indeed, according to *ref.*[45] the heating of **PAn** samples at temperatures $T >$ 293 K (to 373 K) leads to decrease of energy band gap Eg (defined by the edge of the optical reflection) from 3.1 to 2.6 eV, which may be related to the increase of the concentration of charge carriers. At higher temperatures an increase of the energy gap and the fall of conductivity take place, which is also observed by us for films **PAn** at $T > 373$ K (see Fig. 6.4).

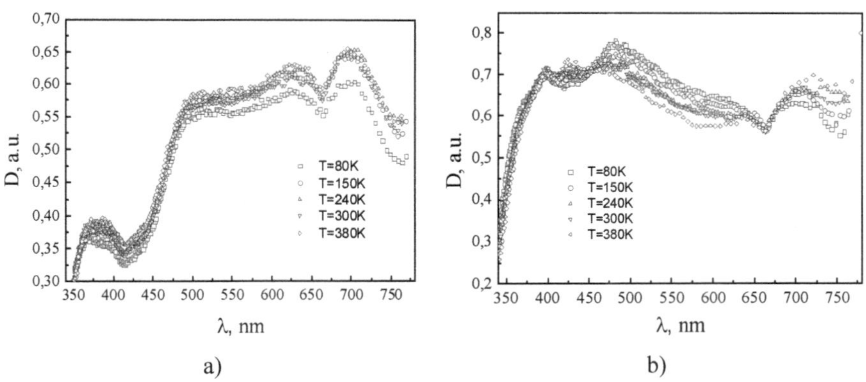

FIGURE 6.7 Optical transmittance spectra for film of **PAn–HCl** (a) and **PAn** doped with $K_3[Fe(CN)_6]$ during 2 h (b) in the visible wavelength range from 340 to 780 nm at different temperatures.

For the films of **PAn** doped by $K_3[Fe(CN)_6]$ an increase of the optical density over the spectral range and a significant shift of the absorption maxima toward the lower energies (so-called, bathochromic shift) are observed. For the films doped for 2–3 h, the absorption maxima are observed at $\lambda = 395$, 425, 490, and 710 nm (see Fig. 6.7 (b)), and for deeper doping times – at 395, 425, 639, and 710 nm, while the band with a peak at $\lambda = 490$ nm almost disappears (see Fig. 6.8). Temperature dependence of the spectral shape and intensity of the absorption lines are different from the previous ones.

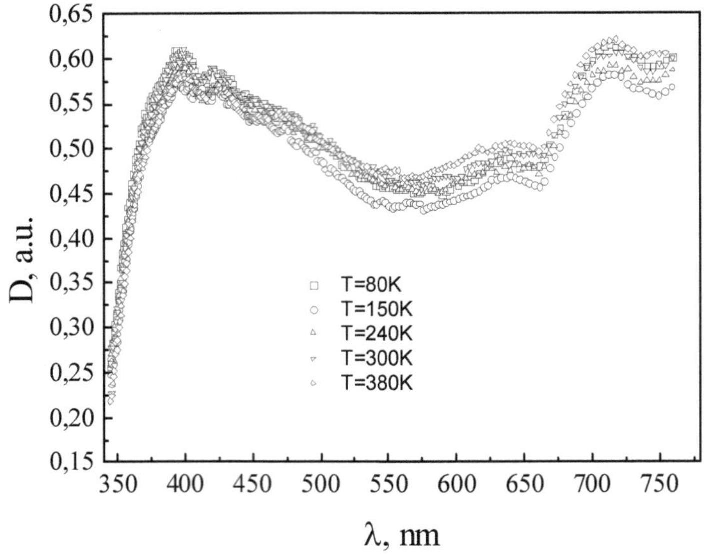

FIGURE 6.8 Optical absorption spectra for **PAn** doped with $K_3Fe(CN)_6$ during 24 h in the visible wavelength range from 340 to 780 nm at different temperatures.

Comparison of the optical absorption of **PAn** films at different content of dopant is shown in Figure 6.9. As can be seen from Figure 6.8, the nature of changes in the intensity of the spectrum depends on the range of the wavelengths (λ). At 80 K the line intensity is minimal for wavelength of 425 nm, while for the wavelength of λ = 490 nm, it is maximum and again is minimum for wavelength of λ = 710 nm. Such behavior was not observed for the sample of **PAn** doped by $K_3[Fe(CN)_6]$ complex with maximum intensity at about 395, 425, 639, and 710 nm (see Fig. 6.2). Doping of the **PAn** film with $K_3[Fe(CN)_6]$ complex is favorable to the increasing intensity of transmittance in spectral region from 340 up to 500 nm, while the increase of the doping time leads to some reducing intensity of the transmittance spectrum, that is well visible in Figure 6.9.

So, the thermochromic effect in films of **PAn** doped with $K_3Fe(CN)6$ is expressed in complex thermo-induced changes in the optical absorption spectra. This impact of temperature on the spectra of the polymer may be connected both with the presence of a rotation in the segments of the polymer chains and with the change of the electronic properties of conjugated polymers.[38,39] At that, the doping of **PAn** by the complex $K_3Fe(CN)_6$ causes the stronger temperature dependence of the optical transition toward the undoped **PAn**.

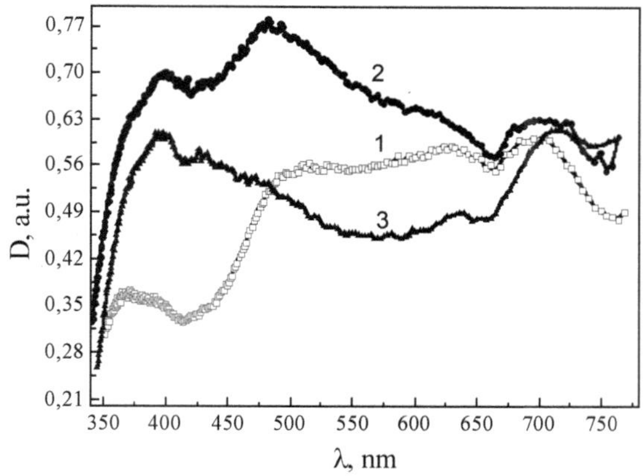

FIGURE 6.9 Optical absorption spectra of **PAn** films with different degree of doping by complex $K_3Fe(CN)_6$: **PAn** (1), **PAn** doped with $K_3Fe(CN)_6$ during 24 h (2), **PAn** doped with $K_3Fe(CN)_6$ during 2 h (3) in the visible wavelength range from 340 to 780 nm at different temperatures.

Sensitivity to the higher temperatures was determined also for the films of **PoT** obtained by electrochemical polymerization on the surface of SnO_2 electrodes and doped with silver nanoparticles.[46] Doping was carried out by exposure of films for some time (15–30 min) in 1% nanosilver solution obtained by the standard method.[47] Thermochromic effect was investigated both for undoped and for composite films of **PoT–Ag** in the temperature range 293–353 K; the corresponding spectra are shown in Figure 6.10.

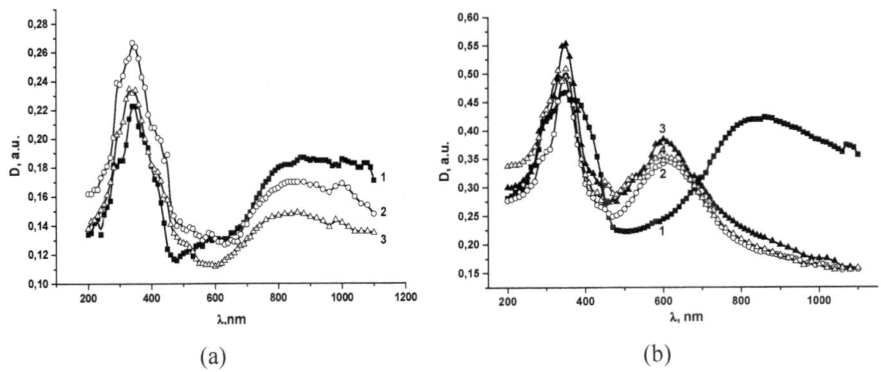

FIGURE 6.10 (a) Optical absorption spectra of un-doped **PoT** film at different temperatures: 1–293 K; 2–323 K; 3–353 K; (b) Optical absorption spectra of **PoT** films, doped by nanoparticles of argentums at different temperatures: 1—un-doped film, $T = 293$ K; 2—**PoT–Ag**, $T = 293$ K; 3—**PoT–Ag**, $T = 323$ K; 4—**PoT–Ag**, $T = 353$ K. The thickness of film is 320 ± 12 nm.

Presented spectra demonstrate the effect of dopant on the optical absorption features of **PoT**. For the undoped film due to heat treatment at temperatures 323 and 353 K the intensity increases and there is some extension of the first absorption band (λ = 370 nm). For a long-wavelength absorption band in the field of 800–1000 nm, there is an inverse relationship – the intensity of the absorption decreases and a peak is shifted toward the lower wavelengths.

For the samples doped with nanosilver the situation is significantly changed, namely, for the first band (λ = 360 nm) along with the increasing of the absorption intensity its width is decreased. Since the absorption at 360 nm caused by π–π^* transitions into band gap, it can be assumed that the doping by nanosilver leads to the greater regulation of the energy structure of the polymer. At this, a broad band with a maximum at 810 nm actually disappears due to the delocalized charge carriers along the polymer chain. We can assume that nanosilver interacts with the charge carriers that are created by imino-benzene fragments of **PAAr**.[21] Reduction of these fragments causes the absorption in the region of 600 nm, which is attributed to the n–π^* transitions in amino-quinoid polymer chains of **PAAr**.[48]

So, the thermo-induced changes in the optical spectra of conjugated **PAAr** are quite complex and to a large-extent depend on the structure of the polymer chain, the presence of substituents, their nature, and the type dopants. However, this gives rises to a target programming of the thermochromic changes in films of conjugated polymers to create on their basis thermochromic windows, sensors, indicators, etc.

KEYWORDS

- **thermochromic effect**
- **polyaniline derivatives**
- **doping**
- **optical spectra**
- **thermal behavior**
- **conformations**
- **conductivity**

REFERENCES

1. Seeboth, A.; Lötzsch, D. *Thermochromic Phenomena in Polymers;* Smithers Rapra Technology Limited: UK, 2008; pp 95.
2. Bamfield, P. *Chromic Phenomena: Technological Applications of Colour Chemistry;* Royal Society of Chemistry: Cambridge, UK, 2001; pp 374.
3. Seeboth, A.; Lötzsch, D. *Encyclopedia of Polymer Science and Technology;* 3rd ed. Kroschwitz, J. I., Ed.; Wiley: New York, USA, 2004; Vol. 12, pp 143.
4. Lee, S. S.; Jin, Y. S.; Son, Y. S. J. Variable Optical Attenuator Based on a Cutoff Modulator with Tapered Waveguides in Polymers. *J. Lightwave Technol.* **1999,** *17* (12), 2556–2561.
5. Franiv, V. A.; Bovgira, O. V.; Girnyk, I. S.; Kushnir, O. S.; Futey, O. V.; Vas'kiv, A. P. Temperature Sensor Based on Crystals Tl_4hgi_6 and Tl_4pbi_6. *Electron. Inf. Technol.* **2013,** *3,* 35–40 (In Ukrainian).
6. Panchenko, T. V.; Strelets, K. Y.; Litovchenko, A. Y. *Thermo induced Optical Absorption of Crystals $Bi_{12}sio_{20}$: Ag.* Вісник Bulletin of Dnipropetrovsk University, *Phys. Electron.* **2009,** *16* (2), 80–85 (In Russian).
7. Yanush, O. V.; Markova, T. S.; Khvostova, N. O. Thermochromic Device (Options). Ru Patent 2449 331.
8. Inoue, T.; Ichinose, M.; Ichikawa, N. Thermotropic Glass with Active Dimming Control for Solar Shading and Daylighting. *Energ. Buildings.* **2008,** *40,* 385–393.
9. Kapustianyk, V.; Korchak, Yu.; Rudyk, V.; Batiuk, A. Influence of the Growing Conditions on Thermocromic Properties of $[NH_2(C_2H_5)_2]_2$cuxzn1–Xcl_4 Crystals. *Funct. Mater.* **2005,** *12* (4), 775–780.
10. Wild, R. L.; Rockar, E. M.; Smith, J. G. Thermocromism and Electrical Conductivity in Doped $Srtio_3$. *Phys. Rev. B.* **1973,** *68,* 3828.
11. Liu, Y.; Tao, H.; Chu, X.; Wan, M.; Bao, J.; Zhao, X. Effects of Addition of Tungsten Chloride on Optical Properties of Vo_2-Based Thermochromic Films Prepared by Sol-Gel Method. *J. Non-Cryst. Solids.* **2014,** *383,* 116–120.
12. Pikin, S. A.; Blinov, L. M. *Liquid Crystals;* Nauka: Moscow, Russia, 1982.
13. Gotra, Z.; Zelinskii, R.; Mikitiuk, Z.; Sorokin, V.; Sushinskii, O.; Phechan, A. *LCD Electronics;* Gotra, Z., Ed.; A priory: L'viv, 2010; pp 532 (In Ukrainian).
14. Ueno, K.; Matsubara, K.; Watanabe, M.; Takeoka, Y. An Electro- and Thermochromic Hydrogel as a Full-Color Indicator. *Adv. Mater.* **2007,** *19* (19), 2807–2812.
15. Sussman, J.; Snoswell, D.; Kontogeorgos, A.; Baumberg, J. J.; Spahn, P. Thermochromic Polymer Opals. *Appl. Phys. Lett.* **2009,** *95* (17), 173116.
16. Bakulin, A. A.; Elizarov, S. G.; Khodarev, A. N. Et Weak Charge-Transfer Complexes Based on Conjugated Polymers for Plastic Solar Cells. *Synth. Metals.* **2004,** *147,* 221–225.
17. Panda, P.; Veldman, D.; Sweelssen, J.; Bastiaansen, J. J. A. M.; Langeveld-Voss, B. M. W.; Meskers, S. C. J. Charge Transfer Absorption for Π-Conjugated Polymers and Oligomers Mixed with Electron Acceptors. *J. Phys. Chem. B.* **2007,** *111* (19), 5076–5081.
18. Carpi, F.; De Rossi, D. Colours from Electroactive Polymers: Electrochromic, Electroluminescent and Laser Devices Based on Organic Materials. *Opt. Laser Technol.* **2006,** *38,* 292–305.
19. `Heeger, A. J. Semiconducting and Metallic Polymers: The Fourth Generation of Polymeric Materials. *Synth. Metals.* **2002,** *123,* 23–42.

20. Aksimentyeva, O. I. *Electrochemical Methods of Synthesis and Conductivity of Conjugated Polymers;* Svit: L'viv, 1998 (In Ukrainian).
21. Molapo, K. M.; Ndangili, P. M.; Ajayi, R. F.; Mbambisa, G.; Mailu, S. M.; Njomo, N.; Masikini, M.; Bakerand, P.; Iwuoha E. I. Electronics of Conjugated Polymers (I): Polyaniline. *Int. J. Electrochem. Sci.* **2012,** *7,* 11859–11875.
22. Rannou, P.; Dufour, B.; Travers, J. P.; Pron, A.; Djurado, D.; Janeczek, H.; Sek, D. Temperature-Induced Transitions in Doped Polyaniline: Correlation between Glass Transition, Thermochromism and Electrical Transport. *J. Phys. Chem. B.* **2002,** *106* (41), 10553–10559.
23. Winokur, M. I.; Spiegel, D.; Kim, Y.; Hotta, S.; Heeger, A. J. Structural and Absorption Studies of Thethermochromic Transition in Poly–(3–Hexylthiophene). *Synth. Metals.* **1989,** *28,* 419–429.
24. Ingunas, O.; Gustafsson, G.; Salaneck, W. R.; Osterholm, J. E.; Laakso, J. Thermochromism in Thin Films of Poly(3-Alkylthiophenes). *Synth. Metals.* **1989,** *28* (1–2), 377–384.
25. Ingunas, O.; Salaneck, W. R.; Osterholm, J. E.; Laakso, J. Thermochromic and Solvatochromic Effect in Poly-(3-Hexylthiophene). *Synth. Metals.* **1988,** *22* (4), 395–406.
26. Yoshino, K.; Nakajima, S.; Onoda, M.; Sugimoto, R. Electrical and Optical Properties of Poly(3-Alkylthiophene). *Synth. Metals.* **1989,** *28* (1–2), 349–357.
27. Stern, R.; Ballauff, M.; Wegner, G. Synthesis and Phase Behavior of Liquid Crystalline Poly(3-N-Alkynl-4-Hydroxybenzoates). *Macromol. Chem. Makromol. Symp.* **1989,** *23,* 373–380.
28. Ahlskog, M.; Paloheimo, J.; Punkka, E.; Stubb, H. Thermochromism in Langmuir–Blodgett Films of Polyakylthiophenes. *Synth. Metals.* **1993,** *57* (1), 3830–3835.
29. Zheng, W. Y.; Levon, K.; Laaakso, J.; Osterholm, J. E. Characterization of Solid-State Properties of Processable N-Alkylated Polyanilines in the Neutral State. *Macromolecules.* **1994,** *27,* 7754–7768.
30. Norris, I. D.; Kane-Maguire, L. A. P.; Wallace, G. G. Thermochromizm in Optically Active Polyaniline Salts. *Macromolecules.* **1998,** *31* (19), 6529–6533.
31. Abella, L.; Pomfreta, S. J.; Adams, P. N.; Monkmana, A. P. Thermal Studies of Doped Polyaniline. *Synth. Metals.* **1997,** *84* (1–3), 127–128.
32. Aksimentyeva, O. I.; Grytsiv, M. Ya.; Konopelnik, O. I. Temperature Dependence of Resistance and Thermal Stability of Doped Polyaniline. *Funct. Mater.* **2002,** *9* (2), 251–254.
33. Aksimentyeva, O. I.; Konopelnik, O. I.; Grytsiv, M. Ya.; Kovorotny, O. I.; Sitar, A. V. *Thermochromic Effect on the Thin Layers of Polyortotoluidine and Polyaniline on the Surface of Tin Oxide;* Reports in Int. Conf. Physics and Technology of Thin Films, Ivano-Frankivsk, Ukraine, May 19–24, 2003, 87–88 (In Ukrainian).
34. Aksimentyeva, O. I.; Konopelnik, O. I.; Zakordonskiy, V. P.; Grytsiv, M. Ya.; Martynyuk, G. V. Thermochromic Effect at the Thin Layers of Conjugated Polyaminoarenes. *J. Phys. Stud.* 2004, *8* (4), 369–372 (In Ukrainian).
35. Konopelnik, O. I.; Aksimentyeva, O. I.; Martynyuk, G. V. Effect of Temperature on the Optical Properties of Conducting Polyaminoarenes and their Composites with Elastic Polymer Matrix. *Molec. Cryst. Liq. Cryst.* **2005,** *427,* 37–46.
36. Aksimentyeva, O. I.; Grytsiv, M. Ya.; Konopelnik, O. I. Temperature Dependence of the Electrical Conductivity and Structure of Amine Contained Polyarylenes. *J. Phys. Stud.* **2002,** *6* (2), 180–184 (In Ukrainian).

37. Kohlman, R. S.; Joo, J.; Epstein, A. J. *Conducting Polymers: Electrical Conductivity, Physical Properties of Polymers Handbook;* AIP Press: New York, Woodbury, 1996; pp 453–478.
38. Ginder, J. M.; Epstein, A. J.; MacDiarmid, A. G. Phenyl Ring Rotation, Structural Order and Electronic States in Polyaniline. *Synth. Metals.* **1990,** *37,* 45–55.
39. Konopelnyk, O. I.; Aksimentyeva, O. I. Thermochromic Effect in Conducting Polyaminoarenes. *Photoelectronics.* **2011,** *20,* 18–22.
40. Zuo, F.; Angelopoulos; M., Macdiarmid, A. G.; Epstein, A. J. Transport Studies of Protonated Emeraldine Polymer: A Granular Polymeric Metal System. *Phys. Rev. B.* **1987,** *36,* 3475–3478.
41. Tung, T. S.; Ho, K. C. Cycling and At-Rest Stabilities of a Complementary Electrochromic Device Containing Poly(3,4-Ethylenedioxythiophene) and Prussian Blue. *Sol. Energ. Mater. Sol. Cells.* **2006,** *90,* 521–537.
42. Huang, L. M.; Wen, T. C.; Gopalan, A. Electrochemical and Spectroelectrochemical Monitoring of Supercapacitance and Electrochromic Properties of Hydrous Ruthenium Oxide Embedded Poly(3,4-Ethylenedioxythiophene)–Poly(Styrene Sulfonic Acid) Composite. *Electrochimica Acta.* **2006,** *51,* 3469–3476.
43. Gomez-Romero, P.; Torres-Gomez, G.; Molecular Batteries: Harnessing $Fe(Cn)_6^{3-}$ Electroactivity in Hybrid Polyaniline–Hexacyanoferrate Electrodes. *Adv. Mater.* **2000,** *12* (19), 1454–1465.
44. Vasyukov, V. N.; Dyakonov, V. P.; Shapovalov, V. A.; Aksimentyeva, E. I.; Szymczak, H.; Piehota, S. Temperature-Induced Change in the ESR Spectrum of the Fe^{3+} Ion in Polyaniline. *Low Temp. Phys.* **2000,** *26* (4), 265.
45. Joshi, G. P.; Saxena, N. S.; Sharma, T. P.; Mishra, S. C. K. Measurement of Thermal Transport and Optical Properties of Conducting Polyaniline. *Ind. J. Pure Appl. Phys.* **2006,** *44,* 786–790.
46. Savytsky, N.; Konopelnyk, O.; Lytvyn, I., Horbenko, Y. The Features of Optical and Electrical Properties of Polyaminoarenes Doped by Inorganic Nanoclaster, Reports Int. Conf., Eureka 2014, Lviv, Ukraine, May 15–17, 2014; pp 82.
47. Mulvaney, P.; Linnert, T.; Henglein, A. Surface Chemistry of Colloidal Silver in Aqueous Solution: Observations on Chemisorption and Reactivity. *J. Phys. Chem. Berlin,* **1991,** *95* (20), 36.
48. Choudhury, A. Polyaniline/Silver Nanocomposites: Dielectric Properties and Ethanol Vapour Sensitivity. *Sens. Actuators. B.* **2009,** *138,* 318.

CHAPTER 7

POLYANILINE IN CHEMO- AND BIOSENSORICS: OVERVIEW

YA. S. KOVALYSHYN and O. V. RESHETNYAK

Department of Physical and Colloid Chemistry, Faculty of Chemistry, Ivan Franko National University of Lviv, 6 Kyryla & Mefodia Str., Lviv 79005, Ukraine

E-mail: kovalyshyn@yahoo.com; reshetniak@franko.lviv.ua

CONTENTS

ABSTRACT

The basic aspects of the electroactive polymers using in general as well as the polyaniline (**PAn**) as classical representative of these polymers in particular under design of chemo- and biosensors are considered in this overview. On example of typical experimental and industrial samples it was analyzed the architecture and performance properties of chemosensors by resistive, amperometric, potentiometric, voltammetric, optical and gravimetric types, their advantages and disadvantages at the analysis of various inorganic and organic substrates. It ihas shown the prospects of application of nanostructured **PAn** and composites on its basis with nano-dispersed metallic, oxide, and mineral fillers at the design of sensor devices. An application of **PAn** in biosensorics is considered in a separate way; in particular, the methods of biological components immobilization on the polymer-modified substrates have been analyzed.

7.1 INTRODUCTION

Information about the state of any system will be reliable only at the availability of reliable data concerning to its composition at given moment of time. Operational information about quantitative contents of the components is important in order to control the chemical–technological process in materials science, environmental, and biomedical diagnostics at the analysis of food and so on. However, in many cases, rapid analysis or continuous monitoring of many chemical or biochemical objects with the use of the methods of classical chemistry is practically impossible. Therefore, in recent years the actuality of the development of direct rapid methods for determination of inorganic and organic (including the biologically active ones) substances is steadily growing. The successes in this area are mainly associated with the modern achievements of materials science, solid-state physics, microelectronics, information technologies, physical chemistry, and molecular biology and so on. An integration of these fields of research in the second half of the 20th century led to the appearance of the independent branch of the scientific researches, so-called *sensorics.*[1] The term "*sensorics*" comes from the *Latin* "*sensus*," which means feeling. Therefore, in the broad sense the conception of **sensorics** includes a set of approaches aimed to the recognition and quantification of particles by different chemical nature in the atmosphere, water or in living organisms. **Chemical sensor** represents by itself the sensor or sensing device (physical device) by a small size, which is designed for detection and quantification of neutral or ionized organic and inorganic particles. Sensors permit to collect, to record, to transmit, to treat, and to distribute the information about the states of physical and chemical systems. This can be information about their chemical composition, shape, structure, position, or dynamics.[2,3] The most important part of the sensor is the *sensitive layer* deposited on a solid substrate, which is changing in a case of the contact with the determined substance (substrate) accompanying by the generation of a signal. In this turn, a signal, or otherwise, the response of sensitive layer to the presence of the particles of substrate in analyte can occur, first of all in the form of the changes in the physical properties of the components of the active layer, such as a change in the conformation of the particles, conductivity, index of refraction, viscosity, increase of weight *etc*. The alternative, in the case of chemical interaction of the substrate with the active layer, is the generation of products, the quantity of which is proportional to the amount of the substrate in the analyte. In both cases, for the fixation of generated in the active layer of the physical and chemical changes they need to be further converted to the form where they can be easily operated, namely, to strengthen, to compare with the standard and finally to display in digital or analog form. Another integral part of the sensor device, which converts the physical and chemical changes into the measurable signals is named as **transducer**. There are electrochemical, electrical, optical, magnetic, gravimetric, and thermometric sensor devices depending on the principle of operation and the nature of the generated signal. The **biosensors** are variety of the chemical sensors; they represent by themselves the combined devices consisting of biochemically or biologically active components (biological component) and electronic transducer. The sensor will perform the analytical function only when the close correlation will exist between the value of signal and the quantity of particles in the analyzed sample. The best is the case when this relationship is directly proportional.

7.2 SENSORY PLATFORMS BASED ON ESPECIALLY OF PAn

Optical, the most importantly, electrical properties of the electroconductive polymers (**ECP**) make them by attractive materials for various sensor devices. The principle structural unit of **ECP** is a linear chain containing the repetitive links of the p-conjugated organic monomers. The systems of this type represent by themselves the insulators in a neutral state. However, chemical or electrochemical reduction (*n*-doping) or oxidation (*p*-doping) leads to the transformation of these structures into the conductors of electrical current. An electrical conductivity is caused by the change of the energy-band structure of polymer and by the formation of the charged carriers. The vast majority of **ECP** belongs to the one-dimensional conductors because the electrons or holes (in a case of *n*- and *p*-doped **ECP,** respectively) migrate along the linear conjugated chains. In terms of solid state physics the **ECP** are interpreted as wide-gap semiconductors with a flat gap, because, unlike to the metallic conductors, the energy gap between the valence and conduction band is available even at room temperature and at its decreasing the conductivity of **ECP** is decreased.

7.2.1 THIN FILMS OF PAn

Sensor platforms based in particular on electroconductive polymers usually are created electrochemically by depositing of the polymer/oligomers on the surface of the working electrode as a result of oxidative polycondensation of starting monomer often in the form of the thin films. For this purpose the potentiostatic or galvanostatic mode is most commonly used. Among other methods it should be also noted that the potentiodynamic (cyclic voltammetry) and pulsed potentiometric mode of synthesis. The values obtained in such way the modified electrode **ECP** serves as the sensor device. By fixing the quantity of electricity which has passed through the electrolyzer during the synthesis–deposition process, we can estimate the thickness of the deposited film. At the same time, on the basis of the final value applied to the electrode potential, it can strictly control the level of **ECP** doping. In addition, this method is suitable for the synthesis of soluble **ECP**, which can be used in particular for the film-forming,[4] which is achieved using the initial monomers with various side substituents.

An alternative method of production of the detectors based on **ECP** is a chemical modification of the polymer precipitated at the expense of various side substituents. Finally, the detectors can be varied by using different types of counterions for compensation of positive charge in a case of *p*-doped polymer which is formed during the oxidation of polymer chains and produces the electroconductive states of the polymer. Eventually, the various sensoric platforms differed by properties can be obtained from the same monomer, since such properties of **ECP** as morphology, molecular weight, length of the conjugation, the order of monomer units connection (microstructure), electroconductivity, band gap, and so forth are determined by the conditions of polymerization, namely by the size of oxidation potential, by the nature of the oxidant and solvent, by temperature, concentration of electrolyte and monomer, others.

PAn is unique among the family of electroconductive polymers since its doping level can be controlled through a non-redox acid doping/base de-doping process.[5] By changing of the doping level, the **PAn** can be modified to suit specific applications. At this, **PAn** in sensor devices can act as both the active layer and transducer. In addition, its good compatibility with many biological components, particularly enzymes, makes it indispensable in the design of biosensors.

The basis of the most sensors on PAn platform is the possibility of the reverse transition between the non-conductive form of **PAn** (emeraldine base)

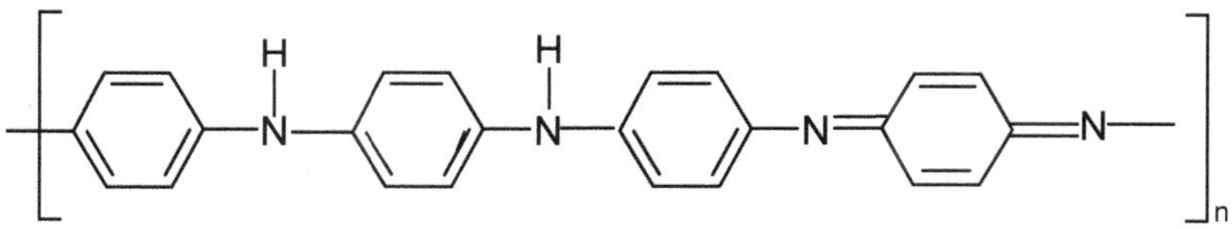

and the conductive protonated form of **PAn** (emeraldine salt):

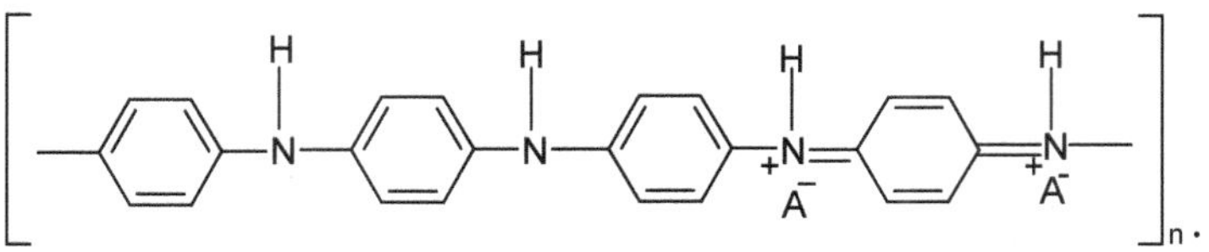

The conductivity of **PAn** is increased with the increasing of doping level (the number of protonated centers). Also, the oxidation or reduction of the emeraldine form of **PAn** is observed in many cases with the formation of pernigraniline or leucoemeraldine, respectively, which in turn may be also in the forms of both the base and salt. All of these forms of **PAn** are different not only on magnitude conductivity, but also on the value of the equilibrium potential, optical characteristics, adsorption capacity, and so forth.

Although the electrochemically deposited layers of **ECP** with the thickness by nanometers belong to the nanoobjects, their physical chemical properties are not always satisfy the requirements for a constructive materials used in molecular electronics. The reason for this is that the physical and chemical properties of **ECP** are also strongly depend on their structure, which in turn depends not only on the type of polymer chains, but also on methods of installation, orientation, and packing density. That is why today in the design of sensor devices the focus increasingly is shifted to the use of nanoparticles of **ECP** or nanostructured **ECP**.

7.2.2 NANOPARTICLES OF PAn

Nanoparticles of **ECP** have the structural morphology with an aspect ratio of approximately 1:1:1 in the all three coordinate axes. As in a case of nanoparticles by different nature, such as metallic ones and metal oxides, these materials have several potential physical and chemical advantages including a large ratio of surface area to volume ratio and high radius of curvature, which dramatically changes the morphology of the surface compared to the bulk polymer structures. This effect, as in ф case of other nanostructures, can lead to significant changes in the activity of the material in electrocatalytic processes. Nanoparticles of **ECP** are obtained as stable suspensions that can be used just as true solutions. This is a significant advantage for production, since the solutions can be prepared from preformed polymeric materials.

The approaches implemented in recent years include the **PAn** synthesis by method *in situ*, the use of sulfoderivatives of **PAn** and the synthesis of water-soluble nano-dispersed **PAn.**[6] Nanodispersions have great promises for use them in sensorics, because they are suitable for inkjet printing, facilitating the application of this conducting polymer directly on the substrate. Small size of the discrete particles of a material is a significant advantage at the implementation of such processes since the likelihood of the nozzles cartridge clogging is reduced. It is shown, that for **PAn** films deposited by inkjet printing method, it can accurately controlled their two-dimensional picture, the thickness and conductivity, ensuring the appropriate level of precision by the capabilities of the inkjet printing.[6] The use of these nanomaterials as ink for inkjet printing offers a new, easy, and economical possibilities of polymeric printed materials using not only in sensorics but also in many other areas, such as energy storage, displays, organic light- emitting diodes. Furthermore, given that inkjet printing technology is suitable for large-scale production, it enables the mass production of devices such as sensors for a wide range of applications.

Nanoparticles of different types can be synthesized by changing of the synthesis conditions. The micellar templates, in which the amphiphilic micelles are formed via polymerization of monomer molecules in a certain area, have a much wide application. An example is the synthesis of **PAn**'s nanoparticles in micelles of dodecylbenzenesulfonic acid (**DBSA**). However, today a limited number of examples of such nanoparticles application in chemical sensorics are described. In particular, nanoparticles of **PAn/DBSA**, deposited by inkjet printing method have been used for the manufacture of printed film sensors and were tested for their ability to detect the ammonium in water and gas media.[7] An amperometric response of the sensor into the flow-through cell[8] is analyzed for detection in the aqueous phase. The best response is obtained at the values of *pH* neared to neutral. The effect of aqueous ammonium leads to the appearance of current response at −0.3 V relatively to Ag/AgCl electrode. Sensor based on **PAn** nanoparticles exhibits an ammonium detection ability in industrially important range from 1 to 100 ppm at the detection limit of 0.54 ppm (3.2 mM) with an accuracy of less than 5%. The nanoparticles of **PAn/DBSA** were used in conductometric sensor with the plicated electrodes produced by the screen printing method[9] for the detection of gaseous-like ammonium. The effects of the sensing layer thickness as well as the operating temperature via the range 1-100 ppm have been also investigated. Regardless the operating temperature, the response time for the films obtained by a single printing was 15 s. However, an increase of the temperature till 80 °C reduces the regeneration time of the sensor to 210 s at the concentration of ammonium 50 ppm. Furthermore, it was shown that the sensitivity to the determination is decreased with the temperature increasing and the sensor showed the response 0.24 for 100 ppm at 80 °C. The background signal of such sensor slightly depends on humidity (in the ranges from 35 to 98%) at room temperature.

7.2.3 NANOSTRUCTURED PAn

At the obtaining of nanostructured **ECP** the regulation of the samples' morphology occurs at a stage of the matrix or as it so-called, **template synthesis**, when the matrixes acts as the peculiar structural directrixes. The "soft templates" such as surfactants, inorganic acids, and complex organic doping agents that promote the self-assembly processes or "hard templates" such as zeolites, porous aluminum oxide, or polymer membranes which due to the one-dimensional nanocanals can program the growth of nanotubes or nanowires (fibers) are usually used as the structural "managers" during the synthesis of nanostructured **ECP**.

In such a way, the nanotubes obtained from the **PAn** doped by $HClO_4$ were characterized by a significant degree of the crystallinity. An ordering of polymer chains gave almost six-fold increase of the conductivity[10] from 1.2 to 7.1 S/cm.

Nanotechnologies give the possibility to do a new approach to the study and application of conductive polymeric materials in chemical sensorics. Control of the structure and morphology of the conductive polymer materials at the nanoscale makes possible the variety of improvements that are impossible in a case of the traditional materials. Nanostructured conductive polymers have several potential advantages compared to their bulk analogues. AFirst, if the synthesis is controlled at the nanoscale, then thanks to the improvement of the ordering and structural reproducibility, the behavior of material is more predictable and reproducible. Secondly, the nanostructured materials provide much more the ratio of the real surface area to volume, thereby greatly increasing the surface area for chemical interactions and electron-transferring processes, thus overcome some limitations of the sensor performance, in particular related to the diffusion. Thirdly, the nanostructurization opens even wider range of composite materials based on **ECP** in general and on **PAn** in particular having new physical, electrical, and electrochemical properties as a result of a large number of new combinations of components. Fourthly, the nanostructured materials due to surface structural and quantum-mechanical features that arise during the nanostructurization, may exhibit the improved catalytic and electrocatalytic properties that do not have their bulk analogues. This should lead to the improvement of the analytical characteristics of the sensor, namely to the lowering of detection limits of substrate visualization, to the improvement of the results reproducibility, reversibility, sensitivity and stability of the sensor, reducing of its sensitivity to the impurities. Finally, the nanostructurization should lead to improved manufacturability of the received materials and devices, their suitability for the production of sensor devices in real industrial conditions.[11]

After all, many methods of nanostructures synthesis permit better and separately to control the synthesis of nanomaterials and processes of their deposition, that is reflected in combination of these advances in the manufacture of sensors. However, an application of conductive polymer nanostructured materials for their using in chemical sensorics is a relatively new area of research, and that is why so many current investigations are focused on general properties of various nanomaterials for sensorics rather than on detailed and thorough assessment of their analytical estimation which should be sustained, if we would understand the real value of these materials.[11]

7.3 NANOCOMPOSITED BASED ON PAn AS SENSITIVE LAYER OF CHEMOSENSORS

Another approach being implemented in the construction of sensors is to use the composites of **PAn** with the nanoparticles of metals, oxides, semiconductors, natural minerals, carbon materials (nanotubes, fullerenes, and others), as well as of others polymers, and so forth. An interaction between the components by different nature often leads to the synergism in such systems, which significantly improves the physical and chemical properties of these materials. At this, polymer (in particular **PAn**) may play a role both of a filler and matrix. The work of Wang et al.[12] can serve as an example of the materials by the first type. These authors have created an organic/inorganic hybrid nanofilm from the chemical vapor deposition of MoO_3, followed by the ion- exchange- induced intercalation of **PAn** between the molybdenum trioxide sheets,[12] bringing about highly sensitive changes in conductivity, particularly to volatile organic carbons. Aniline monomer was intercalated within the MoO_3 film through a cation-exchange process with sodium ions and following polymerized with ammonium persulfate. Conductometric responses were to 50 ppm of formaldehyde in dry air. The **PAn**/MoO_3 hybrid showed the sensitivity to formaldehyde at 50 ppm and to acetaldehyde at 10 ppm with a response ratio of 0.08 and 0.044, respectively (Fig. 7.1).

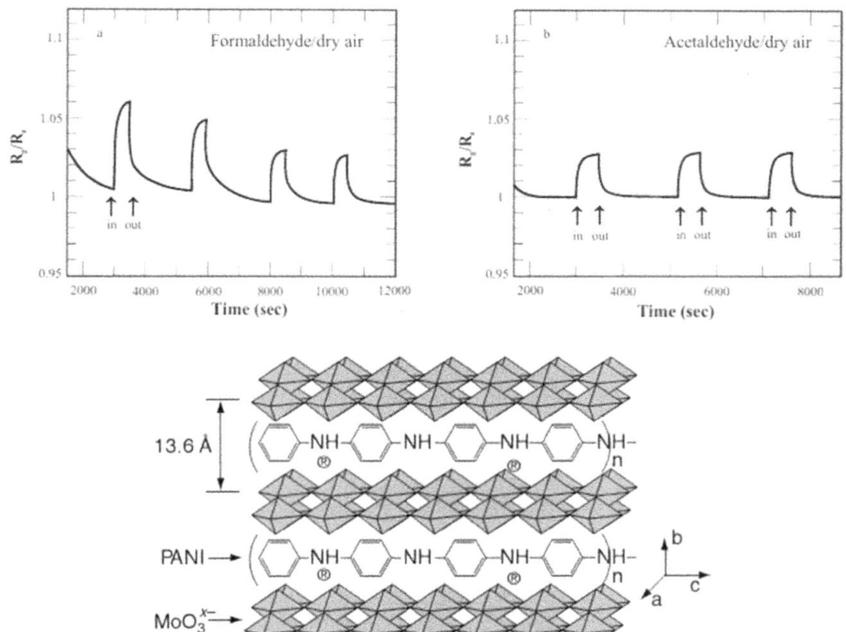

FIGURE 7.1 Conductometric responses of **PAn/MoO$_3$**-based sensor to (a) 50 ppm of formaldehyde and (b) 10 ppm of acetaldehyde in dry air, and (c) structure of **PAn/MoO$_3$** composite.[12]

However, the most widely used sensorics are the composite systems based on nanostructured or unstructured conductive polymers and conductive materials by nano dimensioned degree of dispersion. Films of conducting polymers have long been seen as extremely useful substrates and carriers for such nanoparticles due to their excellent film-formation properties, and good levels of conductivity, as well as their open 3D structures, which allow good diffusion of solution species to the surfaces of the embedded nanoparticles. Composites of these two species thus possess advantageous properties imparted by both materials and have the potential to improve sensor capability. Most attention for chemical-sensing applications has focused on the use of noble and transition metal catalysts.[11] In this case, at the design of electrochemical sensors it should be taken into consideration that the catalytic and redox properties of metal nanoparticles, which will mainly determine the effective functioning of the sensor device, will affect the size, the presence of defects and quantum-–mechanical properties of nanoparticles.[13]

This type of sensor is easy to make using the methods of electrochemistry. The sensor for the determination of hydrogen sulfide,[14] in which the PAn nanowires with 250-320 nm in diameter functionalized by gold nanoparticles (**Au–NPs**) by the cyclic voltammetry were by the sensitive layer, can be as an example. Resistive sensors based on the synthesized composite were characterized by a very low detection limit, exactly ~0.1 ppb, by a wide dynamic range 0.1-100 ppb and by very good selectivity. The operation of the sensor is the interaction of hydrogen sulfide with the surface of gold nanoparticles:

$$H_2S + (Au)_n \rightarrow (Au)_{n-1}AuS + 2H^+ + 2e^-. \quad (7.1)$$

Formation of Au−S bonds lead to an increase in the doping level of **PAn** due to the formation of protons during the reaction, resulting in a significant decrease in the resistance of the composite **Au–NPs/PAn**. In addition, **PAn** and gold nanoparticles act as donor and acceptor accordingly. At this, there is a transfer of an electron from the p-doped **PAn** to partially oxidized ones with hydrogen sulfide **Au–NPs**. On PAn molecules there are ion-radical centers responsible for the charge transfer, the result of which is an increase of the electrical conductivity of the sensitive layer. The sensor has a little response time (<2 min) and regeneration time (<5 min) and at the same time is characterized by a high sensitivity 13.78% per ppb. Response is linear in the range from 0.1 to 500 ppb and the sensor has a high reproducibility.

Another example is the nanocomposite **Cu–NPs/PAn,**[15] which is used for the manufacture of conductometric sensor sensitive to chloroform content in the gas phase. The composite was prepared by filling polymerization, when the pre-**Cu–NPs** were contributed into the polymerization solution containing the aniline, hydrochloric acid, and oxidant (ammonium peroxydisulfate). At the exposure of sensor in chloroform vapor, the resistance of the composite is increased. The sensitivity ($\Delta R/R$) for chloroform concentrations within 10-100 ppm was varied from 1.5 to 3.5. The results of spectroscopy in the long-wave infrared region indicate on the adsorption interaction of chloroform with the metal clusters. At this, the contact of **PAn** with copper is made worse, which leads to increase of the resistance of the composite. Thus, the response of the sensor is formed mainly due to the adsorption–desorption process of chloroform on the surface of copper nanoparticles.

To produce the nanocomposite sensor platforms significantly more complex techniques are also used. Feng et al.[16] have synthesized of **PAn** on the surface of microspheres of polystyrene (**PS**). After removal of **PS** by it dissolving in tetrahydrofuran, the formed hollow spheres of **PAn** were modified by nanoparticles of Au. It was discovered that the significant increase of the electrical conductivity of the composite compared to the unmodified polymer particles. Glassy carbon electrode modified by synthesized composite was used to study the oxidation of dopamine by cyclic voltammetry method. Oxidation peak potential was reduced in the presence of Au and the amperometric response significantly is increased compared with electrodes modified by Au nanoparticles or by the **PAn** only. This is due to the distribution of Au nanoparticles on the surface of hollow spheres of **PAn** and, consequently, by the improvement of the diffusion of dopamine and access to the surface of Au nanoparticles.

The carbon nanotubes (**CNT**) are fillers among the most commonly used in the manufacture of nanocomposites for sensorics. Since their discovery, they have been an object of intense research because of the unique structural electronic, mechanical, optical, and potential opportunities in applied nanotechnologies. High conductivity, high specific surface area, the possibility of chemical modification of their surface by chemical grafting of various functional groups or polymer chains, and biocompatibility make them ideal materials in the construction of electrochemical sensors. In the manufacture of sensors **CNT** are incorporated into the polymer matrix, including PAn one, often by the mechanical mixing or filling polymerization. At this, chemical, electrochemical or photo polymerization, vapor deposition from gaseous phase and other techniques[11] are used. It's proposed that the polymerization in the presence of **CNT** changes the mechanism of the polymerization process that leads to the formation of template structures and orientation of growing polymer chains, resulting in an increase in their degree of crystallinity.

Obtained composites in such a way illustrate a significant higher conductivity compared to pure **ECP**. In particular, **CNT** modified by **Pd–NPs** are sensitive to the presence of hydrogen in the gas phase. The electrical resistance of the sensing layer is abruptly reduced on 10-15% at its exhibiting by air due to the adsorption of oxygen.[17]

It was found that the defects on the surface of single-walled carbon nanotubes play an important role in shaping of the electrical response for a wide range of vapors of organic substances and the sensitivity and chemical selectivity of sensor device can be increased by controlled introductions of the defects.

Thus, the oxidative defects, namely the carboxylated centers, can be obtained by irradiation with ultraviolet light of the sample of carbon nanotube heated to 120 °C in the presence of ozone.[18] It was shown that the carboxylated centers on the surface of carbon nanotube can interact with the molecules in particular of acetone in analyte.[18] At this, along with the electrical conductivity the adsorption capacity of the composite is decreased also. Sensors with modified oxidized nanotubes are also sensitive to methanol, water, hexane, toluene, chlorobenzene, and so forth.

Structurally the sensors on conductive polymer platform generally are similar to the sensor devices with the sensitive elements from the metal oxides, that is they can be in mono-– or multivariate performance. High sensitivity and selectivity usually can be achieved by combining of monosensors into ensembles. Electrochemical response can be repeatedly improved by using of the electrodes with opposite pintle-interdigitated structure.

7.4 THE TYPES OF CHEMOSENSORS BASED ON PAn

PAn into electrical transducer systems is used at least in four types of the electric converters,[19] namely:

- *conductometric converters* with the tracking changes in conductivity;
- *potentiometric converters* with the fixation of potential of the open circle;
- *amperometric converters* with the measurement of current strength at the fixed potential;
- *voltammetric converters,* which are based on the monitoring of the current changes at the potential scanning under given mode.

In addition a significant number of optical gas-sensitive elements as well as sensors that respond to the changes of mass of the sensing element were developed based on **PAn**. As for gravimetric devices, then the basis of their functioning is the piezoelectric sensing effect or registration of the acoustic waves.

Next, based on the specific sensor devices let us consider the general principles of design, principles of operation, limit applications, advantages and disadvantages of each of these types of chemosensors based on **PAn**.

7.4.1 CONDUCTOMETRIC / RESISTIVE SENSORS

Conductometric response is most often used in sensorics. In practice, not the conductivity itself is measured, but the resistance to the passage of electric current R, attributing it to the resistance in the absence of analyte R_0 that is the value of R/R_0. The many of processes that lead to the changes in the density of electric charge carriers and their mobility, change the conductivity. The conductivity of **PAn** depends both on the oxidation state of the main polymer chain and on the degree of protonation on imine sites.[20] Any interaction with **PAn** that alters either of these processes will affect its conductivity. Unlike acids and bases, redox active chemicals and gases can change the conductivity of **PAn** by changing its inherent oxidation state. Interaction of conjugated polymers with acceptors or electron donors leads to the changes in both the density of carriers and their mobility resulting conductivity varies considerably. Electrical properties and, in particular, the resistance to the passage of direct electric current are sensitive to exposure by vapors of various substances, namely acids and bases, organic compounds, hydrogen, and so forth. The magnitude response of the sensor will affect the design, the sensor surface area, porosity, and thickness of the **PAn** fibers. The decrease in porosity and increase of the thickness of the fibrils leading to reduced sensitivity and increased response time of the sensor as analyte harder to penetrate in the polymer on sufficient depth.

A sorption of vapor by the **ECP** induces the physical swelling of the material that affects the electron density of the polymer chains. Therefore, the result of adsorption of particles from the analysis is the change in electrical conductivity $\Delta\sigma$, which is the sum of three components, namely:

$$\Delta\sigma = \left(\Delta\sigma_C^{-1} + \Delta\sigma_n^{-1} + \Delta\sigma_i^{-1}\right)^{-1} \tag{7.2}$$

Here $\Delta\sigma_c$ is the change of the intrachain conductivity of **ECP**; $\Delta\sigma_n$ is the change of intermolecular conductivity by electrons jumping between chains caused by the presence of particles from analyte in the film; $\Delta\sigma_i$ is the change of ionic conductivity between chains under the influence of sorption-party components.

More energy efficient is the conductivity along the polymer chains rather than between them. Therefore, the most changes in conductivity will be caused by the changes of intrachain component of conductivity. For example, highly ordered **PAn** is characterized by higher conductivity compared to the amorphous one. In particular, a significant increase in the electrical conductivity of **PAn**' film doped by dinonylnaphthalene sulfonic acid (**DNNSA**) is observed during the exposure into the ethanol vapor.[21] Obviously, the hydrogen bond between the ethanol and **PAn** promotes the greater convergence of the molecules of ethanol and PAn chains, reducing the energy barrier of the electrons jumping and increasing the mobility of the charge carriers.

PAn is also able to adsorb the hydrogen, but according to the data of different authors the value of the storage capacity ranges from 1-2 to 6-8 by % mass.[22,23] At this, the electrical conductivity used in the design of sensors is significantly increased. Sadek et al.[24] proposed two possible mechanisms to explain the increase of conductivity in this case. The first is that hydrogen can form the bridges between the nitrogen atoms of adjacent chains or partially to protonate the imine nitrogen atoms. As a result of these reactions, the protonation of nitrogen atoms in PAn chain takes place, resulting in a growing number of delocalized carriers (polaron and bipolyarons) and, thus, of the conductivity. According to another mechanism, the change in resistance is due to the fact that during the interaction of hydrogen with **PAn**, a certain fraction of the analyte is atalytically oxidized with the formation of water. The presence of water facilitates the transfer of charge between the **PAn**'s molecules and contributes to the conductivity increase due to reduced resistance both of the doped **PAn** and undoped one. Designed sensor has a small response time (about 30 s) and can be used to detect of the small amounts of hydrogen in air (Fig. 7.2).

The authors of *ref.*[25] report about the chemoresistive sensors made from the monodisperse nanoparticles of **PAn** synthesized by oxidative dispersion polymerization. Polystyrene sulfonic acid (**PSSA**) was used as the stabilizer and doping agent. Transducers based on nanoparticles of **PAn** have been performed by stepwise (layered) spray of the solution containing various concentrations of **PAn**'s nanoparticles and of the multi-walled carbon nanotubes on electrodes by interdigitated structure. This process makes it possible to construct the stable sensors with the reproducible response during the repeated applications. Electrochemical properties of these sensors were tested at successively transmission of the flows of pure nitrogen and nitrogen containing of the volatile organic compounds. Interestingly, that to activate the converters based on nanoparticles of **PAn** can by simply increase of **PAn** content or by adding only of 0.5% of multi-walled carbon nanotubes (**MWCNT**) in order to achieve the resistance lower than 150 Ω. Due to its original conductive architecture, which was established using the atomic force microscopy and represents by itself a double interpenetrating network, composites of **PAn**'s nanoparticles with multi-walled carbon nanotubes exhibit the higher sensitivity and selectivity than the other mixtures compositions, demonstrating the synergism of properties.

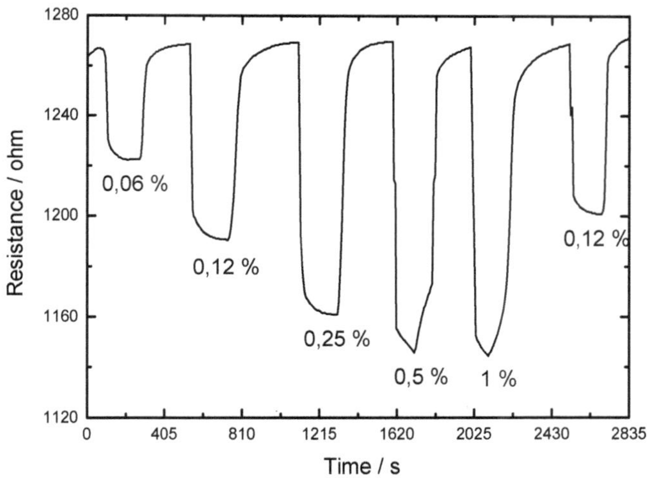

FIGURE 7.2 Dynamic response of the doped **PAn**'s nanofiber based on conductometric sensor to different H_2 gas concentrations in air.[24]

In a case of the presence of conductive network of **PAn**'s particles, as it schematically is shown in Figure 7.3(a), only the polymeric phase is responsible both for the electronic conductivity and for interaction with analytes. Thus, the effect of increase in resistance (which is observed at relatively high concentrations of **PAn**) must be the result of separation of **PAn**'s nanoparticles, due to the adsorption of the solvent molecules on their surface, resulting in the formation of the adsorbed layer and the formation of gaps between the nanoparticles. Conversely, if the molecules of volatile organic compounds are absorbed by the **PAn**'s nanoparticles, resulting in the swelling of polymer

(Fig. 7.3(b)), it leads to an increase in the volume of particles, increasing of the area of contact and mutual diffusion of **PAn**'s macromolecules at the interface between the nanoparticles. The result of this is an increase of the electrical conductivity of the transducer. In addition, according to the theory in the vicinity of the percolation threshold, the level of connection is so weak that the network of **PAn**'s particles consists of a small number of conductive paths, and the most of the particles are not in close contact with each other. Under these conditions, only some electrons may have enough energy to move from the one particle to another by tunneling.

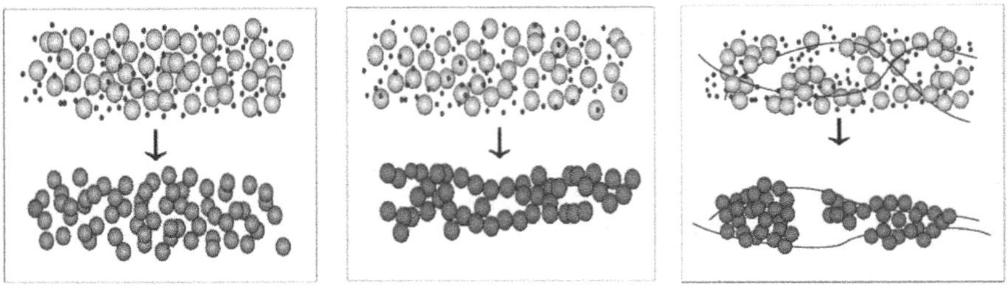

(a) (b) (c)

FIGURE 7.3 Different cases of interaction between the transducer and analyte: (a) is positive response for **PAn**'s nanoparticles; (b) is negative response for **PAn**'s nanoparticles; (c) is positive response for composite of **PAn**'s nanoparticles with multi-walled carbon nanotubes.[25]

Thus, the increase of volume as a result of nanoparticles of the polymer swelling can reduce the average distance between the particles, increasing the conductivity of a sensor and leads to a decrease of the resistance. The relevant model is considered that the **PAn**'s nanoparticles as "smooth" spheres are merged during their expansion; besides, not only their outer shell covered by the **PSSA** contacts, but also the deeper layers of **PAn**, allowing mutually to diffuse to conductive polymeric chains by different nanoparticles.

But how to predict whether resistance will increase or decrease as a result of swelling? It depends on the degree of swelling of the **PAn**, which in turn is a function of the ratio of the number of molecules of volatile organic compound to the number of macromolecules available for swelling on the surface of the nanoparticle. In other words, for a given amount of analyte molecules, the degree of swelling is inversely proportional to the number of **PAn**'s particles. Thus, in transducers with a high concentration of **PAn** (4-5% wt.) the molecules of organic compounds are adsorbed on a surface, while for the systems with a lower content of **PAn** (1-3% wt.), the partial or complete swelling will be take place. After reversing, the positive response of the sensor can be viewed as a transition to another mode of measurement, that is controlled by the swelling degree and, thus, by the size of **PAn**'s particles.

In the third case, a positive response of the sensor that is the resistance with increasing of the analyte concentration is observed. In Figure 7.3(c) the processes that occur in transducer based on composite of **PAn**'s nanoparticles and multi-walled carbon nanotubes are schematically represented. Conductive structure of composite represents by itself a double percolation network of **PAn–NPs** and **MWCNT**. Light net of **PAn**'s nanoparticles, riddled with multiple **MWCNT**, is expanded due to the diffusion of analyte disconnecting the **MWCNT** from the network. The result is an increase of the resistance of a system. Similar effects connected with the expansion of **PAn** are observed for transducers by the same chemical nature but different structure, for example in the case of **MWCNT**, coated with **PAn**. So, this way we can explain the different effects of analytes on conductive characteristics of **PAn**'s composites that are important in the design of conductometric sensors on their basis.

7.4.2 POTENTIOMETRIC SENSORS

Potentiometric transducer is the simplest type of sensor, in which **PAn** plays a role of the sensitive membrane. The changes in the open circuit voltage are fixed as the signal of a sensor. A simple potentiometric sensor based on **PAn** is the sensor to determine acidity of medium. The mechanism of *pH* impact on the value of the equilibrium potential of platinum electrode modified by **PAn**'s film is connected with fact that the electrochemical oxidation of **PAn** leads to an increase of the proportion of quinone diimine fragments in macromolecule of **PAn** and to reduce of the proportion of benzenediamine fragments.[26] Due to the presence of basic amine and imine atoms of nitrogen the **PAn** interacts with protogene acids with the formation of the salts of **PAn**. In the synthesized salts of emeraldine the nitrogen atoms of quinone diimine groups bounded with the anions of organic acids with long or voluminous substituent (2-naphthalenesulfonic acid, camphorsulfonic acid, etc.) that due to the large size and low mobility will block the nitrogen atoms. This decreases the possibility of recovery of the emeraldine to leucoemeraldine, which mainly occurs during the synthesis of **PAn** in the presence of strong mineral acids as doping agents. As the experimental data show, this leads to the improvement of functional characteristics of obtained reverse redox systems, especially at the synthesis of **PAn** from the organic solvents, apparently due to receipt of the oriented polymer films by more ordered structure.

The structure of the macromolecular chain of **PAn** in the form of emeraldine can be represented using the benzenediamine (QH_2) and quinone diimine (Q) groups. The ratio of these groups in the macromolecule determines the degree of oxidation of the polymer, and the degree of protonation of the polymer is a function of pH. Since the electrons and protons take part in the redox transformations of **PAn**, the equilibrium potential of the **PAn**'s electrode is sensitive to the change of *pH*. It also shows that the equilibrium potential is a function of the dissociation constant (**DC**). For example, for the acid–base reaction connected with the quinone group in the macromolecule of **PAn**

$$Q + 2H^{+} \leftrightarrow QH_2^{2+} \tag{7.3}$$

the corresponding dissociation constant for acid can be written as:

$$Kd = \frac{[Q][H^{+}]^2}{[QH_2^{2+}]}. \tag{7.4}$$

Renewal equation for the protonated quinone diimine QH_2^{2+} group is as follows:

$$QH_2^{2+} + 2e^{-} \leftrightarrow QH_2 \tag{7.5}$$

so, the *Nernst* equation becomes:

$$E_{QH_2^{2+}/QH_2} = E^{o}_{QH_2^{2+}/QH_2} + \frac{RT}{2F}\ln\frac{[QH_2^{2+}]}{[QH_2]} \tag{7.6}$$

Substituting the Eq. (7.3) into Eq. (7.5), we obtain

$$Q + 2H^{+} + 2e^{-} \leftrightarrow QH_2. \tag{7.7}$$

Substituting in Eq. (7.6) $[QH_2^{2+}]$ by the expression obtained from Eq. (7.4), we obtain an expression for the Nernst equation:

$$E_{QH_2^{2+}/QH_2} = E^{o}_{QH_2^{2+}/QH_2} + \frac{RT}{2F}\ln\frac{[Q][H^{+}]^2}{K_d[QH_2]} \tag{7.8}$$

The transition of benzene diamine groups into quinone diimine ones requires the cleavage of two protons and two electrons. The concentrations of quinone diimine, benzenediamine groups, and the concentration of hydrogen ions will be included in the *Nernst* equation. All these factors will determine the dependence of the electrode potential on pH medium. Introduction to the structure of **PAn**'s macromolecule of sterically inactive anions of organic acids helps to regulate the electrochemical activity of the **ECP**, limiting it by the processes of transition emeraldine « pernigraniline.

To create a potentiometric chemosensors based on conducting polymers synthesized polymeric products are applied to the surface of an inert working electrode. In model systems the platinum electrodes often are used for such purposes. The platinum electrodes modified by **PAn**'s films under different conditions were tested by the authors of *ref.*[27] as potentiometric electrochemical sensors for the determination of *pH* of aqueous solutions. The results of the *pH* influence on the equilibrium potentials of formed electrodes depending on the conditions of synthesis are shown in Figure 7.4. According to the obtained results (Fig. 7.4(a,b)), for **PAn**-modified Pt-electrode the equilibrium potential change in buffer solutions with pH values ranging 1.81-7.5 is characterized by the linear dependence. In the case of deposition of **PAn** from the dimethyl sulfoxide–aqueous solution (volume ratio of **DMSO**:H_2O = 9:1) and its doping by 2-naphthalenesulfonic acid (**NSA**) (Fig. 7.4(a), lines 4) the deviation of the values of potential from the approximated direct line is quite significant and at low *pH* the electrode loses the sensitivity to the concentration of hydroxonium ions. At the same time, Pt-electrode (modification of its surface has been done by nanoscale films of **PAn** doped with camphorosulfonic acid (**CSA**) by chemical oxidation of aniline with benzoyl peroxide (**BP**) (Fig. 7.4(b), line 1) or electrochemically (Fig. 7.4, straight lines 6,7), is also characterized by a pH-dependent equilibrium potential, reproduced in the pH range 4-8. Modification of the Pt electrode surface by product of polymerization of *N*-(3-(trimethoxysylil)-propyl) aniline (**TMSPA**) also allows us to use this electrode as a *pH* sensor, since, as can be seen from Figure 7.4(a), (straight lines 2,3) in buffers with different pH values and it observed the linear relationship between the potential and value of pH-environment. For the film of **TMSPA** obtained from the organic solvent, this dependence has small deviations from the linearity. It was observed an influence of the substrate (electrode) nature on the value of the potential of sensor (Fig. 7.4(a,b)). In the case of platinum the values of potential are higher than in the case when the **PAn**'s film was deposited on the graphite surface (Fig. 7.4(a), dependence of 7; Fig. 7.4(b), dependence of 2). However, the response of the sensor at the same pH change is almost the same in all cases. Thus, a platinum electrode modified by thin **PAn**'s films synthesized in the presence of **NSA** and **CSA**, is characterized by a pH-dependent equilibrium potential, reproduced in the pH range 1.5-8. Electrodes formed during the electrochemical oxidation of aniline and its derivatives are chemical sensors, sensitive to the changes of pH solutions.

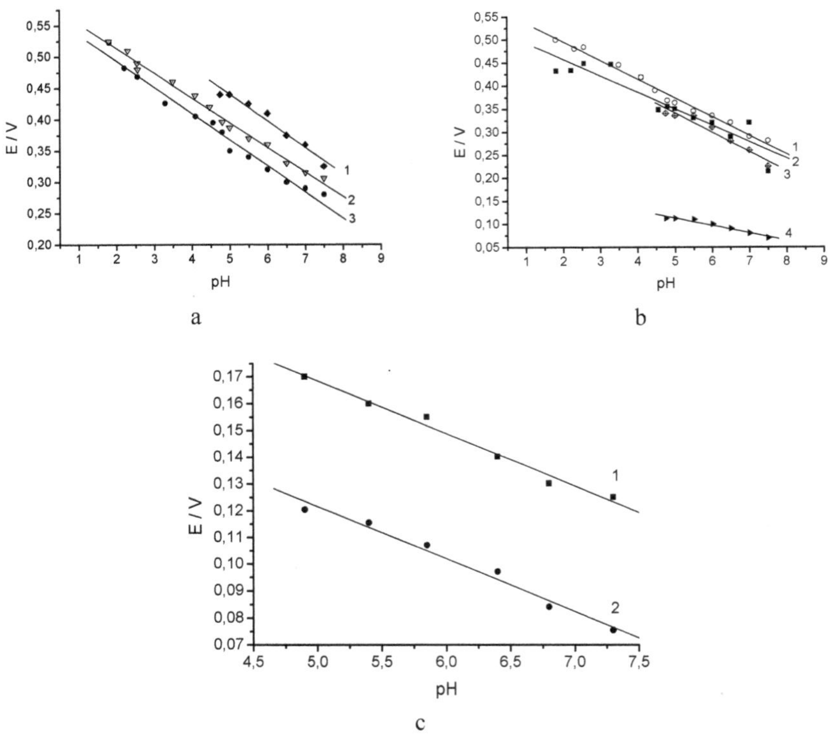

FIGURE 7.4 The change of potential of the electrode modified by electroconductive polymer in the phosphate buffer solution depending on pH of medium. *Polymer synthesis conditions:* *a*: **1** is galvanostatic synthesis on Pt-electrode from 0.1 *M* **An** + 1 *M* $LiClO_4$ solution in acetonitrile; **2** – C(**TMSPA**) = 0.1 M, C (CCl_3COOH) = 1 N in acetonitrile; **3** - C(**NSA**) = 1 M, C(**An**) = 0.1 M, C($LiClO_4$) = 1 M in solution of **DMSO**:H_2O = 9:1; *b*:1 is galvanostatic synthesis on Pt-electrode from solution C(**TMSPA**) = 0.1 M, C(H_2SO_4) = 1 N in water; **2** – C(**NSA**) = 1 M, C(**An**) = 0.1 M, C($LiClO_4$) = 1 M in solution of **DMSO** : acetonitrile = 1:1; **3** - (0.1 M **An** + 1 M $LiClO_4$ + 0.25 M **CSA**); **4** - on graphite electrode (0.1 M **An** + 1 M $LiClO_4$ + 0.25 M **CSA**); *c*: chemical synthesis on Pt (**1**) and graphite (**2**) electrodes from 0.25 M **An** + 0.25 M **BP** + 0.25 M **CSA** solution.[27]

Synthesizing of **PAn** in the presence of different additives, doping agents, or monomers can get the materials that are sensitive to a variety of analytes, not only to pH. For example, authors[28] have been constructed the potentiometric sensor for the determination of Cr (VI) by polymerization of aniline in the presence of hydrochloric acid and diphenylcarbazide. The sensor showed a linear response in semi-logarithmic coordinates ($E = f(lg[Cr(VI)])$) within a concentrations from 10^{-6} to 10^{-1} *M*. Also the potentiometric sensors sensitive to the presence of organic compounds, including saccharose[29], protein[30], and so forth can be constructed based on **PAn**.

7.4.3 AMPEROMETRIC SENSORS

In amperometric sensors on the platform of **PAn** the signal extends from the change of current flowing through the amperometric cell depending on the concentration of the analyte at constant potential of the polymer-modified electrode. In this case, the conjugated polymer plays an active role, participating in the oxidation–reduction processes. Using of pulse technology like to current pulse amperometry or impedance spectroscopy allows us to determine the kinetics of the recall formation.

Amperometric sensor for the determination of catechol has been successfully implemented by Mu.[31] Touch electrode was prepared by electrochemical deposition of copolymer of aniline and *o*-aminophenol on platinum foil by a size 3 ′ 4 mm. Copolymer layer on platinum, as it was turned out, shows the electrocatalytic effect in the oxidation of catechol. At a constant potential of 0.55 V the copolymer is readily oxidized to free radicals

$$\text{PAnOH} \longrightarrow \text{PAnO}^{\bullet} + \text{H}^{+} + \text{e}^{-}, \quad (7.9)$$

which immediately reduce the catechol in the solution with the regeneration of copolymer PAnOH and with the formation of free radical particles $C_6H_4(OH)O\bullet$

$$\text{PAnO}^{\bullet} + \text{C}_6\text{H}_4(\text{OH})_2 \longrightarrow \text{PAnOH} + \text{C}_6\text{H}_4(\text{OH})\text{O}^{\bullet}. \quad (7.10)$$

On the next stage the radicals of hydroquinone are oxidized to the quinone

$$\text{C}_6\text{H}_4(\text{OH})\text{O}^{\bullet} \longrightarrow \text{C}_6\text{H}_4\text{O}_2 + \text{H}^{+} + \text{e}^{-} \quad (7.11)$$

Available in copolymer hydroxyl group plays an important function in the electron transfer between the copolymer electrode and catechol. This reaction also occurs on copolymer electrode and its rate depends on the concentration of catechol in solution. The effect of catechol concentration on the current value of the oxidation can be seen from Figure 7.5. The current response was fixed at the potential 0.55 V, pH 5.0 and temperature 20°C. The linear dependence with the correlation coefficient of 0.997 is observed in the concentrations range 5-80 mM. Thus, this sensor can be used for the determination of catechol content in these concentration limits. It is shown that the sensor is characterized by the sensitivity not only to catechol, but also to phenol, resorcin and hydroquinone. His lifetime was more than 120 times without any significant changes in the magnitude of current response, indicating a high operational stability.

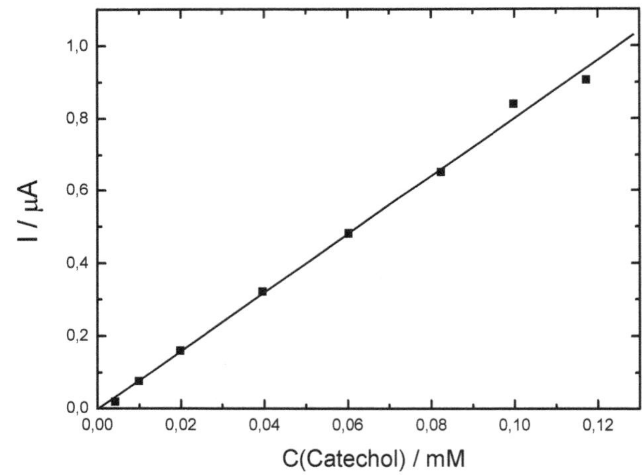

FIGURE 7.5 The ratio between current response and the concentration of catechol within the concentrations range of 5–120 mM.[31]

Use of the ferrocenesulfonic acid (**FSA**) (Fc-Fe^{n+}) as doping agent during the synthesis of **PAn** allows synthesizing the polymer (**PAn–Fc**), which is an effective catalyst for the oxidation of hydrogen peroxide, and on this basis to construct a sensor[32] for determining the concentration of H_2O_2. Iron atom in **FSA** may have the oxidation degrees +2 and +3. Reversible redox process between ions Fe^{2+} and Fe^{3+} plays an important role in the process of H_2O_2 catalytic oxidation. Effective (apparent) activation energy calculated by the log-slope current depends on the inverse temperature is 6 kJ/mol. Such a small value of the activation energy is the reason that the electrode reaction at 0.56 V is very fast and the over current response is formed in a very short time. At 0.56 V, **PAn–Fc** is in oxidation state and the iron in **FSA** is also in a state with a higher degree of oxidation Fe^{3+}. So, when the hydrogen peroxide is in contact with Fe^{3+} in **FSA** the chemical reaction between them takes place with the formation of the intermediate, which subsequently enters in the electrode reaction. The scheme of the H_2O_2 catalytic oxidation can be represented as follows:

$$H_2O_2 + Fc - Fe^{3+} \longrightarrow HO_2^{\bullet} + Fc - Fe^{2+} + H^+ \tag{7.12}$$

$$HO_2^{\bullet} - e^- \longrightarrow O_2 + H^+ \tag{7.13}$$

$$Fc - Fe^{2+} - e^- \longrightarrow Fc - Fe^{3+} \tag{7.14}$$

where Fc–Fe^{2+} and Fc–Fe^{3+} represent two oxidized states of iron in doping agent of **PAn**, namely in **FSA**. At relatively high value of electrode potential formed during the reaction Fc–Fe^{2+} is rapidly oxidized on the electrode to the Fc–Fe^{3+}. The current response of the sensor depends on pH, applied potential, and temperature at given concentration of H_2O_2. Under optimal conditions, the sensor detects the fast (0.2-0.3 min) response to added amounts of hydrogen peroxide, a good operating stability, linear response within the concentrations of H_2O_2 from 4 to 64 mM and a small temperature dependence of the current response.

7.4.4 VOLTAMMETRIC SENSORS

Unlike the potentiometry, in which the current is applied, the voltammetric measurements of the dependence "current–potential" in the electrochemical cell are carried out under non-equilibrium conditions. Therefore, at the voltammetric measurements information about the target analyte enters from the determinations of current value as a function of the applied potential, the magnitude of which is linearly increased during the experiment. At the achievement of a certain value potential on the working electrode, an electrochemical process takes place, which is accompanied by the current flow. The amount of current corresponds to the rate of the electrolytic reaction, which in its simplest form represents by itself the redox half-reaction

$$M^{n+} + e^- \longleftrightarrow M^{(n-1)+} \tag{7.15}$$

Here M^{n+} and $M^{(n-1)+}$ signify the oxidized and reduced forms of the electroactive particle M.

Voltammetric sensors, in the operation process of which the oxidation or reduction current detection of the analyte are entrusted, are highly devices because they are selective and highly sensitive to a wide range of inorganic and organic substances. The benefits of the voltammetric method are confirmed by the Korean researchers who used the glassy carbon coated with electrochemically deposited poly diaminterthiophenol, which was additionally modified by covalently grafted ethylenediaminetetraacetic acid as the working electrode.[33] Polarization of the electrode by linear potential scan or rectangular potentiostatic pulses gave a linear response of reduction current for ions Co(II), Ni(II), Cd(II), and Fe(II) at concentrations ranging from 0.1 to 10 μM and for Cu(II), Hg(II), and Pb(II) with concentration solutions within 0.5–20.0 nM. So, the voltammetric sensors can be confidently recommend for the selective determination of the trace amounts of various heavy metal ions in solution.

Authors of *ref.*[34] used of two types of the composites based on **PAn** and the sodium salt of a copolymer of styrenesulfonate and maleic acid (**PSSMA**) for detection of the ammonium ion in the analyte. Films by the first type of composite, namely **PAn–PSSMA (I)** were prepared by the incorporation of **PSSMA** anions into **PAn**, previously electrochemically polymerized by cyclic voltammetry (**CVA**) on thin film gold electrode at the surface of aluminum oxide (Au/Al_2O_3) plate. Another composite, namely **PAn-PSSMA (II)** was prepared by electropolymerization of aniline by **CVA** method on **PSSMA**/Au/Al_2O_3- electrode, which was prepared by applying of the aqueous solution of **PSSMA** of the appropriate quantity and concentration on the surface of Au/Al_2O_3. For both composites at the increase of the NH^{4+} concentration, a significant increase in the cathodic current peak at ≈−0.3 V is observed (potential scan was performed within (−0.4) ÷ (+0.4) V relative to Ag/AgCl/3M NaCl reference electrode). The high sensitivity of **PAn–PSSMA (I)** and **PAn–PSSMA (II)** are caused by their fibrillar morphology and high porosity. Maximal sensitivity (12 mA/mM) was observed for the film of **PAn–PSSMA (II)** obtained at the applying of 12 ml of 0.5% wt. solution of **PSSMA** and at the applying of **PAn** as a result of 5-fold cyclic potential scanning within (−0.3) ¸ (+1.0) V.

Today the intensive researches concerning to the development of voltammetric sensors based on **PAn** are also carried out. Such sensors can be successfully used for the analysis of biological objects. In particular, there is a high demand for rapid, sensitive and accessible methods of detection of harmful microorganisms, including *E. coli* (*Escherichia coli*), which is a very contagious and potentially deadly pathogen and may occur in food and water. In *ref.*[35] the cells *Escherichia coli* O157:H7 were isolated by immuno magnetic separation and were labeled with bifunction-analyzed electroactive **PAn** (immuno-**PAn**). The labeled cell complexes were deposited on disposable carbon electrode by screen print (**SPCE**) and were pressed to the surface of the electrode by an external magnetic field, in order to enhance the electrochemical signal generated by **PAn**. The **CVA** method was used to identify the **PAn**, and the value of a signal indicates the presence or absence of *E. coli.* A small number of the strain *Escherichia coli* O157:H7 (which corresponds to the original concentration of 70 **CFU**/ml) in seven colony-forming units (**CFU**) were successfully detected by **SPCE** sensor. The analysis requires 70 min from the sampling begin to detect, which is a great advantage compared to conventional methods of cultivation, especially in areas that require high throughput screening of samples and quick results. Another advantage that is not required any biological modification of the surface of sensor. The method can be used to design the compact portable devices.

7.4.5 OPTICAL SENSORS

Although the gas sensitive elements based on conductive polymers are applied primarily with the use of methods for measuring of electrical characteristics, but in the references the information about the optical gas sensitive elements based on **PAn** enough has long appeared.[36] The method of absorption spectroscopy is widely used to detect the ammonia and various acids using the **PAn**'s film. These optically sensitive sensors are characterized with a fast response to analyte and relatively easily are regenerated. If ammonia deprotonates the nitrogen atoms of the imine group, then in the presence of acids the reverse process of their protonation takes place, resulting to decrease or increase of the specific electroconductivity of **PAn** respectively, as well as the changes in the electronic absorption spectrum. This property of **PAn** is used to create the chemical sensors, especially the gas sensitive ones.[37,38]

Scheme of the interaction of ammonia with **PAn** is shown in Figure 7.6. Adsorption of ammonia on the protonated emeraldine base of **PAn**'s films at first has the physical nature. This process is not accompanied by significant change of optical properties, but there is a change in potential, which is related to the change of the polarons mobility. The next step is the chemosorption resulting to the formation of **PAn**–NH_3 complex. Last, the slowest stage is the restructuring of the polymer, which is accompanied by the deprotonation of the imine groups.[39]

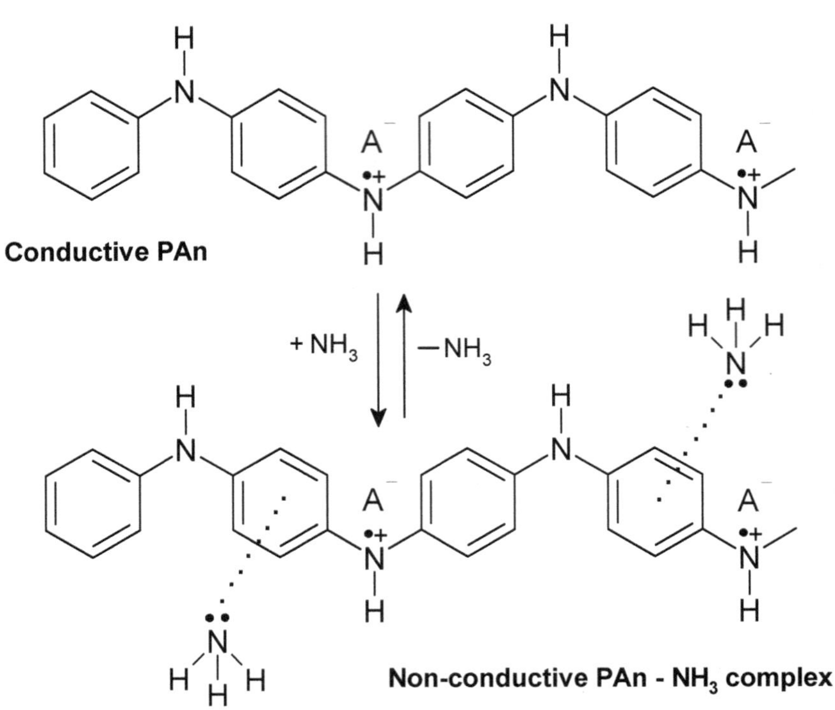

FIGURE 7.6 An interaction of ammonia with **PAn.**[39]

The effect of ammonia adsorbed on the surface of the polymer affects the electronic absorption spectra of **PAn.**[38] Film treated with hydrochloric acid, becomes clear green color with a maximum absorption light band at 800 nm, confirming that the layer of **PAn** is completely protonated. When gaseous nitrogen containing the ammonia is passed through the sensor element coated with **PAn**'s film, there is a distinct color change from green to blue, and the absorption maximum is shifted to 620 nm. Maximum of the absorption bands at ammonia concentration increasing is gradually shifted to shorter wavelength. A change of spectrum is reversible, indicating on the possibility of the **PAn**'s film using as the optically sensitive material for ammonia.

The response of PAn sensitive element to ammonia was tested by measuring of the light absorption intensity in the region of 600 nm. When the concentration of ammonia is increased from 18 to 18,000 ppm, the four-fold increase of the light absorption intensity is observed. Resistance of **PAn**'s films depends on the diffusion of ammonia in the film, which can be expressed by the following equation[40]:

$$R = R_0 \exp\left[(\alpha N)^y\right], \tag{7.16}$$

where R_0 is initial resistance, N is the concentration of ammonia, α and y are constants.

Since the same diffusion process manages by the change of the light absorption intensity, the relationship between light absorption and concentration of ammonia can be expressed by the same equation:

$$A = A_0 \exp\left[(\alpha N)^y\right] \tag{7.17}$$

Here, A_0 is the initial light adsorption and A is light adsorption at concentration N. From the last equation it's calculated log $(\ln(A/A_0))$, namely an expression proportional to logN, in order to register N.

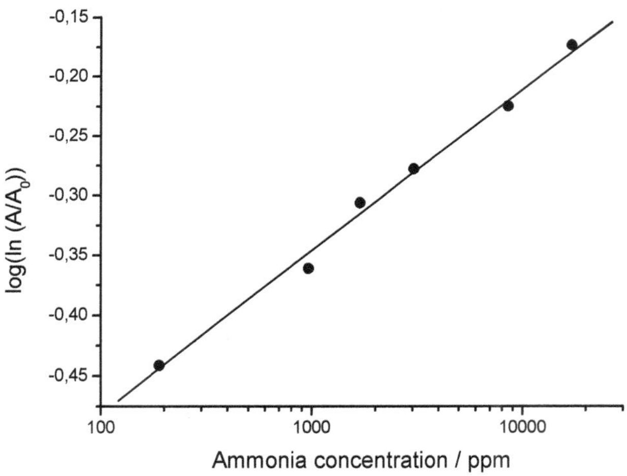

FIGURE 7.7 Calibrating curve of ammonium sensor response on the content of ammonia in reactive mixing.[40]

Typical calibrating curve obtained using the Eq. (7.17) is shown in Figure 7.7. Linearity dependence is observed in the range of 180–18,000 ppm with linear regression coefficient 0.998. The sensitive element of the sensor before experiments usually is treated with 0.1 M hydrochloric acid that leads to complete regeneration of the sensing element. The gaseous nitrogen containing the hydrochloric acid can be used as an alternative method for the regeneration of the sensing element. This method is rapid, simple, and reproducible. Therefore, PAn sensors sensitive for ammonia, based on the spectroscopy method have the advantages over the elements based on the measurement of specific electroconductivity not only in response and time regeneration, but also in ease regeneration of the sensing element. Optical sensors are more suitable for practical use though require the appropriate instrumentation equipment.

Light transmission of the **PAn**'s film will depend on the exposure time and pressure. As a result, for different initial conditions, the response time of sensors on PAn basis at the exposure by gaseous ammonia is varied from 10 to 1000 s. The time dependence of light transmission A at sudden change of pressure is usually given by the sum of two or more exponents[39]:

$$\frac{\Delta A}{A} = \sum K_{A,i}\left[1 - \exp\left(-\frac{t - t_0}{\tau_i^A}\right)\right] \tag{7.18}$$

for adsorption and

$$\frac{\Delta A}{A} = \sum K_{D,i}\left[1 - \exp\left(-\frac{t - t_0}{\tau_i^D}\right)\right] \tag{7.19}$$

for desorption.

Characteristic times τ_i for adsorption and desorption processes are near and depending on pressure can take the values near to 100 and 1000 s, which can be associated with two different adsorption centers on the PAn surface. The activation energies of adsorption centers are near to 0.8 eV. An effect of pressure on the ammonia degree adsorption is described usually by *Langmuir* or *Freundlich* adsorption isotherm. Overall the response time of the sensor will be determined by the rate of physical and chemical sorption process and by the rate of the restructuring of the polymer chains. The last factor will depend on the morphology of **PAn**'s film.

Undoped **PAn** can be used in the design of optical sensors for the determination of acids' vapors, such as formic acid.[41] It were developed the optical sensors based on **PAn** for the analysis of solutions, for example, for determination of pH,[42] for determination of the content of dissolved oxygen.[43] The composite materials based on **PAn** are also used at designing of the optical sensors. The second component of the composite serves as a catalyst for the process in which a substance becomes detectable in acid or base. Last interacting with **PAn**'s layer leads to a change in its color and light-absorption. For example, gold nanoparticles encapsulated by PAn as emeraldine base exhibit the catalytic activity in the oxidation of glucose to gluconic acid.

Gluconic acid dopes the emeraldine base, the process is accompanied by a color change of the **PAn**'s layer from blue to green, and this was used for the colorimetric determination of the glucose in analysis.[44]

7.4.6 PIEZOELECTRIC SENSORS

In piezoelectric devices, the electrical signal is due to vibrations of the crystal, usually of quartz. Electrochemical quartz by crystal microbalance allows to define the mass changes of the electrode at coating levels lower than the monolayer coverage,[45] as well as to investigate the conformational changes of the grafted surface modifiers,[46] and so forth. The operating principle of the quartz sensor is that the increase in weight or thickness of the quartz plate leads to a decrease in the resonant frequency of its oscillations. Change in thickness (Δl_q) of plates (at the expense of deposition/desorption of analyte-substance on/from the surface of quartz) leads to the change of the resonant frequency (Δf_q)

$$\Delta f_q / f_q = -\Delta l_q / l_q \tag{7.20}$$

More mass (ΔM), which is uniformly distributed on the crystal surface area S as a thin film can be considered as equivalent to the change in mass of the quartz plate. This can be expressed by the equation

$$\Delta f = -2,26 \cdot 10^{-6} f_q^{\,2} \frac{\Delta M}{S}. \tag{7.21}$$

Eq. (7.20) and (7.21) hold true for the cases where ΔM is very small. Therefore, the marginal mass change ΔM is directly related to the change of frequency fluctuations Δf. The negative sign in Eq. (7.21) indicates a decrease in the resonant frequency of the resonator with increasing of mass, and *vice versa* an increasing of frequency as a result of desorption processes. Eq. (7.21) gives the most adequate results for thin and hard layers; that is the added weight must not undergo any deformation displacements during the oscillation. If a large mass deposited on the crystal or on the film of the crystal, then the linear dependency provided by this equation was not observed for the frequency changes. Note, that other factors, such as wetting, electric field, and temperature also affect the resonant frequency. That's why it is necessary to calibrate the device for each series of resonators for the quantitative determination of the mass changes. Practical use of piezoelectric quartz crystal resonators as analyzers implies the need to determine of the quantitative relationship between the relative change in resonance frequency and added mass.

In particular, a significant shift of the resonance frequency of quartz plate covered with a layer of **PAn** in the form of emeraldine salt due to the binding of ammonia by **PAn**'s layer allows a quantitative determination of NH_3 in solution or in the gaseous phase. Calculated on the basis of experimental data sensitivity of the sensor is 9642 Hz·l/mol,[47] which could be increased by the introduction of gold nanoparticles to the structure of **PAn**. Incorporation of metallic particles, in turn, also can be controlled by decreasing of the resonant frequency of the sensor (Fig. 7.8). Gold nanoparticles obtained by reduction of $HAuCl_4$ with $NaBH_4$ in the presence of dodecanethiol in the form of dispersion are added to the cell with the resonator, where they arbitrarily make up to polyaninine matrix. At the content of incorporated Au nanoparticles $1.093 \cdot 10^{-6}$ g/cm^2, the sensitivity of such modified sensor was already 53,124 Hz·l/mol (Fig. 7.9). In other words, the PAn films modified by Au–NPs and coated on mass-sensitive receiver are about in five times more sensitive and are characterized by much wider range of the linear response compared to the unmodified ones. The reason for this is obviously more evenly distribution of charge on the surface of the sensor, which contributes to higher adsorption of particles of the substrate, and hence to increase of the sensor sensitivity.

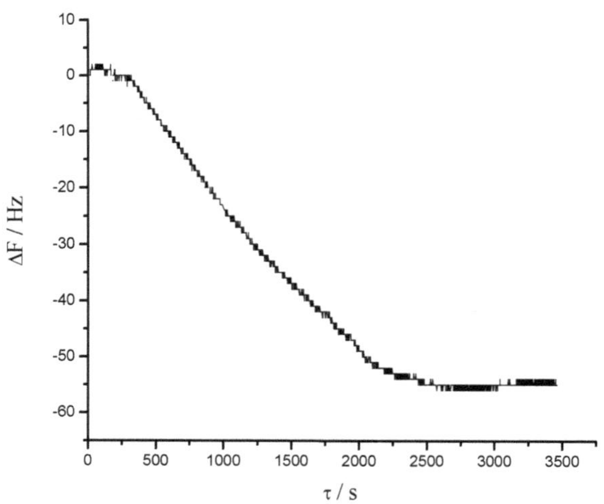

FIGURE 7.8 Frequency response of the piezoelectric sensor depending on the duration of Au nanoparticles incorporation into the **PAn**'s film.

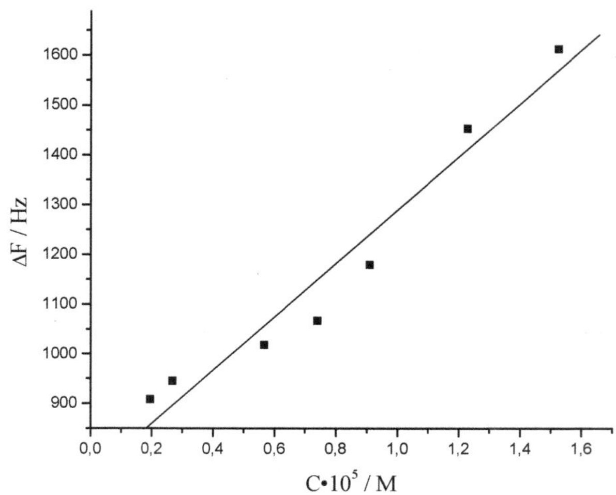

FIGURE 7.9 Frequency response of the mass sensitive sensor modified by Au nanoparticles at the increase of ammonia in the solution and its calibrating curve.

Today a large number of piezoelectric sensors have been designed and studied for the detection of the compounds that are capable physically adsorbed or chemically binded by **PAn**, including: formic acid,[48] carbon tetrachloride, chloroform, dichloromethane, 1,2-dichloroethane vapors, ethyl acetate, ethanol, propanol, hexane, and benzene.[49,50]

7.4.7 ACOUSTICAL SENSORS

Similar to the piezoelectric sensors are the sensors whose operation is based on the measurement of the surface acoustic waves. The acoustic waves are generated on the surface of the piezoelectric crystal at its excitation. Such materials as quartz, tantalate, and lithium niobate[2] are most commonly used as substrate materials in sensors on the acoustic waves. Two systems of opposite-interdigitated transducers, one of which is used for the excitation of wave (transmitter) and the other one for its detection (receiver) is the required element of the sensor. There are sensors on surface and bulk acoustic waves depending on the type of the wave's propagation.

In *ref.*,[51] authors present the one among possible design of the humidity sensor, the principle of which is based on the surface acoustic waves (**SAW**). Sensor was constructed according to the scheme of dual resonator-oscillator with an operating frequency of 300 MHz. **PAn**'s coating is applied along with the resonator of sensor (Fig. 7.10). Al/Au electrodes were used to ensure the corrosion resistance of the sensor chip at the formation of **SAW** resonators. Two SAW resonators were deposited on a quartz substrate and were used as the controls of frequency feedback in the electric circuit of the oscillator. Fine **PAn**'s coating was applied to the resonator of sensor and it served as the sensing material of the sensor for measuring of relative humidity. During the humidity determination due to the adsorption of water molecules, the conductivity of **PAn** is changed resulting in the appearance of the perturbation in the **SAW** propagation. The sensor exhibits a high sensitivity, fast response, good reproducibility, and the stability at room temperature.

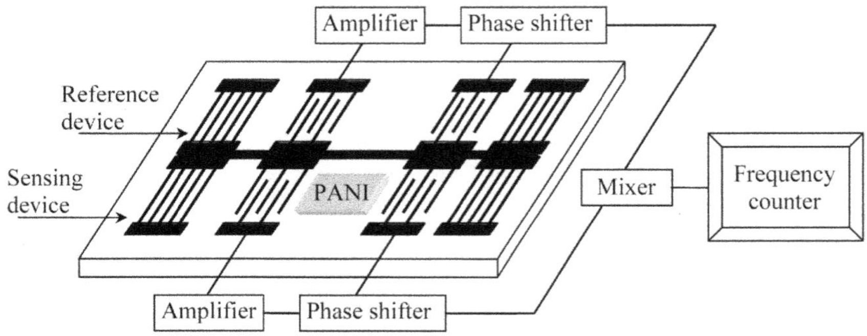

FIGURE 7.10 A scheme of the sensor for the determination of humidity based on SAW.[51]

The **SAW** sensors based on **PAn** for determination of hydrogen,[52,53] humidity,[54] NO_2 and CO[53] have been developed today. At this, at the designing of **SAW** sensors the composites of **PAn** with other polymers, namely polyvinyl alcohol,[54] inorganic materials (with nanoparticles of In_2O_3),[53] and so forth are often used.

7.5 PAn IN BIOSENSORICS

If the biologically active components, such as enzymes, nucleic acids, yeast, liposomes, organelles, bacteria, antibodies-antigens, single-celled organisms, and even entire tissues of higher organisms are contained in the structure of touch sensitive layer device, it can be determined as **biosensor**. Although these are not the fundamental differences between chemo- and biosensors, the last are differed by significantly greater variety, higher selectivity, and sensitivity. Flexible operating conditions make them indispensable in biomedical diagnostics, at the analysis of food quality and environmental protection. In accordance with the accepted interpretation of the term "biosensor," all changes in the composition as well as the process of biochemical recognition occur in the surface layer, the volume of which is very small compared to the volume of the sample of the analyzed solution.

Due to the specificity of the interaction between the particles from analyte and bio-recognizing sensing layer, the detecting physical and chemical changes can be measured by the transducer. Biochemical component acts as a highly selective sorbent keeping the determined component on the surface. Transducer serves as a mechanical support of biochemical component, while registering a signal of biochemical transformations. Just as in the case of chemosensors, depending on the nature of signals the transducers may be electrochemical, optical, mechanical, thermal, acoustic, and so forth type. In addition, they are also classified according to the nature of the biochemical component, distinguishing the enzyme electrodes, immunosensors, DNA-sensors, and so forth.

Regardless of the nature of the biochemical component, the all biosensors can be divided into two groups according to the nature of the signal. The first group is represented by **enzyme sensors (electrodes)**. In such sensors, the measurement of a signal is accompanied by a detectable change in the nature of matter. The signal itself is a measure of the speed of substrate enzymatic reactions transform. The decline in the concentration of the substrate is compensated by its diffusion transfer from the bulk solution, so at the spending of substrate the signal of enzyme sensor should be decreased. However, the degree of conversion of the substrate during the measurement is usually very small, so under such quasi-stationary conditions the signal of the enzyme sensor is stable. Its dependence on the bulk concentration of the substrate as a whole is characterized by sigmoid nature, but in rather wide range of concentrations is approximated by a linear function. These sensors are also called as *kinetic* or *dynamic*. The second group is represented by biochemical sensor, in which the equilibrium is established as a result of biochemical recognition; such equilibrium is not accompanied by chemical transformations of the detectable substance. These biosensors are called as **affine biosensors**. They include **immunosensors**, **DNA-sensors,** and sensors based on supramolecular cellular structures. Maximal value of a signal of the affine biosensors corresponds to the establishment of equilibrium at the interaction of the receptor and determined substance and is determined by the appropriate constant of the equilibrium.[4]

PAn in biosensor devices can serve both as a matrix for the immobilization of enzymes and as the transducer. Also, often these functions are combined. In order to the biosensor was reliably working, the biological material must be surely attached (immobilized) on the surface of the transducer. To ensure an efficient deposition of biomolecules, the next several prerequisites must be satisfied:

- a process of the biological macromolecules immobilization on the surface should be effective and stable;
- the immobilized biomolecules should completely maintain their biological properties;
- they should be compatible and chemically inert concerning to the structure of the host;
- immobilized biomolecules must be accessible to the substrate, which undergoes the biochemical transformation.

Immobilized enzyme has many functional advantages over free enzyme, namely: reusability, increased stability, continuous operation, a quick termination of the reaction, ease of separation of product from the biocatalyst, and lower operating cost. Immobilization of biomolecules on films of conducting polymers can be made using many different procedures while preserving the defining biological properties of biomolecules.

The most common used methods of the immobilization are as follows: an adsorption, microencapsulation, an inclusion, the lacing, the covalent binding and also an electroimmobilization.[2,55]

As noted above, the preferred method of the **PAn**'s films obtaining is the electrochemical method due to its simplicity, the possibility of obtaining of thin polymer layers on electroconductive surfaces of a wide range of configuration, the possibilities for control of the structure and the degree of oxidation of the synthesized polymer and more. Depending on the nature of substituents in the initial monomer the formed polymer coating will be strongly coupled to the surface or will be easily away from it, can be an insulator or a conductor of electric current, less or more heat-resistant, capable to reversible redox, be dense or porous, solid or gel-like, and so forth. Electropolymerization is a simple and a fast procedure that allows to control the thickness of the obtaining biosensitive layer on a basis of value of current or the number of transmitted electricity. Such process is easy automated under the mass production and provides the necessary metrological characteristics of properties of the forming coating. The vast majority of electrochemically deposited polymer films for immobilization of biomolecules belong to electro conducting polymers, such as polymer acetylenes, thiophenes, pyrroles, anilines, indoles, and carbazoles. An important place among these has the **PAn**, although it inferior to the number of relevant studies dedicated to polypyrrole.[4] Thanks to the conductivity of **PAn** the thickness of polymer film can be easily changed and is not to be limited by very thin films in contrast to the non-conducting organic polymers, where the thickness of the polymer film is very small.

Unlike to polypyrrole, the oxidative condensation of which is occurred at the pH near to neutral solutions, **PAn** is synthesized at the oxidation from the strongly acidic aqueous solutions. These conditions are not favorable for the incorporation of enzymes by including of their macromolecules by growing PAn chains via microencapsulation or copolymerization in the presence of one or more electroactive groups in biomolecules. However, a large number of enzymes and other biomolecules can be immobilized by physical adsorption on presynthesized **PAn.**[56,57] This is the simplest method of the enzymes immobilization. Binding between the particles of polymer and enzyme is due to the formation of hydrogen bonds, the *Van der Waals* forces, or ionic interactions. This method almost does not change the native conformation of the enzymes. Adsorbed biomaterial is very sensitive to changes of pH, temperature, ionic strength and the substrate concentration. The disadvantages of this method include the losses of enzyme due to it leaching during the sensor operation. However, an adsorption method can be used for research purposes in order to obtain the sensors, a long life for which will not require. On the other hand, the losses of enzymes can be significantly reduced by applying of the semipermeable membrane on the surface of the sensitive device. Membranes do not hinder to diffusion of the low molecular detectable substances to the sensing layer from the volume of analyte but prevent to the leaching of bioactive components from the surface of the sensor.

Some different approach of the wedging-in of biomolecules in electro synthesized **PAn**'s film is the **electro-immobilization** (*electrochemical doping*). The wedging-in into electroconductive polymer of enzymes enables the efficient contact of the active centers of the enzyme with an electroconductive surface of electrode, which is important in the design of biosensoric devices. This method is quite simple and it permits to precipitate the biologically active molecules on certain parts of the electrode in controlled way. Negatively (or positively) charged enzymes such as glucose oxidase, galactose oxidase or peroxidase can be as doping agents of PAn chains during their oxidation (or reduction).[58] At this, firstly the dedoping of the PAn layer is performed and after that it's doping, but in solution of the corresponding enzyme. For example, **PAn** initially is reduced in phosphate buffer solution at the potential of −500 mV (*versus satur.* Ag/AgCl electrode) to achieve the sustainable current value and then is oxidized at the potential of +650 mV for 20 min in phosphate buffer solution containing of 0.1 mg/ml of the *horseradish peroxidase* (**HRP**).

During this process, the enzyme electrostatically is joined to the **PAn**'s film.[59]

A similar method has been used by the authors of *ref.*[60] Doped **PAn**'s layer with polyvinyl sulfonate on the surface of screen-printed electrode was polarized for 1500 s under nitrogen at −500 mV by pulses duration of 500 ms to restore the surface layer. After that, the electrode was transferred into a phosphate buffer solution containing of 10 mg/ml of **HRP** and at continuous stirring and de-aeration positively was polarized at +700 mV for 1500 s. During oxidation, the protein is electro statically joined to the polymer surface. It was determined the mass of the protein immobilized on the electrode surface. The ratio between the number of electro statically bound protein and its concentration in the solution is described by the expression

$$M = 1,4521 \cdot 10^{-5} \log[\mathrm{HRP}] + 3,375 \cdot 10^{-5} \quad (7.22)$$

where [**HRP**] is the concentration of horseradish peroxidase. The transition from the mass of peroxidase to the number of molecules (M. m. of peroxidase is 44,000 Da)

$$N = 1,9845 \cdot 10^{11} \log[\mathrm{HRP}] + 4,6097 \cdot 10^{10} \quad (7.23)$$

makes it possible to calculate the number of molecules of peroxidase on the electrode surface. It was revealed,[60] that the optimal concentration of protein is 0.75 mg/ml, and the number of protein molecules that immobilized on the surface consists of 4.3618×10^{11}. The number of **HPR** molecules required for the formation of a monolayer is equal to $4.237 \cdot 10^{11}$, if to assume that the area is perfectly flat electrode is $9 \cdot 10^{-6}$ m^2 and the radius of the particle of peroxidase is 2.6 nm. Theoretical number of peroxidase molecules required for the formation of a monolayer, is in good agreement with the experimental data under optimal conditions.

However, an immobilization of biomolecules by electrochemical wedging in conductive or un-conductive polymeric films requires the imposition of relatively high electrical potentials. This can lead to the partial oxidation (reduction) of the biomaterial and loss of its activity. In addition, it is difficult to accurately determine the amount of immobilized enzyme in the polymer net. This complicates the kinetic analysis and the optimization of multilayer enzyme electrodes.

An alternative method of the electrochemical incorporation of enzymes can be the **covalent binding of biomolecules** (*chemical adsorption*) with a surface of pre-synthesized polymeric film containing in the appropriate functional groups. An important advantage of this method is that each stage takes place under optimal conditions. A stage of the polymerization, for example, can occur in highly acidic medium, in organic solvents at high potentials, be accompanied by the generation of monomeric radicals that would destroy the biological molecules, and so forth. In addition, this approach allows us to create the better conditions for the inflow of the analyte to the immobilized biomolecules, facilitates to macromolecular interactions that retain the original properties of polymeric film, for example such as electroconductivity.[61] The main advantage of this method is that in such biosensor the enzyme is strongly fastened to the substrate and not washed during the operation of the sensor. In order to protect the active center of the enzyme during the immobilization, it is often carried out in the presence of substrate. One method of the covalent binding of the enzyme (**E**) is shown[2] in Figure 7.11.

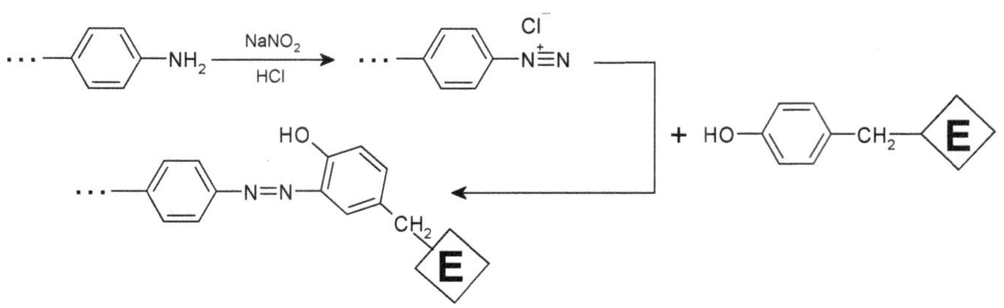

FIGURE 7.11 Covalent bonding of the ferment (**E**) with the substrate by diazotization of the final amine groups of **PAn**.

In cross linking method the bifunctional reagents (namely, glutaraldehyde, hexamethyldiizocyanate, and 1,5-dinitro-2,4-difluorobenzene) are used for the biomaterial immobilization on a solid support. The method is quite useful for the stabilization and for the maintenance of enzymes adsorbed on the surface of the carrier. An example of this method using is given in *ref.*[62] Authors have been constructed the biosensor **PAn/HRP** by dripping method. For this purpose the solution by dissolving of 0.2 mg of **HRP** in 100 ml of 0.05 M of phosphate buffer at pH 7.0 was pre-prepared. Then 3 mg of powder-like **PAn** doped with anthranyl sulfonic acid and 0.2 mg of bovine serum albumin (**BSA**) were added to this enzyme solution, and after that it was stirred with the following addition of the 2 ml (2.5%)

of glutaraldehyde. Then aliquot of the mixture was applied drop wise to the previously pretreated microelectrode and was dried for 2 h. During the drying of a solution the cross linking of the active ingredients of biosensor takes place. This procedure has been used at the manufacture of **PAn**/cytochrome biosensor. The cross linking method is also not completely perfect, since at the construction of sensor there are possible the losses of the enzyme activity and the difficulty of diffusion of the substrate molecules.

A great interest represents by itself the **layered immobilization of enzymes**, although the **PAn** in this method is used quite rarely. To fix the enzyme on the surface of transducer is possible by means of the alternating applying of layers having the oppositely charged ionogenic groups. Layered assembling method, which is based on the electrostatic adsorption of oppositely charged polyions, was firstly proposed[63] by Decher and *Hong* in 1991. Due to the universality and simplicity, the method of layered immobilization is suitable for a wide range of materials, including the proteins, nanoparticles, dyes, and DNA. Polyionic complexes can be obtained by reacting of high macromolecular polyelectrolytes, namely polycations (**PAn**, polythionine, and polythiophenes) and polyanions (nafion, polystyrene sulfonate, and polyvinyl sulfonate). Inclusion of biomolecules takes place in the process of complexes formation during the layered deposition of polyelectrolytes from the solution. These multilayer coatings are characterized by high durability and additionally provide the electrostatic barrier that limits an access of the sample's ions to the electrode that interfere to analysis.[64] The authors of *ref.*[65] were used the method of layered assembly of carboxylated multi walled carbon nanotubes and **PAn** at the manufacture of the biosensor transducer on choline. On a basis formed in such a way they was applied the choline oxidase immobilized by **BSA** and glutaraldehyde. Designed biosensor showed a linear response range of substrate via the concentrations range from $1 \cdot 10^{-6}$ to $2 \cdot 10^{-3}$ M with a correlation coefficient of 0.997. Response time and the detection limit ($S/N = 3$) is 3 s and 0.3 mM, respectively.

There are a number of review articles concerning to **PAn** using in biosensorics[62,66,67] unlike to references dedicated to chemosensors. An interest for using of **PAn** in biosensors primarily is based on fact that it is the effective mediator of the electron transfer in redox enzyme reactions. In Figure 7.12, there is a diagram of electron transfer from the place of biochemical reactions to the electrode through a network of **PAn** in amperometric biosensor. **PAn** offers the many opportunities to combine the analyte receptors interactions and also non-specific interactions in the observed responses. In particular, the transport properties of **PAn**, the electroconductivity, or energy transfer rate enable the increase of the sensitivity of sensors.

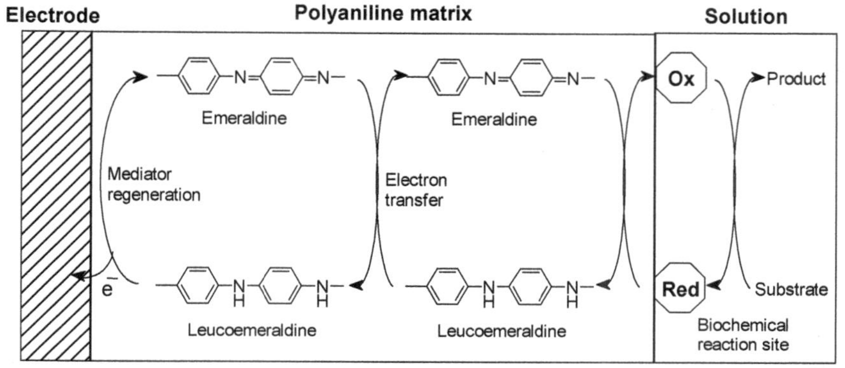

FIGURE 7.12 A scheme of the electron transfer from a place of biochemical reaction on electrode via conductive net of **PAn** in amperometric biosensor.[67]

The ability to reverse oxidation of **PAn** has been applied at the design of biosensors, in which the redox enzymatic reactions take place. First of all, there are sensors based on oxidases: horseradish peroxidase (**HRP**), glucose oxidase (**GOx**), cholesterol oxidase (**ChOx**) and so on. At this, **PAn** often provides the charge transfer between the electrode and the center of biochemical processes, such as in amperometric biosensors. In addition, the ratio between the quantities of quinonediimine and benzenediamine, protonated and non-protonated imine groups such as, for example, in potentiometric, optical, capacitive, and so forth biosensors is changing in **PAn** as a result of the interaction with the products of biochemical processes. The essence of the processes of the response formation remains the same as in chemosensors, but only the component that causes the changes in transducers based on **PAn** is not entered into the system along with the analyte; it is formed or disappears as a result of biochemical reaction. Accordingly, the approaches to the designing of chemo- and biosensors are similar. In particular, a great attention is paid into the use of nanocomposites based on **PAn** as well as metallic and nonmetallic nanoparticles,[68–70] various forms of nanostructured **PAn,**[71] and so forth as the platforms for biosensors. However, in this case the influence of the method of biomaterial immobilization both on operation of the transducer and on biochemical center of the reactions, as well as on ensure of good interaction between them must be taken into account.

7.6 CONCLUSIONS

The ease and the simplicity of the synthesis of thin films of **PAn** on various conductive surfaces, the assemblage of its unique properties, such as the possibility to adjust the electrical conductivity by changing of the doping level, the ability to the charge transfer, the stability, an existence of forms with different oxidation degrees (respectively, different conductivities, and color) and the ease of the transition, a good compatibility with biomolecules and so on, make the PAn by attractive platform at the chemo- and biosensors manufacture. The availability and the potential importance of this **ECP** are increased even more with the possibility of the formation of the various nanostructures (nanospheres, nanotubes, or nanofibers) with its participation, as same as significant improvement of physical and chemical properties at the introduction of nanodispersed particles of metals, oxides, natural minerals, mono- and multi walled carbon nanotubes and more in **PAn** matrix. Already today on a basis of **PAn,** it was created a significant number both of research and industrial designs of the sensors by chemical, medical, food and other appointments, which contains the different principles as for the generation and processing of the analytic signal and on the architecture of the themselves devices. However, the daily increase in the number of objects of analysis, and the need to improve the performance of existing sensor systems, namely their sensitivity, selectivity, operational stability, and so forth makes intensive research in this area. Therefore, it can be foreseen that the number of works devoted to the chemo- and biosensorics, including to **PAn** and materials on its basis will be only to grow.

KEYWORDS

- **polyaniline**
- **composites**
- **chemosensors**
- **biosensors**
- **transducer**
- **biomolecules**
- **immobilization**

REFERENCES

1. Dzyadevych, S. V.; Soldatkin, O. H. *Scientific and Technological Principles of the Design of Miniature Electrochemical Sensors*; Naukova dumka: Kyiv, 2006.
2. Eggins, B. R. *Chemical Sensors and Biosensors*; John Wiley & Sons Ltd.: Chichester, 2002.
3. Turner, A. P. F.; Karube, I.; Wilson, G. S., Eds. *Biosensors: Fundamentals and Applications*; Oxford University Press: Oxford, 1990.
4. Koval'chuk, E. P.; Ostapovych, B. B.; Kovalyshyn, Ya. S. *Chemical and Biological Sensorics*; Ivan Franko National University of L'viv: L'viv, 2012.
5. Huang, W. S.; Humphrey, B. D.; MacDiarmid, A. G. Polyaniline, a Novel Conducting Polymer: Morphology and Chemistry of its Oxidation and Reduction in Aqueous Electrolytes. *J. Chem. Soc. Faraday Trans.* **1986,** *82*, 2385-2400.
6. Crowley, K.; Smyth, M. R.; Killard, A. J.; Morrin A. Printing Polyaniline for Sensor Applications. *Chem. Papers*. **2013,** *67*, 771-780.
7. Morrin, A.; Ngamna, O.; O'Malley, E.; Kent, N.; Moulton, S. E.; Wallace G. G.; Smyth, M. R.; Killard, A. J. The Fabrication and Characterization of Inkjet–Printed Polyaniline Nanoparticle Films. *Electrochim. Acta*. **2008,** *53*, 5092-5099.
8. Crowley, K.; O'Malley, E.; Morrin, A.; Smyth, M. R.; Killard, A. J. An Aqueous Ammonia Sensor Based on an Inkjet–Printed Polyaniline Nanoparticle–Modified Electrode. *Analyst*. **2008,** *133*, 391-399.
9. Crowley, K.; Morrin, A.; Hernandez, A.; O'Malley, E.; Whitten, P. G.; Wallace, G. G.; Smyth, M. R.; Killard, A. J. Fabrication of an Ammonia Gas Sensor Using Inkjet–Printed Polyaniline Nanoparticles. *Talanta*. **2008,** *77*, 710–717.
10. Chen, F.; Tang, W.; Li, C.; Chen, J.; Liu, H.; Shen, P.; Dou, S. Conducting Poly(Aniline) Nanotubes and Nanofibers: Controlled Synthesis and Application in Lithium/Poly(Aniline) Rechargeable Batteries. *Chem. Eur. J.* **2006,** *12*, 3082-3088.
11. Eftekhari, A. *Nanostructured Conductive Polymers*; Wiley: Chichester, 2010.
12. Wang, J.; Matsubara, I.; Murayama, N.; Woosuck, S.; Izu, N. The Preparation of Polyaniline Thin Film Intercalated MoO_3 and its Sensitivity to Volatile Organic Compounds. *Thin Solid Films*. **2006,** *514*, 329-333.
13. Wieckowski, A.; Savinova, E. R.; Vayenas, C. G., Eds. *Catalysis and Electrocatalysis at Nanoparticle Surfaces*; Marcel Dekker: New York, 2003.
14. Shirsat, M. D.; Bangar, M. A.; Deshusses, M. A.; Myung, N. V.; Mulchandani, A. Polyaniline Nanowires-Gold Nanoparticles Hybrid Network Based Chemiresistive Hydrogen Sulfide Sensor. *Appl. Phys. Lett.* **2009,** *94*, 083502(10 p).
15. Sharma, S.; Nirkhe, C.; Pethkar, S.; Athawale, A. A. Chloroform Vapour Sensor Based on Copper/Polyaniline Nanocomposite. *Sens. Actuators. B*. **2002,** *85*, 131-136.
16. Feng, X. M.; Mao, C. J.; Yang, G.; Hou, W. H.; Zhu, J. J. Polyaniline/Au Composite Hollow Spheres: Synthesis, Characterisation, and Application to the Detection of Dopamine. *Langmuir*. **2006,** *22*, 4384-4389.
17. Ma, X.; Li, G.; Wang, M.; Cheng, Y.; Bai, R.; Chen, H. Preparation of Nanowire–Structured Polyaniline Composite and Gas Sensitivity Studies. *Chem Eur. J.* **2006,** *12*, 3254-3260.
18. Robinson, J. A.; Snow, E. S.; Bădescu, S. C.; Reinecke, T. L.; Perkins, F. K. Role of Defects in Single-Walled Carbon Nanotube Chemical Sensors. *Nano Lett*. **2006,** *6*, 1747-1751.

19. Dai, L.; Soundarrajan, P.; Kim, T. Sensors and Sensor Arrays Based on Conjugated Polymers and Carbon Nanotubes. *Pure Appl. Chem.* **2002,** *74*, 1753-1772.
20. MacDiarmid, A. G. Synthetic Metals: A Novel Role for Organic Polymers. *Synth. Met.* **2001,** *125*, 11-22.
21. Svetlicic, V.; Schmidt, A. J.; Miller, L. L. Conductometric Sensors Based on the Hypersensitive Response of Plasticized Polyaniline Films to Organic Vapor. *Chem. Mat.* **1998,** *10*, 3305–3307.
22. Panella, B.; Kossykh, L.; Weglikowska, U. D.; Hirscher, M.; Zerbi, G.; Roth, S. Volumetric Measurement of Hydrogen Storage in HCl-Treated Polyaniline and Polypyrrole. *Synth. Met.* **2005,** *151*, 208-210.
23. Cho, S. J.; Song, K. S.; Kim, J. W.; Choo, K. Hydrogen Sorption in HCl-Treated Polyaniline and Polypyrrole: New Potential Hydrogen Storage Media. Fuel Chemistry Division, 224th National Meeting of the American Chemical Society, Boston, Aug. 18-22, 2002, 47, 790-791.
24. Sadek, A. Z.; Wlodarski, W.; Kalantar–Zadeh, K.; Baker, C.; Kaner, R. B. Doped and Dedoped Polyaniline Nanofiber Based Conductometric Hydrogen Gas Sensors. *Sens. Actuators. A.* **2007,** *139*, 53-57.
25. Lu, J.; Park, B. J.; Kumar, B.; Castro, M.; Choi, H. J.; Feller, J. F. Polyaniline Nanoparticle-Carbon Nanotube Hybrid Network Vapour Sensors with Switchable Chemo-Electrical Polarity. *Nanotechnology.* **2010,** *21*, 255501-255511.
26. Kovalchuk, Y. P.; Ostapovich, B. B.; Turik, Z. L.; Kovalishin, Y. S.; Godovanets, I. M. Kinetics of Oxidative Polycondensation of Aniline in a Methylpyrrolidone Solution. *Ukr. Chem. J.* **2006,** *72*(3-4), 66-71.
27. Kovalchuk, Y .P.; Ostapovich, B. B.; Turik, Z. L.; Kovalishin, Y .S.; Gonchar, M. V. Synthesis and Investigation of Electrically Conductive Polymer Platforms for Biosensors. *Ukr. Chem. J.* **2006,** *72*(11-12), 35-42.
28. Mohammadkhah, A.; Ansari, R.; Fallah Delavar, A.; Mosayebzadeh, Z. Nanostructured Potentiometric Sensors on Polyaniline Conducting Polymer for Determination of Cr (VI). *Bull. Corean Chem. Soc.* **2012,** *33*, 1247-1252.
29. Gupta, N.; Sharma, S.; Mir, I. A.; Kumar, D. Advances in Sensors Based on Conducting Polymers. *J. Sci. Ind. Res. India.* **2006,** *65*, 549-557.
30. Duzgun, A.; Imran, H.; Levon, K.; Rius, F. X. Protein Detection with Potentiometric Aptasensors: A Comparative Study between Polyaniline and Single-Walled Carbon Nanotubes Transducers. *Scientific World J.* **2013,** *2013*, 282756(8 p).
31. Mu, S. Catechol Sensor Using Poly(Aniline-co-o-Aminophenol) as an Electron Transfer Mediator. *Biosens. Bioelectron.* **2006,** *21*, 1237-1243.
32. Yang, Y.; Mu, S. Determination of Hydrogen Peroxide Using Amperometric Sensor of Polyaniline Doped with Ferrocenesulfonic Acid. *Biosens. Bioelectron.* **2005,** *21*, 74-78.
33. Rahman, Md. A.; Park, D. S.; Won, M. S.; Park, S. M.; Shim, Y. B. Selective Electrochemical Analysis of Various Metal Ions at an EDTA Bonded Conducting Polymer Modified Electrode. *Electroanalysis.* **2004,** *16*, 1366-1370.
34. Luo, Y. C.; Do, J. S. Amperometric Ammonium Ion Sensor Based on Polyaniline-Poly(Styrene Sulfonate-co-Maleic Acid) Composite Conducting Polymeric Electrode. *Sens. Actuators. B.* **2006,** *115*, 102-108.
35. Setterington, E. B.; Alocilja, E. C. Rapid Electrochemical Detection of Polyaniline-Labeled Escherichia Coli O157:H7. *Biosens. Bioelectron.* **2011,** *26*, 2208–2214.
36. Agbor, N. E.; Petty, M. C.; Monkman, A. P. Polyaniline Thin Films for Gas Sensing. *Sens. Actuators. B.* **1995,** *28*, 173-179.

37. Kukla, A. L.; Shirshov, Y. M.; Piletsky, S. A. Ammonia Sensors Based on Sensitive Polyaniline Films. *Sens. Actuators. B.* **1996,** *37*, 135-140.
38. Agbor, N. E.; Cresswell, J. P.; Petty, M. C.; Monkman, A. P. An Optical Gas Sensor Based on Polyaniline Langmuir–Blodgett Films. *Sens. Actuators. B.* **1997,** *41*, 137-141.
39. Stamenov, P.; Madathil, R.; Coey, J. M. D. Dynamic Response of Ammonia Sensors Constructed from Polyaniline Nanofibre Films with Varying Morphology. *Sens. Actuators. B.* **2012,** *161*, 989-999.
40. Krutovertsev S. A.; Sorokin, S. I.; Zorin, A. V.; Letuchy, Ya. A.; Antonova, O. Yu. Polymer Film-Based Sensors for Ammonia Detection. *Sens. Actuators. B.* **1992,** *7*, 492-494.
41. Duboriz, I.; Pud, A. Polyaniline/Poly(Ethylene Terephthalate) Film as a New Optical Sensing Material. *Sens. Actuators. B.* **2014,** *190*, 398-407.
42. Jin, Z.; Su, Y.; Duan, Y. An Improved Optical pH Sensor Based on Polyaniline. *Sens. Actuators. B.* **2000,** *71*, 118-122.
43. Li, M.; Liu, W.; Correia, J. P.; Mourato, A. C.; Viana, A. S.; Jin, G. Optical and Electrochemical Combination Sensor with Polyaniline Film Modified Gold Surface and Its Application for Dissolved Oxygen Detection. *Electroanalysis*. **2014,** *26*, 374-381.
44. Majumdar, G.; Goswami, M.; Sarma, T. K.; Paul, A.; Chattopadhyay, A. Au Nanoparticles and Polyaniline Coated Resin Beads for Simultaneous Catalytic Oxidation of Glucose and Colorimetric Detection of the Product. *Langmuir*. **2005,** *21*, 1663-1667.
45. Oyama, N.; Ohsaka, T. Coupling between Electron and Mass Transfer Kinetics in Electroactive Polymer Films – An Application of the in Situ Quartz Crystal Electrode. *Prog. Polym. Sci.* **1995,** *20*, 761-818.
46. Zhang, G. Study on Conformation Change of Thermally Sensitive Linear Grafted Poly(N–isopropylacrylamide) Chains by Quartz Crystal Microbalance. *Macromolecules*. **2004,** *37*, 6553-6557.
47. Koval'chuk, E.; Pereviznyk, O.; Maksymchuk, M.; Makarovs'ka R. Electrochemical Quartz Microbalance Ammonium Sensor. *Visnyk Lviv Univ. Ser. Khim.* **2006,** *47*, 290-294.
48. Yan, Y.; Lu, D.; Zhou, H.; Hou, H.; Zhang, T.; Wu, L.; Cai, L. Polyaniline–Modified Quartz Crystal Microbalance Sensor for Detection of Formic Acid Gas. *Water Air Soil Pollut.* **2012,** *223*, 1275-1280.
49. Ayad, M. M.; El–Hefnawey, G.; Torad, N. L. Quartz Crystal Microbalance Sensor Coated with Polyaniline Emeraldine Base for Determination of Chlorinated Aliphatic Hydrocarbons. *Sens. Actuators. B.* **2008,** *134*, 887–894.
50. Shinen, M. H.; Essa, F. O.; Naji, A. S. Study the Sensitivity of Quartz Crystal Microbalance (QCM) Sensor Coated with Different Thickness of Polyaniline for Determination Vapors of Ether, Chloroform, Carbon tetrachloride and Ethyl acetate. *Chem. Mater. Res.* **2014,** *6,* 7-12.
51. Wang, W.; Xie, X.; He, S. Optimal Design of a Polyaniline-Coated Surface Acoustic Wave Based Humidity Sensor. *Sensors*. **2013,** *13*, 16816-16828.
52. Sadek, A. Z.; Baker, C. O.; Powell, D. A.; Wlodarski, W.; Kaner, R. B.; Kalantar–Zadeh, K. Polyaniline Nanofiber Based Surface Acoustic Wave Gas Sensors – Effect of Nanofiber Diameter on H_2 Response. *Sensors*. **2007,** *7*, 213-218.
53. Sadek, A.Z.; Wlodarski, W.; Shin, K.; Kaner, R. B.; Kalantar–zadeh, K. A Layered Surface Acoustic Wave Gas Sensor Based on a Polyaniline/In_2O_3 Nanofibre Composite. *Nanotechnology.* **2006,** *17*, 4488-4492.
54. Li, Y.; Deng, C.; Yang, M. A Novel Surface Acoustic Wave-Impedance Humidity Sensor Based on the Composite of Polyaniline and Poly(Vinyl Alcohol) with a Capability of Detecting Low Humidity. *Sens. Actuators. B.* **2012,** *165*, 7-12.

55. Ahuja, T.; Mir, I. A.; Kumar, D.; Rajesh. Biomolecular Immobilization on Conducting Polymers for Biosensing Applications. *Biomaterials.* **2007,** *28*, 791-805.
56. Verghese, M. M.; Ramanathan, K.; Ashraf, S. M.; Mathotra, B. D. Enhanced Loading of Glucose Oxidase on Polyaniline Films Based on Anion Exchange. *J. Appl. Polym. Sci.* **1998,** *70*, 1447–1453.
57. Chaubey, A.; Paude, K. K.; Singh, V. S.; Malhotra, B. D. Co-immobilization of Lactate Oxidase and Lactate Dehydrogenase on Conducting Polyaniline Films. *Anal. Chim. Acta.* **2000,** *400*, 97-103.
58. Yang, Y.; Shaolin, M. Bioelectrochemical Respoused of the Polyaniline Horseradish Peroxidase Electrodes. *J. Electroanal. Chem.* **1997,** *432*, 71-78.
59. Nomngongo, P. N.; Ngila, J. C.; Msagati T. A. M. Indirect Amperometric Determination of Selected Heavy Metals Based on Horseradish Peroxidase Modified Electrodes. In *Biosensors - Emerging Materials and Applications*; Serra, P. A., Ed.; InTech: Rijeka, 2011; pp 569-589.
60. Morrin, A.; Guzman, A.; Killard, A. J.; Pingarron, J. M.; Smyth, M. R. Characterization of Horseradish Peroxidase Immobilization on an Electrochemical Biosensor by Colorimetric and Amperometric Techniques. *Biosens. Bioelectron.* **2003,** *18*, 715-720.
61. Cosnier, S. Biomolecule Immobilization on Electrode Surface by Entrapment on Attachment to Electochemically Polymerized Films. A Review. *Biosens. Bioelectron.* **1999,** *14*, 443-456.
62. Michira, I.; Akinyeye, R.; Somerset, V.; Klink, M. J.; Sekota, M.; Al–Ahmed, A.; Baker, P. G. L.; Iwuoha, E. Synthesis, Characterisation of Novel Polyaniline Nanomaterials and Application in Amperometric Biosensors. *Macromol. Symp.* **2007,** *255*, 57-69.
63. Decher, G.; Hong, D. J. Building of Ultrathin Multilauer Films by a Selfassembly Proces. II. Consecutive Adsorbtion of Anionic and Cationic Bipolar Amphiphies and Polyelectrolytes on Charged Surfaces. *Ber. Bunsenges Phys. Chem.* **1991,** *95*, 1430-1433.
64. Cabaj, J.; Sołoducho, J. Layered Biosensor Construction. In *State of the Art in Biosensors - General Aspects*; Rinken, T., Ed.; InTech: Rijeka, 2013; pp 37-65.
65. Qu, F.; Yang, M.; Jiang, J.; Shen, G.; Yu, R. Amperometric Biosensor for Choline Based on Layer-By-Layer Assembled Functionalized Carbon Nanotube and Polyaniline Multilayer Film. *Anal. Biochem.* **2005,** *344*, 108-114.
66. Wei, D.; Ivaska, A. Electrochemical Biosensors Based on Polyaniline. *Chem. Anal.* **2006,** *51*, 839-852.
67. Dhand, C.; Das, M.; Datta, M.; Malhotra, B. D. Recent Advances in Polyaniline Based Biosensors. *Biosens. Bioelectron.* **2011,** *26*, 2811-2821.
68. Zhai, D.; Liu, B.; Shi, Y.; Pan, L.; Wang, Y.; Li, W.; Zhang, R.; Yu, G. Highly Sensitive Glucose Sensor Based on Pt Nanoparticle/Polyaniline Hydrogel Heterostructures. *ACS Nano.* **2013,** *7*, 3540–3546.
69. Davydenko, N.; Kovalyshyn, Ya.; Ostapovych, B. Glucose Biosensor Based on Polyaniline-Silver Composite Platform. *Visnyk Lviv Univ. Ser. Khim.* **2013,** *54*, 414-419.
70. Demydchuk, I.; Kovalyshyn, Ya. Glucosooxidase Sensor Based on Metal-Polymeric Platform. *Visnyk Lviv Univ. Ser. Khim.* **2012,** *53*, 416-424.
71. Hao, Y.; Zhou, B.; Wang, F.; Li, J.; Deng, L.; Liu, Y. N. Construction of Highly Ordered Polyaniline Nanowires and Their Applications in DNA Sensing. *Biosens. Bioelectron.* **2014,** *52*, 422-426.

CHAPTER 8

CORROSION PROTECTION OF ALUMINUM AND AL-BASED ALLOYS BY POLYANILINE AND ITS COMPOSITES

O. V. RESHETNYAK and M. M. YATSYSHYN

Department of Physical and Colloid Chemistry, Faculty of Chemistry, Ivan Franko National University of L'viv, 6 Kyryla & Mefodia Str., L'viv 79005, Ukraine

E-mail: reshetniak@franko.lviv.ua; M_Yatsyshyn@franko.lviv.ua

CONTENTS

ABSTRACT

The main aspects of aluminum and aluminum alloys protection by corrosion coatings based on various forms of polyaniline (**PAn**) were considered in the presented chapter. The advantages and disadvantages of modern chemical and electrochemical methods of application of **PAn** protective coatings on aluminum-containing substrates were analyzed as well as the relationship between the conditions of application of protective layers and their protective properties was shown. The modern approaches to the use of **PAn** in protective corrosion coatings, such as doping by polymeric acids, the formation of double-strand **PAn** complexes, the use of **PAn** as the pigment filler in paint coating, as well as a major component of various polymer–polymer or polymer-inorganic hybrid composites were considered. Proposed today mechanisms of protective action of protonated and deprotonated forms of **PAn** were in detail analyzed concerning to the corrosion of aluminum-containing substrates by forming of the protective oxide or salt passivation layers, as well as the inhibition of redox processes by **PAn** doping anions was considered. A special attention has been paid into the negative role of the intermetallic surface inclusions concerning to the initiation of corrosion and to the use of conversion cerium coatings to eliminate of this problem.

8.1 INTRODUCTION

Since the beginning of active research (1974) of a new class (at that time) of electroconductive polymers (**ECP**), one among their representatives, namely polyaniline (**PAn**), gained the practical importance. It is used individually as a powder and films, in composition of composite materials as filler of crystalline structures having the developed porous or layered structure and, at the same time, as the polymer matrix for other polymers, micro- or nanoparticles of metals, oxides of different nature, natural minerals, etc. Today the commercial production of **PAn** and its composites under the trademarks *Corrpassiv*, *Version*, *Panipol CX*100*X*03, and *PANDA,* etc. has been established.

Among the modern industries, which use the significant amounts of **PAn**, it should be selected the one concerning to the coatings on a basis of **PAn** for the protection of metals and alloys against the corrosion.[1–4] Corrosion of metals annually causes the huge material and economic losses worldwide.

All metallic constructions, starting from a variety of vehicles and ending by usual spike or by building blocks for construction are exposed to the corrosion. The corrosion of various communications and equipment of chemical, petrochemical, and other industries is separate question. Therefore, the search for new constructive materials and means of corrosion protection is an important area of scientific and of applied investigations in chemistry. One of the new areas of exploration and of development of the means of corrosion protection of metal products is the use of electroconductive polymers, in particular of **PAn** that can provide effective protection in acidic and alkaline, aqueous, and non-aqueous media. Many studies have shown that the **PAn** is effective protective anticorrosive coating (**PACC**) not only for iron and for various steels,[5–9] but also for non-ferrous metals, including the alluminium.[2,10]

Aluminum and its alloys (**AA**) have a much wide application as constructive materials in various industries; they are widely used for food packaging and in electronics, in the building, petroleum, and chemical, especially in the aerospace and machine-building industries. However, these materials are quite corrosive–sensitive to environments containing chloride ions. Until recently, the protection against the corrosion of **AA** was carried out with the use of coatings containing the hexavalent chromium (Cr^{6+}). However, chromates are harmful to human health and are environmentally hazardous. Therefore, at the end of the twentieth century there was a request to ecologically pure treatment of surface of aluminum and its alloys in creation of **PACC**. Investigation of non-toxic treatments to improve the stability of aluminum corrosion was the subject of interest of large industrial companies and as a result, led to the abandonment of toxic phosphate–chromate coatings in favor of **PAn**-containing coatings in particular. Active searches for the optimal synthesis routes and also new directions of application of **PAn** and its derivatives as both main and auxiliary components of active corrosion protection of metals and alloys based on them are ongoing today. Technologies of use of protective coatings of aluminum alloys based on **ECP** are promising alternative of the technology of chromate coating due to good corrosion protective properties of their relatively low cost and good environmental compatibility. Among recent achievements can be identified the creation of so-called "smart" corrosion inhibiting coatings[9,11] based on **ECP** in general and on **PAn** in particular, which define the modern application of these polymers as **PACC**.

8.2 GENERAL CHARACTERISTICS OF ANTICORROSIVE COATING BASED ON ECP

Corrosion is the result of chemical or electrochemical reaction of metal (metal alloy) with the environment or aggressive component of this environment. For a long time the most common and the most economical method of corrosion protection of metal objects and constructions are the coatings of organic nature. Such coatings primarily serve by physical superficial barrier for corrosive ingredients, which can cause corrosion of metal structures or parts.

Solving the problem of protection against the corrosion by organic coatings is impossible without finding out another question, namely: "How and why the coatings control the corrosion?" In accordance with G. P. Bierwagen,[12] this question contains several components of sub-questions that constantly arise as revealing of the essence of corrosion control using organic coatings, namely:

- *How organic coatings protect metal substrates from the corrosion?*
- *How to form a high-quality corrosion-resistant coating based on organic polymers?*
- *How and why organic coatings are destroyed in the corrosion process?*
- *How describe the corrosion protection by organic coating?*
- *Do we really know everything about the corrosion protection of metals by organic coatings?*

These questions become particularly relevant in the application of protective organic coatings of **ECP**, including **PAn**, polypyrrol (**PPy**), their mixtures, etc., since these polymers even today are still under study and systematization of obtained knowledge about their structure and physical–chemical properties.

ECP in general and **PAn** in particular are chemically inert and stable polymers in different environments and do not contain harmful components or impurities. Unlike many other polymeric substances **ECP** can be used as primary **PACC**, as matrix-base of **PACC** as well as additives to various anticorrosive paints and varnishes. Ones of the first attempts to apply of **PAn** for corrosion protection of metals was realized by D. DeBerry (1985),[13] which showed that **PAn**, electrodeposited on passivated steel increased its corrosion resistance in highly acidic medium of H_2SO_4. As was shown by A. MacDiarmid,[14] the mechanism of this effect is similar to anodic protection of metals against the corrosion. Later B. Wessling (1998) in first time proposed the paints containing the **PAn** and showed their effectiveness on corrosion protection.[15] Over the next five years reproducible results in the creation of **PACC** based on dispersions of pure **PAn**[16] were obtained.

Since the release of the first published works devoted to **PACC** based on **ECP**, their number increased significantly. However, often there are controversial results, which make it impossible to state unequivocally about the mechanism action of anticorrosive conductive materials in different environments.

The beginning of serious research investigations concerning to protection of metals from the corrosion is considered by T. P. McAndrew[17] the 1991 year, when were published the first works devoted to protection against corrosion of carbon steel, made in the American scientific laboratories *LANL*, and *NASA*. The first attempts to summarize the achievements concerning the problem of corrosion protection of metals by **ECP** were done by J. D. Stenger-Smith[18] and B. Wessling.[15] The works[1,19] were devoted to general questions of the metals protection against the corrosion, including the **ECP**. Authors of *ref.*[7] summarized the results on the mechanism of catalytic action of doped and un-doped **PAn** on steel surface. In particular, it was shown that the protective layers based on **PAn** serve as active electronic barrier to corrosion. Several diverse overview works, in which presented the results of experimental and theoretical studies of corrosion processes and participation in them of **ECP** and, **PAn** in particular, have been published today. The questions raised by the authors of these studies are still relevant today. The most of these works have been devoted first of all to corrosion resistance of steel products, but the results were partially generalized also for corrosion resistance of aluminum and of **AA**. So, in *ref.*[2] the state of research concerning to use of **PAn** and other polymers for corrosion-resistant coatings of nonferrous metals, including the aluminum and alloys on its basis is considered just as it was done in short overview in *ref.*[20] All this is by evidence that the interest in the use of **PAn** for the corrosion protection of aluminum and aluminum alloys in the future does not decrease but only will be increased.

8.3 PAn: SHORT CHARACTERISTIC AND PHYSICO–CHEMICAL PROPERTIES

The structure of **PAn**'s macromolecule can be represented as the set of repeated alternations of reduced (di(benzeneamine) [–(C_6H_4)–N(H)–(C_6H_4)–N(H)–]) and oxidized (benzeneamine–quinoimine [–(C_6H_4)–N=(C_6H_4)=N–])

fragments (see Fig. 8.1). With the use of *IR*, *UV/Vis*, *ESR*, *NMR*, Raman, and *X*-ray photoelectron spectroscopy the existence of three forms of **PAn**[21–25] was confirmed depending on the synthesis conditions and subsequent treatment, namely leucoemeraldine, emeraldine, and pernigraniline (see Schemes 8.2-8.3), which are differed by their physical and chemical properties. The most stable form of **PAn** isemeraldine, in which the ratio between oxidized and reduced fragments of macromolecular chain is 1:1 (see Fig. 8.2). As a result of redox processes, the emeraldine can move into leucoemeraldine or pernigraniline,[40] which represent by themselves the wholly reduced and oxidized forms of **PAn,** respectively (see Figs. 8.2–8.3). In the presence of inorganic and organic acids, the nitrogen atoms of amine (-NN-) or imine (-N=) groups of the main chain of **PAn** can be quaternizated due to the protonation to $-N^{+}H_{2}-$ or $-N^{+}{=}N$ groups. As a result, unprotonated forms of **PAn**, namely leucoemeraldine bases (**LB**), emeraldine bases (**EB**), and pernigraniline bases (**PnB**) pass in the form of the salts - **LS**, **ES,** and **PnS**, respectively (see Scheme 8.2). In particular, as a result of the transition **EB** → **ES** the electrical conductivity of the polymer can be changed in the range from 10^{-10} S/cm up to 100 S/cm and higher (depending on the doping level),[26,27] because the protonation of nitrogen atoms provides the higher conductivity of **PAn** due to the contribution of ionic component (see Scheme 8.2). Since both electrochemical and chemical synthesis of **PAn** is carried out in acidic medium, it is usually formed its **ES** – the substance of green color. **LS** and **PnS** are relatively unstable forms of **PAn** due to their strong reducing and oxidizing properties, respectively. That's why they relatively easy can move in **EB** or **ES** forms under the action of ambient environment.[28]

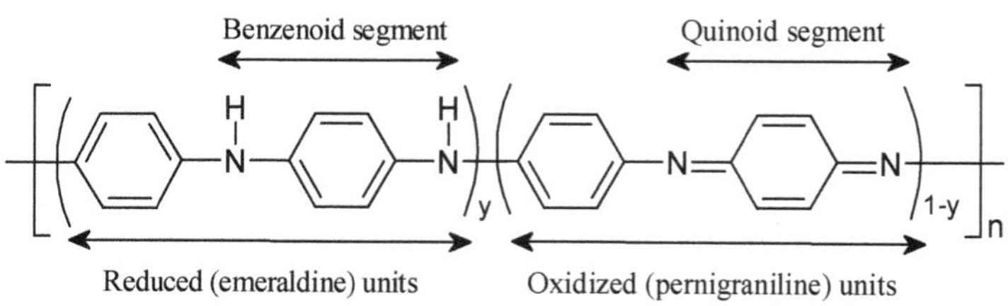

FIGURE 8.1 The structure of emeraldine ($y = 0.5$) base of **PAn**.

Mutual transitions of **PAn**'s forms is depicted by authors of *ref.*[3] in the form of *Pourbaix* diagram in the coordinates "potential – pH" (see Fig. 8.4). Change of the potential or *pH* of medium changes the forms of **PAn**, which is accompanied by chemical and physical processes in the polymer macromolecules. In particular, emeraldine base form of **PAn**, which consists of

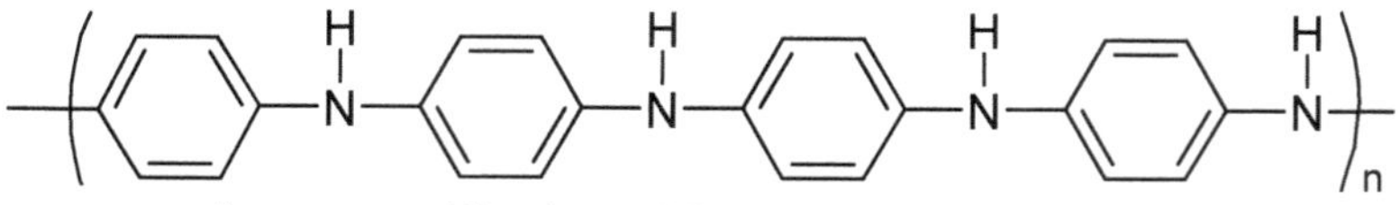

Leucoemeraldine base (LB), y = 1, colorless, insulator

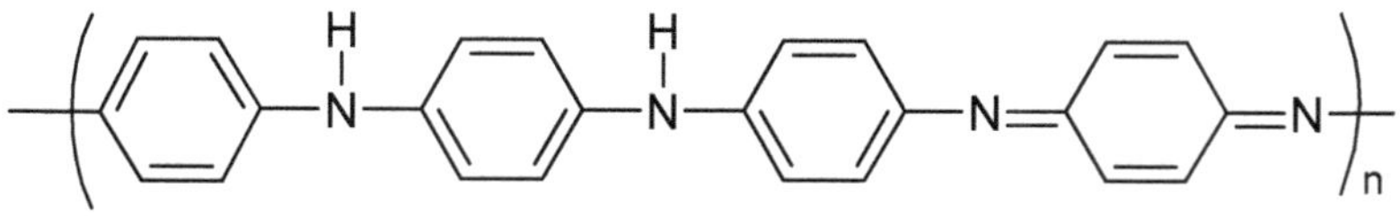

Emeraldine base (EB), y = 0.5, blue, insulator

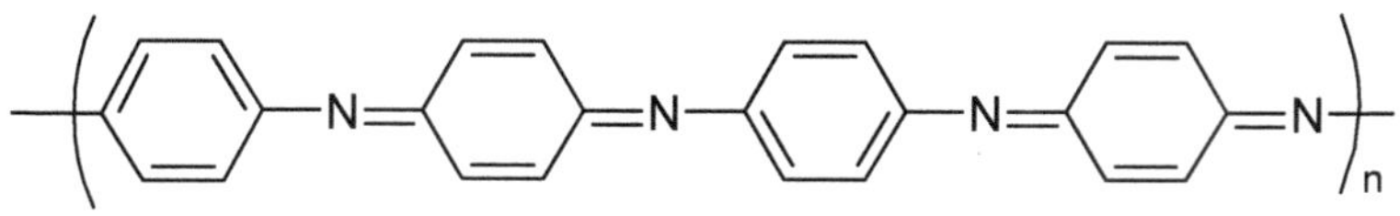

Pernigraniline base (PnB), y = 0

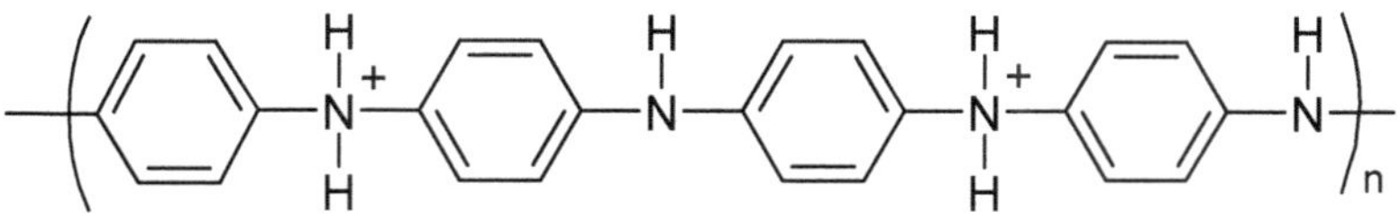

Leucoemeraldine salt (LS), colorless, insulator

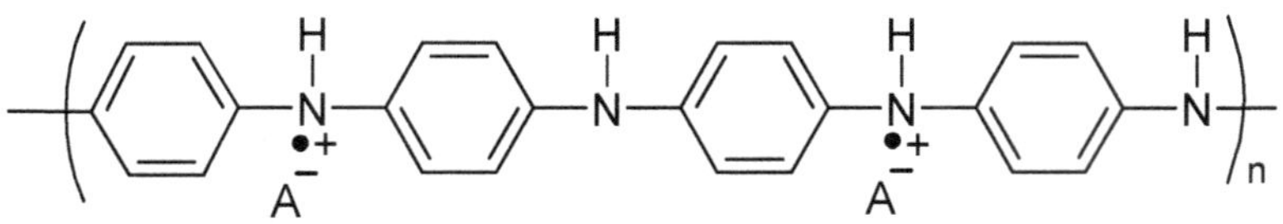

Emeraldine salt (ES) with separated polarons, green, conducting

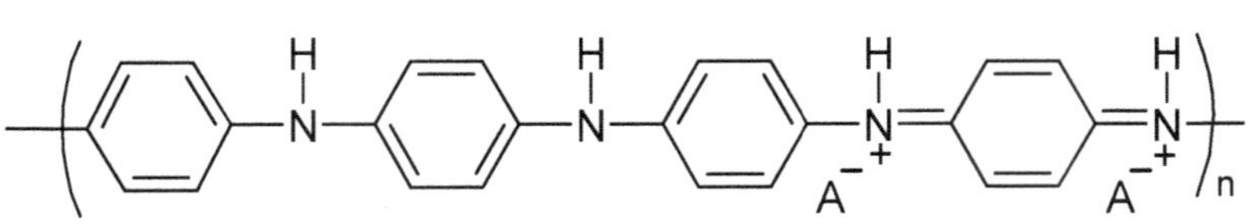

Emeraldine salt (ES) - bipolaron form

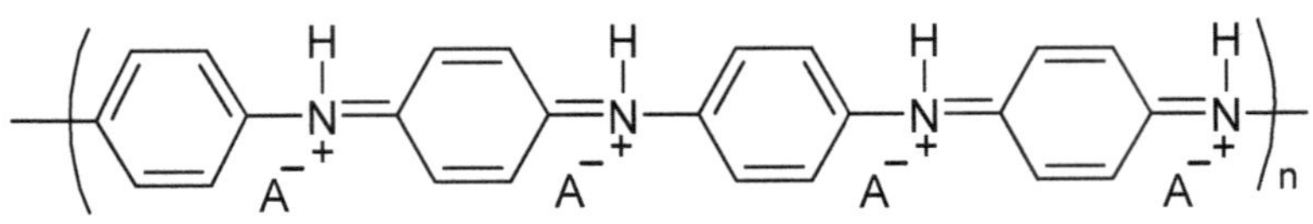

Pernigraniline salt (PnS).

FIGURE 8.2 The forms of **PAn** (see also Scheme 8.1), where A^- is Cl^-, ClO_4^-, HSO_4^-, etc. counter-ion.

alternating **PnB** and **LB** units is converted to the electronically conducting form **ES** by protonation in acidic media ($pH < 4$). From the authors[3] point of view, it can be an important factor for a protective layer since there are local changes in *pH* during the corrosion processes. For example, the results of corrosion tests of samples of alloy **AA** 5182 showed that deposited on their surface by spin coating film of **PAn** (**EB**) by thickness of 2-3 microns reduces the corrosion current of **AA** samples more than on order of magnitude. At this, a potential of the samples is shifted significantly toward more positive values. Acting as "H^+–ion trap" as a result of the possible **EB** → **ES** transition, and characterizing by low electronic conductivity, such layer significantly reduces the diffusion of proton and reduction of oxygen on the substrate. In addition, so there is a delay local acidification in the early stages of corrosion of aluminum surfaces coated with **EB**, which is also a favorable factor to increase the efficiency of corrosion protection.

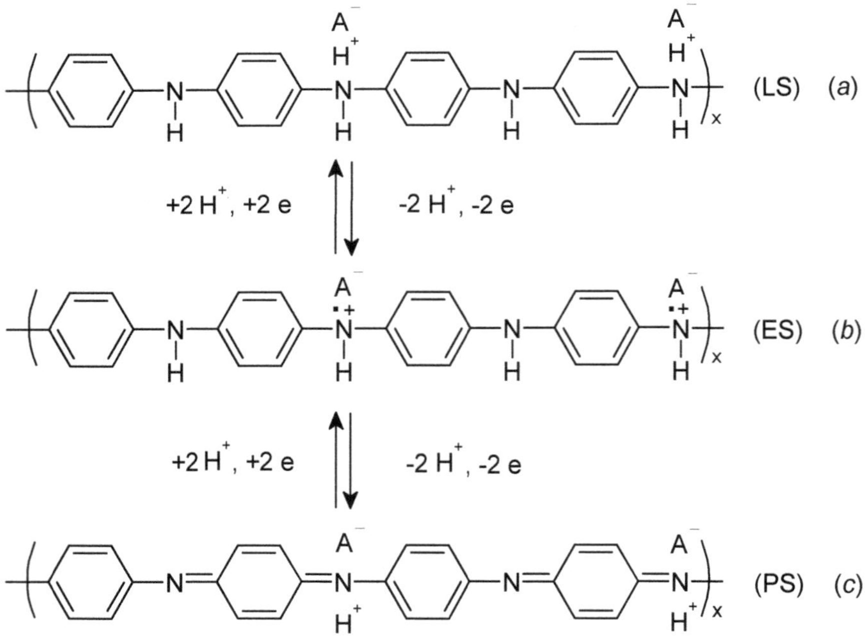

FIGURE 8.3 The change of **PAn**'s form as a result of redox processes: (a) leucoemeraldine salt (**LS**); (b) emeraldine salt (**ES**); (c) pernigraniline salt (**PnS**).

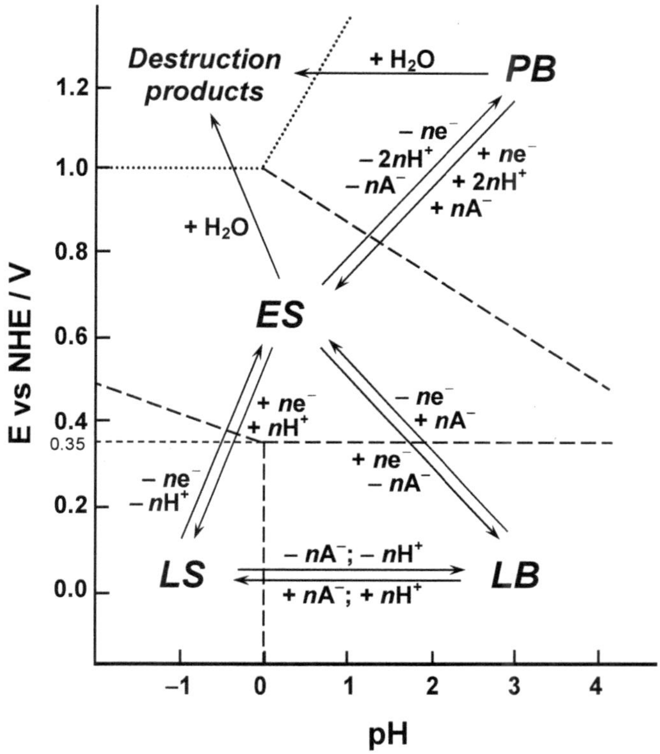

FIGURE 8.4 Diagram in the coordinates "potential (relatively to normal hydrogen electrode) – pH" for the forms of **PAn** in aqueous solutions. (The neutral form of **ES**, exactly **EB**, is not presented due to its lack of electroactivity), according to Cecchetto and co-workers.[118]

Electrochemical oxidation of aniline (**An**) in aqueous solution of 0.1 M H_2SO_4 on Al and Al–Pt electrodes (platinized Al-electrode) proceeds in different ways. An oxidation of **An** on Al–Pt electrode proceeds similarly as on the Pt electrode[29] while on a purely Al-electrode process is much slower, and density currents move *LE/Em* for the same number of cycles are ~12 and 40 mA/cm^2 for Al and Al–Pt electrode, respectively. An oxidation process of **An** on the Al electrode is characterized by an induction period,[29] which depends on the concentration of monomer.[30] The presence of the induction period as well as the difference in the values of current densities are caused by oxidation of just–polished surface of Al electrode during anodic polarization branch and by the formation of the oxide film, which leads to much lower currents of **An** oxidation, which is started at E » + 0.8 V. In addition, such films are not adhesive to the surface of the Al electrode. Therefore, an increase of peak currents of oxidative transition *LE/E* begins not immediately, but only after dilution (partial or complete) of surface film of Al_2O_3 (in the case of Fig. 8.5 (a) and Fig. 8.6 (a) after 10 cycles of potential scan).

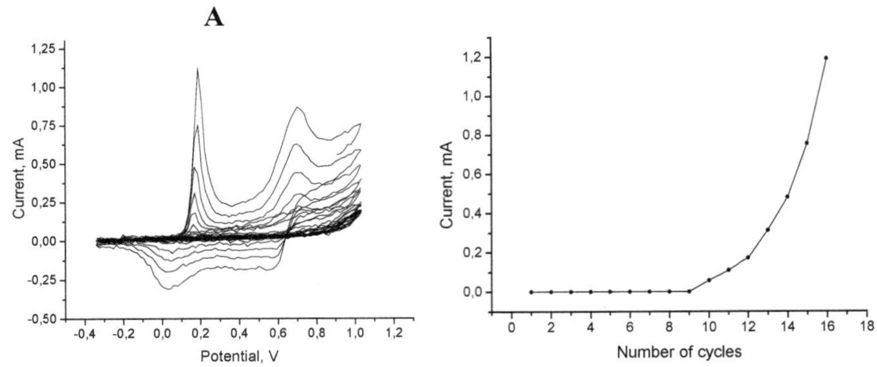

(a) (b)

FIGURE 8.5 Cyclic voltammograms of previously mechanically decontaminated Al electrode in 0.5 M solution of **An** in 0.5 M H_2SO_4 (*a*) and kinetic curves of the formation of the first anodic peak (**A**) at $E = \sim (+0.1)\text{-}(+0.3\ V)$ (b). The potential scanning rate (s_E) was equal to 50 mV s^{-1}.

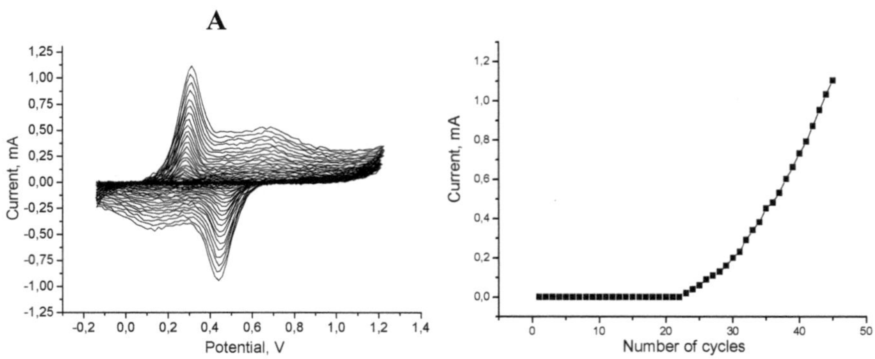

(a) (b)

FIGURE 8.6 Cyclic voltammograms of Al electrode, the surface of which was not previously mechanically decontaminated, in 0.25 M solution of **An** in 0.5 M H_2SO_4 (*a*) and kinetic curves of the formation of the first anodic peak (**A**) at $E = \sim (+0.2)\text{-}(+0.4)$ V (b) (s_E = 50 mV s^{-1}).

Interestingly, that the authors of *ref.*[29–31] did not observe characteristic intermediate anodic peak at E » 0.5 V on the cyclic voltammograms of

PAn-modified Al electrode in a solution of sulfuric acid, while in *ref.*[32] practically under the same conditions for the electrochemical deposition of **An**, the authors traced of this peak. The reason of this may be the differences in terms of advance preparation of the working electrode. In particular, we found that the redox-formation of **PAn**'s films without intermediate peak, but with the induction period proceeds on mechanically purified disk Al electrode (see Fig. 8.5). An oxidation peak of **An** at using of Al electrode is not seen so clearly as on cyclic voltammograms of Pt electrode[33] due to the influence of a protective oxide film. Proof of this fact is the cyclic voltammograms obtained by us during comparative studies of **An** oxidation on pre-unprepared Al electrode (without prior mechanical treatment of the surface from superficial film of native oxide) in a plate (see Fig. 8.6). In this case, on the cyclic voltammograms intermediate peak also was not observed and there is a presence of the induction period (over a much larger number of cycles > 20). At this, the first anodic peak at ~ 210 mV, which corresponds to the transition of leucoemeraldine/emeraldine, with increasing number of cycles is shifted toward the anode, which may indicate some difficulties of polymerization process proceeding.

Conductive coatings, including also **PAn**, belongs to the category of "smart" or ideal coatings that inhibit the corrosion so that they can generate an inhibitor only upon the occurrence of the corrosion. The mechanism of their functioning can be regarded as the work of nanodevice that can influence on the corrosion process. In aerated electrolyte, metallic aluminum, as was shown by the authors,[11] theoretically provides the potential near ~1.7 V, which is sufficient for the potentiation of chemical or physical nanodevice that produces a corrosion inhibitor on demand. The analogue of such a device can serve a natural passivation of steel in concentrated nitric acid, where the "corrosion inhibitor" is a compact oxide film that spontaneously is formed on the metal surface. Functional nanocoatings can provide "smart" feature, using, where appropriate, the surface energy of the system to produce the inhibiting material.[11] Similar formation of the corrosion inhibitors – compact oxide films inherently in practically for all active metals (Al, Fe, Ni, Cu, etc.) in the presence of molecular oxygen in the system. Preparation and properties of oxide films on aluminum surface particularly is described in detail in *ref.*[34] In addition it is necessary to take into account that at the anodic polarization of electrode (see Fig. 8.1) coated with **PAn** film, the hydrated acid anion is drawn into the polymer layer; the doping process takes place, which leads to high osmotic pressures in the film. Conversely, the dedoping process takes place at the cathodic polarization. These cyclic changes may also affect the adhesiveness of the polymer film to the surface of the aluminum substrate.

8.4 THE METHODS OF APPLYING OF PACCs BASED ON ECP

ECP can be applied to metal surfaces by many methods, among which the most important and the most common are mechanical, chemical, and electrochemical. Hybrid coatings containing the **ECP** are also deposited on the metal samples via spin coating method.[35]

Mechanical (physical) method consists in the deposition of ready paints or varnishes on properly prepared metal surface. Deposition is carried out by immersion in a solution, lacquer, paint, by spray of solution, with the use of brush or roller, etc. with following evaporation of the solvent. However, in the case of individual **PAn** this approach is practically implemented rather rare (e.g., in *ref.*[3,36,37]) because of low adaptability (the ability for processing) of the polymer. **EB** (un-doping form, insulator) of **PAn** is only partially soluble in the few organic solvents (dimethylformamide, N-methylpyrrolidone, and others),[3,36] while doped form of **PAn** (conductive form) is practically insoluble in almost all solvents with the except of partial solubility in concentrated sulfuric acid and N-methylpyrrolidone. Therefore, chemically synthesized **PAn** is often used as a pigment/filler that is a part of the protective coatings based on other polymers or epoxy resins (**ERs**).[4] However, it should be noted that the application of thin and uniform coatings of this type on complex surfaces causes significant difficulties.

More promising is the application of **ECP** layers *in situ* during electrochemical synthesis of the polymer. Electrochemical methods are more multifaceted and have the broader perspectives, especially given the possibility of modification of various upon nature conductive surfaces of conductive polymers by homogeneous and adhesive thin films.[38] Therefore, electrochemical methods are actively used in technology for *in situ* passivation and coating by facing protective layer of structural metals, including also the aluminum.[8]

The films of **ECP** can be easily obtained by electrodeposition from the solutions of the corresponding monomers. The structure, morphology, and corrosion behavior of deposited coatings depends not only on the method of electrochemical deposition of **ECP**, but also from electrochemical parameters of deposition, in particular the current density and the applied potential as well as the monomer concentration, nature, and the concentration of the electrolyte, etc.[8] Electropolymerization has certain advantages over other methods of applying of anticorrosion coatings, namely: highly automated, relatively low cost, gives the possibility of deposition from aqueous solutions, the ability to realize the small-sized coating, the ability to form a polymer on the irregular objects of complex form in one stage, the ability to control the properties of **ECP** by changing of the electrochemical deposition parameters and to obtain of **ECP** in one of its inherent forms. However, considering the benefits should to be said about the disadvantages of electrochemical deposition methods. The main one is the presence of water in the film of **ECP**, which can be removed by drying. However, based on the structure of the **ECP**, they always will absorb moisture from the environment to achieve some of its equilibrium concentration.

Important for **PACC** based on polymers is a thickness, adhesion, porosity, structure, surface morphology, and other properties of film created by any method. In particular, the question of **PAn**'s coatings adhesion to the surface of aluminum and steel, which is important for the effective functionation of the corrosion protection, is described in detail in *ref.*[39] This is especially true at the electrochemical deposition of organic coatings on metallic substrates. Ensuring of good protective properties of polymer coatings requires the compulsory prior mechanical, chemical, or electrochemical treatment of metal surface. In some cases, this involves the formation of so-called conversion coatings on metallic surfaces due to the interaction of metal surface with the components of the environment and due to the formation of stable films comprising a metal of a protected construction. However, in most cases, the pre-treatment of metal products before applying of anticorrosion coatings is to eliminate the superficial oxide films. In the case of aluminum and **AA** both physical and chemical, electrochemical, or combined methods[29,40] are used for this purpose.

In turn, an important is question also about the state of the active metal surface, including aluminum, after applying of **PACC**. Here it important for the formation of adhesive coatings also has an oxide film that is formed either in the process of coating, including the electrochemical method, or is formed on the surface of the metal interphase/protective coating during the operation. It is obvious that for aluminum and its alloys important parameter is the maximal thickness of the oxide film, which completely prevent the formation even under corrosion-resistant coating is virtually impossible.

8.5 CHARACTERISTICS OF PAn's FILM PROTECTIVE COATINGS

An important characteristic of organic **PACC**, including on the basis of **PAn**, is the surface morphology of polymer films, because it determines the protective properties of the coating as a physical barrier. The surface morphology of polymer films and the resulting corrosion behavior of coatings based on **PAn**, in turn, is dependent on the mode of electrochemical deposition (galvano- or potentiostatic, potentiodynamic), concentration of monomer (**An**),[41] and at the copolymerization also on the ratio of monomers.[8] The potential scanning rate and the limits of scanning have an essential impact on the morphology of the deposited films during electrochemical synthesis of **ECP** under conditions of potentiodynamic mode.[42] The main factor that affects on the surface structure of the films of **ECP** and, respectively on their physical and chemical characteristics is the magnitude of the applied potential,[8,41] which determines the polarization current density of electrode[8] at potentiostatic mode of the electrochemical deposition. For example, the electrosynthesis of **PAn** in potentiostatic conditions at low potentials (+ 0.32 V *rel.* Ag/AgCl – minimum potential of **An** polymerization) leads to the formation of a dense films, and at high potentials [3] + 0.84 V of porous loose films.[43] At the same time, authors of *ref.*[44] reported that at higher current density in galvanostatic deposition of **PAn** from aqueous solutions of oxalate acid on aluminum alloy **AA**3004 the compact more adhesive **PAn** coatings are formed. So, obvious is the impact of composition and nature of the electrolyte on the kinetics of electrochemical deposition. In particular, authors of *ref.*[42] indicate the importance of impurities in solution of alkali metal cations.

Feature of electrochemically modified surfaces by **PAn** layers of aluminum or **AA** is that the areas with different thickness of polymer films, and even uncovered by **PAn** areas coexist on them.[45] One of the main reasons that lead to this, there is different initial thickness of the oxide film of Al_2O_3 on aluminum-containing surfaces. It should be noted that the thickness of **PAn**'s film deposited on metal-electrode is not only geometric parameter. It also determines the physical and chemical characteristics of **ECP**. In particular, in very thin films the length of the chain conjugation increases, which can significantly affect on the spectral and electrochemical properties of the polymer, and the most importantly in terms of **PACC** quality on its conductivity.[46] In particular, Martins et al.[47] related the deposition of **PAn** films on **AA**6061–T6 alloy in 0.5 mol L^{-1} H_2SO_4 and 0.5 mol L^{-1} **An** using cyclic voltammetry and potentiostatic polarization. In both methods, the films obtained were adherent and presented a cauliflower structure. However, a slight increasing of the corrosion potential for **PAn** coated substrate as compared with bare metal indicates that no significant protection is achieved by the polymer coating. This result was attributed exactly to the conductive character of the **PAn** films. In addition, the formation of conductive **PAn** is possible only when it is deposited from strongly acidic media, whereas in media with higher *pH* the **PAn** with low conductive is obtained. Electrosynthesis of **ECP** proceeds in accordance with the ion-radical mechanism with their simultaneous doping. It is clear, that the nature of the solvent and of the electrolyte-dopant[48–50] will also affect on the oxidation of the monomer and on the formation of the macromolecular structure of the films. In particular, it was shown that the nature of the electrolyte anion affects on the nucleation of the polymer chain on the surface of the metal substrate.[51] On the other hand, the formation of a conducting polymer is made possible in a solvent with a low-donor number, which keeps both the **An** and the **An** radical protonated.[52] As for the nature of the electrolyte during the electrochemical deposition of **PAn**, it is clear that the classical method of its electrosynthesis in the medium of HCl is inadmissible in view of the well-known corrosive properties of chloride ion. At the same time, as was shown by the authors of *ref.*[53] **PAn** coated aluminum electrodes synthesized on aluminum electrode from aqueous solution of 0.25 M **An** and 0.2 M sodium benzoate initially provide corrosion protection of aluminum, decreasing the corrosion current density at least in 15 times.

Often results presented in references on corrosion resistance of **PAn** coatings obtained practically under identical conditions are somewhat contradictory. Particularly, the **PACC** were obtained at the electrodeposition of **PAn** on aluminum alloy **AA**3004 from an aqueous solution containing the oxalate acid; such coatings had excellent anticorrosive properties in aqueous 3.5% NaCl.[42] The corrosion current decreases significantly from 6.55 mA cm^{-2} for uncoated alloy to 0.158 mA cm^{-2} for **PAn**-coated **AA**3004. The corrosion rate of the **PAn**-coated **Al** is found to be 5.17×10^{-4} mm $year^{-1}$, which is ~40 times lower than that observed for bare Al. The positive shift E_{corr} = ~0.11 V in potential corrosion (to E_{corr} = ~0.9 V vs. **SCE**) indicated the protection of the Al surface by the **PAn** coatings.[44] At the same time, **PAn**'s films, formed via galvano- and potentiostatic methods on samples **AA2024T6** alloy in a medium of 0.5 M aqueous solution of oxalate acid showed lack of corrosion resistance in 1% NaCl solution due to the presence of pores in film.[54]

The nature of the electrode, on which the polymerization is carried out,[55] determines the structure of **PAn**'s films, since the first stage of polymerization is the formation of the adsorbed layer on the surface of electrode. Dissolving of the metal that occurs at the anode potential sweep in the first anodic cycle, does not prevent the electrochemical oxidation of **An** on the surface of such active metals as Fe, Ni, etc. However, the adsorption of monomer molecules on the metal surface at the initial stage of electrochemical oxidation of **An** slows of electrochemical dissolution of the electrode-substrate and subsequent deposition of the polymer, which begins at ~ +0.9 V (see Fig. 8.5 and 8.6) almost completely blocks the metal surface from the dissolving in the following anodic cycles.[32]

Electrodeposition of **PAn** from acidic solutions to spontaneously passivated electrodes, such as Al is not as simple as the Fe and Ni, due to the existence of formed dense dielectric film of Al_2O_3 (in contrast to semiconductor films of Fe_2O_3 or NiO). Factors that can lead to effective deposition of **PAn** on such surfaces as a result, firstly, reducing of the induction period during the polymerization are as follow: previous cathodic or anodic treatment of such surfaces;[56] the treatment with the use of chelate (complexone) additive (e.g., alizarine),[31] which blocks the reaction of hydrogen elimination on aluminum; and finally, the use of the concentration of monomer [3] critical concentration of 0.4 M.

To improve the protective properties of the film of **ECP,** it was proposed simultaneous anodization of aluminum surface, thereby increasing the stability of **PAn**' films as for penetration of ion Cl^- to the surface of metal, and electrodeposition of **PAn**.[57] Another approach to increase the protective properties of the **PAn**'s coatings is deposition of the so-called top-coat layers, such as layers of **ER** on **PAn**-modified surface. Non-pigmented epoxy coatings were obtained on the mild steel[44] by cathodic deposition from an **ER** emulsion at constant voltage (~250 V) and stirring conditions. After coating for 3 min, the samples were rinsed with distilled water and cured for 30 min at 180 °C. The average measured thickness of the epoxy coatings was 30 mm.[58] As shown by the authors of *ref.*,[53] these doubly coverages modifying the surface of **PAn–ER** are promising for corrosion protection also of Al and **AA**.

The double-strand **PAn** can be considered as one option for top-coat layers. The results show that such coverages are not surrendered (in particular, in 0.5 M solution of NaCl) per anticorrosive properties to conversion chemical coatings obtained with the use of chromates, and in some cases (double corrosive solution of salt and acid, *pH* 3.6), even exceeding of them.[59] Double-strand **PAn** is a molecular complex of linear polymers that are non-covalently bonded in a side-by-side fashion. The second outer layer is forming by polymers containing the anionic functional groups. Examples of the polyanions are poly(acrylic acid) (**PAA**),[60] poly(styrenesulfonic acid) (**PSSA**), poly(methylvinylether-*alt*-maleic acid), poly(methylacrylate–*co*–acrylic acid),[59] poly(acrylamide–*co*–acrylic acid)[61], etc. For example, when an aqueous solution of **An** and the **PAA** is mixed, the **An** monomers bind to the **PAA** backbone due to various intermolecular forces. After the solution of precursor adducts is acidified, the **An** is polymerized by an oxidant, such as sodium persulfate or hydrogen peroxide. The resulting product is a green colored solution.

The advantage of double-strand complexes first of all is due to the fact that polyanione takes the function of doping agent, the doping ions are not washed out even after long term exposure to a corrosive environment. Thus, high stability of the complex as a whole and the stability of the conductive form of **PAn** in particular are provided. In addition, with proper choice of polymeric dopant, the conductive polymer can be dispersed easily in solvent (unlike to individual **PAn**) to be used as a coating material, which can be deposited in mechanical way.[59] It should be noted that the presence of functional groups, such as -COOH in polymer-dopant is important not only because of the possibility of appearance of electrostatic forces between the chains dragging, but also of hydrogen bonds (see Fig. 8.7). Moreover, they also provide an opportunity of covalent grafting of the coating to aluminum-containing substrates in the presence on their surface of the oxide layer. A similar phenomenon in particular has been used for a long time for the formation of self-assembled layers of fatty acids on oxidized metal surfaces, including the aluminum.[62–64]

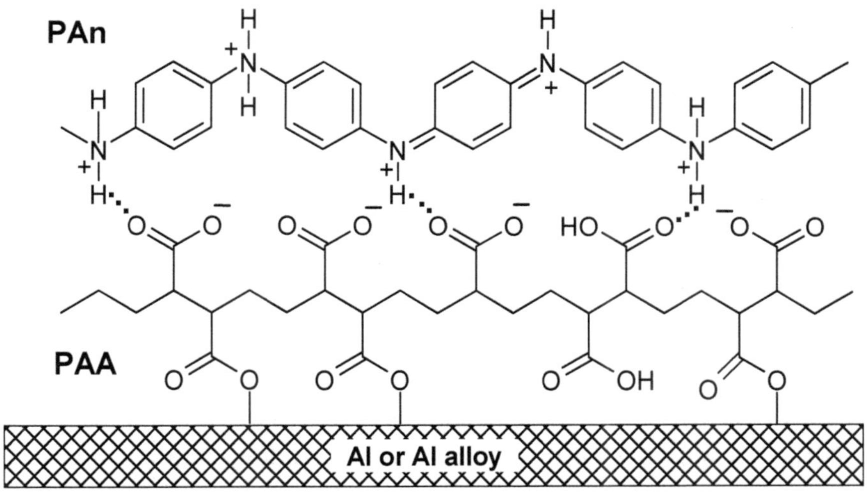

FIGURE 8.7 Formation of double-strand **PACC** on aluminum or **AA** on example of **PAn,** and poly(acrylic acid) (possible hydrogen bonds are indicated by three points).

The above-mentioned advantages of double-strand **PAn** are partially implemented in the case of simple doping of chemically synthesized **PAn** by polymer acids. In particular, by authors of *ref.*[37] the **PAn** doped by poly(methylmethacrylate–*co*–acrylic acid) was prepared using the template-guided synthesis method. The investigations of mechanically deposited on alloy **AA**3104–**H**19 films of such polymer from saturated ethyl acetate solution showed that the corrosion resistance of the obtained samples in 0.5 mol L^{-1} NaCl aqueous solution is significantly better compared to the protective coatings based on **ER**.

Concerning to the above-mentioned self-assembled layers, their pre-deposition on the surface of Al and **AA** can significantly improve the anticorrosive resistance of **PAn**-coatings. In particular, as was shown for the alloy **AA**5182,[65] its pre-treatment before deposition of **PAn** led to that obtained **PACC** showed increased corrosion resistance in neutral salt spray compared with the samples pre-treated with H_3PO_4 solutions. The reason for this effect is the best physico–chemical affinity between the polymer layers and chemically modified by 3-aminopropil–*tert*–oxylane surface, which significantly improves the adhesion to it of the **PAn** film (**EB**). Similar results were also obtained using the surface treatment of aluminum **AA**1050 by aminopropyl trimethoxy silane and vinyl trimethoxy silane before applying of the clear epoxy coatings,[66] which can be used in the formation of combined **PAn–ER** protective coatings.

Further development of this approach can be considered the using of silane compound containing the monomers of conducting polymers as a functional group. In particular, as a primer for corrosion inhibition of aluminum the *N*-(3-trimethoxysilylpropyl) (**AnSi**)[67] was used. Combining the monomers of conducting polymers and silane functionalities at a molecular level demonstrated the synergistic effect in the result of the formation of improved barrier action. Experimental results indicated that macro-agglomerates composed by siloxane and **An** oligomers, the latter mainly stabilized as H–complexes, are formed in **AnSi** methanol/water (95:5) solution with *pH* adjusted to 4.6 by adding of the acetic acid. Upon surface modification of **AA** specimens by single and multiple immersions in the hydrolyzed **AnSi** solutions, crosslinking of self-assembled macromolecules is further promoted, thus obtaining the polymeric **AnSi** hybrid network. The studies conducted on both the silane solution and the resulting films, jointly with several corrosion tests in naturally aerated 0.6 M NaCl solutions (*pH* 6.5 ± 0.2), strongly support the protection performance of such hybrid films.

To improve the manufacturability of **PAn**, it is often used in mixtures with other polymers.[68,69] This approach is another method to obtain the qualitative **PAn**-containing **PACC** on active metals. In the case of electroactive monomers for the synthesis of polymeric composites may be used the electrochemical (galvano– or potentiostatic) methods of synthesis, particularly as in the case of deposition of composite films **PAn**-polypyrrol (**PPy**) on **AA** of **AA**2024–**T**3 grade from aqueous solutions of oxalic acid.[8] The results of investigations on anticorrosion stability showed that the use of such conductive composite coating to protect the substrate leads to a significant reduction in the rate of corrosion (even four orders of magnitude depending on the deposition conditions, such as capacity and duration of polarization, the molar ratio of monomers, etc.). At this, the corrosion potential is shifted in the anode region up to 600 mV, and the corrosion currents (I_{corr}) are decreased from 1.703 mA for uncovered **AA**2024–**T**3 substrate to 0.543 mA and 0.0025 mA in the case of substrates protected by **PAn** and **PAn/PPy** protective films, respectively.

8.6 AN APPLICATION OF PAn AS THE PIGMENT IN PROTECTIVE COATINGS

Given the already aforementioned poor solubility of **PAn**, which complicates its application from the solutions, an alternative way is often used, namely the introduction of **PAn**' dispersions (5-30 wt.%) in the composition of epoxy or polymer protective coatings that can be used as a paint coating. In the first case, pre-synthesized doped **PAn** (**ES** form) is dispersed in **ER** with various applications of inorganic additives, which further is cured on the protecting surface. An example of this pigment can serve the **PAn** doped with *ortho*–phosphate acid (**PAn**–H_3PO_4)[4] or lignosulfonate (**PAn–LGS**),[70] which with varying success have been used as the pigment additives to **ER** for the formation of corrosion resistance layers on **AA** of **AA**2024**T**3 grade.

If 30 wt.% addition of **PAn**–H_3PO_4 to composite based on **ER** resulted in passivation of surface of metal substrate in the medium 0.5% NaCl through the formation of insoluble in water aluminum phosphate precipitate on defective areas of coating, the **PAn–LGS**/epoxy coating is no longer an effective corrosion resistant coating after 30 days immersion in 0.6 M NaCl due to loss of the dopant ions.

In the second case, the doped **PAn** is dispersed in the solution of polymer in organic solvent and applying of the protective coating is reduced to an evaporation of the solvent from coated on the surface of such polymer blend. An example is a mixture of **PAn** doped by *p*–toluene sulfonic acid (**PAn–TSA**) and polyvinylbutyral–*co*–vinylalcohol-*co*-vinylacetate, dissolved in ethanol.[71] It was established that such blend effectively inhibits the filiform corrosion of **AA**2024–T3 aluminum alloy. A somewhat different approach was used at the obtaining of polymeric composite based on doped **PAn** by camphorosulfonic acid (**PAn–CSA**) and poly(vinyl acetate–*co*–butyl acrylate).[72] In this case the **An** solution was added to co-polymer latex while stirring, followed by slow addition of oxidant (ammonium persulfate solution) under stirring, resulting in a dark green polymer suspension. Synthesized **PAn–CSA**/co-polymer powder is soluble in dichloromethane, and such solution was tested for corrosion protection of aluminum alloys **AA**2024–**T**3.

But the mixtures of dedoped **PAn** (**EB** form) with polymer acids are more commonly used. In this case, in the manufacture of blends the repeated doping of **PAn** takes place already by polymeric anion, and such mixtures acquire the properties similar to the aforementioned double-strand **PAn**. By similar properties possess for example, the blends of **PAn–CSA** and poly(methylmethacrylate) (**PMMA**), obtained by discharging of the solutions of these polymers in *m*–cresol.[73] Seegmiller et al. [73] reported that **AA**2024–**T**3 alloy coated with such blend provided better corrosion protection than **PMMA** system.

One may expect that the methods of application of **PAn** as a part of **PACC** to protect the Al and **AA** may be soon expanded. This is confirmed by the results obtained by H. Kukačkova and A. Kalendova concerning to the protection of steel.[74] Unlike to classic approach when exactly the dispersions of **PAn** is one of the components of polymer or epoxy protective layers, in this Chapter **PAn** was used to modify the surface of dispersed inorganic fillers of **PACC**. In particular, in chemical way it were modified the surfaces of the fillers with non-isometric particle shape based on silicates-talc, calcined kaolin, and the filler based on natural graphite. To evaluate the anticorrosive efficiency of series of model paints, fillers, and pigments were dispersed in a medium molecular weight waterborne **ER**. The value of pigment volume concentration of fillers and comparative pigment (zinc–aluminum phosphate) was constant in all the paints (10 vol.%). Testing results shown that surface modification of the lamellar particles of filler contribute to enhance the barrier and anticorrosive efficiency. In particular, the paint films containing lamellar talc covered by **PAn**-layer provided better anticorrosion efficiency than coatings with reference pigment (zinc–aluminum orthophosphate) in all test corrosive environments (in a neutral salt spray chamber, in a condensation chamber with content of sulfur dioxide and immersion test). Moreover, authors summarized that the combination of lamellar shape of filler particles with the action of conducting polymer could be suitable for the eventual application in eco-friendly paints instead of traditional anticorrosion pigments.

8.7 PROTECTIVE COATINGS BASED ON HYBRID PAn-CONTAINING COMPOSITES

Together with the polymer–polymeric composites the hybrid coatings have a much wider application. At electrochemical synthesis of **PAn** in the presence of sols or suspensions of different metals as well as of their oxides or natural minerals the incorporation of the dispersed phase into deposited polymeric film proceeds with the formation of composite protective layer. The enhanced corrosion protection effect of such composites relative to pure **PAn** in the form of coating on metallic surface was attributed to the combination of the redox catalytic property of **PAn** and the barrier effect of the inorganic filler dispersing in the composite. Given the prospects of such coatings, their obtaining, properties, and applications have been the subject of detailed studies in a number of reviews, for example.[75,76] At the same time, an attention of the most researchers is focused on composites of **PAn** with metal particles of varying degrees of dispersion (from nano- to micro-) with noble (Au, Pt, Pd, and Ag) and non-noble transition metals (Cu, Ni, Co, Fe, Ta, Zn, Mn, Zr, and rare earth metal Ce). However, most of these coatings, mainly because of their value have no practical value for use of them as the protective anticorrosion coatings. Therefore, more important is the use of non-metal particles as a filler in composites, namely such as carbon materials, especially carbon nanotubes and graphene, oxides and mineral dispersions. However, in this case the amount of fillers-materials that were tested as a part of composites based on **PAn** as **PACC** exactly of aluminum and aluminum alloys is relatively limited.

The use of carbon materials in **PACC** also has some reservations due to the fact that their introduction into the polymer matrix in amounts higher than the percolation threshold leads to an abrupt increase in the electrical conductivity of the coatings. Therefore somewhat aloof in this group of materials is nonconductive nanodiamond (**ND**). In particular, it was demonstrated that **PAn–ND** fibers are conductive and that the insertion of the insulating **ND** particles inside the **PAn**' matrix does not significantly affect the electrical properties of **PAn** but induces a noticeable increase of the thermal stability and a decrease of the temperature-induced decomposition of the **PAn** backbone.[77] The films of composites **PAn(EB)/ND** (diameter of **ND** particles was 5-10 nm) have been deposited by authors of *ref.*[78] via spin-on method from the dimethylformamide solutions and were tested as **PACC**. The results showed that these composites shown excellent corrosion inhibitor characteristics in the case of steel in 1 M NaCl, 0.1 M HCl, and 0.1

M H_2SO_4 aqueous solution, but the authors argue that these films have to behave well also on the protection of aluminum.[78]

However, more often the oxide materials, such as titanium (IV) oxide are used as the fillers in the synthesis of **PACC** based on **PAn**. The reason for the interest to this compound at the creating of **PACC** are especially well developed methods of nanodispersed TiO_2 obtaining.[79] Due to very high values of specific surface area of TiO_2 nanopowders it can be achieved a much better improvement of physical and chemical properties of the composite by the substantially lower filler content. An example of TiO_2–**PAn** composites as **PACC** can serve the results of *ref.*[57] The results of corrosion studies have shown that as a result of the incorporation of oxide nanoparticles with an average size 5-10 nm into deposited on the surface of **AA**3105 alloy layer of **PAn** its corrosion resistance is significantly improved compared with pure **PAn**' films. Improved protection of such coverages is due to the fact that TiO_2 nanoparticles block the pores of the anodic oxide film on the surface of alloy thereby preventing the contact of surface of the alloy and aggressive medium.[80] Comparison of similar properties (thickness of the composite films 100–250 nm, the average size of incorporated nanoparticles $\leq$ 10 nm) of composites **PAn**/TiO_2 and **PAn**/ZrO_2 showed that the composites of the first type provide significantly better protection of aluminum alloys,[80] which indicates the significant impact of nature of the filler material on the corrosion resistance of polymer-oxide protective coatings. Another mineral nanofiller used in **PACC** on the Al-containing surfaces is SiO_2. In particular, the **PAn (EB)** was dissolved by the authors of *ref.*[36] in N–methylpyrrolidinone (**NMP**) solvent and formed solution was added to the silica sol-gel liquid and stirred to form the **PAn** doped-sol. The different ratio (2.5-10% of **PAn**) coatings were then applied to **AA2024–T3 AA** surface using a spray method and oven dried in air. Obtained hybrid coatings by thickness of 5-13 mm successfully protected **AA**2024 in both acidic (HCl, *pH* 3.5) and neutral (3.5% NaCl) solutions for long periods up to 2 and 24 months, respectively.

Another example of hybrid Si-containing coating can be the composite *Ormosil*–$Ni_{0.5}Zn_{0.5}Fe_2O_4$/**PAn**.[35] Composite $Ni_{0.5}Zn_{0.5}Fe_2O_4$/**PAn** was obtained via polymerization of **An** *in situ* in toluene in the presence of particles Ni–Zn ferrite and $FeCl_3$ used as initiator at different ratios inorganic filler: An (1:1, 1:2, and 1:5). Modification of the obtained composite by *Ormosil* (organically modified silicates) was carried out in 0.05 M HNO_3 aqueous solutions with the use of 3–glycidoxypropyltrimethoxysilane, tetraethoxysilane, and curing agent tetraethylenepentamine as the precursors.[35] Hybrid coatings were deposited on metallic samples via spin method. I_{corr} values for untreated samples **AA** (4.13–4.35)$\times 10^{-6}$ A cm^{-2} and treated by *Ormosil*/

hybrid coatings are much lower (6.02–7.88) × 10^{-8} A cm^{-2} and are lower than for coatings treated by pure *Ormosil* (9.24–9.48) × 10^{-8} A cm^{-2}). Such result is primarily associated with the formation of denser films that provide an effective physical barrier against the penetration of water molecules and thus better protect against localized pitting-formation. Electrochemical and water–salt aerosol investigations of *Ormosil*' films on **AA6061–T6** and **AA**2024–**T**3 showed that barrier and anticorrosive properties of the investigated films are improved exactly through the use of $Ni_{0.5}Zn_{0.5}Fe_2O_4$/**PAn** hybrid system.

However, from economical point of view, namely the cost, the most promising fillers in composites based on **ECP** are native minerals. In particular, the *clay* minerals (nontronite, smectite, hectorite, halloysite, kaolin, bentonite, montmorillonite), mordenite and other zeolites, perovskite, glauconite–silica, etc. have been used widely as fillers of **PAn** matrix in such composites. However, just the montmorrilonite clay (**MMT**) as components of **PACC** of aluminum and **AA** were tested almost exclusively. This is because the use of **MMT** in polymer nanocomposites requires an initial step of purification of the clay, followed by the interchange of the interlayer inorganic cations with short chain organic ones, such as alkylammonium ions. This step increases the interlayer spacing as well as improves the compatibilization with the polymer. The incorporation of clay platelets into polymer matrix, in the form of organic-based coatings, may effectively enhance the corrosion protection performance of polymers on metallic surface. In particular, homogeneous and adherent **PAn–MMT** nanocomposite coatings were electrosynthesized on aluminum alloy **AA**3004 by using the galvanostatic polarization method in the of 0.5 M aqueous oxalic acid solution.[81] The FTIR technique confirms the incorporation of nanoclay particles in the **PAn**' matrix during electropolymerization. The corrosion protection effect of the coatings was studied electrochemically in 3.5 wt.% aqueous NaCl electrolytes. The I_{corr} values decreased from 6.55 A cm^{-2} for uncoated to 0.102 A cm^{-2} for nanocomposites-coated **AA** under the optimal conditions, while the corrosion rate of the **PAn–MMT** nanocomposites-coated Al was found to be about 190 times lower than that observed for uncoated Al. **PAn–CSA–MMT** clay nanocomposites also has been used as the filler of **ER** paints coatings.[82] As in the previous case, it was found that **PAn-MMT-ER** coatings have much better protective properties compared to **PAn-ER** layers in 3.5% NaCl aqueous solution at 65 °C. At this, the corrosion resistance of **PAn-MMT-ER** coatings is increased with **MMT** content increasing up to 5 wt.%, confirming the important role of mineral filler to create the qualitative **PACC**.

8.8 MECHANISMS OF PROTECTIVE ACTION OF PAn AS TO CORROSION OF ALUMINUM-CONTAINING SUBSTRATES

Organic coatings based on polymers act primarily as a physical barrier that prevents the penetration to the surface of metal substrate of the components of an environment, especially water, molecular oxygen, and in case of contact with solutions of such corrosive and aggressive ions as Cl^- and etc. One of the main tasks of such coatings is to prevent the cathodic reaction of oxygen reduction:

$$2\,H_2O + O_2 + 4\,e \rightarrow 4\,OH^-, \tag{8.1}$$

which can occur under protective layer on the surface of metal? In the case of aluminum, due to the amphoterism of the metal and its oxide, the increasing of *pH* as a result of the process proceeding is probably the most important factor that accelerates the corrosion. However, the effectiveness of this type of coatings eventually is deteriorated via time due to prolonged contact with an aggressive medium. Under the film of organic coating it may develop so-called local corrosion (filiform, pitting, stomach, etc.). It begins in local areas and further is developed on the surface of metal and penetrates into its depths. The formation of bubbles, holes and cavities, "swelling" of the superficial layer are the demonstrations of local corrosion. According to the presented results, the permeability, and as a result, also corrosion resistance of polymeric coatings, including on the basis of **PAn**, can be significantly improved by the introduction of various mineral fillers.[35,36,79,81]

Generally organic protective coatings usually have a good adhesion to the surfaces of metals. However, for long duration of polymer coating operation or under changes in operating conditions as a result of both corrosive environment action and physical factors (temperature, light, mechanical loads, etc.) can be traced the bundles on the edge surface of the metal/coating, that is, the "delamination" of **PACC** from the surface of metal. Ultimately this leads to a complete loss by coating of the properties of a barrier.[83] All these features are typical also for **PACC** based on **PAn**.

Meantime, the feature of protective coatings based on **ECP** in general and on **PAn** in particular is their ability to inhibit further corrosion in the presence on a surface of coating of small point defects or when it is mechanically damaged (the presence of scratches). The question of mechanisms of corrosion inhibiting of aluminum-containing substrates by **PAn** protective layer or by protective layers on its basis is extremely interesting and versatile. Primarily this is due to various forms of **PAn**, which it can take, to the possibility of reversible transitions under the influence of various factors, to physical and mechanical properties of **PAn**' films and composites on its basis. The mechanism of the active action of **PAn**-containing coating can be substantially differed. In particular, *Torresi* and co-authors[73] indicate three different mechanisms of inhibition in the case of **PACC** damage, namely: 1) the formation of a protective oxide passivation coating; 2) an inhibition of redox processes by doping **PAn** anions, which pass into solution as a result of dedoping of **PAn**; 3) the formation at the site of damage of protective film of insoluble compound which is formed by aluminum cations and again the doping **PAn** anions (see Fig. 8.8). The authors of *ref.*[5,6] three time phases of the anticorrosion action of **PACC** mark out on the basis of protonated **PAn**, namely: 1) the passivation of the metal surface; 2) active corrosion protection with the participation of **PAn** in the form of salt (doped form); 3) the transition to the protective barrier after un-reverse dedoping of **PAn** (the transition to the form of base). The total duration of all three periods will determine the effectiveness of protective coating fully.

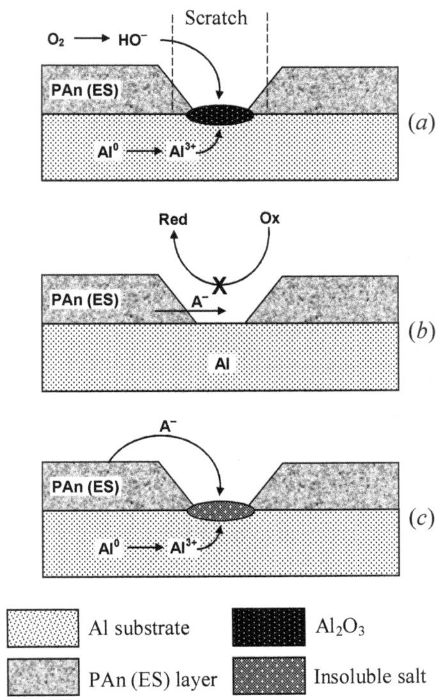

FIGURE 8.8 Three different mechanisms by which **PAn** can provide the corrosion protection, according to Torresi and co-workers,[73] (where **Ox** might include H^+, H_2O, and O_2 and **Red** might include H_2 and OH^-).

Application of **PAn**-containing coating can lead to a shift of the equilibrium potential of the metal in its passivation region (see Fig. 8.8 (a)), which is typical first of all for the ferrum.[13] Many authors have shown that thanks to its high redox potential and its electronic conductivity, **PAn** in the **ES** form also is able to passivate aluminum, which is evidenced firstly by the significant shift in the anodic potential region of Al-containing electrodes at the application on them of **PAn**-containing **PACC**.[8] The shift of potential depends both on the conditions of obtaining and on the composition of the coating and corrosive-active environment. Therefore, the shift of the potential in the anodic region is not the cause but the consequence of the passivation of surface of metal (iron or aluminum). If to damage the protective coating on metal, the foil of **ECP** becomes the cathode in relation to the metal surface, generally of anodic. In general, the interaction of electroconducting polymers with active metal-substrate (Me), based on the results of *ref.*,[84,85] can be described by following sequence of the transformations, where A^- is the one-valent anion doping the **ECP**:

$$(1/n)\mathrm{Me} + (1/m)\mathrm{ECP}^{m+}(\mathrm{A}^-)_m + (y/n)\mathrm{H_2O} \rightarrow (1/n)\mathrm{Me(OH)}_y^{(n-y)+} + (1/m)\mathrm{ECP}^0 + (y/n)\mathrm{H}^+ + \mathrm{A}^-; \tag{8.2}$$

$$(m/4)\mathrm{O_2} + (m/2)\,\mathrm{H_2O} + \mathrm{ECP}^0 + m\mathrm{A}^- \rightarrow \mathrm{ECP}^{m+}(\mathrm{A}^-)_m + m\mathrm{HO}^-. \tag{8.3}$$

Eq 8.2-8.3 represent two consecutive redox-processes, namely the reduction of *p*-doped electroconductive polymer (oxidized form) by metal–substrate with its subsequent oxidation of the obtained reduced form by molecular oxygen. Authors of *ref.*[70] consider that in the case of aluminum under immersion conditions, oxidation of metal can act as a trigger for the reduction of the **PAn**' coating, leading to the formation of a passive, protective aluminum oxide–hydroxide layer. As a result, coverage based on **ECP** is insensitive (from the corrosion point of view) to pin holes and small scratches.

Detailed mechanism in the case of **PAn**-coating on Al-substrate with taking into account the fact that its surface at the site of injury is hydrated as a result of contact with the electrolyte solution is shown[86] in Figure 8.9. In acidic or neutral medium in the presence of molecular oxygen the form **ES** of **PAn** reversibly is reduced to **LB** form (see Fig. 8.9 (a)). In fact, there is a cage galvanic circuit, where anodic process is the corrosion of the metal substrate and the cathodic process is reduction of molecular oxygen directly on the border of **PAn**|scratch. After coating of the defect by corrosion products, that in the case of aluminum can be represented by the equations:

$$Al^0 \rightarrow Al^{3+} + 3e^-; \quad (8.4)$$

$$2Al^{3+} + 3H_2O \rightarrow Al_2O_3 + 6H^+, \quad (8.5)$$

cathodic and anodic processes become by spatially-separated; the reaction of electrochemical reduction of oxygen is moving to the outer surface of the protective coating. This leads to a decrease in the diffusion of water and hydroxyl ions to the metal surface (scratch/defect), to gradual increase in the concentration gradient of molecular oxygen and to the subsequent decrease of the electrochemical activity of the metal. The reduction of oxygen is displaced to the periphery, where there is a local increase of pH at the expence of the ions HO^-, which are formed, leading to reversible dedoping of **PAn**. The electron transfer from the metal to the polymer is mediated by the presence of **ES,** which can be formed from an acid–base interaction between the hydrated substrate and the polymer, according to the data of works.[87,88] Generally, the corrosion process lasts until it will be achieved a certain critical thickness of oxide–hydroxide layer, and the rate of these processes will reduced to zero value.

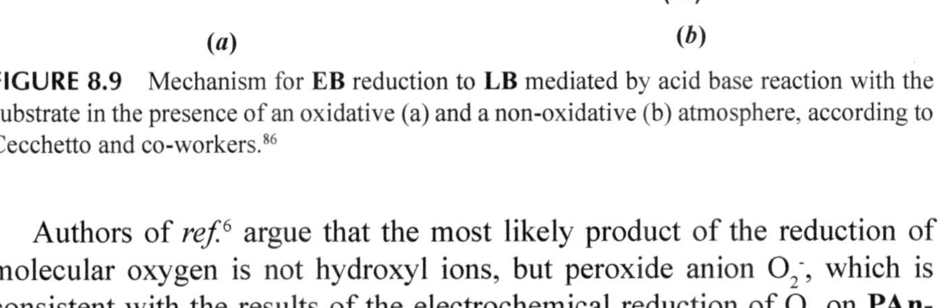

FIGURE 8.9 Mechanism for **EB** reduction to **LB** mediated by acid base reaction with the substrate in the presence of an oxidative (a) and a non-oxidative (b) atmosphere, according to Cecchetto and co-workers.[86]

Authors of *ref.*[6] argue that the most likely product of the reduction of molecular oxygen is not hydroxyl ions, but peroxide anion O_2^-, which is consistent with the results of the electrochemical reduction of O_2 on **PAn**-modified electrode in physiological solution.[89] Moreover, Otsuka and co-workers reported in *ref.*[90] that when **PAn** powder was added to water superoxide is generated. The formation of superoxide was confirmed by using ESR, amperometry and enzyme techniques. Peroxide anion, which is formed, next diffuses to the metal surface and further, the transformation taking place just as at the formation of metals oxides.[91] They can be divided into three consecutive stages: physical adsorption of peroxide anion O_2^- that goes into its dissociative chemisorption; penetration of oxygen from chemosorbed layer into the superficial layer of the metal and finally the formation of oxide. It has been suggested that the participation of the anion O_2^- facilitates the penetration stage, since as compared with molecular oxygen the bond O-O in peroxy-anion is weakened, which facilitates the dissociation of the molecule. This mechanism and appropriate scheme have been proposed to form a passivation coating on the surface of iron.[6] In our opinion; such a mechanism can be realized also on aluminum-containing surfaces. Somewhat modified by us scheme of transformations with taking into account that the corrosion of metal and the formation of the passivating oxide layer take place exactly on the defects or damages of **PAn**'s protective coating is shown in Figure 8.10.

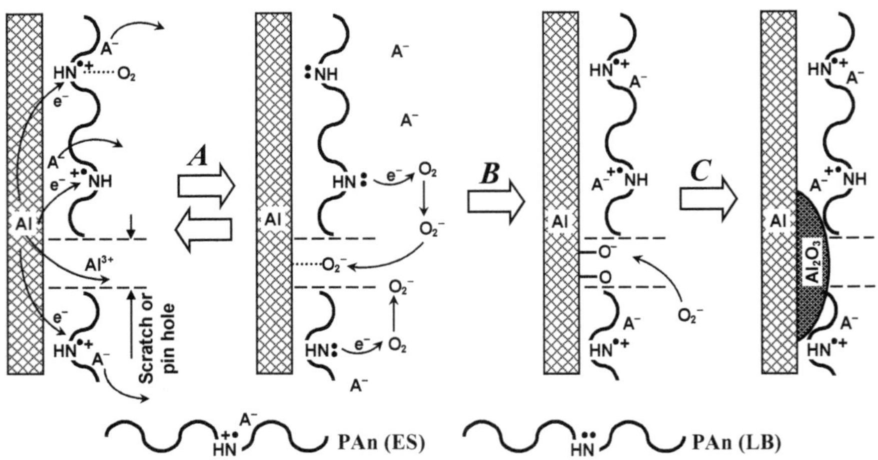

FIGURE 8.10 The scheme of catalytic formation of passivating oxide layer on the surface of aluminum in the place of the defect of **PAn**'s (**ES**) **PACC**: the stage (**A**) is the formation of complex **PAn**-molecular oxygen, an oxidation of aluminum substrate by **PAn** with reduction of **ES** to **LB**, the displacement of anions of dopant A^- into volume of electrolyte; the stage (**B**) is oxidation of **LB** by molecular oxygen to **ES** with the formation of peroxide ion O_2^- and its physical adsorption on the surface of aluminum substrate; the stage (**C**) is dissociative chemisorption of peroxide ion on the surface of metal and the formation of passivating film of Al_2O_3.

A somewhat different picture is observed in air-free neutral solutions. In this case, the reduction processes and dedoping of **PAn** are irreversible (see Fig. 8.9 (b)), and **ES** is converted into a mixture of **EB** and **LB**, that even visually is evident in the change of color of the protective film near the deposited scratch. Here we must stop on the protective properties of undoped **PAn**, because the views of researchers regarding the relationship between the efficiency of corrosion protection of these **ES** and **EB** forms of **PAn** are still disagreed. In some cases, the using exactly of **EB** form of **PAn** is much better compared with the coatings based on **ES.**[92] For example, authors of *ref.*[93] showed that the highest corrosion inhibition factor for **AA**3003 alloy in aqueous 3.5% NaCl and 0.1 N HCl solutions was obtained for undoped **PAn** being equal to 12 and 4.4 in neutral and acidic media, respectively. The reason of this is observations reported elsewhere[92–94] that **EB** membranes exhibit a significantly lower permeability to both halide anions and protons, compared to the acid-doped **ES** form. The basis of differences in physical and chemical properties of **ES** and **EB** is primarily the difference in electronic conductivity between these forms of **PAn**, and also acid–base balance between them. It is clear that the effectiveness of corrosion resistance of two forms of **PAn** primarily will depend on *pH* of medium. As a result of doping the polymer chain of **PAn** acquires a positive charge (becomes by polyelectrolyte), which leads to the hydrophysics of **PAn**' surface and facilitates the diffusion of hydrated ions from the bulk aqueous electrolyte into volume of doped film (resulting in a lower value of pore resistance) compared to film based on undoped or dedoped **PAn**, which poses no charge. In addition, as mentioned above, **EB**-coating, acting as H^+–ion trap, inhibits local acidation the overall rate of the corrosion process.

The above-described mechanism of corrosion due to the formation of passivating oxide layer, however, cannot explain the inhibitory effect of coatings on the basis of **PAn**, which is observed in media containing the chloride-ions. In such solutions, the corrosion of aluminum-containing materials would only increased, especially in the anodic polarization of aluminum since the anions Cl^- have the ability to destroy the oxide layer, which causes primarily of the filiform corrosion of aluminum. As a result of the reduction of macromolecule of **ECP** (**PAn**) and oxidation–ionization of superficial atoms of metal in the defect on the nitrogen atoms of **PAn** it is created the excess negative charge, thereby pushing of the doping anion from the macromolecule. This process proceeds similar to the process that occurs at the cathodic polarization of **ECP** films including **PAn** on different electrode surfaces. *Kendig* et al. proposed the concept,[11] which is based on the assumption that the doping of **PAn** anions which begin to be stand out in the electrolyte at the damage of protective layer possess by the properties inhibiting of the corrosion redox process (see Fig. 8.11). Proposed theoretical model was confirmed by the authors as an example of corrosion inhibition of **AA2024–T3** in neutral NaCl solutions.[11] This laid the foundations for constructing of the so-called "smart" coatings ("smart–release" coatings), which are able to provide the corrosion inhibitors "on demand" - only at the initiation of corrosion resulting from the damage of coating or in the presence of defects in it.

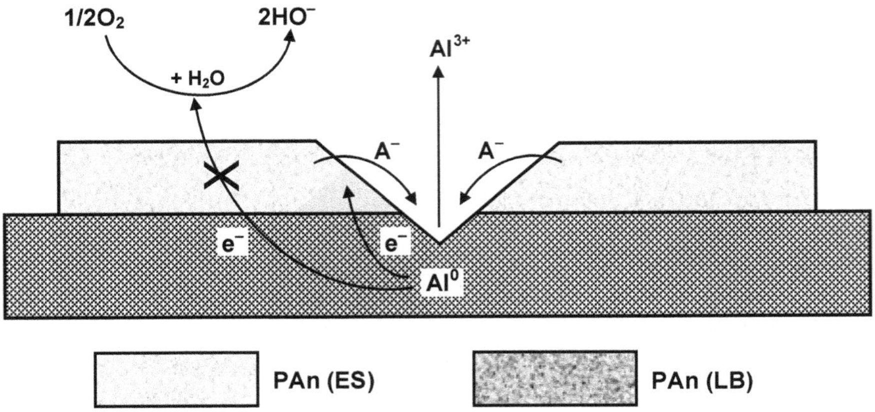

FIGURE 8.11 Scheme of the protection of a scribe by conducting polymer doped with a corrosion inhibiting anion A^-, according to Kendig and co-workers.[11]

With this in mind, as to the effectiveness of corrosion inhibiting of metal substrate the first is the nature of doping anion, which in particular is noted in *ref.*[71,95,96] Since the galvanic circuit is realized in corrosion system, the first arising question is what kind of process, namely cathodic or anodic the anions of dopant should inhibite. The answer to this question was obtained due to the results of the authors' works,[97,98] which have shown that the inhibitors of the oxygen reduction reaction are important. One possible result of this is to prevent cathodic delamination of the coating as caused by hydroxide generation in the result of Reaction 8.1 at the metal surface (see Fig. 8.8 (b)).[84,99] A related mechanism would be the prevention of direct damage to the metal surface due to cathodic corrosion induced by HO^- generation and the consequent activation of the metal toward further O_2 reduction. Such result as to the importance of inhibition exactly of molecular oxygen reduction reaction is logical in view on the known catalytic activity of **PAn** at the electrochemical reduction of O_2.[100–103] For the visualization of catalytic function of **PAn** in the system **PAn**-Al there are the all necessary prerequisites, namely: it provides 1) the oxygen exchange (**PAn** reversibly adsorbs of oxygen[104] and transforms it into an active form by the formation of weak (leaky) complex O_2-P,[105] where P is unpaired electron polarons of **ES**), with its following removal as a part of the reaction products (oxide layer); 2) electronic exchange (reversible redox transitions of the forms of **PAn**), and also 3) the bonding of the reagents and intermediates (oxygen and dopant).

The result of the displacement of the ions' dopant from the polymer layer can be the realization of the third mechanism of corrosion inhibition of the metal substrate (see Fig. 8.8 (c)). At using of **PAn** doped for example, with phosphate[4] or oxalate acids[42,54] the corresponding anions-dopants at the site of injury of a protective layer form the insoluble compounds by the interaction with the ions (Al^{3+}) of corroded substrate.

Considering the protection of aluminum and aluminum alloys from the corrosion, it is impossible to avoid another important aspect. It is well known that even the industrial samples of aluminum (so-called 1×××series) contain up to 1 wt.% of impurities. The main alloying elements of aluminum alloys are Cu (2×××series), Mn (3×××series), Si (4×××series), Mg (5×××series), Mg and Si (6×××series), Zn and Mg (7×××series), and other elements (Li is an e.g., 8×××series).[106] In this case, in addition to the main alloying element in the composition of alloy always there are equally significant amounts of other elements (see Table 8.1), which may lead to crystallization of different (depending on a grade of alloy) intermetallic particles of aluminum with alloying elements mainly on the surface of the alloys during the metallurgical processes of their obtaining. As shown by many researchers[3,73,107-109] exactly the presence of trace amounts of superficial cathodic intermetallic particles leads to the formation of pitting and rupture of **PACC** continuity under active corrosive conditions. In particular, scanning electrochemical microscopy was combined with scanning electron microscopy-energy-dispersive spectroscopy elemental mapping to show that heterogeneous cathodic reactivity (corrosion of separate areas of surface) at **AA**2024 surfaces is correlated with the locations of intermetallic particles.[110] This is confirmed by the fact that, for example, improved protective properties of **PAn (EB) PACC** have been obtained by removing iron–manganese particle from the substrate (aluminum alloys **AA**5182) before the coating. The reasons for this influence of alloying applications, but rather inclusion of their particles of intermetallic compounds in the surface layer of the aluminum substrate are: 1) less thickness of passivating oxide film on these areas, and because they are more sensitive to the corrosive effects; 2) intermetallic inclusions (particles) in the most cases are more noble (inert) compared to aluminum matrix of alloy, and therefore

TABLE 8.1 Composition of Industrial Samples of Aluminum and Aluminum' Alloys (in Accordance with the *ISO* 209–1: 1989).[106]

ISO designation	IRR	Content alloy additives/wt.% (minimal limit/maximal limit)										Al
		Si	Fe	Cu	Mn	Mg	Cr	Zn	Ti	Other		
										Each	Total	
Al 99.5	1050 A	-	-	-	-	-	-	-	-	-	-	99.50
		0.25	0.40	0.05	0.05	0.05	-	0.07	0.05	0.03	-	-
Al Cu4SiMg	2014	0.50	-	3.9	0.40	0.20	-	-	-	-	-	Remainder
		1.2	0.7	5.0	1.2	0.8	0.10	0.25	0.15	0.05	0.15	
Al Cu4Mg1	2024	-	-	3.8	0.30	1.2	-	-	-	-	-	
		0.50	0.50	4.9	0.9	1.8	0.10	0.25	0.15	0.05	0.15	
Al Mn1Mg1	3004	-	-	-	1.0	0.8	-	-	-	-	-	
		0.30	0.7	0.25	1.5	1.3	-	0.25	-	0.05	0.15	
Al Mn0.5Mg0.5	3105	-	-	-	0.3	-	-	-	-	-	-	
		0.6	0.7	0.3	0.8	0.20	0.40	0.15	-	0.05	0.15	
Al Mg4.5Mn0.7 (A)	5183	-	-	-	0.50	4.3	0.05	-	-	-	-	
		0.40	0.40	0.10	1.0	5.2	0.25	0.25	0.15	0.05	0.15	
Al Mg1SiCu	6061	0.40	-	0.15	-	0.8	0.04	-	-	-	-	
		0.8	0.7	0.40	0.15	1.2	0.35	0.25	0.15	0.05	0.15	
Al Mg0.7Si	6063	0.20	-	-	-	0.45	-	-	-	-	-	
		0.6	0.35	0.10	0.10	0.9	0.10	0.10	0.10	0.05	0.15	
Al Mg0.7Si(A)	6063 A	0.30	0.15	-	-	0.6	-	-	-	-	-	
		0.6	0.35	0.10	0.15	0.9	0.05	0.15	0.10	0.05	0.15	
Al Si1MgMn	6082	0.7	-	-	0.40	0.6	-	-	-	-	-	
		1.3	0.50	0.10	1.0	1.2	0.25	0.20	0.10	0.05	0.15	
Al Zn5.5MgCu	7075	0.30	0.15	-	-	0.6	-	-	-	-	-	
		0.6	0.35	0.10	0.15	0.9	0.05	0.15	0.10	0.05	0.15	

*Note: **IRR** is International Registration Record.

will show the cathodic character; the reduction of hydrogen or oxygen on the surface will proceed on them leading to a local increase of *pH* of the aqueous medium, which further activates the pitting corrosion of adjacent with intermetallic inclusion areas. The mechanism of initiation of pitting corrosion around the particles is described in *ref.*[108,111]

Negative role of intermetallic superficial inclusions significantly is increased in the case of deposition of insufficiently dense **PACC** (e.g., pure polymeric without the use of fillers, etc.). However, it was found that the corrosion resistance of such coatings, for example of **PAn**' films doped with oxalate acid on samples of alloy **AA2024T6**, can be improved to 90% by post-treatment in cerium ion containing (1000 ppm) hot (60°C) solution at for 30 min.[54] The authors attribute this to the fact that in the presence of ions of Ce(III) on defective areas of coating is formed the cerium conversion coatings (**CCC**) based on oxides of cerium (III) and (IV), according to the following equations:[54]

$$Ce^{3+} + 3HO^{-} \rightarrow Ce(OH)_3; \quad (8.6)$$

$$2Ce(OH)_3 \rightarrow Ce_2O_3 + 3H_2O; \quad (8.7)$$

$$2Ce(OH)_3 + 2HO^{-} \rightarrow 2CeO_2 + 4H_2O + 2\ e^{-}. \quad (8.8)$$

It is obvious that in the latter transformation the **PAn** in the form of **ES** takes part as an oxidizing agent. A similar mechanism was proposed by *Wang* et al. for a Gr/Al composite surface.[112] However, one can not dismiss another possibility, namely, the deposition–intercalation in the pores of **PAn** exactly in a salt medium of insoluble cerium oxalate precipitate with self–formation of polymer–inorganic composite that can also improve the corrosion resistance of such coatings. The result of the deposition of oxide coatings Ce(III) and Ce(IV) is slowing of the oxygen reduction reaction and hence decrease the corrosion rate. Similar results for **CCC** were obtained also for other **AA**, including **AA**7075,[45,113] **AA**2024,[114] **AA**2014,[115] and **AA6082**,[116] including in absence of **PAn' PACC**. In particular, Mishra and co-workers[117] investigated the inhibitory effect of $LaCl_3$ and $CeCl_3$ added to NaCl solution against corrosion, concluding that, for all the concentrations studied, $CeCl_3$ was the better corrosion inhibitor. Generally, of the rare earth compounds, cerium shows maximum corrosion protection efficiency. An important reason for the superior efficiency of cerium is due to Ce(IV), which can be formed in high *pH* value environments.[117]

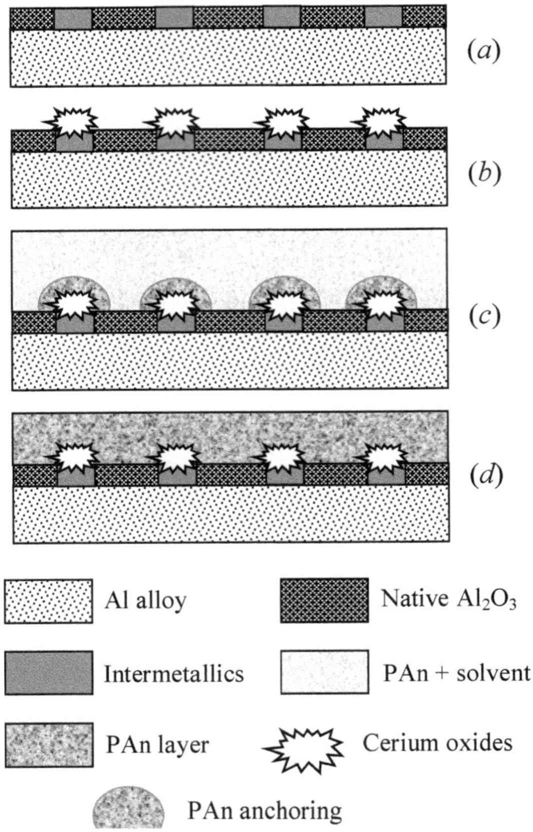

FIGURE 8.12 Scheme of the formation of duplex (**CCC/PAn**) on the surface of **AA**, according to Johansen and co-workers:[118] (a) initial sample of aluminum alloy with intermetallic regions and native oxides; (b) sample after cerium electroless deposition, (c) initial **PAn** anchoring on the **CCC**-modified intermetallic region; (d) **PAn** layer after solvent evaporation, representing the final treatment stage for the duplex coating.

A new approach to the creation of protective coatings based on **CCC** has been proposed by *Johansen* et al.,[118] who conducted a comparative analysis of the corrosion protection of **AA**6063 aluminum alloy by three different coatings, namely: cerium conversion, PAn–**EB,** and duplex (**CCC/PAn**) coatings. The scheme of deposition of duplex coatings is shown in Figure 8.12. Its peculiarity was that the **CCC** does not eliminate the defects of **PAn** coating, as in the *ref.*[56] and was served as its substrate. For the conducting polymer coating, a saturated solution of chemically synthesized undoped **PAn** in N–methyl–pyrrolidone was obtained. The samples with deposited previously cerium conversion coating were then painted using a paint brush with the **PAn** casting solution and dried at 50 °C under vacuum. The results of investigations showed that the AA coated with the duplex coating is corroded at a lower rate than with cerium or **PAn** coatings individually primarily due to the fact that application of **CCC** graded corrosive effect of intermetallic particles on the surface of **AA**. Moreover, it was found that the resulting surface after the deposition of serium conversion coating is rough and electrochemically active, providing all the conditions of anchorage and reactivity necessary for the formation of more adherent **PAn** films.

8.9 CONCLUSIONS

Summarizing the above-presented results, we can conclude that the **PAn** and composites on it basis are effective and, as a result, promising **PACC** also of aluminum and aluminum alloys. At the same time, we note that currently available data are quite difficult for analysis and generalization. The reason for this is primarily that the majority of research is one-sided because the attention of researchers focused almost exclusively on the protective role of **PACC**. In turn, the choice of material coverage depends not only on the conditions in which this coverage will continue to work, but also on the nature of the substrate material, because it also affect on the morphology and on adhesion of deposited protective coating, especially under electrochemical deposition conditions. As noted above, both qualitative and quantitative composition of aluminum alloys are significantly differed depending on series and mark. However, the results of the comparative analysis of peculiarities of morphology and interaction of material-substrate with a protective layer, its anticorrosive properties for different aluminum-containing substrates and of the same applied coating are practically absent in references, similar to the conclusions, and recommendations as to benefits of using of any type of **PAn**-containing **PACC** for protection of **AA** by given series.

At the same time, analysis shows that exceptionally **PAn**' coatings are significantly inferiored in anticorrosive properties to variety of modern composites on their basis. This leads to the fact that significantly are intensified the researches devoted to the development and testing of such materials where **PAn** on the one hand can be dispersed phase in various varnishing and painting compositions or dispersion medium in which impose a variety of materials–fillers. Today a little attention is paid to the study of the influence of the degree of dispersion of the filler material such as oxide materials, natural minerals and so on corrosion resistance of composites on their basis. It is well known that the transition from micro– to nanodimension level of dispersion often drastically changes the physical and chemical properties of virtually identical qualitative composition of materials. Therefore, we can expect that the transition to the nanoscale level will lead to a qualitative leap in the evolution of research dedicated to corrosion protection of metal products including aluminum and aluminum alloys.

KEYWORDS

- **aluminum**
- **aluminum alloys**
- **protective anticorrosion coating**
- **corrosion protection**
- **mechanism**

REFERENCES

1. Brooman, E. W. Modifying Organic Coatings to Provide Corrosion Resistance. Part III: Organic Additives and Conducting Polymers. *Met. Finish.* **2002,** *100,* 104-110.
2. Tallman, D. E.; Spinks, G.; Dominis, A.; Wallace, G. G. Electroactive Conducting Polymers for Corrosion Control. Part 1. General Introduction and a Review of Non-Ferrous Metals. *J. Solid State Electrochem.* **2002,** *6,* 73–84.
3. Cecchetto, L.; Ambat, R.; Davenport, A. J.; Delabouglise, D.; Petit, J. P.; Neel, O. Emeraldine Base as Corrosion Protective Layer on Aluminium Alloy AA5182, Effect of the Surface Microstructure. *Corr. Sci.* **2007,** *49,* 818-829.
4. Sathiyanarayanan, S.; Azim, S. S.; Venkatachari, G. Performance Studies of Phosphate-Doped Polyaniline Containing Paint Coating for Corrosion Protection of Aluminium Alloy. *Appl. Polymer Sci.* **2008,** *107,* 2224–2230.
5. Schauer, T.; Joos, A.; Dulog, L.; Eisenbach, C. D. Protection of Iron against Corrosion with Polyaniline Primers. *Prog. Org. Coat.* **1998,** *33,* 20–27.
6. Ogurtsov, N. A.; Shapoval, G. S. Catalytic Effect of Polyaniline for Anticorrosion Protection of Mild Steel. *Catalysis Petrochem.* (Kataliz i neftekhimia), **2001,** *9-10,* 5-12. (in Russian).
7. Spinks, G.; Dominis, A.; Wallace, G. G.; Tallman, D. E. Electroactive Conducting Polymers for Corrosion Control. Part 2. Ferrous Metals. *J. Solid State Electrochem.* **2002,** *6,* 85–100.
8. Iroh, J. O.; Zhu, Y.; Shah, K.; Levine, K.; Rajagopalan, R.; Uyar, T.; Donley, M.; Mantz, R.; Johnson, J.; Voevodin, N. N.; Balbyshev, V. N.; Khramov, A. N. Electrochemical

Synthesis: A Novel Technique for Processing Multi-Functional Coatings. *Prog. Org. Coat.* **2003,** *47,* 365–375.

9. Willner, I.; Basnar, B.; Willner, B. From Molecular Machines to Microscale Motility of Objects: Application as "Smart Materials," Sensors, and Nanodevices. *Adv. Funct. Mater.* **2007,** *17,* 702–717.
10. Twite, R. L.; Bierwagen, G. P. Review of Alternatives to Chromate for Corrosion Protection of Aluminum Aerospace Alloys. *Progr. Org. Coat.* **1998,** *33,* 91–100.
11. Kendig, M.; Hon, M.; Warren, L. "Smart" Corrosion Inhibiting Coatings. *Prog. Org. Coat.* **2003,** *47,* 183–189.
12. Bierwagen, G. P. Reflections on Corrosion Control by Organic Coatings. *Prog. Org. Coat.* **1996,** *28,* 43-48.
13. DeBerry, D. W. Modification of the Electrochemical and Corrosion Behavior of Stainless Steels with an Electroactive Coating. *J. Electrochem. Soc.* **1985,** *132,* 1022-1026.
14. Mac Diarmid, A. G. *Short Course on Electrically Conductive Polymers*; New Paltz: New York, 1985.
15. Wessling, B.; Posdorfer, J. Corrosion Prevention with an Organic Metal (Polyaniline): Corrosion Test Results. *Electrochim. Acta.* **1999,** *44,* 2139-2147.
16. Wessling, B. Passivation of Metals by Coating with Polyaniline: Corrosion Potential Shift and Morphological Changes. *Adv. Mater.* **1994,** *6,* 226-228.
17. McAndrew, T. P. Corrosion Prevention with Electrically Conductive Polymers. *Trends Polym. Sci.* **1997,** *5,* 7-12.
18. Stenger–Smith, J. D. Intrinsically Electrically Conducting Polymers. Synthesis, Characterization, and Their Applications. *Prog. Polym. Sci.* **1998,** *23,* 57-79.
19. Grundmeier, G.; Schmidt, W.; Stratmann, M. Corrosion Protection by Organic Coatings: Electrochemical Mechanism and Novel Methods of Investigation. *Electrochim. Acta.* **2000,** *45,* 2515–2533.
20. Gvozdenovic, M. M.; Grgur, B. N. Electrochemical Polymerization and Initial Corrosion Properties of Polyaniline–Benzoate Film on Aluminium. *Org. Coat.* **2009,** *65,* 401–404.
21. Ohsaka, T.; Ohnuki, Y.; Oyama, N. IR Absorption Spectroscopic Identification of Electroactive and Electroinactive Polyaniline Films Prepared by the Electrochemical Polymerization of Aniline. *J. Electroanal. Chem.* **1984,** *161,* 399–405.
22. Syed, A. A.; Dinesan, M. K. Review: Polyaniline - a Novel Polymeric Material. *Talanta.* **1991,** *38,* 815–837.
23. Kaplan, S.; Conwell, E. M.; Richter, A. F.; MacDiarmid, A. G. Ring Flips as a Probe of the Structure of Polyanilines. *Macromolecules.* **1989,** *22,* 1669–1675.
24. Łapkowski, M.; Geniés, E. M. Evidence of Two Kinds of Spin in Polyaniline from in Situ EPR and Electrochemistry: Influence of the Electrolyte Composition. *J. Electroanal. Chem.* **1990,** *279,* 157–168.
25. Lee, H. T.; Wang, C. C. Effects of Interactions among Polyaniline, Camphorsulfonic Acid and Silica on the Structure and Properties of Their Conductive Hybrids. *Pol. Engineer. Sci.* **2008,** *48,* 439-447.
26. Malinauskas, A. Chemical Deposition of Conducting Polymers. *Polymer.* **2001,** *42,* 3957-3972.
27. Yatsyshyn, M.; Koval'chuk, E. Polyaniline: Chemical Synthesis, Synthesis Mechanism, Structure and Properties, Doping. *Proc. Shevchenko Sci. Soc. Chem. Biochem.* **2008,** *21,* 87-102 (in Ukrainian).

28. Plesu, N.; Ilia, G.; Pascariu, A.; Vlase, G. Preparation, Degradation of Polyaniline Doped with Organic Phosphorus Acids and Corrosion Essays of Polyaniline–Acrylic Blends. *Synth. Met.* **2006,** *156,* 230–238.
29. Pournaghi–Azar, M. H.; Habibi, B. Electropolymerization of Aniline in Acid Media on the Bare and Chemically Pre–Treated Aluminum Electrodes a Comparative Characterization of the Polyaniline Deposited Electrodes. *Electrochim. Acta.* **2007,** *52,* 4222–4230.
30. Khrushch, V.; Panzyga, O.; Yatsyshyn, M.; Koval'chuk, E. *Production and Properties of the Films of Conducting Polymers on the Aluminium Surfaces*, 12th Scientific Conference "Lviv Chemical Readings – 2009": Proceedings. L'viv, Ukraine, June 1-4, 2009; Publishing Centre of Ivan Franko National University of L'viv: L'viv, 2009. Ф21 (In Ukrainian).
31. Huerta–Vilca, D.; de Moraes, S. R.; de Motheo, A. J. Aspects of Polyaniline Electrodeposition on Aluminium. *J. Solid State Electr.* **2005,** *9,* 416–420.
32. Prasad, K. R.; Munichandraiah, N. Potentiodynamic Deposition of Polyaniline on Non-Platinum Metals and Characterization. *Synth. Met.* **2001,** *123,* 459-468.
33. Yatsyshyn, M. M.; Boichyshyn, L. M.; Demchyna, I. I.; Nosenko, V. K. Electrochemical Oxidation of Aniline on the Surface of an Amorphous Metal Alloy $Al_{87}Ni_8Y_5$. *Russ. J. Electrochem.* **2012,** *48,* 502-508.
34. Lohrengel, M. M. Thin Anodic Oxide Layers on Aluminium and Other Valve Metals: High Field Regime. *Mater. Sci. Engineer. R.* **1993,** *11,* 243-294.
35. Wu, K. H.; Chen, P. H.; Yang, C. C.; Ho, W. D.; Liu, C. I. Infrared Stealth and Anticorrosion Performances of Organically Modifited Silicate–NiZn Ferrite/Polyanilin Hybrid Coatings. *J. Polym. Sci.: Part A: Polym. Chem.* **2008,** *46,* 926–935.
36. Akid, R.; Gobara, M.; Wang, H. Corrosion Protection Performance of Novel Hybrid Polyaniline/Sol–Gel Coatings on an Aluminium 2024 Alloy in Neutral, Alkaline and Acidic Solutions. *Electrochim. Acta.* **2011,** *56,* 2483–2492.
37. Oliveira, M. A. S.; Moraes, J. J.; Faez, R. Impedance Studies of Poly(Methylmethacrylate–*Co*–Acrylic Acid) Doped Polyaniline Films on Aluminum Alloy. *Prog. Org. Coat.* **2009,** *65,* 348–356.
38. Breslin, C. B.; Fenelon, A. M.; Conroy, K. G. Surface Engineering: Corrosion Protection Using Conducting Polymers. *Mater. Design.* **2005,** *26,* 233–237.
39. Motheo, A.; de Jesus, Bisanha, L. D. Chapter 2. Adhesion of Polyaniline on Metallic Surfaces. In *Aspects on Fundamentals and Applications of Conducting Polymers*; de Jesus Motheo, A., Ed.; In Tech: Rijeka, 2012; pp 19–40.
40. McGovern, W. R.; Schmutz, P.; Buchheit, R. G.; McCreery, R. L. Formation of Chromate Conversion Coatings on Al–Cu–Mg Intermetallic Compounds and Alloys. *J. Electrochem. Soc.* **2000,** *147,* 4494-4501.
41. Conroy, K. G.; Breslin, C. B. The Electrochemical Deposition of Polyaniline at Pure Aluminium: Electrochemical Activity and Corrosion Protection Properties. *Electrochim. Acta.* **2003,** *48,* 721-732.
42. Andrade, G. T.; Aguirre, M. J.; Biaggio, S. R. Influence of the First Potential Scan on the Morphology and Electrical Properties of Potentiodynamically Grown Polyaniline Films. *Electrochim. Acta.* **1998,** *44,* 633-642.
43. Volfkovich, Yu. M.; Sergeev, A. G.; Zolotova, T. K.; Afanasiev, S. D.; Efimov, O. N.; Krinichnaya, E. P. Macrokinetics of Polianiline Based Electrode: Effects of Porous Structure, Microkinetics, Diffusion and Electrical Double Laer. *Electrochim. Acta.* **1999,** *44,* 1543-1558.

44. Shabani–Nooshabadi, M.; Ghoreishi, S. M.; Behpour, M. Electropolymerized Polyaniline Coatings on Aluminum Alloy 3004 and Their Corrosion Protection Performance. *Electrochim. Acta.* **2009,** *54,* 6989–6995.
45. Kamaraj, K.; Sathiyanarayanan, S.; Venkatachari, G. Electropolymerised Polyaniline Films on AA 7075 Alloy and Its Corrosion Protection Performance. *Prog. Org. Coat.* **2009,** *64,* 67–73.
46. Goldenberg, I.; Petty, M.; Monkman, A. A. Comparative Study of the Electrochemical Properties of Dip-Coating, Spun and Langmuir–Blodgett Films of Polianiline. *J. Electrochem. Soc.* **1994,** *141,* 1573-1576.
47. Martins, N. C. T.; Moura e Silva, T.; Montemor, M. F.; Fernandes, J. C. S.; Ferreira, M. G. S. Polyaniline Coatings on Aluminium Alloy 6061–T6: Electrosynthesis and Characterization. *Electrochim. Acta.* **2010,** *55,* 3580-3588.
48. Tang, H.; Kitani, A.; Shiotani, M. Effects of Anions on Electrochemical Formation and Overoxidation of Polyaniline. *Electrochim. Acta.* **1996,** *41,* 1561-1567.
49. La Craix, J. D.; Diaz, A. F. Electrolite Effects on the Switching Reaction of Polyaniline. *J. Electrochem. Soc.* **1988,** *135,* 1475-1463.
50. Krylov, V. A.; Kurys', Ya. I.; Pokhodenko, V. D. Influence of the Medium on the Electrochemical Behavior of Polyaniline in Aprotic Electrolytes. *Theor. Exp. Chem.* **1993,** *29,* 154-157.
51. Mandić, Z.; Duić, L.; Kovačiček, F. The Influence of Counter-Ions on Nucleation and Growth of Electrochemically Synthesized Polyaniline Film. *Electrochim. Acta.* **1997,** *42,* 1389-1402.
52. Naudin, E.; Gouerec, P.; Belanger, D. Electrochemical Preparation and Characterization in Non-Aqueous Electrolyte of Polyaniline Electrochemically Prepared from an Anilinium Salt. *J. Electroanal. Chem.* **1998,** *459,* 1-7.
53. Gvozdenović, M. M.; Grgur, B. N. Electrochemical Polymerization and Initial Corrosion Properties of Polyaniline-Benzoate Film on Aluminium. *Progr. Org. Coat.* **2009,** *65,* 401–404.
54. Karpagam, V.; Sathiyanarayanan, S.; Venkatachari, G. Studies on Corrosion Protection of Al2024 T6 Alloy by Electropolymerized Polyaniline Coating. *Curr. Appl. Phys.* **2008,** *8,* 93–98.
55. Maia, D. J.; De Paoli, M. A.; Alves, O. L. Growth of Linear Polianiline Chains in a Layeder Tin(IV) Phosphonate Host. *Synth. Met.* **1997,** *90,* 37-40.
56. Huerta–Vilca, D.; De Moraes, S. R.; De Motheo, A. J. Anodic Treatment of Aluminum in Nitric Acid Containing Aniline, Previous to Deposition of Polyaniline and Its Role on Corrosion. *Synth. Met.* **2004,** *140,* 23–27.
57. Zubillaga, O.; Cano, F. J.; Azkarate, I.; Molchan, I. S.; Thompson, G. E.; Cabrai, A. M.; Morais, P. J. Corrosion Performance of Anodic Films Containing Polyaniline and TiO_2 Nanoparticles on AA3105 Aluminium Alloy. *Surf. Coat. Tech.* **2008,** *202,* 5936-5942.
58. Grgur, B. N.; Gvozdenović, M. M.; Mišković–Stanković, V. B.; Kačarević–Popović, Z. Corrosion Behavior and Thermal Stability of Electrodeposited PANI/Epoxy Coating System on Mild Steel in Sodium Chloride Solution. *Progr. Org. Coat.* **2006,** *56,* 214–219.
59. Sun, L.; Liu, H.; Clark, R.; Yang, S. C. Double-Strand Polyaniline. *Synth. Met.* **1997,** *84,* 67-68.
60. Yang, S. C.; Brown, R.; Racicot, R.; Lin, Y.; McClarnon, F. Electroactive Polymer for Corrosion Inhibition of Aluminum Alloys. *ACS Symp. Ser.* **2003,** *843,* 196-206.

61. Gupta, G.; Birbilis, N.; Khanna, A. S. An Epoxy Based Lignosulphonate Doped Polyaniline– Poly(Acrylamide Co–Acrylic Acid) Coating for Corrosion Protection of Aluminium Alloy 2024–T3. *Int. J. Electrochem. Sci.* **2013,** *8,* 3132-3149.
62. Thompson, W. R.; Pemberton, J. E. Characterization of Octadecylsilane and Stearic Acid Layers on Al_2O_3 Surfaces by Raman Spectroscopy. *Langmuir.* **1995,** *11,* 1720–1725.
63. Tao, Y. T. Structural Comparison of Self–Assembled Monolayers of N–Alkanoic Acids on the Surfaces of Silver, Copper, and Aluminium. *J. Am. Chem. Soc.* **1993,** *115,* 4350–4358.
64. Tao, Y. T.; Lee, M. T.; Chang, S. C. Effect of Biphenyl and Naphthyl Groups on the Structure of Self-Assembled Monolayers: Packing, Orientation, and Wetting Properties. *J. Am. Chem. Soc.* **1993,** *115,* 9547–9555.
65. Cecchetto, L.; Denoyelle, A.; Delabouglise, D.; Petit, J. P. A Silane Pre–Treatment for Improving Corrosion Resistance Performances of Emeraldine Base-Coated Aluminium Samples in Neutral Environment. *Appl. Surf. Sci.* **2008,** *254,* 1736-1743.
66. Mohseni, M.; Mirabedini, M.; Hashemi, M.; Thompson, G. E. Adhesion Performance of an Epoxy Clear Coat on Aluminum Alloy in the Presence of Vinyl and Amino–Silane Primers. *Prog. Org. Coat.* **2006,** *57,* 307–313.
67. Flamini, D. O.; Trueba, M.; Trasatti S. P. Aniline-Based Silane as a Primer for Corrosion Inhibition of Aluminium. *Progr. Org. Coat.* **2012,** *74,* 302–310.
68. Anand, J.; Palaniappan, S.; Sathyanarayana, D. N. Conductiong Polyaniline Blends and Composites. *Prog. Polym. Sci.* **1998,** *23,* 993-1018,
69. Pud, A.; Ogurtsov, N.; Korzhenko, A.; Shapoval, G. Some Aspects of Preparation Methods and Properties of Polyaniline Blends and Composites with Organic Polymers. *Prog. Polym. Sci.* **2003,** *28,* 1701-1753.
70. Gupta, G.; Birbilis, N.; Cook, A. B.; Khanna, A. S. Polyaniline-Lignosulfonate/Epoxy Coating for Corrosion Protection of AA2024–T3. *Corr. Sci.* **2013,** *67,* 256–267.
71. Williams, G.; McMurray, H. N. Polyaniline Inhibition of Filiform Corrosion on Organic Coated AA2024–T3. *Electrochim. Acta.* **2009,** *54,* 4245–4252.
72. Gustavsson, J. M.; Innisa, P. C.; Heb, J.; Wallace, G. G.; Tallman, D. E. Processable Polyaniline–HCSA/Poly(Vinyl Acetate–*Co*–Butyl Acrylate) Corrosion Protection Coatings for Aluminium Alloy 2024–T3: A SVET and Raman study. *Electrochim. Acta.* **2009,** *54,* 1483–1490.
73. Seegmiller, J. C.; da Silva, J. E. P.; Buttry, D. A.; de Torresi, S. I. C.; Torresi, R. M. Mechanism of Action of Corrosion Protection Coating for AA2024–T3 Based on Poly(aniline)–Poly(methylmethacrylate) Blend. *J. Electrochem. Soc.* **2005,** *152,* B45-B53.
74. Kukačková, H.; Kalendová, A. Investigation of Mechanical Resistance and Corrosion–Inhibition Properties of Surface–modified Fillers with Polyaniline in Organic Coatings. *J. Phys. Chem. Solids.* **2012,** *73,* 1556–1561.
75. Ćirić–Marjanović, G. Recent Advances in Polyaniline Composites with Metals, Metalloids and Nonmetals. *Synth. Met.* **2013,** *170,* 31–56.
76. Li, X.; Sun, J.; Huang, M. Preparation and Properties of Nanocomposites of Polyaniline and Metal Nanoparticles. *Prog. Chem.* **2007,** *19,* 787-795.
77. Tamburri, E.; Guglielmotti, V.; Orlanducci, S.; Terranova, M. L.; Sordi, D.; Passeri, D.; Matassa, R.; Rossi, M. Nanodiamond-Mediated Crystallization in Fibers of PANI Nanocomposites Produced by Template–Free Polymerization: Conductive and Thermal Properties of the Fibrillar Networks. *Polymer.* **2012,** *53,* 4045-4053.

78. Gomez, H.; Ram, M. K.; Alvi, F.; Stefanakos, E.; Kumar, A. Novel Synthesis, Characterization, and Corrosion Inhibition Properties of Nanodiamond–Polyaniline Films. *J. Phys. Chem. C.* **2010,** *114,* 18797- 18804.
79. Volkov, S. V.; Koval'chuk, E. P.; Ogenko, V. M.; Reshetnyak, O. V. Nanochemistry, Nanosystems, and Nanomaterials. Naukova dumka: Kyiv, 2008; pp 127-130 (In Ukrainian).
80. Zubillaga, O.; Cano, F. J.; Azkarate, I.; Molchan, I. S.; Thompson, G. E.; Skeldon, P. Anodic Films Containing Polyaniline and Nanoparticles for Corrosion Protection of AA2024T3 Aluminium Alloy. *Surf. Coat. Techn.* **2009,** *203,* 1494–1501.
81. Shabani-Nooshabadi, M.; Ghoreishi, S. M.; Behpour, M. Direct Electrosynthesis of Polyaniline–Montmorrilonite Nanocomposite Coatings on Aluminum Alloy 3004 and Their Corrosion Protection Performance. *Corr. Sci.* **2011,** *53,* 3035-3042.
82. Hosseini, M. G.; Jafari, M.; Najjar, R. Effect of Polyaniline–Montmorillonite Nanocomposite Powders Addition on Corrosion Performance of Epoxy Coatings on Al 5000. *Surf. Coat. Technol.* **2011,** *206,* 280–286.
83. Zarras, P.; Anderson, N.; Webber, C.; Irvin, D. J.; Irvin, J. A.; Guenthner, A.; Stenger–Smith, J. D. Progress in Using Conductive Polymers as Corrosion–Inhibiting Coatings. *Radiat. Phys. Chem.* **2003,** *68,* 387–394.
84. Kinlen, J.; Silverman, D. C.; Jeffreys, C. R. Corrosion Protection Using Polyaniline Coating Formulations. *Synth. Met.* **1997,** *85,* 1327-1332.
85. Deng, Z.; Smyrl, W. H.; White, H. S. Stabilization of Metal-Metal Oxide Surfaces Using Electroactive Polymer Films. *J. Electrochem. Soc.* **1989,** *136,* 2152-2157.
86. Cecchetto, L.; Delabouglise, D.; Petit J. P. On the Mechanism of the Anodic Protection of Aluminium Alloy AA5182 by Emeraldine Base Coatings. Evidences of a Galvanic Coupling. *Electrochim. Acta.* **2007,** *52,* 3485–3492.
87. Sato, F.; Sakurai, T.; Hoshino, K. Effect of Degreasing Treatments on the Passivity of Pure Aluminium Sheet. *Mater. Sci. Forum.* **1996,** *217–222,* 1629-1634.
88. Chehimi, M. M.; Watts, J. F.; Eldred, W. K.; Fraoua, K.; Simon, M. XPS Investigations of Acid–Base Interactions in Adhesion. Part 4. - Use of Trichloromethane as a Molecular Probe for the Quantitative Assessment of Polymer Basicity. *J. Mater. Chem.* **1994,** *4,* 305-309.
89. Kawashima, N.; Takamatsu, M.; Morita, K. Superoxide Generator Using Polyaniline Catalyst. *Colloid. Surface. B.* **1998,** *11,* 297-299.
90. Otsuka, S.; Saito, K.; Morita, K. Generation of Superoxide by Adding Polyaniline to Water. *Chem. Lett.* **1996,** *25,* 615–616.
91. Roberts, M. W.; McKee, C. S. Chemistry of the Metal-Gas Interface. Clarendon Press: Oxford, 1978; pp 594.
92. Wen, L.; Kochierginsky, N. M. Doping-Dependent Ion Selectively of Polyaniline. *Synth. Met.* **1999,** *106,* 19-27.
93. Ogurtsov, N. A.; Pud, A. A.; Kamarchik, P.; Shapoval G. S. Corrosion Inhibition of Aluminum Alloy in Chloride Mediums by Undoped and Doped Forms of Polyaniline. *Synth. Met.* **2004,** *143,* 43–47.
94. McAndrew, T. P. Corrosion Prevention with Electrically Conductive Polymers. *Trends Polym. Sci.* **1997,** *5,* 7-12.
95. Kinlen, P.; Liu, J.; Ding, Y.; Graham, C. R.; Remsen, E. E. Emulsion Polymerization Process for Organically Soluble and Electrically Conducting Polyaniline. *Macromolecules.* **1998,** *31,* 1735-1744.

96. de Souza, S.; Pereira, J.; de Torres, S. C.; Temperini, M.; Torresi, R. Polyaniline Based Acrylic Blends for Iron Corrosion Protection. Electrochem. *Solid-State Lett.* **2001,** *4,* B27-B30.
97. Ilevbare, G. O.; Scully, J. R. Oxygen Reduction Reaction Kinetics on Chromate Conversion Coated Al–Cu, Al–Cu–Mg, and Al–Cu–Mn–Fe Intermetallic Compounds. *J. Electrochem. Soc.* **2001,** *148,* B196-B207.
98. Clark, W. J.; Ramsey, J. D.; McCreery, R. L.; Frankel, G. S. A Galvanic Corrosion Approach to Investigating Chromate Effects on Aluminum Alloy 2024–T3. *J. Electrochem. Soc.* **2002,** *149,* B179-B185.
99. Tallman, D. E.; He, J.; Gelling, V. J.; Bierwagen, G. P. Scanning Vibrating Electrode Studies of Electroactive Conducting Polymers on Active Metals. *ACS Symp. Ser.* **2003,** *843,* 228-253.
100. Malinauskas, A. Electrocatalysis at Conducting Polymers. *Synth. Met.* **1999,** *107,* 75–83.
101. Barsukov, V. Z.; Khomenko, V. G.; Katashinskii, A. S.; Motronyuk, T. I. Catalytic Activity of Polyaniline in the Molecular Oxygen Reduction: Its Nature and Mechanism. *Russ. J. Electrochem.* **2004,** *40,* 1170-1173.
102. Cui, C. Q.; Lee, J. Y. Effect of Polyaniline on Oxygen Reduction in Buffered Neutral Solution. *J. Electroanal. Chem.* **1994,** *367,* 205-212.
103. Khomenko, V. G.; Barsukov, V. Z.; Katashinskii, A. S. The Catalytic Activity of Conducting Polymers toward Oxygen Reduction. *Electrochim. Acta.* **2005,** *50,* 1675-1683.
104. Aasmundtveit, K.; Genoud. F.; Houźe, E.; Nechtschein, M. Oxygen-Induced ESR Line Broadening in Conducting Polymers. *Synth. Met.* **1995,** *69,* 193-196.
105. Kang, Y. S.; Lee, H. J.; Namgoong, J.; Jung, B.; Lee, H. Decrease in Electrical Conductivity upon Oxygen Exposure in Polyanilines Doped with Hcl. *Polymer.* **1999,** *40,* 2209-2213.
106. International standard ISO 209–1. (First edition 1989-09-01). Wrought Aluminium and Aluminium Alloys - Chemical Composition and Forms of Products - Part 1: Chemical Composition. International Organization for Standartization: Geneva, Switzerland, 1989.
107. Park, J. O.; Paik, C. H.; Huang, H.; Alkire, R. C. Influence of Fe–Rich Intermetallic Inclusions on Pit Initiation on Aluminum Alloys in Aerated NaCl. *J. Electrochem. Soc.* **1999,** *146,* 517-523.
108. Foley, R. T. Localized Corrosion of Aluminum Alloys - A Review. *Corrosion (NACE).* **1986,** *42,* 277-288.
109. Leth–Olsen, H.; Nordlien, J. H.; Nisancioglu, K. Formation of Nanocrystalline Surface Layers by Annealing and Their Role in Filiform Corrosion of Aluminum Sheet. *J. Electrochem. Soc.* **1997,** *144,* L196-L197.
110. Seegmiller, J. C.; Buttry, D. A. A SECM Study of Heterogeneous Redox Activity at AA2024 Surfaces. *J. Electrochem. Soc.* **2003,** *150,* B413-B418.
111. Nişancioğlu, K. Electrochemical Behavior of Aluminum-Base Intermetallics Containing Iron. *J. Electrochem. Soc.* **1990,** *137,* 69-77.
112. Wang, C.; Wu, G.; Zhang, Q.; Jiang, L. Characterization and Corrosion Protection Properties of Cerium Conversion Coating on $Gr_{(f)}$/Al Composite Surface. *J. Mater. Sci.* **2008,** *43,* 3327-3332.
113. Kamaraj, K.; Karpakam, V.; Sathiyanarayanan, S.; Venkatachari, G. Electrosynthesis of Polyaniline Film on AA 7075 Alloy and Its Corrosion Protection Ability. *J. Electrochem. Soc.* **2010**, *157,* C102-C109.

114. Johansen, H. D.; Brett, C. M. A.; Motheo, A. J. Corrosion Protection of Aluminium Alloy by Cerium Conversion and Conducting Polymer Duplex Coatings. *Corr. Sci.* **2012,** *63,* 342-350.
115. Aldykewicz, A. J., Jr.; Isaacs, H. S.; Davenport, A. J. The Investigation of Cerium as a Cathodic Inhibitor for Aluminum–Copper Alloys. *J. Electrochem. Soc.* **1995,** *142,* 3342-3350.
116. Mishra, A. K.; Balasubramanian, R. Corrosion Inhibition of Aluminum Alloy AA 2014 by Rare Earth Chlorides. *Corros. Sci.* **2007,** *49,* 1027–1044.
117. Decroly, A.; Petitjean, J. P. Study of the Deposition of Cerium Oxide by Conversion on to Aluminium Alloys. *Surf. Coat. Technol.* **2005,** *194,* 1–9.
118. Yasakau, K. A.; Zheludkevich, M. L.; Karavai, O. V.; Ferreira, M. G. S. Influence of Inhibitor Addition on the Corrosion Protection Performance of Sol-Gel Coatings on AA2024. *Prog. Org. Coat.* **2008,** *63*, 352–361.

CHAPTER 9

SYNTHESIS AND PHYSICO–CHEMICAL PROPERTIES OF COMPOSITES OF CONJUGATED POLYAMINOARENES WITH DIELECTRIC POLYMERIC MATRICES

O. I. AKSIMENTYEVA[1], O. I. KONOPELNYK[2], G. V. MARTYNIUK[3], and O. M. YEVCHUK[1]

[1]Department of Physical and Colloid Chemistry, Faculty of Chemistry, Ivan Franko National University of Lviv, 6 Kyryla & Mefodia Str., Lviv 79005, Ukraine. E-mail: aksimen@ukr.net

[2]Department of General Physics, Faculty of Physis, Ivan Franko National University of Lviv, 8 Kyryla & Mefodia Str., Lviv 79005, Ukraine. E-mail: konopel@ukr.net

[3]Department of Methodology of Teaching of Physics and Chemistry, State Humanitarian University of Rivne, 31 Astafona Str., Rivne 33000, Ukraine. E-mail: galmart@ukr.net

CONTENTS

ABSTRACT

The synthesis conditions and physico–chemical properties of conductive composites of conjugated polyaminoarenes (**PAArs**) – polyaniline (**PAn**), polyorthotoluidine (**PoT**), polyorthomethoxy-aniline (**PoMA**) in polymer matrixes of polymethylmethacrylate (**PMMA**), polyvinyl alcohol (**PVA**), polybuthylmethacrylate (**PBMA**), polyacrylic acid (**PAA**), and polymetacrylic acid (**PMA**) have been studied. It was found that the dependence of specific conductivity on the conducting polymer content has a percolation character with extremely low "threshold" (2–5% vol.). Based on the temperature dependence of EPR in the temperature interval of 4.2–300 K, a stronger delocalization of charge carries in composites compared with individual polyaminoarenes has been developed. It is established a connection between electrical, mechanical, and thermomechanical properties of polymer–polymer composites studied. On the basis of study optical changes in film composites under gas action the method of obtaining the flexible color indicators for express control of ammonia content in gas environment has been developed.

9.1 INTRODUCTION

Conductive polymers and composites cause an increased scientific interest in connection with the development of new fields of science and technology, including nanotechnologies, display products, sensory, power converters by new type.[1–5] However, most of conductive polymers have the conjugated system of p–electronic bonds, namely polyaniline (**PAn**) and its derivatives- polyaminoarenes (**PAArs**),[3–5] are difficultly given to technological processing due to the low solubility, poor mechanical and thermomechanical properties, which hinders of their widespread practical use. One way to provide the flexibility to conducting polymers is their application onto flexible or textile fabrics, or the use of special dopants.[6–8]

An improvement of physical and chemical properties of conducting polymers can be achieved by the creation of composites with hyperelastic or thermoplastic polymers, which mainly are dielectrics. In these composites conductive polymer fillers can provide the transition "insulator–conductor" at much lower levels compared to the metal and carbon fillers; at this, a filler and dielectric matrix have similar values of the specific density that provides the resistance to separation of such systems.

Conductive composites based on conjugated polyarenes (**PArs**) and elastomeric polymeric matrices are representatives of a new type of composite materials, in which unlike to traditional metal fillers the conductive conjugated polymers (such as polypyrrol (**PPy**) or **PAn**) are used as conductive components.[9,10]

Positive results can be expected at using of conducting polymers composites with thermoplastic materials, including the well-known polymers, namely polyacrylate, polycarbonate, polyvinyl chloride, and others.[6–10] Composites based on polyvinyl alcohol (**PVA**),[7,11,12] polymethylmethacrylate (**PMMA**)[13–15] which are the main component of "organic glass" call of particular interest in recent years. Given the fact that conducting polymers unlike to metallic powders or carbon fillers[16–18] can provide not only the conductivity but also interesting optical and electro-optical properties of composites[3,5] they can be considered as promising materials for organic electronics.[19]

The most important areas of conductive polymer–polymeric composites application are antistatic materials; some of them have been used in the manufacture of screens.[19] Another important application of conductive polymer–polymeric composites is their ability to absorb the electromagnetic field that can be used for antiradar protection.[6,8] Promising is the use of such composites in the electrotechnical, electronic, radio engineering, and other industries of production for obtaining of the conductive adhesives, sealants, potting compounds, and antistatic coatings on metallic and non-metallic surfaces.

At this time, physical–chemical properties of composites of conducting polymers with dielectric polymer matrices studied insufficiently; for an example, the questions concerning to the relationship between the composition of composites (content and chemical structure of components) and their properties (conductivity, mechanical and thermal deformative characteristics), the question of an influence of the dielectric matrix on the regularities of the charge transfer as well as on the change of the optical properties of composites under the influence of external factors demand a detailed study.

The aim of this work was to study the mutual influence of conjugated polyaminoarene and dielectric polymer matrix on the regularities of the charge transfer, mechanical, thermal mechanical, and optical properties of polymer–polymeric composites, the determination of relationship between the physical and chemical properties of composites and chemical structure and content of the components, the development on this basis of the effective methods for polymeric composites with conductive polymer filler.

9.2 EXPERIMENTAL

Dielectric polymeric matrixes of **PMMA**, **PVA**, polybutyl methacrylate (**PBMA**), polyacrylic acid (**PAA**) and polymethacrylic acid (**PMA**) have been used for the investigations. **PAn** and its derivatives, namely poly-o-toluidine (**PoT**) and poly-o-methoxyaniline (**PoMA**) were applied as electroconductive polymeric fillers; the structure of elementary link is represented in **Chaper 3**.

Composites in the form of dispersed samples and film composites were obtained by thermal pressing method,[11,20] by oxidation and electrochemical polymerization of aminoarenes (**AAr**) into the polymer matrixes,[21–23] by ultrasonic blending of polymers in common solvent.[14] To study the structure and morphology of polymer films, the scanning electron microscope *ISI–DS*–130 and an optical microscope with digital camera *Nicomed* "*Nicon*-2500" have been used. *ESR* spectra were recorded using the *X*–ray spectrometer with a frequency of field n = 9.756 ± 0.001 GHz at T = 4.2–300 K (Institute of Physics, Polish Academy of Sciences, Warsaw). To study the structure of the composites the Raman spectroscopy (T64000 spectrometer, Ar laser 514.5 nm, Łódź Polytechnic, Łódź) was used.

The electrical properties of the composites were studied by measuring of the electrical resistance (R) with subsequent of its recalculation on specific resistance, based on the size of the samples. Determination of the resistivity and temperature dependence of conductivity was performed using the two-probe method,[24] the chromel–Copel thermo couple, the electrical resistance was recorded by digital voltmeter–ohmmeter *VK* 2–20 and teraohmmeter *EB*–13A. Optical absorption spectra were recorded with the use of spectrophotometer *SF*–46 in the wavelength range 360–1100 nm at T = 293 K. The mechanical properties of the composites were studied by measuring of the microhardness using the *Heppler* consistometer according to the technique described in *ref.*[25] Thermomechanical properties of composites were measured using a modified *Vick's* device under simultaneous heating (2 deg/min) and loading (1 kg) as it is described in *ref.*[25,26]

9.3 RESULTS AND DISCUSSION

9.3.1 TEMPLATE SYNTHESIS OF COMPOSITES OF POLYAMINOARENES WITH DIELECTRIC POLYMERIC MATRIXES

The most promising method for providing of the elasticity and thermoplasticity of **PAAr** is to create the electroconductive polymeric composites in which the mixing of the components is occurred at the molecular level.[5,27] In chemical synthesis, in particular at the synthesis of nanosized polymeric composites, the technology of template or matrix synthesis with the so-called "soft" templates is used.[27] Soft matrix synthesis is implemented under chemical and spatial matching (complementarity) of monomers in the chain on the one hand, and of the matrix from other hand; at this, the elementary acts are carried out between the monomers and growing macromolecules (and also with oligomers during matrix polycondensation) bonded with the matrixes.[28] Such approaches have been used by us for the obtaining of composites of conjugated **PAAr**s with polymeric matrices (**PVA**, **PAA**, and **PMA**) when the polymerization of **AAr** is actually proceeded in polymeric gel, macromolecules of which are "soft" templates.[29,30]

It was established that the obtaining of nanodispersed **PAn** (particle size consists of 200–500 nm) is possible under conditions of oxidative polymerization of aniline in the presence of dissolved matrice of polyvinylpyrrolidone or **PVA** or other polymers.[10,31] However, data on the kinetic regularities of polymerization of **AAr** into the polymer matrices obtained not enough. In this work to study the kinetics of oxidative polymerization of **AAr** in solutions of **PVA** and polymeric acids, the spectrophotometric method was chosen.[32]

The process of polymerization of **AAr** (aniline (**An**), o-toluidine etc) is quite complex and proceeds via the mechanism of oxidative combination or via condensation of aromatic amines.[2] First step is an oxidation of the monomer with the break of one electron and following formation of cation–radical, which can be isomerizated into quinoid cation–radical (Fig. 9.1).

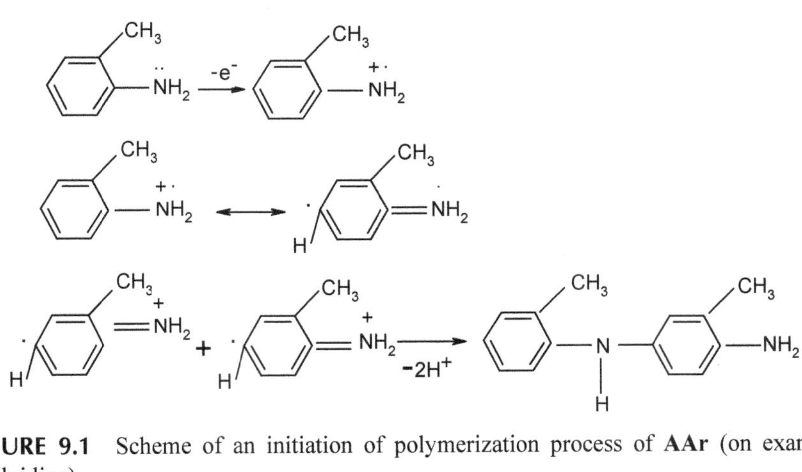

FIGURE 9.1 Scheme of an initiation of polymerization process of **AAr** (on example, o-toluidine).

The combination of such primary and isomerized radicals based on principle "head–tail" leads to the formation of the dimer that is accompanied by the deprotonation. Next, the successive processes of dimer oxidation and oxidative combination with monomer, which determine the growth of the polymer chain, take place.

The reaction product, namely **PAn** or **PoT** is colored substance and gives the absorption bands at 380–420 nm (π–π* is transition in the bandgap), 600–650 nm (n–π transition in amino–quinoid system) and 700–800 nm (an absorption of the carriers in the polaron band).[1,2,22,23]

According to the *Beer–Lambert–Bouguer* law,[32] the concentration of colored substances is directly proportional to the optical density of the solution (*D*):

$$D = \varepsilon \cdot C \cdot l \tag{9.1}$$

where ε is molar absorption coefficient, *C* is the concentration of substance, *l* is the thickness of the cell. If the absorbance density is changed due to the formation of the reaction product, and in the initial moment of time the concentration of the product, and accordingly, optical density $D = 0$, the reaction kinetics can be described by the following kinetic equation:

$$\frac{dD}{dt} = k\frac{dx}{dt}, \tag{9.2}$$

where *x* is the concentration of the product of reaction in the moment of time *t*.

To determine the rate constant we have used the semi-logarithmic dependence

$$\ln D = k \cdot t \tag{9.3}$$

where *k* is the rate constant.

It was found that the accumulation of oxidative polymerization products, namely **PAAr** into reaction mixture as a function of time is described by *s*-shaped kinetic curve which is typical for complex polymerization processes, including the autocatalytic ones.[2] Presentation of the obtained data in semi-logarithmic coordinates reveals the three plots of the kinetic curve – initial, transitional, and final with the exit in the "plateau" (Fig. 9.2).

The linear dependence *lnD - t*, which is observed at the initial and final section of the kinetic curve, gives the grounds for a formal description of the reaction rate of polymerization of aniline at the initial and final stages by the kinetic equation of the first order. In the presence of **PVA** in the reaction solution the overall view of kinetic curves practically does not change, but on the linear ares of semi-logarithmic dependencies can be the marked the change of the inclination (Fig. 9.3).

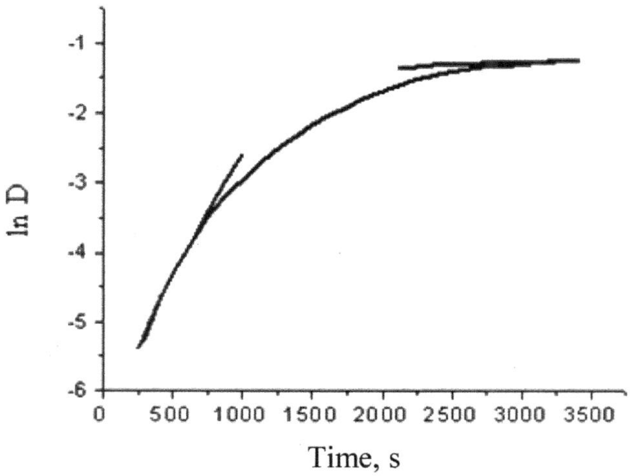

FIGURE 9.2 Dependence of logarithm of optical density of reactive mix on time for **An** polymerization ($C = 2.5 \cdot 10^{-2}$ M) in the presence of $2.5 \cdot 10^{-2}$ M $(NH_4)_2S_2O_8$ in 0.5 M H_2SO_4 at T = 293 K. Concentration of **PVA** = 0.

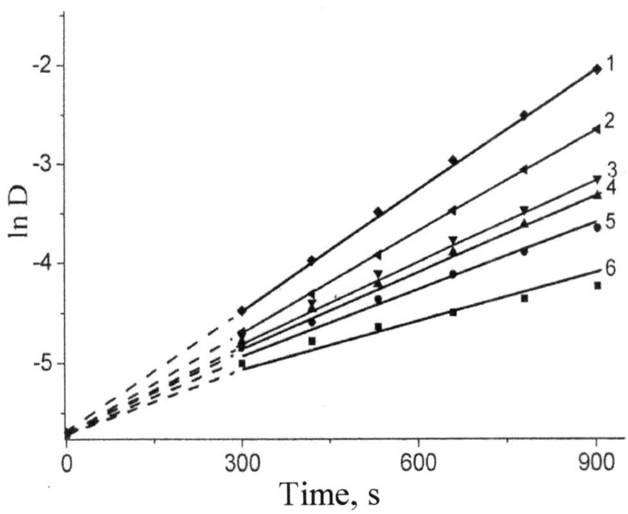

FIGURE 9.3 The change of the logarithm of optical density of reactive mix in the presence of **PVA**, mass.%: 1–0; 2–0.125; 3–0.375; 4–0.563; 5–1; and 6–3.

As can be seen from the data presented in Figures 9.2 and 9.3 as well as in Table 9.1, the polymer matrice has a significant influence on the kinetics of the process, slowing down the first, initial stage of polymerization, to which corresponds the rate constant k_1, and has a little effect on the final stages of polymerization, in particular, on the constant k_2 (Table 9.1). Therefore, the matrix of **PVA** inhibits the rate of processes of initiation and growth of chains of **PoT** and **PAn**.

TABLE 9.1 The Constants of the Reaction Rate of Polymerization of o-Toluidine and **An** in Solution of **PVA.**

Concentration of PVA, %	o-Toluidine		An	
	$k_1 \cdot 10^3$, s^{-1}	$k_2 \cdot 10^3$, s^{-1}	$k_1 \cdot 10^3$, s^{-1}	$k_2 \cdot 10^3$, s^{-1}
0	5.5 ± 0.20	0.7 ± 0.15	3.80 ± 0.15	0.8 ± 0.1
0.125	3.4 ± 0.15	0.7 ± 0.15	3.25 ± 0.15	0.8 ± 0.1
0.375	3.1 ± 0.15	0.9 ± 0.15	3.04 ± 0.15	0.9 ± 0.2
0.563	2.3 ± 0.20	0.8 ± 0.15	2.25 ± 0.15	0.8 ± 0.1
1.00	2.8 ± 0.15	0.5 ± 0.15	1.88 ± 0.15	0.8 ± 0.1
3.00	1.6 ± 0.10	0.8 ± 0.15	1.03 ± 0.15	0.9 ± 0.1

Probably, the fixation of monomer proceeds on the macrochains of **PVA** containing of residual acetate groups causing the immobilization of monomer and conceived chains on the surface of the fibrils. This reduces the mobility and changes the spatial orientation of the reactants, which reduces the rate of oxidation and oxidative combination of **AAr**.

The processes of polymerization (polycondensation) of **AAr** in the presence another polymer of high molecular weight, in particular such as **PVA**, under certain conditions can be considered as matrix synthesis.[28] Using the method of matrix the polymer–polymeric composites having more ordered structure than obtained by simple mixing of the solutions of polymers, as well as the nanoscale composites, which cannot be obtained from the finished polymers due to insolubility of them, can be synthesized. **PAArs** that are not soluble in water and in the most well-known organic solvents are belonging to these polymers. Typically, monomers and oligomers are reversibly binded with the matrix via rather weak intermolecular interactions, namely electrostatic, donor–acceptor. **PVA**, which is used in this paper, is characterized by a high content of residual acetate groups; this foresees both of electrostatic and chemical interactions with amino groups both of monomers and nucleated polymer chains.

The macromolecules of polymeric electrolytes, namely of **PAA** and of polymetacrylic acid (**PMMA**), which are capable to form the fairly stable molecular complexes with **AAr**[33,34] and also to act as the doping agents of conjugated polymer chain may be also used as the soft templates for the synthesis.

For the synthesis of such composites, the reaction mixture containing of **PMMA** or **PAA** (gel), monomer compound (aniline, o-toluidine), oxidant (e.g., ammonium peroxydisulfate) dissolved in 0.5 M of aqueous sulfuric acid is used.

The polymerization process begins after proceeding of certain induction period (1–5 min), the duration of which increases with the concentration of the polymer matrix increasing and the concentration of monomer and oxidant decreasing. The presence of this period is the fact that the process of oxidative polymerization of aromatic amines is characterized mostly by autocatalytical nature,[2] and some time is required for the synthesis of pernigraniline oligomers which are the catalyst of the process.

Spectrophotometric study of the kinetics of polymerization of aniline complexes with **PAA** showed the significant effect of the polymer matrix on the rate of a process. Increasing of the induction period in 4–5 times and the slowing down of the initial stages of polymerization are the most pronounced for the system **PMMA–PAn.**[29] Probably, the immobilization of monomer and nucleated macrochains on the surface of templates reduces the mobility and changes the spatial orientation of the reactants, which affects the rate of oxidative conjunction of aniline. At that, the nanodispersions of **PAn** stable to sedimentation for several months are formed. The formation of ordered polymeric ensembles during the process of film formation on hard surfaces leads to a significant increase of electroconductivity of films obtained from the synthesized nanodispersions compared with other methods of their formation at the improved morphology of coatings.[35]

The resulting polymer–polymeric materials form flexible, transparent films (Fig. 9.4(a)). Research of the morphology of obtained film composites showed of their fairly ordered structure (Fig. 9.4(b)).

View of optical spectra of film samples (thickness 0.7 ± 0.05 mm) greatly depends on the type of polymer matrix used as film-forming material (Fig. 9.5). Thus, for composites **PoT–PAA** two absorption bands are observed: the first is in the field of wavelength $\lambda = 400–420$ nm ($\pi–\pi^*$ transition in the bandgap of **PAAr**) and the second one at $\lambda = 750–850$ nm (the absorption in polaron zone). For composites **PoT–PMA** the second band is observed in the field $\lambda = 550–600$ nm, which is typical for electronic transitions in benzoquinoid fragments of **PAArs**.[32]

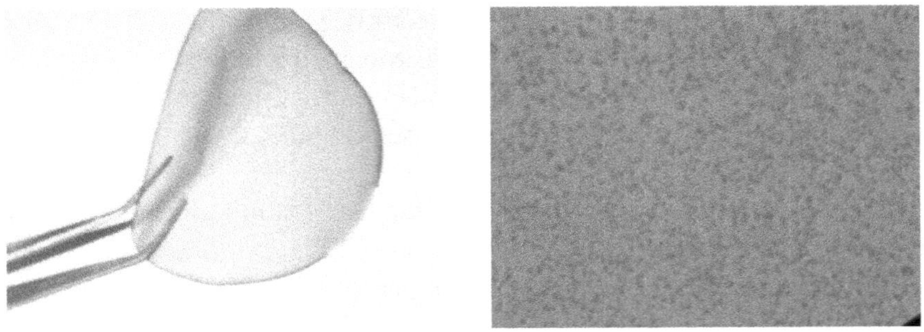

(a) (b)

FIGURE 9.4 (a) Images of the film composites based on **PMA** and **PoT** at the content of **PoT** equal to 2.5 mass.%. (b) Magnification is equal to 600.

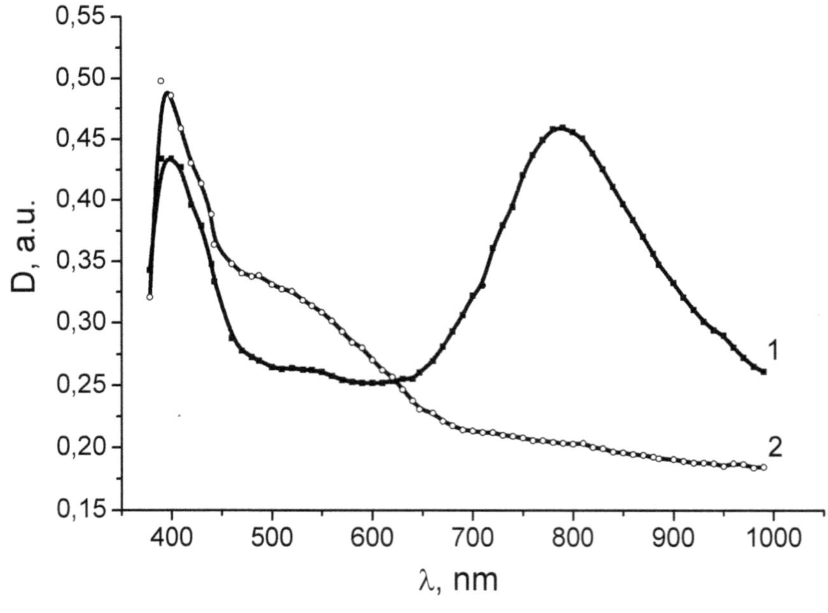

FIGURE 9.5 Absorption light-spectra for the film composites of **PoT** with the polymeric matrixes of **PAA** (1) and **PMA** (2) at the content of electroconductive polymer equal to 2.5 mass.%.

For electroconductive polymeric composites obtained under conditions of electrochemical polymerization of **AAr** in matrix of **PVA** previously deposited on the optically transparent electrode, the clear absorption bands of eletroconductive polymer (**PAn**) in the range of 380–420 nm (π–π* transition), 600–650 nm (n–π transitions in benzoquinoid fragments) and at λ = 750–850 nm (absorption in polaron zone) are observed. Polymerization is accompanied by the increasing of absorption intensity and by some displacement of the band position maxima (Fig. 9.6).

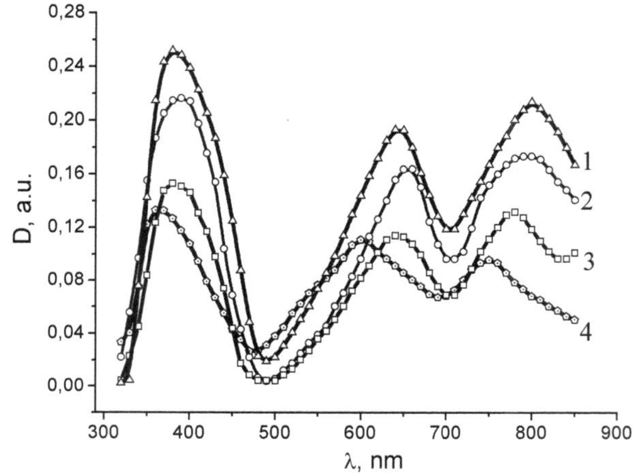

FIGURE 9.6 Absorption light-spectra for the film composites of **PVA–PAn** depending on time of the electropolymerization (1–30; 2–20; 3–15; and 4–10 min) of aniline in the film of **PVA** (the thickness is 8 μm).

Thus, in the optical spectra of composites of conjugated **PAAr**s and high-polymeric matrices there are the main features typical for the absorption of a charge's carriers inherent to organic semiconductors based on conjugated polymers. This allows us to assume that the mechanism of a charge transfer in composite materials will be determined primarily by the properties of electroconductive polymeric fillers.

9.3.2 THE NATURE OF ELECTROCONDUCTIVITY IN COMPOSITES WITH ELECTROCONDUCTIVE POLYMERIC FILLERS

The conductivity of polymeric materials which contain the electroconductive fillers will depend on how the filler particles are placed one against the other and from the contact resistance between the adjacent particles.[16–18] The term "*conductivity mechanism*" designates the mechanism of a charge transfer from one particle of the filler to another one.[36,37] The specific resistivity of polymeric compositions depends on the resistance of pure fillers, which in turn, will be determined by the totality of contact resistances between the particles.[18]

The filler is distributed in a polymer matrix with the formation of various structures, namely: matrix, statistical, oriented, and stoistic. In addition, the filler's particles can be distributed between the granules or globules of polymer. All structures, except for the matrix one, can meet in real polymeric materials, matrix structure inherent to nanoscale polymer–polymeric composites.

The mechanism of electroconductivity depends on the type of electroconductive filler, its concentration, dispersion method and degree of dispersion of the filler in the polymer, the temperature, and other factors. Charge transfer in the composite is carried out along the chain, which consists of filler particles between which there is a direct contact.[37] It is known, that in polymer–polymeric systems which form the interpenetrating polymer networks[33] (**IPN**), the conductivity can manifest itself at the content of the electroconductive filler equal to 1–2%.

For scientific description of such systems, it was developed the theory, which was called the "*percolation theory*". This theory makes it possible to describe the processes of diverse nature under conditions when at the gradual change of one parameter (e.g. concentration), the properties of the system are changed abruptly.[37] The word "*percolation*" in English means the "*flow*"; that is why sometimes along with the term "*percolation theory*" can come across the term "*flowing theory*".

In chemistry, the percolation theory is used for the description of polymerization processes. Percolation process may also lead to the self-organization and to the formation of ordered structures. Objects that are formed during the percolation are fractals. Until such time when there is not a chain of conductive islands which bind the whole volume of the sample, the specimen is an insulator (Fig. 9.7(a)). At the appearance of electroconductive circuit, the properties of material are changed abruptly and there is a phase transition "insulator–conductor"; in other words, the sample becomes conductive. It follows, that the component, which is responsible for the conductivity in formed composite should be fully delocalized across the dielectric matrix. As a result of this delocalization, the continuous cluster of the conductivity is formed. Unlike the theory of temperature phase transitions, accordingly to which the transition between two phases is occurred at a critical temperature, the percolation transition is geometrical phase transition. Percolation threshold or critical concentration separates two phases: in the one phase there is a finite cluster, and in the second one infinite cluster. It was proved that in area lower than percolation threshold the infinite cluster not exists and in the area above of percolation threshold, a cluster exists and is only one. Schematically, the formation of percolating cluster is shown in Figure 9.7.

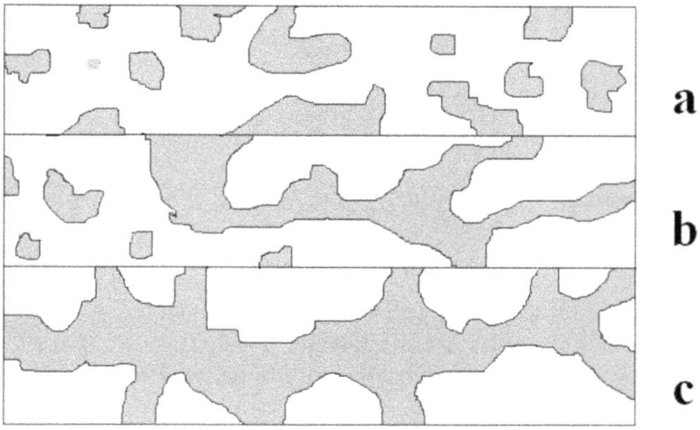

FIGURE 9.7 Schematic representation of the process of the formation of percolating cluster.[38]

To determine the structure of percolation cluster it was proposed several geometric models; one among them is a model of *Scale–Shklovsky–de Gene.*[36,37]

On the other hand, authors[36] consider that the conductivity is realized mainly by the emission of electrons through the deepenings between the particles. As the theory of electrical contacts shows, the passage of current is possible not only through the direct contact of two conductors, but also when there is aerial deepening between them or film of the dielectric.[38] Depending on the size of the deepening, applied voltage and temperature are different possible mechanisms of electrons hopping through the potential barrier, one of which is a tunnel effect. At this, continuous conductive channels begin to be formed and the resistance is decreased.

We found that the composites of conjugated **PAAr**s are conductive only after achievement of a certain threshold concentration of conjugated polymer in a dielectric matrix, then the specific conductivity of the obtained composites grows on 6–8 orders.[39–41]

The dependence of electroconductivity of film composites on the content of the electroconductive polymer has a percolation character with the percolation "threshold" value (P, vol.%) depending on nature of **PAAr**: $P = 2.1\%$ (**PAn**), 1.7% (**PoMA**), and 2.8% (**PoT**) for composites based on **PVA**. In the case of composite **PMMA–PAn** the percolation threshold is achieved at $P = 2\%$ vol. (Fig. 9.8(a,b)).

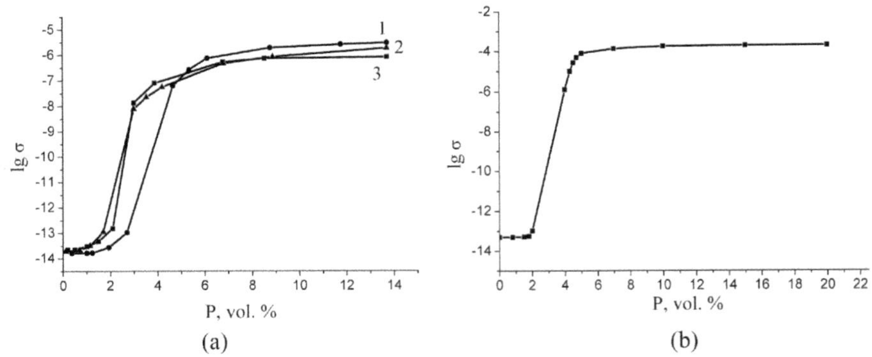

FIGURE 9.8 (a) Dependence of logarithm of specific conductivity of composites based on **PVA** on the content of electroconductive polymer: 1—**PoT**, 2—**PoMA**, and 3—**PAn**; (b) Dependence of logarithm of specific conductivity of composites **PMMA–PAn** on the content of **PAn**.

TABLE 9.2 An Impact of the Content of Electroconductive Polymer (**ECP**) on Specific Conductivity and Order of Anisotropy of Polymeric Composites Based on **PVA**.

The content of ECP, vol.%		**The thickness of film, мм**	**Specific conductivity, $\sigma_{\perp}\cdot 10^8$, Ohm^{-1} cm^{-1}**	**Specific conductivity, $\sigma_{II}\cdot 10^8$, Ohm^{-1} cm^{-1}**	**The order of anisotropy, $\sigma_{II}/\sigma_{\perp}$**
0		0.26 ± 0.02	10^{-14}	–	–
PAn	2.3	0.32 ± 0.03	1.0 ± 0.1	1.1 ± 0.2	1
	4.6	0.29 ± 0.01	0.1 ± 0.1	14 ± 2	13
	6.9	0.32 ± 0.01	1.2 ± 0.1	29 ± 1	24
	9.2	0.33 ± 0.05	2.2 ± 0.2	76 ± 2	35
	13.8	0.33 ± 0.07	3.6 ± 0.2	210 ± 4	58
PoMA	2.3	0.36 ± 0.01	2.9 ± 0.1	3.0 ± 0.2	1
	4.6	0.36 ± 0.01	3.9 ± 0.1	21 ± 2	5
	6.9	0.35 ± 0.01	8.7 ± 0.1	150 ± 5	17
	13.8	0.42 ± 0.01	29.0 ± 0.1	823 ± 8	28

Obtained on Teflon or glass substrate film composites based on **PVA** have a significant anisotropy of conductivity - the resistance values measured in a direction parallel to the plane of the film σ_{II} is much smaller compared to the resistance measured in the perpendicular direction $\sigma_{\perp}$ (Table 9.2). The degree of anisotropy, which is a measure of the ratio $\sigma_{II}/\sigma_{\perp}$, is increased with the increasing of concentrations of conductive polymer. The obtained results permit to do assumption about the possibility of ordering of electroconductive macrochains mainly in parallel to the plane of the film.[40]

Obviously, the anisotropy of the conductivity of investigated film composites is stipulated by features of the structure of electroconductive polymers, their ability to be in the low order state and can be considered as their main feature. In such systems with a 1 D electronic structure the conductivity is defined by high mobility of the carrier along the polymer chain.[1,2,42] The electroconductivity becomes three-dimensional when the probability of diffusion of the electron between adjacent chains is increased, that is realized via movement between the defects of chains.[42–44] The presence of such a strong anisotropy testifies to the assumption that in the result of the polymerization of **AArs** in **PVA** matrix the formation of polymeric composites proceeds that includes the interaction of functional acidic (acetate) and basic (amine) groups.

If the polymer matrix is polymeric electrolyte, then the dependence of the specific resistivity of composites on the content of the electroconductive polymer has a rather complicated nature (Table 9.3). If the specific resistivity of "dry" matrixes **PAA** and **PMMA** is quite high and in the absence of impurities consists of r = 10^{14}–10^{15} Ohm×m, then in the presence of about 2% of **PAAr** the resistance is sharply decreased on 8–10 orders. With increasing of the concentration of **PAAr** the specific electroconductivity of composites (s = 1/r) firstly is increased peaking at 6.8% content of **PoT** (Table 9.3), and then is decreased. It is likely, that at such content of electroconductive component in the composite the proper contact between the particles with the formation of a continuous cluster of conduction is provided.[36] Reducing of s with increasing of **PoT** content is associated with the deterioration of the contact due to the loss of mechanical properties of the composite, since the conjugated **PAAr** has a loosening effect.[1]

As can be seen from the data represented in the Table 9.3, the formation of polymer–polymeric composites with ion-conducting matrixes in some cases causes an increasing of the specific conductivity in 1.5–2 times compared to the polymer synthesized in the absence of **PAA** or **PMMA** matrixes. Thus, the polymeric electrolytes serve by additional alloying agents of **PAAr** that was observed previously for the composites of **PAn** with **PMMA.**[39,41]

Exploring the temperature dependence of the specific resistivity normalized to the resistance measured at room temperature (ρ/ρ_{293}), it was revealed that for the all investigated composites, like the most organic semiconductors, on the initial part of the temperature increasing the specific resistivity of the samples is decreased exponentially. Linear dependence lg (ρ/ρ_{293})–$1/T$

in a certain temperatures range (Fig. 9.9) allows us to calculate the value of activation energy of the charge transfer (E).

TABLE 9.3 The Dependence of Specific Resistance of Composites on the Content of **PoT.**

Polymeric matrix	The content of PoT *mass.%*	*ρ, Ohm*m* (T = 294 K)	$\sigma \cdot 10^3$, *S/m*
PAA	2.2	530	1.9
	4.2	204	4.9
	6.2	184	5.4
	8.1	413	2.4
	9.9	472	2.1
PMA	2.5	584	1.7
	4.8	434	2.3
	7.1	224	4.5
	9.3	300	3.3
	11.4	493	2.0
Without matrix	100.0	418	2.4

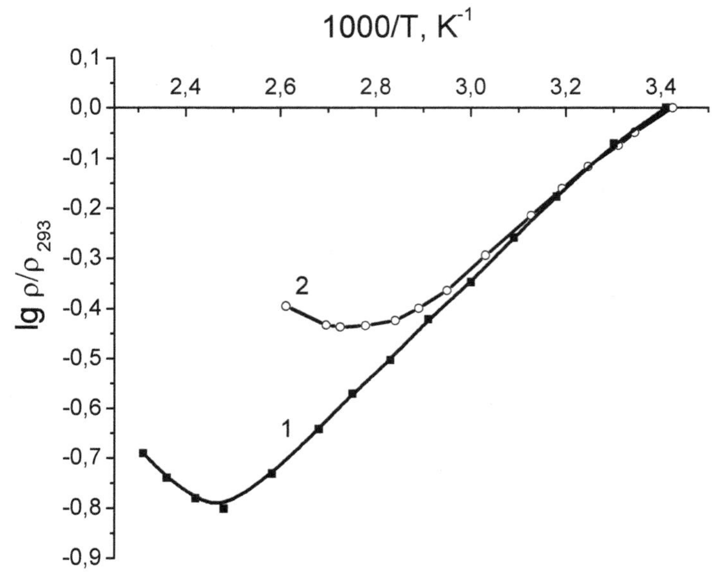

FIGURE 9.9 Temperature dependence of specific resistance of **PAn** (1) and film composite **PVA–PAn** (2). The content of electroconductive filler (**PAn**) consists of 13.8 vol.%.

As we can see from Figures 9.9 and 9.10, the semiconductor character of the electroconductivity of conjugated polymer is not disturbed in composites, but there is a significant impact of the structure of the polymer matrix on the activation parameters of the charge transfer. The value of activation energy determined for the temperatures range T = 293–358 K for composites of **PAn** with **PAA** is E = 0.147 eV, while for the composites of **PMA** at the same content of **PAn** E = 0.06 eV. For the composites based on **PoT** the values of E are somewhat higher and consist of 0.173–0.305 eV depending on the composition of the composite. An availability of dielectric **PVA** matrix even slightly improves the indexes of the charge transport. The value of E in the range of T = 293–358 K for composite **PVA–PAn** consists of 0.47 ± 0.1 eV, while for the obtained under similar conditions **PAn** value consists of E = 0.49 ± 0.01 eV.

Thus, the presence of dielectric and ion-conducting matrixes does not change the semi-conducting nature of electroconductivity and optical absorption inherent to the electroconductive polymers based on conjugated **PAArs** at the same time providing the value of specific conductivity as 10^{-3} S/m, which is sufficient for the use of synthesized composites at the manufacture of antistatic coatings, screens, antirad protection, and so forth.

One of the methods for obtaining of nanoscale polymer–polymeric composites is the dispersion of the solutions of different polymers in a common solvent with its following evaporation.[8] In this case we talk about the formation of polymer "*blends*" or molecular composites. This technique was used by us for obtaining of polymeric blends **PMMA–PAn** using a mixture of organic solvents, namely dioxane, toluene, and dimethylformamide.[39,41] Ultrasonic dispersion of mixed in the requisite proportion of solutions of **PMMA** and **PAn** doped with sulfuric acid in a mixture of dimethylformamide–chloroform (1:1) was carried out for 2 h. Composites were obtained after organic solvent evaporation in a vacuum at temperatures 343–363 K for several days.

As in the case of the film composites based on electroconducting polymers, the concentration dependence of the electroconductivity of composites **PMMA–PAn** has the percolation character with the value of "threshold" percolation equal to 0.8–2 vol.% (Fig. 9.3). The value of specific conductivity at room temperature (σ_{293}) is within $5.0 \cdot 10^{-8}$–$2.8 \cdot 10^{-5}$ $Ohm^{-1} \cdot m^{-1}$ at the content of **PAn** equal to 0.8–10 vol.% (Table 9.4). At the study of the temperature dependence of the specific resistivity of composites **PMMA–PAn** was found that the effective activation energy of conductivity greatly depends on the content of **PAn** and is within (0.38–2.16) ± 0.04 eV.

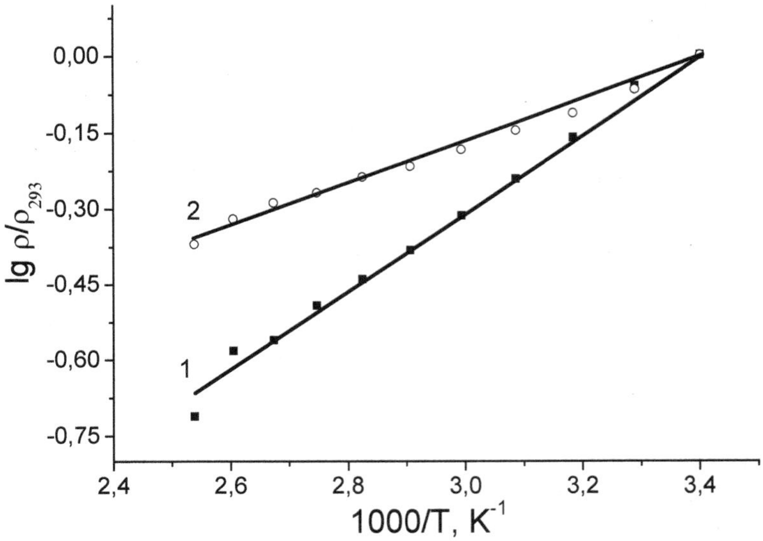

FIGURE 9.10 Temperature dependence of the specific resistance of composites based on **PAA** (1), **PMA** (2) and **PAn** (the content of **PAn** is 2.5 mass.%).

TABLE 9.4 An Impact of the Content of **PAn** on the Parameters of Conductivity of Composites **PMMA–PAn.**

The content of PAn, %	σ_{293}, Ohm^{-1}·m^{-1}	ρ_0, Ohm·m	ε_σ, eV	Temperature range, K
0.0	10^{-14}	–	–	–
0.3	$1.7 \cdot 10^{-7}$	-	–	–
0.8	$5.03 \cdot 10^{-6}$	$3.98 \cdot 10^{-6}$	2.16	298–365
2.0	$8.12 \cdot 10^{-5}$	$2.69 \cdot 10^{-3}$	1.91	298–351
10.0	$2.76 \cdot 10^{-5}$	$3.24 \cdot 10^{-2}$	1.30	305–397
20.0	$2.16 \cdot 10^{-6}$	$3.47 \cdot 10^{-2}$	0.26	306–355
100.0	$3.6 \cdot 10^{-7}$	$6.23 \cdot 10^{-6}$	0.38	298–363

We can assume that with increasing of the **PAn** content in the composite the number of percolation bridges is growing, thus the additional channels to transfer of the charge are created that causes the reduction of the activation energy of the conductivity and an increase of the free path of the charge carriers. As can be seen from the Table 9.4, the specific conductivity of the composites **PMMA–PAn** in some cases exceeds the value of specific conductivity defined for "pure" polymeric filler. This fact is described in the *ref.*[45] for composites **PMMA–PAn** at the content of **PAn** equal to 40%. In our studies an increasing of the specific conductivity of "blend" was observed at the content of **PAn** equal to 2 vol.%.

According to the *Raman* spectroscopy for the composite **PMMA–PAn** the characteristic peaks of benzene (1670 cm^{-1}) and quinoid (1165 cm^{-1}) rings are observed, as well as the absorption band of imine groups –C = N– attached to the quinoid rings (1490 cm^{-1}) and other absorption bands of **PAn** and **PMMA**, the intensity and form of which depend on the content of **PAn** in composite (Fig. 9.11).

An analysis of the obtained spectra showed that at the increase of **PAn** in composite the most significant changes have the *Raman* bands, that is typical for nitrogen binded with the molecules of **PAn** (2500–3500 cm^{-1}) and for etheric oxygen characteristic for **PMMA** (1270–1200 cm^{-1} and 2180 cm^{-1}). This allows us to assume the existence of intermolecular interactions in composites **PMMA–PAn**, which is due to these functional groups.[46] Perhaps, the additional acidic doping of **PAn** by residues of **PMMA** takes place due to such interactions that causes a high conductivity of composites **PMMA–PAn** compared with the starting **PAn**.

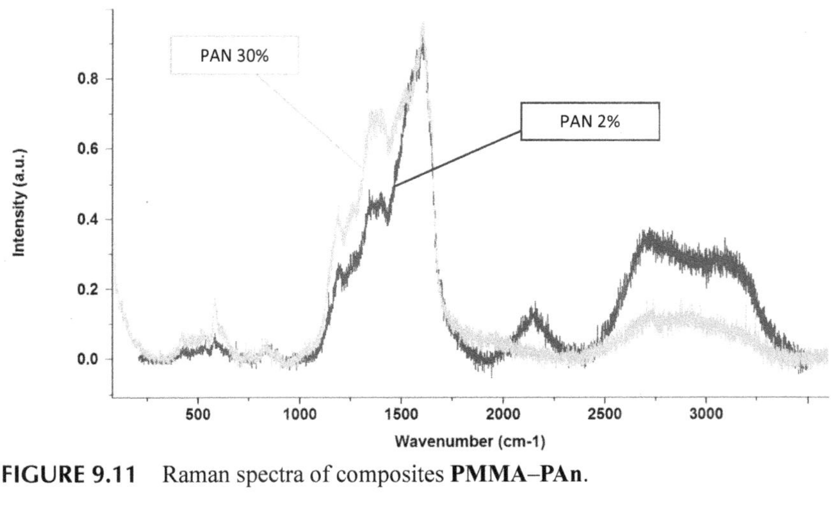

FIGURE 9.11 Raman spectra of composites **PMMA–PAn**.

On the other hand, the presence of such interaction causes the formation of "anchor" points on the macromolecular chain of **PMMA**, on which the fixing of **PAn** macrochains as well as their "deployment" with the transition of conformation of twisted helix to deploy one can take place. This phenomenon is known as the effect of "secondary doping"[2] and it most likely is the reason for increasing of the conductivity of composites.[41,46]

Found peculiarities of the electroconductivity are characterized by a number of features that allow us to include the composites of conjugated **PAArs** and elastic high-polymeric matrixes to the systems in the formation of which the mechanisms of matrix polymerization are involved.[28]

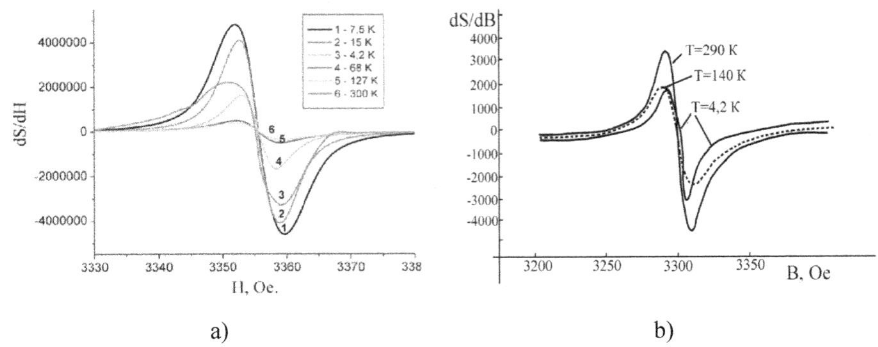

FIGURE 9.12 *ESR*–spectra for composite (a) **PMMA–PAn** (10%) and (b) **PAA–PAn** (10%) at different temperatures.

Proof of this fact is extremely low values of percolation threshold for these composites (near 2 vol.%), the anisotropy of conductivity, the degree of which is increased with the concentration of electroconductive polymer increasing. Probably, there is shown a structural matrix effect, which is the ability of the polymer matrix to influence on the length and chemical structure of the polymer chains of derived polymer (**PAAr**), including on their spatial structure. The presence of the structures of this type provides the safety both of the properties inherent to high-polymeric matrices and semiconductor nature of electroconductivity inherent to conjugated **PAArs**.

For a more detailed study of the impact of the dielectric matrix on the nature of interaction of components of investigated composites it was used the method of electron paramagnetic resonance (**ESR**). **ESR** spectra of composites **PMMA–PAn** at filling above the percolation threshold (10%) were studied in the temperature range 4.2–300 K. It was found that the shape of the **ESR** line for composites in general is similar to the **ESR** spectra of pure **PAn** and this is the unit line without hyperfine structure characteristic for the most electroconductive **PAArs** (Fig. 9.12).

As can be seen from the data represented in Table 9.5, the value of g-factor for composite **PMMA–PAn** is fairly stable throughout the temperatures range and is closer to the g-factor of a free electron in comparison with pure **PAn** ($g = 2.0036$).

TABLE 9.5 Temperature Dependence of *ESR*–Parameters for Composite **PMMA–PAn.**

Temperature, *T, K*	The linewidth *ΔHpp, mT*	The constant of spin–spin interaction, $T_2 \cdot 10^8$, *s*	Asymmetry of the signal *A/B*	$g \pm 0.0002$
7.5	8.21	2.18	1.001	2.0025
15	5.87	1.41	1.002	2.0025
28	5.15	1.37	1.003	2.0025
68	5.05	1.33	0.950	2.0025
97	6.81	1.81	1.050	2.0025
127	7.44	1.98	1.070	2.0025
227	7.76	2.06	1.075	2.0025
245	8.05	2.14	1.081	2.0026
300	8.29	2.20	1.080	2.0027

It was found, that the shape of a line of the **ESR** signal is almost symmetrical at low temperatures (Fig. 9.12(a)), but with increasing of T the symmetry of a line is slightly broken. The degree of asymmetry signal measured as the ratio of the height of the peaks (*A*/*B*) reaches a value equal to 1.08 passing through a minimum at $T = 68$ K. For **PAn** doped by sulfuric acid the degree of asymmetry is 1.5. For polymeric composites, **PMMA–PAn** the value of the width of signal (*ΔHpp*) is much higher in comparison with pure **PAn** (3.1 mT), that confirms the significant delocalization of the charge carriers in the polymeric composite.

A similar shape of temperature dependences for the width of *ΔHpp* line and of the degree of its asymmetry *A/B* as well as the constant of spin–spin interaction (T_2) defined in accordance with *ref.*[47] can be a sign of the presence of a phase transition in the field of $T = 45$–68 K.

TABLE 9.6 Parameters of *ESR*–Spectra of **PAn** in Polymeric Matrix of **PAA.**

Parameters of ESR–spectrum	PAn	PAn–PAA
g-factor, $T = 298$ K	2.0026	2.0008
Ns, 1/*g*	$4.5 \cdot 10^{19}$	$6.1 \cdot 10^{20}$
ΔHpp, *G*	3.6	18

At the obtaining of **PAn** composite in **PAA** matrix the similar picture is observed, namely an extension of a line of signal and an increasing of concentration of paramagnetic centers (Table 9.6), which indicates on a greater delocalization of charge in the composite, which explains the fact of the specific conductivity increasing in such systems due to additional doping by polymeric matrix.

Based on the data of **ESR**–spectroscopy it can be assumed that at the formation of composites of **PAArs** with the polymers **PMMA**, **PAA**, **PMAA**, and **PVA.** The electronic structure of the material is undergoing to significant changes, which with extremely low percolation threshold may indicate on the formation of nanosized polymer–polymeric composite by matrix type.

9.3.3 PHYSICO–MECHANICAL, THERMOMECHANICAL AND ELECTRICAL PROPERTIES OF COMPOSITES OF POLYAMINOARENES AND DIELECTRIC POLYMERIC MATRICES

The use of composite polymeric materials in various fields of technology requires the knowledge of their physical and mechanical properties, in particular such as mechanical strength, capacity for thermoplastic deformation, electrical conductivity, and so on. The properties of composites are determined not only by the characteristics of components that make up the composite, but also to a large extent by the interaction of components between themselves, by method of composite manufacturing, by degree of the composite dispersion or the by degree of its crystallinity.

Despite the extensive study of electrical, physical, thermal, and mechanical properties of composites with electroconductive fillers,[16–18] for the polymer–polymeric systems have received a little information about the impact of polymeric fillers on physical and mechanical properties as well as on thermoplastic characteristics of composites and also about their relationship with the electrical properties.

Polymeric composites based on dielectric matrix **PVA**, **PMMA**, **PBMA** and dispersed polymeric filler obtained by the method of thermal pressing can be regarded as a system of "high-molecular polymer matrix - dispersed phase". At this, general regularities of the impact of filler on temperature of phase transitions should be correlated with inorganic fillers. The electrical properties of conventional polymeric composites with carbon or metal fillers are mainly determined by a value and by type of conductivity of inorganic dispersed phase, and at the concentrations that exceed the threshold, the specific resistivity of the composite is determined by the resistance of electroconductive filler[16]. At the formation of polymeric composites by method of two polymeric dispersed phases (conductive and non-conductive) pressing, the specific conductivity of the composite as same as with the inorganic fillers, has also the percolation nature.[11,41] In this paper the physical and mechanical as well as thermomechanical and electrical properties of composites **PVA, PMMA,** and **PBMA** with electroconductive polymers of **PAn** and its derivatives obtained by thermal pressing method are studied.

At the study of electrical, physical, and mechanical properties of composites based on the matrixes of **PVA** obtained by thermal pressing method, it was found that the introduction of the polymer filler significantly affects their microhardness and thermomechanical characteristics, as well as the specific conductivity of the composites, and the nature of this influence largely depends both on the type of a polymer and from its content.

We found, that in the presence of **PAArs** in the **PVA** matrix, the changes of the limited value of conical point of fluidity (*F*) or microhardness (typical dependences are shown in Fig. 9.13) take place. The microhardness was calculated in accordance with the formula:

$$F = \frac{4 \cdot G \cdot 10^4}{\pi \cdot s^2} \tag{9.4}$$

where F is the microhardness, N/m^2, G is the load, N; S is the depth of the cone penetration, cm.

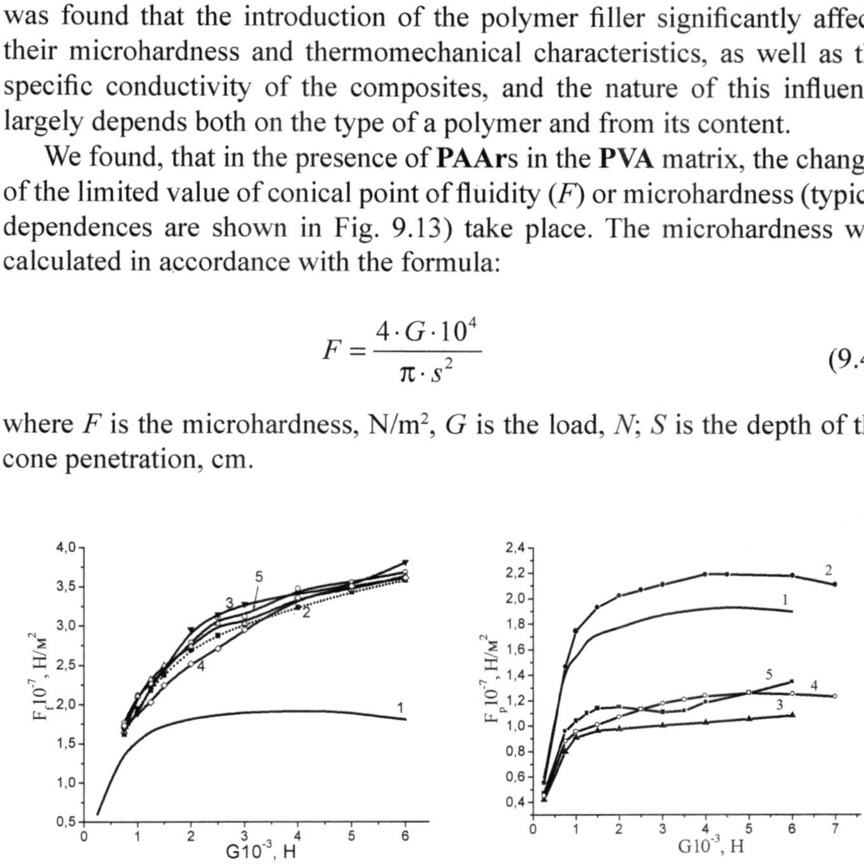

(a) (b)

FIGURE 9.13 The dependence of conical point of flowability on the loading for the composites (a) **PVA–PoMA**; (b) **PVA–PAn**, the content of polyaminoarene, %: 1–0; 2–3.2; 3–5.7; 4–10.7; 5–15.1; and 6–18.2.

The nature of the impact is largely determined by the chemical structure of the electroconductive polymer used as the filler. An introduction of fillers, which have a substituent in the aromatic ring (**PoT** and **PoMA**), leads to an increase of the microhardness on average in 1.3–1.5 times. In the largest extent, the microhardness of composite is increased in the presence of **PoT** in the **PVA** matrix (Table 9.7). For composites, **PVA–PAn** the nature of impact is more complex: small concentrations of filler cause some increase of the microhardness while at $\varphi > 5\%$ the conical point of fluidity of composites (F) is reduced. Starting from the data concerning to microhardness it can be assumed that in this case the packing density of the composite is reduced. The nature of the dependence of specific conductivity of the composite **PVA–PAn** on the content of the electroconductive polymer (Table 9.7) may also indicate on the above said statement. At the content of **PAn** equal to 10–15 vol.% the specific conductivity consists of $\sigma = 2.4\times10^{-3}$–$6.7\times10^{-4}$ S/cm whereas the increase of the concentration of the **PVA** till the 18.2% causes the decrease of the specific conductivity almost on order.

TABLE 9.7 An Impact of Polymeric Filler on Specific Conductivity and Microhardness of Polymeric Composites Based on **PVA.**

The content of filler, (vol). %	PVA–PAn		PVA–PoT		PVA–PoMA	
	Specific conductivity, σ, S/cm	Micro-hardness, $F_f \cdot 10^{-7}$, N/m²	Specific conductivity, σ, S/cm	Micro-hardness, $F_f \cdot 10^{-7}$, N/m²	Specific conductivity, σ, S/cm	Microhardness, $F_f \cdot 10^{-7}$, N/m²
0	10^{-14}	1.9	10^{-14}	1.9	10^{-14}	1.9
1.8	$5.7\cdot10^{-7}$	–	$2.1\cdot10^{-9}$	–	$1.1\cdot10^{-8}$	–
3.2	$2.0\cdot10^{-5}$	2.1	$7.2\cdot10^{-8}$	3.2	$5.2\cdot10^{-7}$	2.3
5.7	$5.5\cdot10^{-5}$	1.5	$3.3\cdot10^{-7}$	2.5	$1.9\cdot10^{-6}$	2.8
10.7	$2.4\cdot10^{-3}$	1.2	$1.8\cdot10^{-7}$	3.2	$1.9\cdot10^{-6}$	2.6
15.1	$6.7\cdot10^{-4}$	1.1	$9.4\cdot10^{-8}$	3.1	$1.0\cdot10^{-6}$	2.9
18.2	$3.7\cdot10^{-5}$	1.0	$5.4\cdot10^{-8}$	2.8	$1.8\cdot10^{-7}$	2.7
100.0	$2.1\cdot10^{-3}$	–	$1.4\cdot10^{-5}$	–	$3.6\cdot10^{-4}$	–

Study of thermomechanical properties of polymer composites showed that the introduction of the polymer filler does not change the overall look of thermomechanical curves (Fig. 9.14), but highly affects the temperature of elastic deformation (T_e), the value of hyperelastic module ($E\infty$) and molecular weight of the kinetic segment (Mc) compared with an individual **PVA** (Table 9.8).

Starting from the experimentally obtained values of relative compression (ε), we can say that in a case of the composites **PVA–PoT** the high elastic

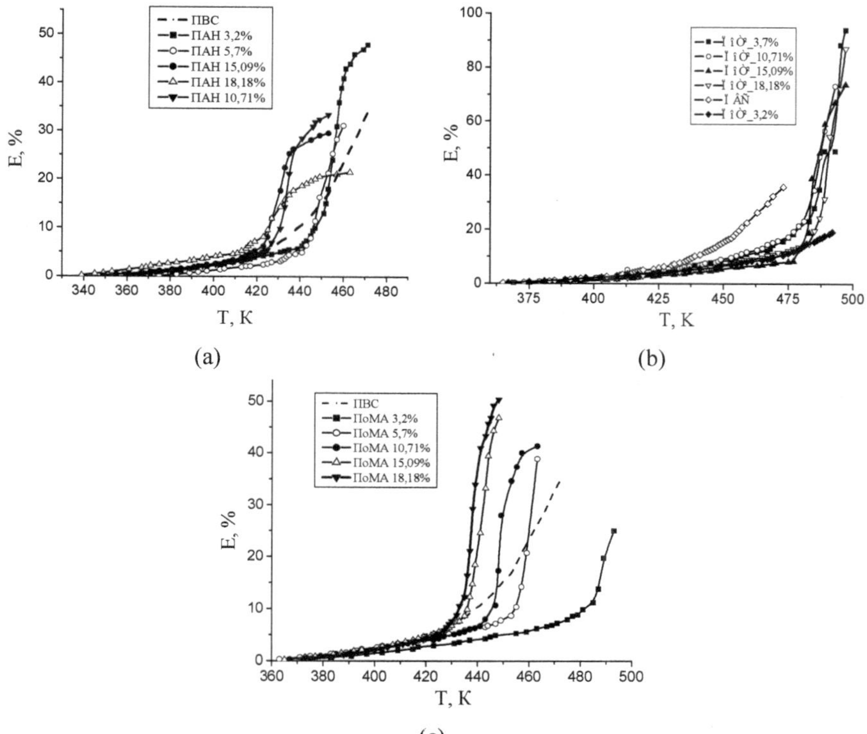

FIGURE 9.14 Thermomechanical curves of composites **PVA–PAn** (a), **PVA–PoT** (b), and **PVA–PoMA** (c).

deformation is decreased that may indicate on increasing of mechanical strength of the composites compared with pure **PVA**. In the case of **PAn** and **PoMA**, the similar effect of filler takes place till the values φ » 10.7% after achievement of which the relative deformation of the composite is increased. Found effect suggests the possibility of interaction of **PAArs**' particles with high elastic **PVA** matrix. At this, fine filler to some extent may play a role of the adsorbent, on the surface of which the segments of the **PVA** macromolecule are adsorbed. Formed in this way highly ordered adsorption layers increase the mechanical strength of the polymer composite.[48] In the case of **PVA** composites with conjugated **PAArs**, the presence of interaction between residual acetate groups of polymeric matrix and amine groups of the polymeric filler is not excluded. The presence of such interaction causes a significant impact on the temperature of hyperelastic state of polymeric composition and also on the value of flow temperature (Table 9.8). The calculated values of the length (molecular weight) of kinetic segment in the presence of filler slightly reduced compared to the **PVA**, which may indicate on increased frequency of crosslinks (*N*). For fillers **PoMA** and **PoT** inherent stronger interaction with functional groups of **PVA**, as evidenced by the shorter kinetic segment and the corresponding increase of the frequency of crosslinks which for the degree of filling $\varphi = 10.7\%$ consists of 2.18×10^3 (**PoT**), 2.12×10^3 (**PoMA**), and 1.54×10^3 (**PAn**).

TABLE 9.8 Thermomechanical Characteristics of Polymeric Composites.

The content of electroconductive polymer, % (vol.)	E, %	$T_{h.e.}$, K	E_∞, 10^{-5} N/m²	M_c. kg / mole
PoT 3.20	3.79	465	26.38	549
5.67	4.09	453	24.45	578
10.71	4.14	443	24.15	572
15.09	2.68	475	37.31	397
18.18	3.38	461	29.58	486
PAn 3.20	4.2	438	19.04	461
5.26	2.68	441	29.85	717
10.71	4.91	424	16.29	811
15.09	6.22	419	12.86	1016
18.18	8.13	416	9.84	1318
PoMA 3.20	2.91	459	34.36	416
5.26	4.52	449	22.12	633
10.71	4.01	439	24.94	549
15.09	5.18	429	19.31	693
18.18	4.69	427	21.32	625
PVA (100%)	5.29	433	15.1	894

In the case of composite **PVA–PAn**, after a slight increase the lowering of the flow temperature is observed, which becomes visible at the content of **PAn** $\varphi > 5.3\%$. These data are in good correlation with the values of the microhardness of composites (Table 9.7), namely after a small increase at $\varphi > 5.7\%$ the microhardness of composite is decreased. Perhaps, in this case the fine particulate filler (**PAn**) forms its own polymer net with the formation of infinite cluster of conductivity as in the case of metal and carbon fillers.[16,36] In particular, this is evidenced by higher values of the specific conductivity of **PAn** composites on 2–3 times compared to other polymeric composites. When $\varphi \geq 10\%$, the growth of the module of hyperelastic deformation is observed as same as significant increase of the length of the kinetic segment. This points to the appearance of possible movement of the chains of **PVA** and **PAn** as a whole due to the destruction of relatively fragile crosslinking bonds.[49] Obtained data concerning to the study of microhardness and thermomechanical properties of composites of **PAAr**s and **PVA** confirm that the materials that retain the unique properties of electroconductive polymers (namely, high electroconductivity, low specific weight, and a semiconductor character of the charge transfer) acquire the new features such as mechanical strength, capacity for hyperelastic deformation, which are inherent to the traditional polymeric materials.

For the composites based on **PBMA** an introduction of **PAn** filler increases the microhardness from 4.51×10^9 N/m^2 (**PBMA**) till 7.5×10^9 N/m^2 at the content of conductive polymer equal to 15% (Table 9.9). The determined effect indicates that the **PAn** acts as reinforcing component in the composite despite the fact that the microhardness of pure **PAn** is less than the microhardness of pure **PBMA**. The combination of these two polymers improves the mechanical properties of composites, possibly due to intersegmental interaction between the filler and polymer matrix resulting in compaction of composite and increase of its microhardness almost in 1.7 times.[20]

The obtained thermomechanical curves and also characteristic transition temperatures (Table 9.9) show the obvious impact of the filler (**PAn**) on physical and mechanical properties of composite samples. This is especially noticeable for the flow temperature (T_f), hyperelastic deformation and to a lesser extent for the glass temperature (T_g).

The strength and the relaxation properties both of the polymers and of the composites on their basis depend on the crosslinking density (concentration of the effective units both of chemical and physical nature) and on the length of interstitial fragment of spatial net (M_c). Similar peculiarities explain the formation of the physical net, the nodes of which are molecules of the filler.[48,49] The creation of such a net leads to the limited mobility of kinetic segments of the polymer matrix, and thus is a factor that leads to increasing of the flow temperature. An increase of the **PAn** content leads to increasing of the quantity of polymer–polymeric links and thus leads to the limitation of the segments mobility.

Comparing the results of thermomechanical studies it should be noted that the increasing of the content of the conductive filler **PAn** over 7% leads to increase of the module of high elasticity and to decrease of the molecular weight of interstitial segment. Relegation of the M_c parameter indicates an increase of intermolecular contacts and the formation of more dense compositional net.

In good correlation with these features, there is dependence of the specific conductivity of composites **PBMA–PAn** on the content of the electroconductive filler (Table 9.9). It was found that only at the content of **PAn** more than 7–10% it is observed a significant increase of the conductivity. It should be noted that there is an optimum area of the ratio **PAn–PBMA** (10–15% of **PAn**), at which the improvement of mechanical and thermomechanical properties of composites is observed. Further, increase of the content of the conductive filler difficults the formation of samples.

TABLE 9.9 Temperature and Thermomechanical Parameters of Composites **PBMA–PAn.**

The content of PAn, %	T_g, °C	T_f, °C	F_∞*10^{-9}, N/m^2	M_c, g/mole	Specific conductivity, Ohm^{-1}cm^{-1}
0	44	46	4.51	–	$2.0 \cdot 10^{-14}$
5	42	45	5.53	–	$8.4 \cdot 10^{-14}$
7	48	87	5.73	510	$1.5 \cdot 10^{-8}$
10	45	80	7.22	360	$3.2 \cdot 10^{-7}$
15	46	85	7.52	310	$3.1 \cdot 10^{-3}$

PMMA–PAn composites are preferably obtained by the method of thermal pressing[15,45,50] at temperatures that are equal to the temperatures of softening and flow of **PMMA**. At this, the content of the electroconductive filler (**PAn**) is varied within 30–60 wt.%. Data on the effect of fillers on physical–mechanical and thermomechanical properties of such composites are almost absent. Given the fact that we was determined a low threshold of percolation for composites **PMMA–PAn** obtained by ultrasonic dispersion of polymer solutions, we have studied the properties of composites at much lower content of **PAn** than described in the references, namely within 1–10%.

As can be seen from the presented values of experimentally obtained and calculated density of the composites, at the content of **PAn** 1–2% there is a slight decrease in the density of the material that is consistent with the defined limited microhardness (Fig. 9.15). With increasing of **PAn** content the experimentally determined density of the composite is increased slightly compared with the calculated due to the possible interactions of **PAn** with the functional groups of **PMMA**, particularly with acidic ones which are formed during the partial saponification of **PMMA** at high temperatures.[45]

TABLE 9.10 Physical–Chemical Characteristics of Composites **PMMA–PAn.**

The content of PAn, W%	Density, exper., ρ, g/cm^3	Density, calc., ρ, g/cm^3	Interfacial microhardness $F_\infty \cdot 10^{-8}$, N/m^2	Loading, G, N	Glass temperature, T_g, K	Flow temperature, T_{flow}, K
0	1.19 ± 0.01	1.190 ± 0.001	2.05 ± 0.10	25 ± 1	383 ± 1	412 ± 1
1	1.19 ± 0.01	1.190 ± 0.001	1.87 ± 0.10	36 ± 1	381 ± 1	414 ± 1
2	1.21 ± 0.01	1.191 ± 0.001	1.88 ± 0.10	41 ± 1	379 ± 1	420 ± 2
4	1.22 ± 0.01	1.192 ± 0.001	1.81 ± 0.10	44 ± 1	382 ± 1	425 ± 2
10	1.22 ± 0.01	1.195 ± 0.001	1.83 ± 0.10	60 ± 1	384 ± 1	428 ± 2

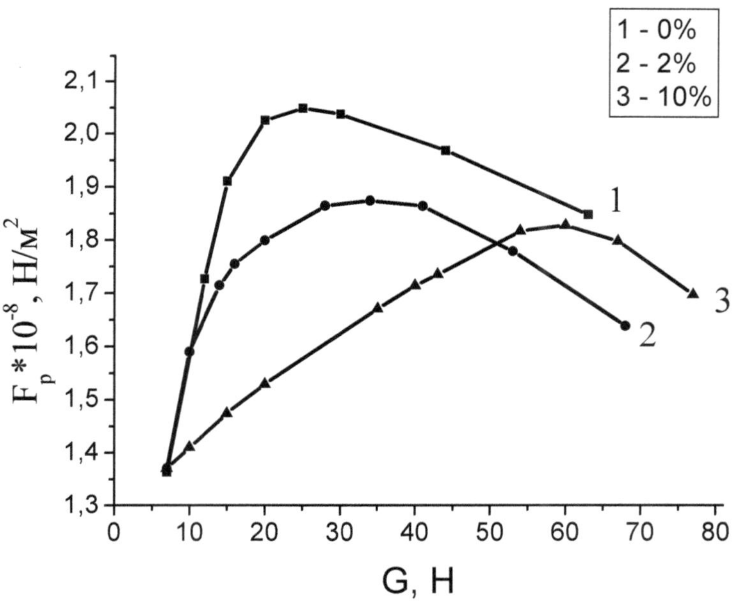

FIGURE 9.15 The dependence of microhardness on the loading for the composites **PMMA–PAn** at different content of **PAn**.

Thermomechanical curves of composites **PMMA–PAn** presented in Figure 9.16 in general have a shape similar to pure **PMMA**. As can be noted from the shape of the thermomechanical curves for these composites, the area of the hyperelasticity is not defined. This gives grounds to assume that the composites, as same as the **PMMA** itself, have a linear structure of the macromolecules.[48]

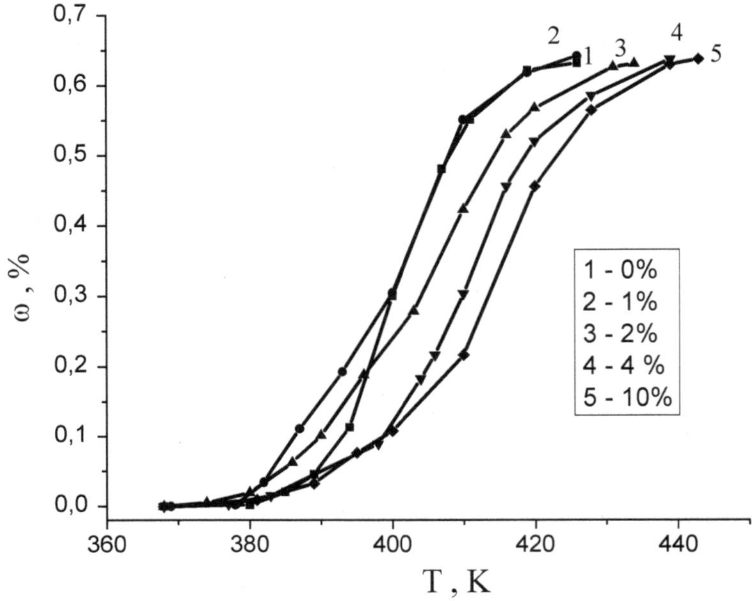

FIGURE 9.16 Dependence of the deformation on temperature for the composites **PMMA–PAn**. The content of **PAn** is indicated on insertion.

Analyzing the obtained data, it can be noted fairly stable values of the glass temperature for the composites **PMMA** (Table 9.10) regardless of the content of **PAn**. At the increasing of the filler's content, the uniform increase of the flow temperature of the composites compared to **PMMA** is observed. The obtained data give the reasons to assume that at temperatures higher than 410 K, where **PMMA** is in a fluid state, a chemical interaction between the two polymers can occur. This interaction may have the character of additional doping, in other words small amounts of **PMA** which are formed during the partial saponification of **PMMA**[45,51] may be alloying agent (or dopant). On the other hand, the presence of such molecules can have the "plasticizing" action on the structure of **PAn** globular coil, turning it into the conformation of the "deployed spiral" as it takes place under the action of camphorosulfonic acid or *m*–cresol[52] on **PAn** and is regarded as the "secondary doping".

According to the results of thermomechanical measurements obtained composites of conjugated **PAAr**s and **PMMA** have the ability to fluidity that makes them suitable for thermoplastic processing and gives reason to use as electroconductive thermoplastic materials for thermal extrusion of products - antistatic protective screens, electronic engineering parts, and so forth.

9.3.4. FLEXIBLE ELEMENTS OF GAS SENSORS BASED ON POLYMER–POLYMERIC COMPOSITES

Using of sensors for monitoring of the environment for the determination of gases in the air or the pH level in water currently is in dire need due to environmental degradation and its impact on human health and safety. At the same time, the existing touch devices that are manufactured using the energy intensive vacuum technologies and expensive semiconductor materials[53] cannot provide the individual needs of each person in a reliable and inexpensive methods of the environment monitoring (water and air) in the “field” conditions, since they require the external sources of energy, cost and complex equipment for the information processing. It is important that the information signal of such sensor could be perceived by human visually (e.g., by color change). In this regard, the sensors based on conjugated polymers with its own electronic conductivity, which can change the optical and electrical properties under the influence of external conditions, namely the temperature,[22] acidity,[54] adsorption of chemicals[55] are promising. There is known the sensitivity of conjugated polymers, in particular, of **PAn** and its derivatives to the action of polar gases (ammonia, nitrogen dioxide, and phosphine) and vapors of organic solvents, namely ethanol, acetone, benzene, and others.[55–58] Unlike to the well-known oxide or ceramic sensors,[55,57] the use of polymeric films does not require of high operating temperatures and the films themselves can be obtained via nonaerated chemical methods using the domestic raw materials including the **PVA**, which is widely used for making of films, it is characterized by high transparency, the flexibility and by the strength.[51] We have studied the regularities of the formation and the properties of flexible elements of optical sensors based on conjugated **PAAr**s (**PAn** and **PoT**) formed in **PVA** matrix for rapid control of the ammonia content in the air.

As it was found in previous studies,[56] thin films of conjugated **PAAr**s obtained by electrochemical deposition on glass plates with a transparent conductive layer **ITO** or SnO_2, easily change their optical spectrum and therefore color under the action of low pressure of ammonia; that’s why it was used to develop the visual indicators of the freshness of the animal products. However, for the practical implementation more accessible and cost indicators must be offered; for this purpose expedient is to obtain the flexible indicator films based on composites of conjugated **PAArs** with elastic polymeric matrix, in particular, with such famous film-former as **PVA**.

An obtaining of the free flexible film which would as composite of conjugated **PAAr** and **PVA** was carried out by oxidative polymerization of 0.01–0.025 M solution of **AAr** in an aqueous gel of **PVA** with the concentration from 0.125 to 5.0 wt.% accordingly to the developed method.[21] The samples have been prepared by the method of watering of the composition on the surface of Teflon or organic glass and by the monolithization of film for 48 h at room temperature and 4 h in a thermostat at 323–333 K. After separation from the substrate, it was obtaining an uniform, flexible green film, which was used for further researches.

As can be seen from the photographs presented in Figure 9.17, the dielectric polymeric matrix and conjugated **PAAr** form an integral composite structure. The comparison of the morphology of "pure" **PAn**'s film obtained by polymerization of aniline on the surface of SnO_2 (Fig. 9.17 (a)) and composite film obtained by polymerization of aniline in the **PVA** matrix (Fig. 9.17 (b)), indicates on a significant impact of the matrix on the structure of film. One can note the presence of certain ordering (self- ordering) in the obtained composite with the formation of nearly regular hexagons built with globules of **PVA**, inside of which there are particles of conjugated **PAAr**.

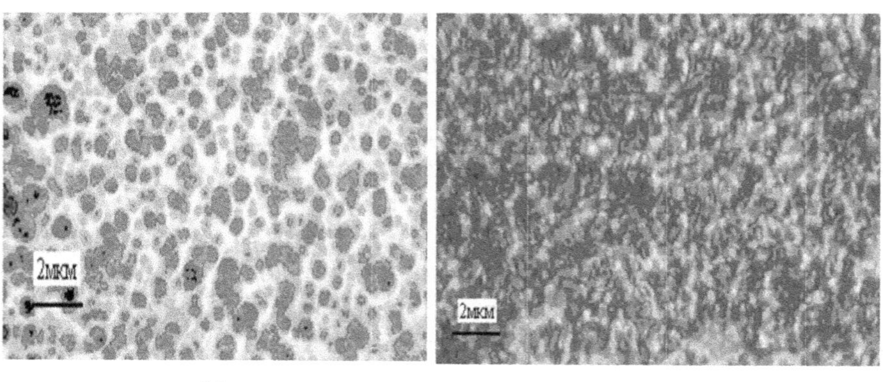

(a) (b)

FIGURE 9.17 Photomicrographies for: (a) **PAn**'s film obtained by electrochemical polymerization of aniline on the surface of SnO_2; (b) composite **PVA–PAn** obtained by oxidizing polymerization of 0.025 M aniline in 2.5% gel of **PVA**. The thickness of films is 0.05 mm (a) and 0.15 mm (b) correspondingly.

Study of thermomechanical properties of the obtained polymeric composites showed that at the formation of **PAAr** in **PVA** matrix there are kept the all termodeformative properties of the polymeric matrix, in particular, the elasticity inherent to **PVA** (Section 9.3.3).

Film composites of **PAArs** in hyperelastic polymeric matrices under the action of polar gases, including ammonia, exhibit the gas-chrome effect, general laws of which are as same as observed for the films of **PAArs**.[56]

The absorption spectra of free films of **PVA–PAAr** (Fig. 9.18) are characterized by the presence of two main bands at 360–390 nm (π–π* transition) and 750–830 nm (an absorption in polaron zone) inherent to conjugated polymeric systems. Under the action of alkaline gas (ammonia) there is a significant change in the optical spectrum, namely, a general increase in the intensity of absorption and the appearance of the band in the range 580–620 nm characteristic for imine–quinoid structure of polymers that is accompanied by visual changes of color of free films.[59,60]

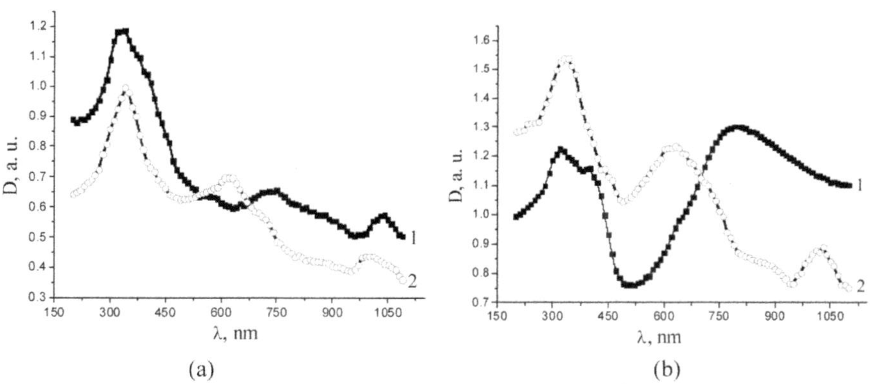

FIGURE 9.18 Optical absorption spectra of free films of composites **PVA–PAn**: *a* - air, *b* - ammonia's atmosphere ($P = 6.4$ kPa). The thickness of films is 0.15 mm (a) and 0.25 mm (b) correspondingly. The content of **PAn** in composite consists of 7.2%.

The gas-chrome effect in composite films is caused by the dedopping process of acid-doped **PAn** with the molecules of ammonia (Fig. 9.19) due to the withdrawal of a proton from a nitrogen atom with the formation of ammonium cation[56] NH_4^+. The process is accompanied by a restructuring of the electronic structure of **PAn** from the conductive green emeraldine salt to the low-conductive intensely blue base.[61]

The proposed sensor of visual control of the ammonia content[60] consists of a polymeric matrix within which there is an indicator substance **(PAAr)**, fixing stand and the defense grid. It's admitted the use of the sticky tape, through which the indicator is attached to the pipeline or walls of the production facilities, or the inside of the packaging product.

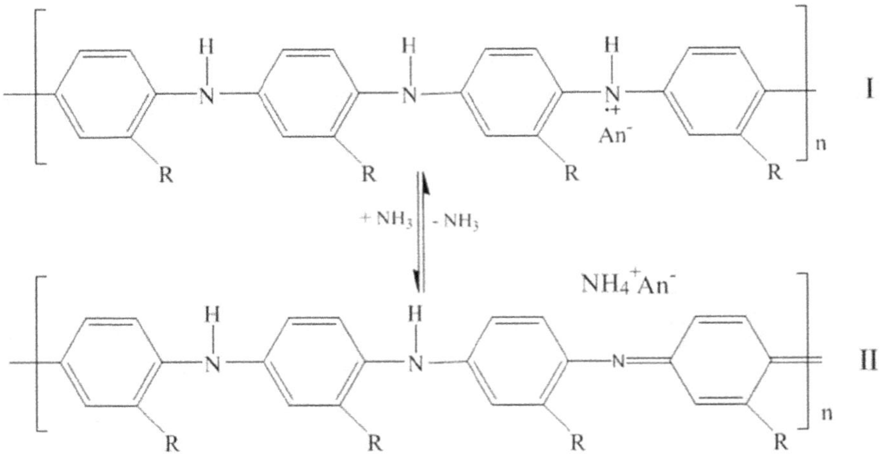

FIGURE 9.19 The structure of elementary link of emeraldine salt of **PAn** (**I**) before the action of ammonia and of the emeraldine base (**II**) formed after the action of ammonia (An^- is alloying anion).

At the starting point, the film has an intense green color and stores it in the absence of ammonia in the air, or throughout the shelf life of the products livestock under condition of the fresh food. In the presence of small amounts of ammonia (partial pressure P = 0.01–0.03 kPa) the color change of the indicator from green to blue is visually observed. At higher concentrations of ammonia, the color of film is considerably weakened to light blue. The speed of the sensor action depending on the nature of the polymeric layer and on the temperature is shown in Table 9.11.

Time of the enter on stationary value of the optical density of composite film is higher than of **PAAr'** one and depends on temperature, pressure of gas, and type of indicator substance embedded into a matrix of **PVA**, but in general is ranged from several seconds to several minutes. After termination of contact with the ammonia medium the back processes are occurring, namely the decomposition of unstable ammonium cation on hydrogen ion and ammonia, the ammonia desorption and recovery of the properties of the film. However, the rate of desorption of ammonia from the composite film is quite slow (from 25–30 min to 2 h) and at higher partial pressures of ammonia (more than 10 kPa) the process is irreversible.

TABLE 9.11 The Results of the Testing of Flexible Sensors with Different Indicated Substances. The Thickness of Free Composition Films is 0.15 mm.

The pressure of ammonia, kPa	Time of color change of sensor, seconds			
	PAn (t = 18–20 °C)	PVA–PAn (t = 18–20 °C)	PoT (t = 0–1 °C)	PVA–PoT (t = 0–1 °C)
1	0.5–1	5	2	12
0.1	2	15–20	5	60
10^{-2}	30–40	190	65	360
10^{-3}	60	300	100	450

The obtained results indicate on the possibility of composite films using as disposable sensors of ammonia, which can be used in personal protective of employees (pipelines and chemical industry), as well as monitoring of freshness of packaged products of animal origin, the deterioration of which is accompanied by evolving of ammonia. The obtaining of a sensitive matter on flexible polymeric carriers allows to simplify the manufacturing technology and significantly to reduce the cost of the sensors, to make them available to every consumer. The developed method of obtaining of flexible color indicators can be used for express control of the ammonia content both in the air and under the packaging of animal products.[60]

9.4 CONCLUSIONS

Based on the complex study both of regularities of obtaining and physicochemical properties of composites of conjugated **PAAr**s and dielectric polymeric matrices, it was found that the presence of electroconductive polymeric fillers in matrices of high polymers essentially influences on the all comlex of the characteristics of the composites. The nature of this influence largely depends both on the type of **PAAr** and on its content; at that, the polymer matrix affects the kinetics of the initial stages of the process of composites formation. On the basis of spectroscopic studies and thermodeformative researches, it was confirmed the intermolecular interaction of polymeric matrixes with electroconductive polymeric filler. It was shown that the dielectric polymer matrix keeps the properties inherent both to high polymers (hyperelasticity and thermoplasticity) and to semiconductivity of conjugated electroconductive polymers. The obtained results were the basis for the development of sensitive gas sensors elements on flexible carriers.

KEYWORDS

- **conjugated polyaminoarenes**
- **dielectric polymer matrix**
- **conductivity**
- **percolation**
- **EPR spectroscopy**
- **thermomechanics**
- **flexible gas sensor**

REFERENCES

1. Heeger, A. J. Semiconducting and Metallic Polymers: The Fourth Generation of Polymeric Materials. *Synth. Met.* **2002,** *123,* 23–42.
2. Aksimentyeva, O. I. Electrochemical Methods of Synthesis and Conductivity of Conjugated Polymers. Svit: L'viv, Ukraine, 1998.
3. Konopelnik, O. I.; Aksimentyeva, O. I.; Tsizh, B. R.; Chokhan, M. I. Physical and Technological Properties of the Sensor Materials Based on Conjugated Polyaminoarenes. *Phys. Chem. Solid State.* **2007,** *8*(4), 786–790.
4. Ćirić–Marjanović, G. Recent Advances in Polyaniline Research: Polymerization Mechanisms, Structural Aspects, Properties and Applications. J. *Synthetic Metals.* **2013,** *177,* 1–47.
5. Molapo, K. M.; Ndangili, P. M.; Peter, M.; Ajayi, R. F.; Mbambisa, G.; Mailu, S. M.; Njomo, N.; Masikini, M.; Baker, P.; Iwuoha, E. I. Electronics of Conjugated Polymers (I): Polyaniline. *Electrochem. Sci.* **2012,** *7,* 11859–11875.
6. Laska, J.; Zak, R.; Pron, F. *Conducting Blends of Polyaniline with Conventional Polymers*; Proceeding of Icsm'96: Praha, 1996; pp 117–118.
7. Grosh, M. A.; Barman, S. K.; Chatteriee, De, S. Low Temperature Electrical Conductivity of Polyaniline – Polyvinyl Alcohol Blends. *Solid State Commun.* **1997,** *103*(11), 629–633.
8. Namazi, H.; Kabiri, R.; Entezami, A. Determination of Extremely Low Percolation Threshold Electroactivity of the Blend Polyvinyl Chloride/Polyaniline Doped with Camphorsulfonic Acid by Cyclic Voltammetry Method. *Eur. Polym. J.* **2002,** *38,* 771–777.
9. Tarkuc, S.; Sahin, E. ; Toppare, L. ; Colak, D. ; Cianga, I. ; Yagci, Y. Synthesis, Characterization and Electrochromic Properties of a Conducting Copolymer of Pyrrole Functionalized Polystyrene with Pyrrole. *Polymer.* **2006,** 47, 2001–2009.
10. Dispenza, C.; Lo Presti, C.; Belfiore, C.; Spadaro, G.; Piazza, S. Electrically Conductive Hydrogel Composites Made of Polyaniline Nanoparticles and Poly (N–Vinyl–2–Pyrrolidone). *Polymer.* **2006,** *47,* 961–971.

11. Ukrainets, A. M.; Aksimentyeva, O. I.; Martiniyk, G. V.; Konopelnyk, O. I.; Evchuk, O. M. Thermomechanical and Electrical Properties of Composites of Conjugated Polyaminoarenes and Polyvinyl Alcohol. *Problems of Chemistry and Chemical Technology.* **2004,** *3,* 132–135.
12. Mirmohseni, A.; Wallace, G. G. Preparation and Characterization of Processable Electroactive Polyaniline–Polyvinyl Alcohol Composite. *Polymer.* **2003,** *44,* 3523–3528.
13. Dubey Rama ; Bag, D. S.; Varadan, V. K.; Lal, D.; Mathur, G. N. Polyaniline Coating on Glass and PMMA Microspheres. *React. Funct. Polym.* **2006,** *66,* 441–445.
14. Aksimentyeva, O. I.; Tsizh, B. R.; Konopelnik, O. I.; Ukrainets, A. M.; Shapovalov, V. A. Metallic Behaviorr of the Conductivity in the Polyaniline – Poly-(Methylmethacrylate) Nanoscale Blends. *Nanosyst. Nanomater. Nanotechnol.* **2007,** *5*(1), 87–92.
15. Angappane, S.; Srinivasan, D.; Rangarajan, G.; Prasad, V.; Subramanyam, S.; Wessling, B. Transport and Magneto-Transport Study on Some Conducting Polyanilines. *Physica B.* **2000,** *284,* 1982–1983.
16. Mamunya, Ye. P. Electrical Properties and Structure of Polymer Composites with Conductive Fillers; Dsci Dissertation: Kyiv, Ukraine. 2003.
17. Pomogaylo, A.; Rozenberg, A.; Ufland, I. Metal Nanoparticles in Polymers. Chemistry: Moscow, Russia, 2000.
18. Hao, X.; Gai, G.; Yang, Y.; Zhang, Y.; Nan, C. W. Development of the Conductive Polymer Matrix Composite with Low Concentration of the Conductive Filler. *Mater. Chem. Phys.* **2008,** *109,* 15–19.
19. Saxena, V.; Malhotra, B. D. Prospects of Conducting Polymers in Molecular Electronics. *Curr. Appl.Phys.* **2003,** *3,* 293–305.
20. Ukrainets, A. M.; Melnyk, G. M.; Evchuk, O. M.; Aksimentyeva, O. I. Physical and Mechanical Properties of the Polybutylmethacrylate and Polyaniline Composites. *"Scientific Notes" of Ternopil National Pedagogical Univ. Series: Chemistry;* Ternopil National Pedagogical University: Ternopil,Ukraine, 2009; Vol. 15, pp 64–67.
21. Aksimentyeva, O. I.; Ukrainets, A. M.; Konopelnyk, O. I.; Evchuk, O. M. Process for the Preparation of Conductive Polymer Composites. Ua Patent No 53159a.
22. Konopelnyk, O. I.; Aksimentyeva, O. I.; Martyniuk, G. V. Effect of Temperature on the Optical Properties of Conducting Polyaminoarenes and their Composites with Elastic Polymer Matrix. *Molec. Cryst. Liq. Cryst.* **2005,** *427,* 37–46.
23. Martyniuk, G. V.; Aksimentyeva, O. I.; Konopelnyk O. I.; Poliovyi, D. Electrochemical Synthesis and Optical Properties of Conjugated Polyaminoarenes and Polymethylmethacrylate Composites Bulletin of L'viv University, *Chem.* **2010,** *51,* 366–371.
24. Aksimentyeva O. I.; Grytsiv, M. Ya.; Konopelnik, O. I. Temperature Dependence of Resistance and Thermal Stability of Doped Polyaniline. *Funct. Mater.* **2002,** *9*(2), 251–254.
25. Martyniuk, G. V. Physical and Chemical Properties of Composites of Conjugated Polyaminoarenes and Dielectric Polymer Matrices. Phd Dissertation: Lviv, Ukraine, 2008.
26. Zakordonskiy, V. P.; Skladanyuk, R. V. Thermomechanical Properties and Structure of Epoksyamin Grids Formed in the Presence of Colloidal Silica. Bulletin of L'viv University, *Chem.* **2000,** *39,* 304–310. (In Ukrainian).
27. Liu, T.; Christian, B.; Chu, B. Nanofabrication in Polymer Matrixes. *Prog. Polym. Sci.* **2003,** *28,* 5–26.
28. Kabanov, V. A.; Papisov, I. M. Complex Formation between the Complementary Polymers and Oligomers in Dilute Solutions. *Macromol. Compounds. A.* **1979,** *21*(2), 243–281 (In Russian).

29. Aksimentyeva, O. I.; Boyko, V. V.; Gavenko, S. F.; Evchuk, O. M. Template Synthesis of Nano Dispersions Conductive Based on Polymer Electrolyte and Polyaniline. Ukrainian Xii Conference of Macromolecular Compounds, Kyiv, Ukraine, 2010.
30. Aksimentyeva, O. I.; Martyniuk, G. V.; Martyniuk, I. V.; Skoreyko, N. T. The Study of the Kinetics of Oxidative Polymerization of Aniline in Solution of Polyvinyl Alcohol. *"Scientific Notes" Of Ternopil National Pedagogical Univ. Series: Chemistry*; Ternopil National Pedagogical University:Ternopil,Ukraine, 2013; Vol. 17, 64–67.
31. Somani, P. R. Synthesis and Characterization of Polyaniline Dispersion. *Mater. Chem. Phys.* **2002,** *77,* 81–85.
32. Sverdlova, O. V. Electronic Spectra in Organic Chemistry. Chemistry: Leningrad, Russia. 1985. 33. Smirnov, M. A.; Bobrova, N. V.; Dmitrie, I. Yu.; Bukolšek, V.; Elyashevich, G. K. Electroactive Hydrogels Based On Poly(Acrylic Acid) And Polypyrrole. *Poly. Sci. Ser. A.* **2011,** *53*(1), 67–74.
34. Hoa, C. H.; Liub, C. D.; Hsieha, C. H.; Hsiehb, K. H.; Lee, S. N. High Dielectric Constant Polyaniline/Poly(Acrylic Acid) Composites Prepared by in Situ Polymerization. *Synth. Met.* **2008,** *158,* 630–637.
35. Evchuk, O.; Aksimentyeva, O. I.; Horbenko, Y. Optical and Electrical Properties of Composites of Conjugated Polyaminoarenes and Polymeric Electrolytes. Bulletin of L'viv University, *Chem.* **2012,** *53,* 352–356 (In Ukrainian).
36. Tarasevich, Y. Y. *Percolation Theory, Applications, Algorithms*; Chemistry: Moscow, 2002.
37. Shklovski, B. I.; Efros, A. L. Percolation Theory and Conductivity of Strongly Inhomogeneous Media. *Ufn.* **1975,** *117*(3), 401–436 (In Russian).
38. Aneli, J.; Zaikov, G.; Mukbaniani, O. Physical Principles of the Conductivity of Electrical Conducting Polymer Composites (Review). Ch & Ch T. **2011,** *5*(1), 75–87.
39. Aksimentyeva, O. I.; Konopelnyk, O. I.; Martiniyk, G. V.; Yurkiv, V. V.; Shapovalov, V. A. Percolation Phenomena and Spin Dynamics in Pani–Pmma Blends. *Molec. Cryst. Liq. Cryst.* **2007,** *468,* 309–316.
40. Aksimentyeva, O. I.; Konopelnik, O. I.; Ukrainets, A. M.; Grytsiv, M. Ya.; Martyniuk, G. V. Anisotropy of Conductivity and Percolation Phenomenon in Film Composites of Polyaminoarenes and Polyvinyl Alcohol. J. *Phys. Chem. Solids.* **2004,** *5*(1), 142–146 (In Ukrainian).
41. Yurkiv, V.; Aksimentyeva O.; Ukrainets, A.; Martyniuk, G.; Opaynych I. The Electrical Properties of Polyaniline Composites and PMMA. Bulletin of L'viv University, Ukraine, *Chem.* **2006,** *47,* 307–311.
42. Coleman, J. N.; Curran, S.; Dalton, A. B.; Davey, A. P.; Mccarthy, B.; Balu, W. J. Percolation-Dominated Conductivity in a Conjugated Polymer–Carbon Nanotube Composite. *Phys. Rev. B.* **1998,** *58*(12), 7492–7498.
43. Kohlman, R. S.; Joo, J.; Epstein, A. J. Conducting Polymers: Electrical Conductivity. *Physical Properties of Polymers Handbook;* Mark, J. E., Eds.; Amer. Inst. Phys, Woodbury: New York, 1996; pp 453–478.
44. Prigodin, V. N.; Epstein, A. J. Nature of Insulator-Metal Transition and Novel Mechanism of Charge Transport in the Metallic State of Highly Doped Electronic Polymers. *Synth. Met.* **2002,** *125,* 43–53.
45. Srinivasan, D.; Natarajan, T. S.; Bhat, S. V.; Wessling, B. Electron Spin Resonance Absorption in Organic Metal Polyaniline and its Blend with PMMA. *Solid State Comm.* **1999,** *18,* 503–508.

46. Aksimentyeva, O. I. ; Konopelnyk, O. I.; Opaynych, I.; Tzish, B.; Ukrainets, A.; Ulansky, Y. Interaction of Components and Conductivity in Polyaniline–Polymethylmeth Acrylate Nanocomposites. *Rev. Adv. Mater. Sci.* **2010,** *23,* 30–34.
47. Etkins, P. *Physical Chemistry;* Mir: Moscow, Russia, **1980,** Vol. 2.
48. Tager, A. A. *Physical Chemistry of Polymers;* Chemistry: Moscow, Russia., 2012.
49. Lipatov, Y. S. *Physical Chemistry of Filled Polymers;* Chemistry: Moscow, Russia, 1977.
50. Makela, T.; Hactainen, T.; Ahopelto, J.; Isotalo, H. Imprinted Electrically Conductive Polyaniline Blends. *Synth. Met.* **2001,** *121,* 1309–1310.
51. Encyclopedia of Polymers. Soviet Encyclopedia: Moscow, Russia, 1972–1977.
52. Min, Y.; Xia, Y.; Mac.Diarmid, A. G.; Epstein, A. J. Vapor Phase Secondary Doping of Polyaniline. *Synth. Met.* **1995,** *69,* 159–160.
53. Dorozhkin, L. M.; Rozanov, I. A. Chemical Gas Sensors in the Environment Diagnosis. *Sensor.* **2001,** *2,* 2–9.
54. Aksimentyeva, O. I.; Hlushyk, I. P.; Stakhira, P. Y.; Martyniuk, G. V.; Fechan, A. V. New Ph Optical Sensor for Medicine and Environment Monitoring Application Based on the Thin Conducting Polymer Films //Ix Polish–Ukrainian Symposium "*Theoretical and Experimental Studies of Interfacial Phenomena and their Technological Application*" 5–9 September, **2005,** Sandomerz, Poland. Proc. pp 1.
55. Wilson, S. A.; Jourdain, R. P.; Zhang, Q.; Dorey, R. A.; Bowen, C. R.; Willander, M.;Qamar Ul Wahab, Al-hilli, S. M.; Nur, O.; Quandt, E.; Johansson, C.; Pagounis, E.; Kohl, M.; Matovic, J.; Samel, B.; Wijngaart, D. W. van.; Edwin Jager; Carlsson, D.; Djinovic, Z.; Wegener, M.; Moldovan, C.; Abad, E.; Wendlandt, M.; Rusu, C.; Persson, K. *New Materials for Micro-Scale Sensors and Actuators. An Engineering Review. Materials Science and Engineering R: Reports*; Linköping University Electronic Press: Ostergotland, 2007; Vol. 56 (1–6), pp 1–129.
56. Tsizh, B. R.; Chokhan, M. I.; Aksimentyeva, O. I.; Konopelnyk, O. I.; Poliovyi, D. O. Sensors Based on Conducting Polyaminoarenes to Control the Animal Food Freshness. *Mol. Cryst. Liq. Cryst.* **2008,** *497,* 586–592.
57. Tsizh, B. R.; Aksimentyeva, O. I.; Chokhan, M. I.; Lazorenko, V. Y. Structure and Gas Sensitivity of the Zno Sensor of Ethanol. *Solid State Phenomena.* **2013,** *200,* 305–310.
58. Choudhury, A. Polyaniline/Silver Nanocomposites: Dielectric Properties and Ethanol Vapour Sensitivity. *Sens. andActuators B.* **2009,** *138,* 318–325.
59. Aksimentyeva, O. I.; Konopelnyk, O. I.; Tsizh, B. R.; Evchuk, O. M.; Chokhan, M. I. Flexible Elements of Optical Sensors Based on Conjugated Polymer Systems. *Sens. Electron.* and*Microsyst.Technol.* **2011,** *2*(8), 34–39 (In Ukranian).
60. Aksimentyeva, O. I.; Tsizh, B. R.; Chokhan, M. I.; Evchuk, O. M. Sensor of Visual Control of Ammonia Content. Ua Patent No 65401, 2011.
61. Tsizh, B. R.; Aksimentyeva, O. I.; Vertsimakha, Ya. I.; Lutsyk, P.; Chokhan, M. I. Effect of Ammonia on Optical Absorption of Polyaniline Films. *Mol. Cryst. Liq. Cryst.* **2014,** *589,* 116–123.

CHAPTER 10

POLYANILINES: GENERATION OF ELECTROCHEMILUMINESCENCE

O. V. RESHETNYAK

Department of Physical and Colloid Chemistry, Faculty of Chemistry, Ivan Franko National University of Lviv, 6 Kyryla & Mefodia Str., Lviv 79005, Ukraine. E-mail: reshetniak@franko.lviv.ua

CONTENTS

ABSTRACT

The sources of the electrochemiluminescence (**ECL**) during the electrochemical synthesis of polyaniline (**PAn**) and copolymers of aniline (**An**) and luminol (**Lum**) by different composition at high electrode potentials have been investigated and analyzed. It was proposed that the generation of low intensity of luminosity in acidic aqueous solutions at the potentials over + (1.3–1.6) V (respectively to the silver/silver chloride saturated electrode) is related with irreversible destruction of the **PAns** in the form of pernigraniline as a result of their interaction with free radical intermediates of oxidation of solvent (water) and anions of the base electrolyte. It was found, that in a case of **An** and **Lum** copolymerization the optimal conditions for obtaining of copolymers with fluorescent properties is the using of aqueous–dimethylsulfoxide (**DMSO**) medium at volumetric ratio of the solvents 9:1. It was shown that obtained by varying the ratio of monomers in the initial polymerization mixture copolymers exhibit the electrochemiluminescent activity in aqueous alkaline solutions ($pH = 10.5$). The feature of **ECL** of copolymers is the presence of two waves of the appearance of luminosity, namely, in the intervals of the potentials + (0.4–1.0) V and + (1.0–2.5) V. It was suggested, that the source of the ECL is a direct electrochemical oxidation (the first wave) and chemical oxidation by free radicals of electrochemically generated intermediates (so-called, the second wave) of luminolic fragments of copolymeric chain.

10.1 INTRODUCTION

Electrochemiluminescence (or electrochemically generated luminescence (**ECL**)) is the phenomenon of electromagnetic radiation in the visible or ultraviolet region of the spectrum, induced by electrochemical transformations. The source of the radiation is the ray deactivation of particles in excited triplet or singlet states, which, in turn, are formed on the electrodes or in the near-electrode space as a result of electrochemically initiated reactions.

At the initial stage of the researches the interest to the **ECL** was caused by declared possibility to use of this phenomenon in display technologies,[1] for electrochemical pumping of the lasers,[2–4] for the study of kinetics of chemical processes, in particular polymerizing ones[5], etc. However, afterwards it appeared that the most promising, in view of practical application, is the analytical aspect of this method of the investigations.[7–10] Today developed **ECL** methods of quantitative analysis not only of inorganic ions, or fairly simple organic substances, but also of pharmaceuticals, explosives, alkaloids, amino-acids, etc. Electrochemically generated luminescence is underlying in electro-optical chemo- and biosensors.[11–14] Especial interest is the biological aspect of **ECL** use, in particular at the analysis of DNA hybridization[15–16] as well as at the immunological analysis. Using of the **ECL** labels for biological objects, such as an antigen or antibody[17–19] gives the possibilities to develop very sensitive and, most importantly, fully automated methods of analysis of the bioassays.

ECL is characterized by significant advantages over classical chemiluminescence (**CL**), among which three are the most essential. Firstly, short-lived intermediates (usually free or charged (anion- or cation-) radicals) that are involved in the formation of luminescence-active product are generated electrochemically directly in the reactive medium, and therefore the intensity of luminescence less is limited by a small lifetime of the intermediates. The second is the process of **ECL** flow is controlled by the potential of electrode, which provides the possibility to carry out the selective determination of the components of the solution without previous separation of the investigated mixture. Thirdly, the emitters of radiation are concentrated near the surface of the electrode, and are not scattered throughout the volume of the solution as in the case of **CL**. This gives the possibility at properly selected optical scheme of **ECL**-installation to fix considerably weaker light flows, increasing the sensitivity of the measurements.

The localization of a process into adsorbed on the surface of the electrode layer determines the impact of the surface excess of electroactive components and state of the electrode surface on the intensity of **ECL**. Moreover, as in any heterogeneous process, the first stage of the electrochemical transformation is the adsorption of electroactive particles on the electrode surface. The size and type of the adsorption equilibrium determine not only the directions but also the features of the following transformations. In particular, between the ability of molecules of organic substances to be adsorbed on electrode surface and the intensity of **ECL** there is a relationship.[1] Therefore the use of **ECL** allows to analyze not only the composition of the solutions (homogeneous **ECL**-analysis), but also the state of the electrode surface (heterogeneous **ECL**-analysis)[9] and the speed of adsorption–desorption balance attainment on its surface.

The most famous and the most studied electrochemiluminescent luminophors are metal-organic complexes or clusters, containing the ions of Ag, Al, Au, Cd, Cr, Cu, Eu, Hg, Ir, Mo, W, Os, Pd, Pt, Re, Ru, Tb, or Tl and also polycyclic aromatic compounds that have the significant fluorescence. The mechanism implemented in this case at the radiation appearance is an annihilation of cationic and anion radicals, formed at the cathode and anode electrode polarization pulses, respectively. The features of **ECL** in such systems fairly well studied, particularly the overviews[7–8,10,20–22] devoted to them. At the same time, the electrochemical generation of luminescence is possible also in galvano- or potentiostatic conditions, where the radiation emitter is formed mainly in accordance with the recombination mechanism in result of the interaction of electrochemically generated free radicals. Peroxydisulfates, luminol (**Lum**), lucigenin, acridone,[23–24]etc. belong to such **ECL**-systems which are becoming increasingly practical importance.

ECL is also observed for polymers[25–29], which today is widely used in relevant technologies such as light-emitting electrochemical cells (**LECs**)[25,30–33] which are alternative to well-known light-emitting diodes (**LEDs**), fast-response displays,[34] sensors,[35–36] chemical imaging,[37] etc. As same as in a case of low molecular organic compounds, the generation of **ECL** of polymers is possible at the formation of a conjugated network of p-electrons in a polymer chain. Such semiconductor polymers the most often contain carbazole, fluorene, *p*-phenylene, phenylenevinylene, and thiophene and other links in the π-conjugated polymeric chain[26] and require the use of alternating current for generation of **ECL**. At the same time we have shown that in the case of polymeric anilines possible generation of luminosity can be observed also at the use of direct current electrolysis mode.

10.2 EXPERIMENTAL DETAILS

To register the voltammetric characteristics in potentiodynamic conditions (speed sweep potential $s_E = 20$ and 50 mVs^{-1}) the potentiostate/galvanostate *PI*–50–1 controlled by computer has been used. The registration of the results took place automatically with the frequency of the record of 25 points/1 s. Polished platinum discs with the area of 1 cm^2 were used as the working and auxiliary electrodes. The all values of the potentials in presented paper are given regarding the saturated Ag/AgCl electrode, which served as the reference electrode. H_2SO_4 or HCl were used as base electrolyte.

During the ECL investigations the unit for the study of weak radiation was applied.[38] Potential at the time of **ECL** investigations under potentiodynamic conditions was scanned with the speed of 5 mVs^{-1} in the range from (−0.2) to –(+3.0 V or in the range from (0.0) to (+2.8) depending on the studied system.

For the luminescent analysis the **Lum** ("*Fluka*", ≥ 98%) was used as received. An aniline (**An**) ("*Sigma–Aldrich*", > 95%) was previously distilled under vacuum. To prepare the working solutions the twice-distilled water was used. The remaining reagents (analytical grade) were used without further purification. Both the electrochemical synthesis of polymeric samples and **ECL**-study was conducted after 15 min bubbling of prepared solutions with argon to remove the dissolved oxygen.

Electrochemical synthesis of thin layers of polyluminol and its copolymer with **An** on the surface of the working platinum electrode was carried out in potentiostatic mode at +1.2 V for 2, 5, 10, and 15 min from 0.5 M aqueous or 0.05 M aqueous-dimethylsulfoxide (**DMSO**) (volume ratio 9:1) solutions of H_2SO_4. The concentration of **Lum** in the initial solution was 0.001 or 0.01 M (respectively in water and aqueous-**DMSO** media), the concentration of **An** – 0.001; 0.01; 0.1, and 0.25 M. After the synthesis, the electrode with coated polymeric (copolymeric) layer was repeatedly washed with distilled water and for study of its **ECL** properties was transferred into alkaline solution with $pH = 10.5$. Constant acidity of the medium was supported using the borate (H_3BO_3–Na_3BO_3) buffer solution.[39]

10.3 GENERATION OF LUMINESCENCE DURING ELECTROCHEMICAL DEGRADATION OF POLYANILINE

As it is know, the electrochemical synthesis of polyaniline (**PAn**) on platinum electrode is occurred at the potentials of working electrode in about 0.8–1.0 V regarding the saturated Ag/AgCl electrode.[40–41] In order to determine the behavior of **PAn** electrode at high values of electrode potential, which in particular may arise at the charging of **PAn** secondary chemical sources of electricity, we carried out the cyclic voltammetric investigations of sulfuric- and hydrochloric aqueous solutions of **An** in a much broader range of potentials such as 0.0–2.8 V on the stationary platinum electrode.[42]

Preliminary results of cyclic voltammetric studies have shown that in the potential range of 0.0–1.3 V in the medium both of sulfuric and hydrochloric acids, at the cyclic voltammograms there are peaks corresponding to **An** oxidation and to oxidation–reduction of its polymeric forms. In the range of the potentials 1.3–1.6 V there is observed a sharp increase of the oxidation currents (see Fig. 10.1). The results of comparison of the cyclic voltammograms of aqueous solutions of H_2SO_4 and HCl in the presence (see Fig. 10.1) and in absence (see Fig. 10.2) of **An** suggest that this fact is related with the oxidation of base electrolyte. Proof of this is the presence of two waves of oxidation, namely at 1.1 and ≈ 1.6 V at the voltammogram of HCl solution (see Fig. 10.2, curve 1) corresponding to oxidation of Cl^- anions and water, respectively, while for the sulfuric acid solution it was fixed only decomposition of water (see Fig. 10.2, curve 2).

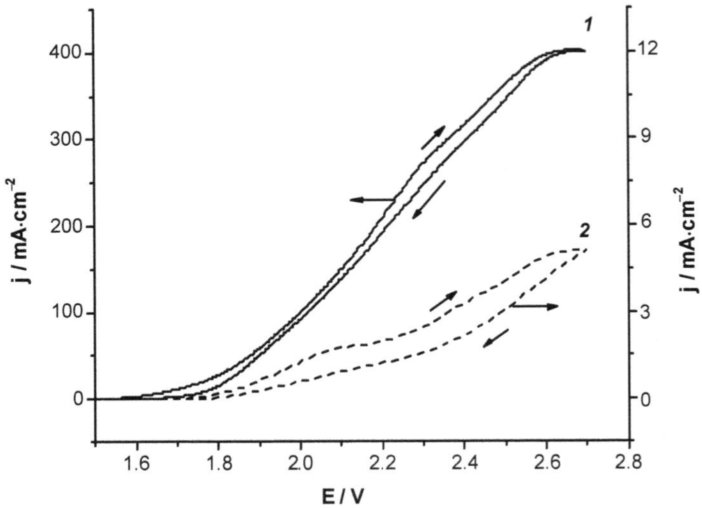

FIGURE 10.1 Cyclic voltammograms of platinum working electrode in 0.1 M acidic aqueous solution of aniline during the first scanning cycle of the potential in the range of the potentials +(1.5–2.8) V. Base electrolyte: 0.5 M H_2SO_4 (1) and 1.0 M HCl (2); s_E = 20 mVs^{-1}.

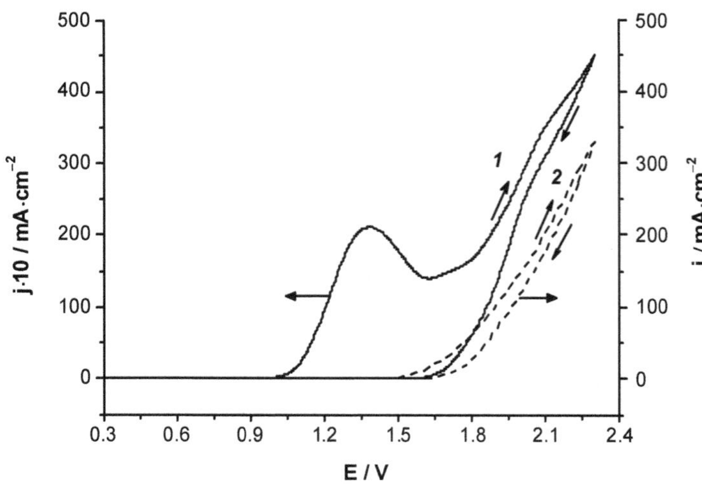

FIGURE 10.2 Polarization curves of platinum working electrode in aqueous solutions (1) HCl (1.0 M) and (2) H_2SO_4 (0.5 M); s_E = 20 mVs^{-1}.

It was found that at the potentials more than +1.2 V the polymerization of **An** is occurred mainly in the volume of the electrolyte with the formation of the oxidized form of polyaniline – pernigraniline. However, the main feature of this process is weak ECL which in potentiodynamic conditions was recorded at the potentials of working electrode ≥ 16 V (see Fig. 10.3). In the case of oxidation of aqueous solutions of base electrolytes (HCl and H_2SO_4) without **An** the electrochemiluminescence is not observed. It was found that the intensity of the **ECL** in sulfuric acid medium is somewhat higher than in the hydrochloric solutions, so the all subsequent studies were conducted only in the medium of H_2SO_4.

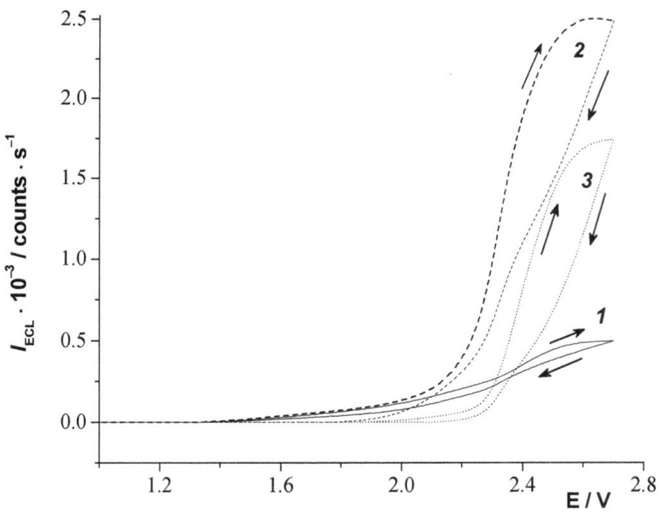

FIGURE 10.3 Dependence of the intensity of electrochemiluminescence (I_{ECL}) at the Pt electrode in 0.5 M solution of H_2SO_4 on the applied potential at the concentration of aniline, M: $1 \cdot 10^{-3}$ (1); $1 \cdot 10^{-2}$ (2); $1 \cdot 10^{-1}$ (3); $s_E = 5$ mVs^{-1}.

The phenomenon of weak ECL during the electrochemical polymerization of **An** can be related with the process of the polymer chain growth since it occurs at the potentials of working electrode that significantly exceed the value corresponding to the formation of polymer.[40–41] Moreover, it was found that in the presence of reduced forms of **PAn** (leucoemeraldine and emeraldine) at the surface of electrode the radiation also is not observed, because their primary oxidation during the potential scanning takes place. Proof of this is the change in color of **PAn** synthesized at different potentials from dark green to brown and black with the increase of the electrode potential.[43–44] Therefore, a possible source of **ECL** can be irreversible degradation (overoxidation) of **PAn**. Since the gap of polymeric chain is possible in any place that is with the formation of the products of different molecular weight, it explains the low intensity of the ECL. As it is known, in a case of the formation of excited particles with high molecular weight, the likelihood of their decontamination without radiation is increased. This means, that for the generation of radiation is necessary the formation of active particles (radicals or contrasts (CRs)) with low molecular weight (perhaps, even of monomeric nature). Their following interaction between themselves, in particular dimerization, can lead to the formation of luminescent active products in an excited state, and accordingly, to the radiation of the photon of light due to the ray decontamination.

Participation of **PAn** in the processes leading to the generation of luminescence is proved by **ECL**-investigations in potentiostatic conditions at $E = +2.0$ V with the use of the electrode on which the various amounts of **PAn** (form of the pernigraniline salt) in the form of the thin films have been previously electrochemically deposited. For this purpose, the electrode was previously polarized in 0.1 M solution of **An** for 1–10 min at the potential +1.0 V, that is corresponding to the formation of the pernigraniline at the electrode surface. It was found, that in the presence of the polymer at the surface of electrode both the intensity and duration of **ECL** are increased, that is the total luminosity ($Q_{ECL} = \int I_{ECL}\, dt$), compared with the case when the previous polarization was not performed (see Fig. 10.4).

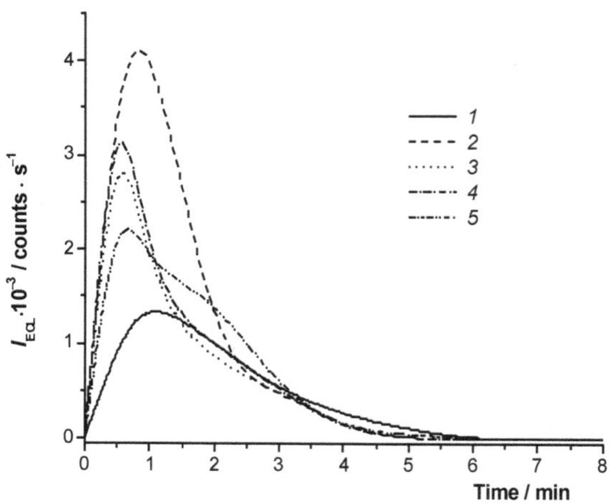

FIGURE 10.4 The dependence of the total radiation Q_{ECL} during the polarization of platinum electrode in a solution of 0.1 M **An** on the applied potential (1) and on the duration of the previous polymerization of **An** on the surface of the electrode (2) (the value of potential during the previous polymerization and during the study of luminescence: +1.0 and +2.0 V, respectively; the base electrolyte: 0.5 M H_2SO_4).

It is well known, that the oxidative degradation of **PAn** takes place at the potentials +(0.8–0.9) V as a result of nucleophilic attack of cation-radical centers in the polymer chain (emeraldine form)[45] by the molecules of solvent.

Therefore, at the potentials corresponding to the emergence of **ECL** (over +1.6 V) the formation of a luminescence active products is possible only due to the degradation of **PAn** in the form of pernigraniline under the action of high-reactive intermediates generated in the system in parallel with the electrochemical oxidation of **PAn**, namely due to the oxidation of chloride-anions (in particular, $Cl^{\bullet}$–radicals) or water (for example, $HO^{\bullet}$–radicals). One of the main products of electrochemical degradation of **PAn** in aqueous solutions is a *p*-benzoquinone, for which as is well known[46–47] the phosphorescence in the *UV–vis* spectral region is characteristic. So, obviously exactly its ray decontamination at the transition from the excited into the main electronic state is by the source of the **ECL** under such conditions. The formation of this radiation emitter in system during the polymerization of **An** (the presence of monomeric molecules in polymerization solution) is also quite possible as a result of recombination of CRs of **An**- intermediates at the formation of **PAn**[41] primarily with the hydroxyl radicals, which are also generated at the electrode under the same conditions. Making an analogy between electrochemical[48] and chemical destruction of **PAns**[49] also can be suggested that the luminescence active products in this system can also be by other potential products of **PAn** degradation, namely dimers of CRs $C_6H_5NH_2^{+\bullet}$, *p*-diaminobenzene, quinoneimines, quinondiimine, etc.[50] which are the low molecular weight substances, in which a system of conjugated p-bonds exists. At the same time accurately to determine the nature of the radiation emitter was not failed because of too low intensity of generated luminescence and the resulting inability to examine its spectral composition.

According to the results shown in Figure 10.2, in the medium of H_2SO_4 the density of currents of oxidation is more than on one order exceeds the density of currents in hydrochloric acid. This fact explains the higher values of **ECL** intensity in sulfuric acid solutions since the number of active intermediates decomposition of water ($HO\bullet$–radicals, H_2O_2, singlet molecular oxygen, etc.) is growing and, correspondingly the number of luminescent-active particles formed by their interaction with **PAn** or low-intermediate oxidation products of the initial monomer is increased also. In a case of the previous deposition of **PAn**'s films on the surface of the working electrode it was established that the maximum of the radiation intensity can be traced at the electrode, where the duration of the previous polarization was 3 min. With further increase of the duration of preliminary polymerization both the intensity of **ECL** and its duration are decreased. In particular, the duration of **ECL** after 3 min of application was about 8 min, whereas after 7 min – only 3.5 min. At the same time, the maximum value of the total luminosity was recorded at 5 min application of **PAn** (see Fig. 10.4, curve 2). Since the amount of polymer at the electrode surface is increased, then the likelihood of electro-active particles formation in volume of polymer and their participation mainly in the cross-linking processes of polymer chains are increased via time leading to the luminescence quenching. Proof of this is much better adhesion of the polymer to platinum after polarization at high values of the electrode potential.

In potentiostatic conditions the ECL was recorded already at the potentials over +1.3 V. This difference in the value of the potential, at which the luminescence is appeared compared to potentiodynamic conditions, is explained by too low intensity of **ECL** and relatively high potential rate sweep (inertness of system). The change of **ECL** intensity during time in potentiostatic conditions (see Fig. 10.5) is classical: the intensity of luminescence is increased during the first minute of electrolysis, then gradually is decreased accordingly to exponential dependence. This kind of dependence is related with a sharp decrease of quantity of electroactive reagent (**An**) in the near-electrode layer due to its low concentration in the initial solution and through partial blocking of the surface of electrode synthesized by **PAn** that was recorded visually. The maximal intensity of radiation was fixed at 2.0 V, while the maximum of total luminosity (total number of photons that are emitted during the electrolysis) at the potential 1.5 V (see Fig. 10.4, curve 1).

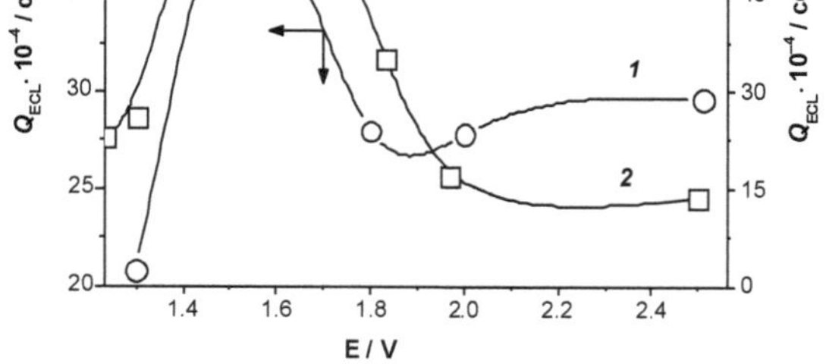

FIGURE 10.5 The change via time of the electrochemiluminescence intensity I_{ECL} during the polarization of platinum electrode in 0.1 M solution of **An** at the applied potential, V: +1.3 (1); +1.5 (2); 1.8 (3); +2.0 (4); 2.5 (5); base electrolyte: 0.5 M H_2SO_4.

This experimental fact can be explained by the **ECL** intensity increasing with the increase of the potential since the density of polarization current is increased and, consequently, the rate of the luminescence active products of electrolysis accumulation is increased also. However, in the case of even higher potentials of electrode the likelihood of their participation in polycondensation reactions is increased resulting in reduction both of the radiation intensity and total quantity of emitted photons. This explains also the fact that the maximal intensity of the luminescence was recorded at the concentration of **An** 1×10^{-2} M (see Fig. 10.3). With the increase or decrease of the concentration of **An** on one order, the radiation intensity is decreased, though, as the results of the cyclic voltammetry evidenced the oxidation currents are practically independent on the concentration of **An**.

10.4 ELECTROCHEMILUMINESCENCE OF THE COPOLYMERS OF ANILINE AND LUMINOL

5-Amino-2,3-dihydrophthalazine-1,4-dione (or luminol) due to the presence in its structure an aromatic cycle bounded to the amine group can also be attributed to **An**'s derivatives. **Lum** is the most well-known and probably the most commonly used luminescent reagent. For the first time the chemiluminescence during the oxidation of **Lum** was recorded by *Albrecht*[51] in 1928, and a year later *Harvey*[52] reported about generating of ECL of **Lum** in alkaline solutions at the anode potential 2.8 V. Today chemiluminescence[53–55] and ECL[56–61] of **Lum** is the subject of numerous publications due to primarily analytical aspects of this phenomenon application which are now shifting toward the analysis of objects of biological origin. The feature of **Lum** is that it also creates the polymer[62–66] and electroconductive (with the system of conjugated p-bonds) copolymers of **Lum** with other monomers, including the **An**[67,68], which also have the luminescent properties. It enables effectively to use thin films of such polymeric materials in (electro)chemiluminescent sensor devices.[69–73] Copolymer with luminescent polyluminol fragments (links) can simultaneously play a role not only sensitive layer, but also a platform for the application of biologically active substances in the design of biosensors. For example, in a case of modification of electrode surface by polymeric **Lum** such platform was used for chemi- or electrochemiluminescent definition of flavin,[62] ephedrine[64] and H_2O_2,[74] and in combination with another layer (namely, poly(Fe(II)*tris*(5-aminephenanthroline) also glucose.[65]

However, obtaining the significant quantities of polyluminol and its copolymers is restricted primarily by low solubility of **Lum** in acidic aqueous solution. This problem is eliminated by the use of water–organic solvents[66] including the **DMSO**. That is why we have investigated the features of electrochemical synthesis of copolymers of **An** with **Lum** as well as electrochemiluminescent activity of obtained products.[75]

To determine the conditions of synthesis of the copolymer of **Lum** and **An** we have conducted the cyclic researches in the volt-range potentials (−0.2)–(+1.2) V in 0.5 M solution of H_2SO_4. The concentration of **Lum** was 5×10^{-4} M, the concentration of **An** was varied in the ranges from 1×10^{-2} to 5×10^{-2} M. As shown in Figure 10.6, and on a cyclical voltammogram of the platinum electrode in an acidic aqueous solution of **Lum**, two peaks of oxidation are observed, namely at +0.65 and ~ (+0.95) V. Since only the first process is reversible, this means that the oxidation of polymer which is formed at the electrode, corresponds just to this peak whereas the maximum at +0.95 is related with the oxidation of initial monomer. Introduction of the **An** into the reaction mixture (see Fig. 10.6(b)) leads to the appearance on the curves of cyclical voltammograms another two peaks of oxidation. Additional studies have shown, that in a case of the sweep potential within −0.2–+0.9 V the cyclical voltammogram was obtained, which is almost identical as in the absence of **An**. Therefore, we can say, that the oxidation of monomeric **Lum** is occurred at lower potentials compared to the **An**. In this case, irreversible oxidation at $E = +1.1$ V corresponds to oxidation of **An** and reversible peak at ~ (+0.3) V is related with the oxidation of **PAn** units in the copolymer, which is formed on the electrode (see Fig. 10.5(b)). Note, that the introduction of **An** leads to a shift of the oxidation/reduction potentials of polyluminol fragments in the area of smaller potentials. For example, if into solution of **Lum** the current maxima of oxidation/reduction were observed at the potentials +(0.65/0.58) V (see Fig. 10.6(a)), then in the presence of **An** at the concentration of 0.01 M these values were as follows: +(0.58/0.50) V (see Fig. 10.6(b)). Despite the fact that *C*(**Lum**) is in twice smaller than the *C*(**An**) (and, as a result, the currents of oxidation of **An** monomer significantly higher compared with the currents of **Lum** oxidation), the currents of oxidation/reduction of **Lum** fragments in the copolymer are higher compared to the currents of oxidation/reduction of **An** fragments. In our opinion, this is an indication that the intermediate products of oxidation of **Lum** are characterized by higher reactivity compared with CRs of **An**, and therefore they are mostly involved in the formation of copolymer.

An increase of the **An** concentration till 0.025 mol L^{-1} (molar ratio **Lum**:**An** = 1:5) leads to a decrease of the difference between the heights of the currents of oxidation of **Lum** and **An** fragments of copolymer (see Fig. 10.6(c)). In addition, there is another reversible maximum of currents of oxidation/reduction at +0.8/+0.7 V, which similar to the results of **PAn** research can be attributed to the secondary transformations of polymer. In particular, they can be cross-linking of polymeric chains through the nitrene-cation[76] or the oxidation of products of partial destruction of copolymer formed in the case of polarization of electrode at the potentials over +1.2 V.[48] The currents of maxima of oxidation/reduction of **Lum** and **An** fragments are almost aligned at the molar ratio **Lum**:**An** = 1:10 (see Fig. 10.6(d)). However, in this case the height of the maximum corresponding to electrochemical conversion of products of the copolymer degradation is significantly increased.

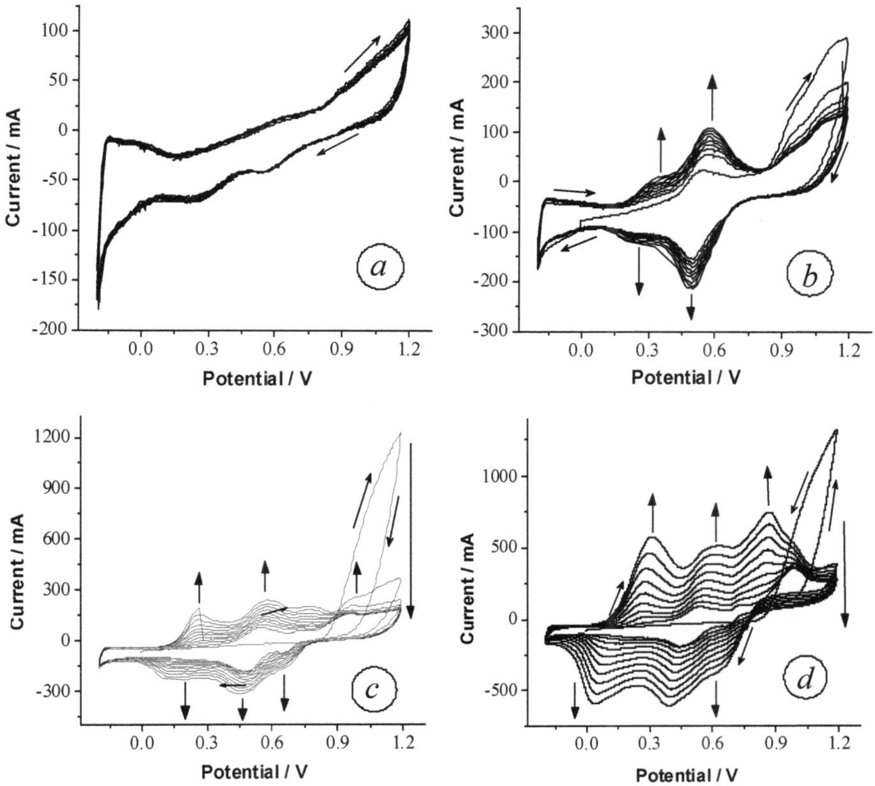

FIGURE 10.6 *CVA*-curves of Pt-electrode in 0.5×10^{-3} M solution of **Lum** (a) and after insertion of 1.0×10^{-2} (b), 2.5×10^{-2} (c) and 5.0×10^{-2} M (d) additives of **An**. Base electrolyte: 0.5 M aqueous solution of H_2SO_4; potential scanning limits: (−0.2)–(+1.2) V; $s_E = 50$ mVs^{-1}.

Electrochemiluminescent properties of **Lum** and **An** copolymer were investigated in potentiodynamic mode. The synthesis of the polymer sample, given the results of previous voltammetric studies, was carried out under potentiostatic conditions at the electrode potential +1.2 V for 2–15 min in 0.5 M aqueous solution of H_2SO_4 containing the starting monomers. Following investigations of the **ECL** activity of the polymeric (copolymeric) layer were carried out in aqueous borate buffer solution at *pH* = 10.5. The choice of medium for research of the **ECL**-properties was caused by the fact that exactly under these conditions the **ECL** of **Lum** is generated at the platinum electrode.[77] However, due to very small amounts of copolymer that was formed at the electrode, the **ECL** of copolymer was not recorded. Therefore, in order to increase the concentration of **Lum** the water-**DMSO** solution was used in the initial solution for the synthesis of polymeric samples similar to the chemical synthesis of polyluminol.[66] Introducing of 10% of **DMSO** into the starting solution and reducing of the concentration of H_2SO_4 to 0.05 M (on one order versus CVA-researches) made it possible to increase the concentration of **Lum** to 0.01 mol L^{-1}.

ECL under these conditions is detected during the synthesis of polyluminol and its copolymerization with **An**. As shown in Figure 10.7, the luminescence occurs at ~E ≈ (+2.0) V and rapidly is increased with the potential increasing. A similar phenomenon was observed earlier during the electrochemical synthesis of **PAn**.[42] Therefore, we can say that in this case, the source of the **ECL** is the destruction of polymer that was previously (at lower values of electrode potentials) formed on the electrode. The peculiarity of this phenomenon is that in a case of **An** introduction (see Fig. 10.7, curves 2 and 3) in the starting polymerizing solution the **ECL** intensity is significantly reduced, while at *C*(**An**) = 0.1 M it was not fixed.

In a case of polyluminol the **ECL** in borate alkaline (*pH* 10.5) buffer aqueous solution occurs at the potential ≥ +0.3 V (see Fig. 10.8). Under similar conditions there is **ECL** of monomeric **Lum** at Pt-electrode[77] and therefore, the hydrazine group of **Lum** not changes and passes into the polymer intact during the polymerization.

The maximal intensity of radiation during the electrochemical oxidation of polyluminol in potentiodynamic mode increases with the increase of amount of polymer on the electrode surface (the duration of polymerization). At the same time, the potential of the appearance of ECL ($E_{\mathbf{ECL}}$) shifts to more positive potentials at the decrease of the duration of polymerization. In particular, for the duration of polymerization = 2 min the value of $E_{\mathbf{ECL}}$ already exceeds +0.5 V. At the same time, it is clearly visible that at the potential > +1.0 V on the potentiodynamic curve the **ECL** the second weak

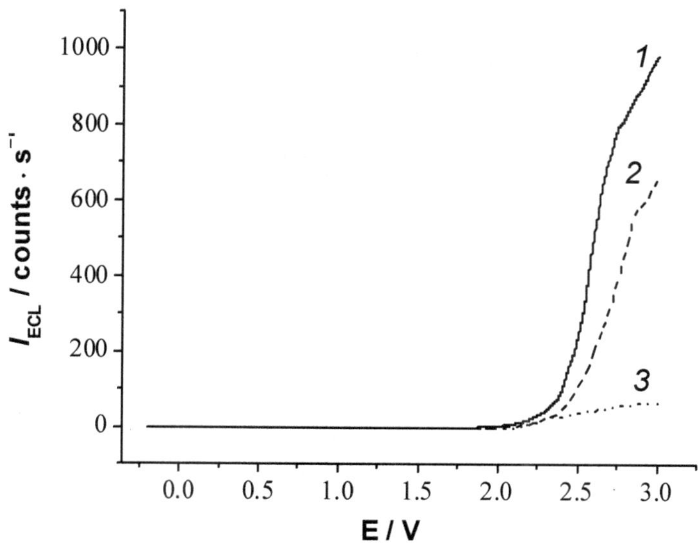

FIGURE 10.7 Potentiodynamic (s_E = 5 mVs^{-1}) curve of *ECL* generation during the electrochemical polymerization of **Lum** from its 0.01 M solution in 0.5 M H_2SO_4 aqueous-**DMSO** (9:1) solution in the absence (1) and in the presence of 0.001 (2) and 0.01 M (3) of aniline.

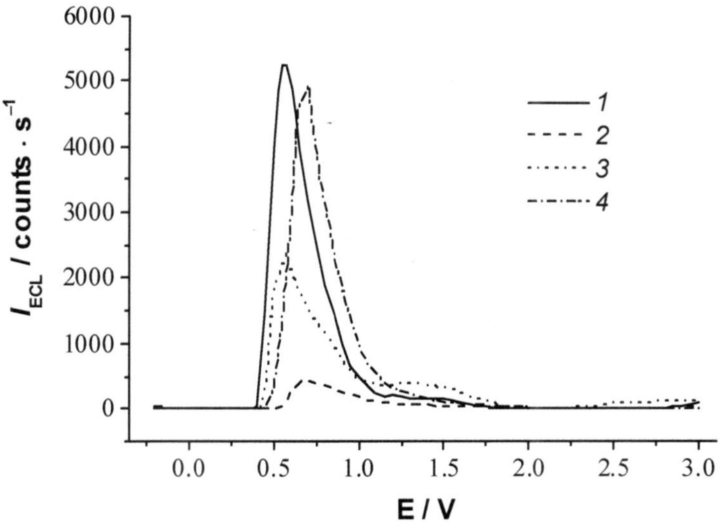

FIGURE 10.8 Potentiodynamic (s_E = 5 mVs^{-1}) curve of *ECL* generation of polylumynol in borate buffer solution (*pH* = 10.5). The duration of electrochemical synthesis of polymeric sample from 0.05 M of aqueous-**DMSO** (9:1) solution of **Lum** at the potential of electrode *E* = +1.2 V, min: 15 (1); 10 (2); 5 (3); 2 (4).

wave of radiation is appeared. Therefore, we can assume that the first wave of **ECL** is related with oxidation of **Lum** fragments due to their direct full electrochemical oxidation in accordance with the mechanism proposed in *ref.,*[78] or with the participation of molecular oxygen,[23] since the initial solutions were not previously deaerated. At the same time, the second wave of **ECL** is related with the oxidation of **Lum** fragments under the action of products of electrochemical oxidation of ions of the base electrolyte (OH^- ions) – $HO^\bullet$–radicals. The introduction of **An** into the starting polymerizing mixture at the molar ratio **Lum**:**An** = 10:1 (see Fig. 10.9), 1:1 (see Fig. 10.10) and 1:10 (see Fig. 10.11) practically does not change the nature of the **ECL**-curves. The luminescence is occurred in the same ranges of the potentials, namely from +0.4 to +1.0 V but the **ECL** intensity is significantly higher than for pure polyluminol (see Fig. 10.8). In this case **ECL** is decreased with the duration of preliminary polymerization increasing. Comparison of the results of **ECL** studies with **CVA**-curves allows to assert that this effect is related primarily with large currents of the **Lum** oxidation in the presence of **An**. Therefore, at the same duration of polymerization the total number of the **Lum** fragments into polymer obtained on the electrode surface increases, which leads to an increase of the intensity of **ECL**.

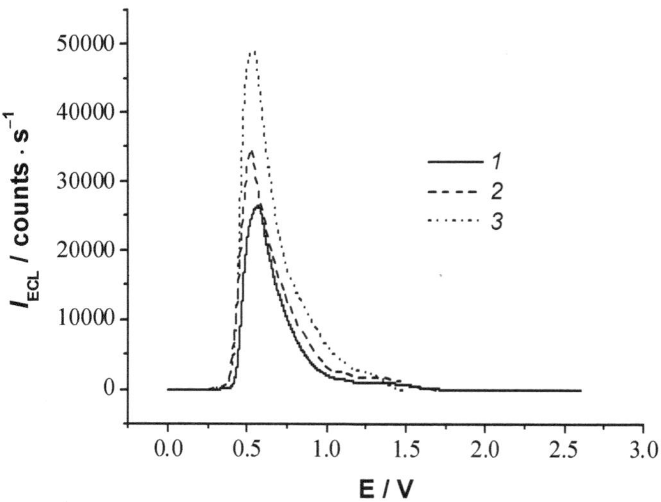

FIGURE 10.9 Potentiodynamic (s_E = 5 mVs^{-1}) dependence of the *ECL* intensity for copolymer of poly(luminol–aniline) in aqueous borate buffer (*pH* = 10.5) from the value of applied electrode potential at the duration of electrochemical synthesis of the polymer sample, min: 15 (1); 10 (2); 5 (3). The conditions of the polymerization: 0.05 M aqueous-**DMSO** (9:1) solution of H_2SO_4; the potential of electrode E = +1.2 V; C_0 (**Lum**) = 0.01 M; C_0 (**An**) = 0.001 M.

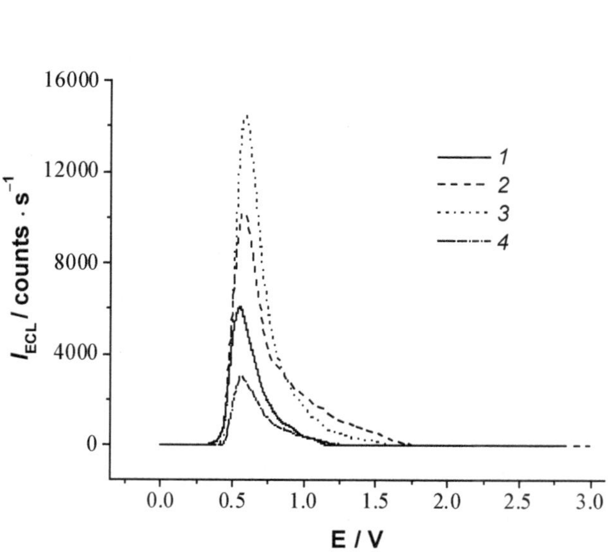

FIGURE 10.10 Potentiodynamic (s_E = 5 mVs^{-1}) dependence of the *ECL* intensity for copolymer of poly(luminol–aniline) in aqueous borate buffer (*pH* = 10.5) from the value of applied electrode potential at the duration of electrochemical synthesis of the polymer sample, min: 15 (1); 10 (2); 5 (3); 2 (4). The conditions of the polymerization: 0.05 M aqueous-**DMSO** (9:1) solution of H_2SO_4; the potential of electrode E = +1.2 V; C_0 (**Lum**) = 0.01 M; C_0 (**An**) = 0.001 M.

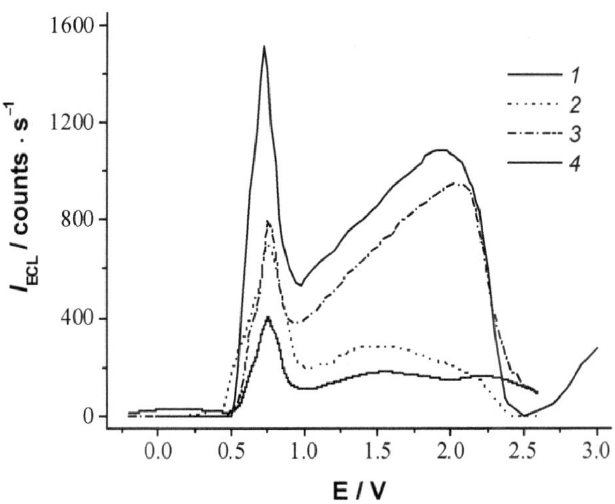

FIGURE 10.11 Potentiodynamic (s_E = 5 mVs^{-1}) dependence of the *ECL* intensity for copolymer of poly(luminol–aniline) in aqueous borate buffer (*pH* = 10.5) from the value of applied electrode potential at the duration of electrochemical synthesis of the polymer sample, min: 15 (1); 10 (2); 5 (3); 2 (4). The conditions of the polymerization: 0.05 M aqueous-**DMSO** (9:1) solution of H_2SO_4; the potential of electrode E = +1.2 V; C_0 (**Lum**) = 0.01 M; C_0 (**An**) = 0.01 M.

At the same time, an increasing of the concentration of **An** from 0.001 to 0.1 M (see Fig. 10.9 and 10.10) significantly reduces the intensity of the **ECL**. Therefore, we can assume that in this case due to the formation of thicker polymeric film, the part of the **Lum** fragments of the copolymer is not oxidized resulting the reducing of the radiation intensity. Proof of this is also already mentioned previously effect of higher **ECL** intensity in the case of reducing the duration of the polymerization.

For a large content of **An** (C(**An**) $\geq$ 1.0 M) in the starting polymeric mixture the nature of potentiodynamic **ECL**-curves is changed significantly. If in a case of the molecular ratio **Lum**:**An** = 1:10 (see Fig. 10.11) additionally there is a second wave in the range of the potentials **ECL** +(1.0–2.5)V, then at the ratio **Lum**:**An** = 1:25 (see Fig. 10.12) an intensity of this wave grows almost on order and it becomes only on the **ECL**-curve. In addition, the dependence of the I_{ECL} on the duration of the preliminary polymerization is the same as for pure polyluminol (see Fig. 10.8), namely the intensity of radiation is increased again with the increasing of the density of copolymer deposited on the electrode.

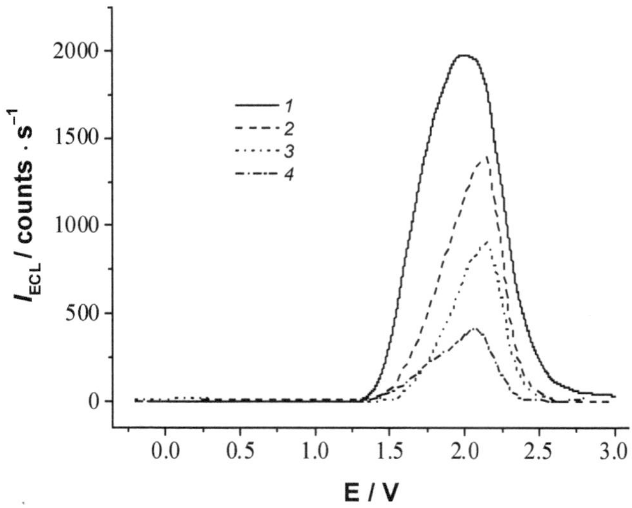

FIGURE 10.12 Potentiodynamic (s_E = 5 mVs^{-1}) dependence of the *ECL* intensity for copolymer of poly(luminol–aniline) in aqueous borate buffer (pH = 10.5) from the value of applied electrode potential at the duration of electrochemical synthesis of the polymer sample, min: 15 (1); 10 (2); 5 (3); 2 (4). The conditions of the polymerization: 0.05 M aqueous-**DMSO** (9:1) solution of H_2SO_4; the potential of electrode E = +1.2 V; C_0 (**Lum**) = 0.01 M; C_0 (**An**) = 0.25 M.

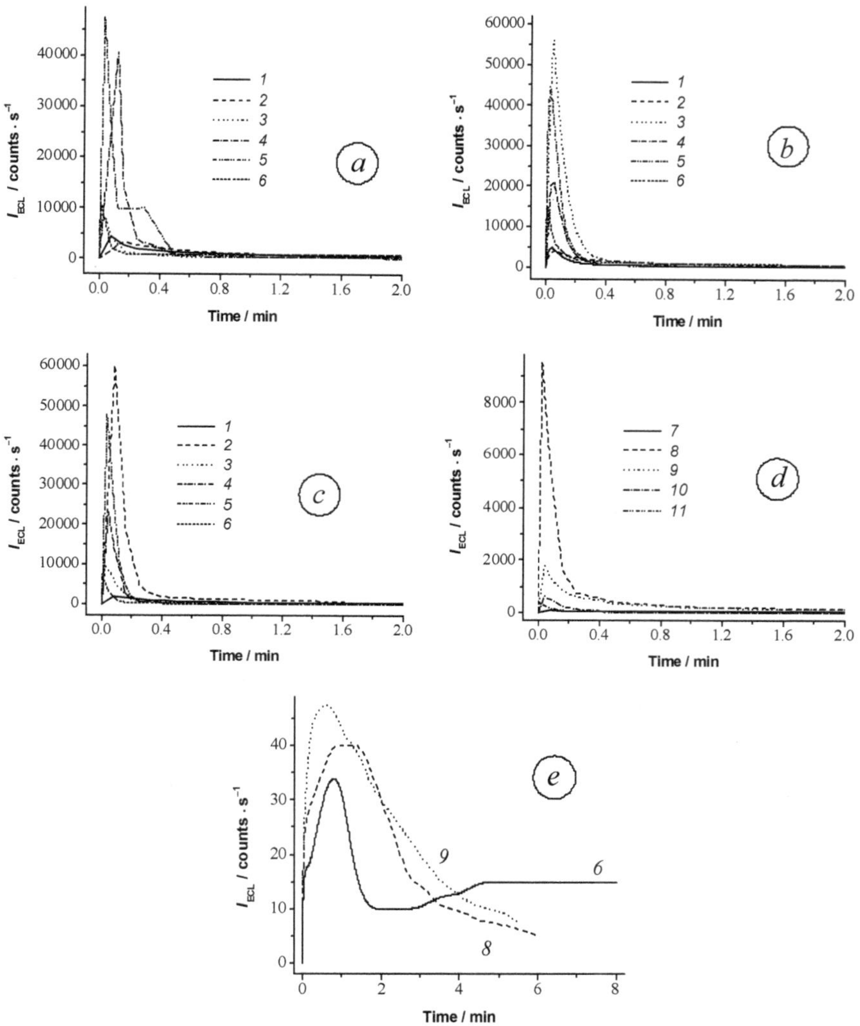

FIGURE 10.13 The change via time of the light emission during the electrochemical oxidation of polyluminol (a) and copolymer poly(aniline–luminol) (b–e) in aqueous borate buffer solution ($pH = 10.5$) under potentiostatistical conditions at the value of the potential of working Pt-electrode, V: 0.6 (1); 0.8 (2); 1.0 (3); 1.2 (4); 1.4 (5); 1.6 (6); 1.8 (7); 2.0 (8); 2.2 (9); 2.4 (10); 2.6 (11). The conditions of the electrochemical polymerization: the potential of electrode $E = +1.2$ V; the duration of synthesis 5 (e) and 10 (*a–d*) min; 0.1 M **Lum** + (0.0 M (a); 0.001 M (b); 0.01 M (c); 0.1 M (d) and 0.25 M (e) **An**) + 0.05 M H_2SO_4 of aqueous-**DMSO** (9:1) solution.

If to draw an analogy between **ECL** of **Lum**[77] and polyluminol (see Fig. 10.8), it can be suggested that the second wave of radiation is related with the oxidation of **Lum** fragments by the intermediates of oxidation of OH^- ions, such as HO•–radicals that can penetrate into the all volume of polymer film on the electrode surface. In this case, the first wave of **ECL** is related only with direct electrochemical oxidation of **Lum** fragments that come into contact with the surface of the electrode, which leads to a significantly lower radiation intensity of polyluminol compared to the copolymer.

Since for (electro)chemically synthesized **PAns** or their copolymers the most likely form is emeraldin salt,[41,49,78] and the oxidation of **Lum** is occurred easier compared to **An**, it is possible to suggest that the synthesized copolymer will contain the following fragments:

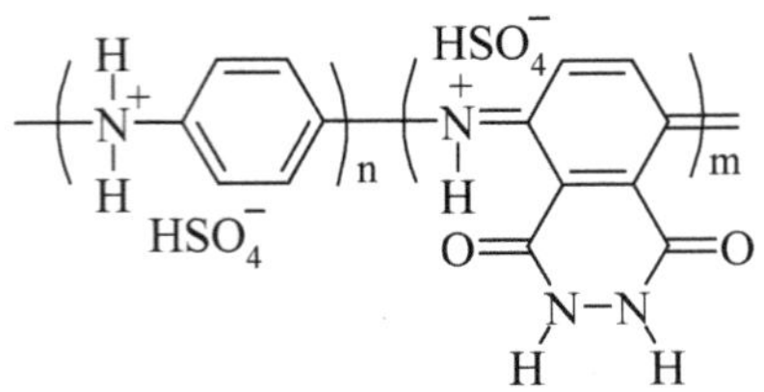

At the same time, since **ECL** is generated in an alkaline medium, then in this process the copolymer will participate in a form of emeraldin base. In this case, the scheme of luminescence occurrence for the fragment of copolymer containing one **Lum** fragment can be represented as follow (here the asterisk (*) denotes electronically excited link):

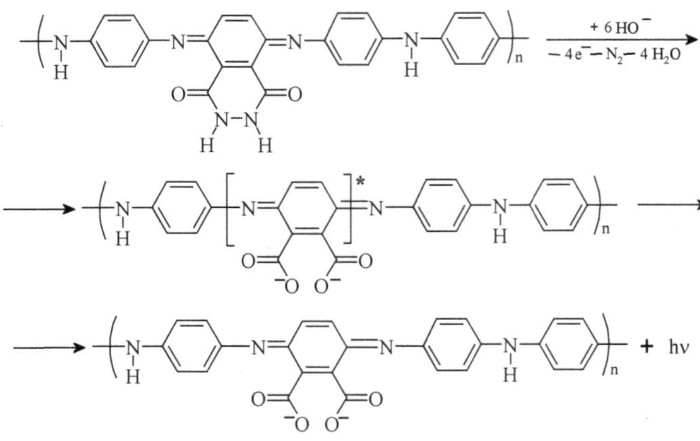

Under potentiostatic conditions the nature of change of the radiation intensity with time is identical for the all synthesized samples. As Figure

10.13 (a–d) illustrates, the intensity of **ECL**, which corresponds to the first wave of the luminescence appearance abruptly increases to the maximal value during the first 2–3 s starting from the polarization of the electrode. Similar character has also a radiation intensity that corresponds to the second wave of **ECL** (see Fig. 10.13(e)). But in this case it is on two to three orders of magnitude lower than the intensity of the first wave. In addition, there is also much smaller the duration of radiation. If for the first wave of **ECL** it was over 15 min, then for the second one it was only about 5 min.

10.5 CONCLUSIONS

Therefore, the electrochemical synthesis of **PAn**s from the acidic aqueous solutions at high values of electrode potential is accompanied by secondary transformations of oxidized form of **PAn** (pernigraniline), namely its irreversible oxidative destruction (overoxidation) and cross-linking of polymeric chains. High reactive intermediates of the oxidation of solvent (water) and base electrolyte are the oxidizing agent. At the potentials over +1.3 V the electrochemical degradation of **PAn** is accompanied by weak luminescence and the emitter of radiation at this likely is the *p*–benzoquinone or other low-molecular products of degradation with a system of conjugated p–bonds. This makes it possible to use the **ECL** method for monitoring of the state of **PAn** coatings and for the analysis of the processes on their surfaces in devices for various purposes.

At the same time, copolymers of **An** and **Lum** show the **ECL**-activity in alkaline solutions at the expense of the presence of **Lum** fragments in the polymeric chain. Proof of this is the fact that the optimal value of the *pH* medium for the generation of radiation is 10.5 like for **ECL** of 5–amine–2,3–dihydrophthalazine–1,4–dion. Using of mixed aqueous-organic, namely water-**DMSO** solvent, allows to increase the concentration of **Lum** in the initial polymerization solution that permits to obtain the copolymer "enriched" by **Lum** fragments, which may be a promising material for development of fluorescent sensors. At the same time, an analysis of the conditions of **ECL** appearance for copolymer of **An** and **Lum**, namely the presence of two waves of the radiation generation, showed that the best for this purpose is the use of the first wave of **ECL** in the range of potentials +(0.4–1.0) V. In this case, in order to achieve the maximal intensity of generated radiation light the ratio **An**:**Lum** in the initial polymerization mixture should not exceed the molar ratio of 10:1.

KEYWORDS

- **electrochemiluminescence**
- **polyaniline**
- **electrochemical destruction**
- **luminol**
- **copolymerization**
- **cyclic voltammetry**

REFERENCES

1. Schaper, H.; Köstlin, H.; Schnedler, E. New Aspects of D-C Electrochemiluminescence. *J. Electrochem. Soc.* **1982,** *129,* 1289–1294.
2. Measures, R. M. Prospects for Developing a Laser Based on Electrochemiluminescence. *Appl. Opt.* **1974,** *13,* 1121–1133.
3. Heller, C. A.; Jernigan, J. L. Electrochemical Pumping of Laser Dyes. *Appl. Opt.* **1977,** 16, 61–66.
4. Horiuchi,T.; Niwa,O.; Hatakenaka, N. Evidence for Laser Action Driven by Electrochemiluminescence. *Nature.* **1998,** *394,* 659–661.
5. Koval'chuk, E. P.;Ganushchak, N. I.; Prysyazhnyj, V. M.; Obushak, N. D. Electrochemiluminescence during Electrolysis of Diazonium Salts in the Presence of the Acceptors of Free Radicals. *Ukr. Khim. Zhurnal+.* **1982,** *48,* 491–493 (In Russian).
6. Koval'chuk , E. P.; Reshetnjak, O. V.*Electrochemiluminescent Studies of the Polymerization on the Metallic Cathode;Abstracts of the Int. Symposium "Polymers at the Phase Boundary"*; L'viv univ:Ukraine, 1994; pp 8.
7. Fähnrich, K. A.; Pravda, M.; Guilbault, G. G. Recent Applications of Electrogenerated Chemiluminescence in Chemical Analysis. *Talanta.* **2001,** *54,* 531–559.
8. Knight, A. W. A Review of Recent Trends in Analytical Applications of Electrogenerated Chemiluminescence. *Trends Anal. Chem.* **1999,** *18,* 47–62.
9. Kukoba, A. V.; Bykh A. I.; Svir I. B. Analytical Applications of Electrochemiluminescence: An Overview. *Fresenius J. Anal. Chem.* **2000,** *368,* 439–442.
10. Greenway, G. M. Analytical Applications of Electrogenerated Chemiluminescence. *Trends Anal. Chem.* **1990,** *9,* 200–203.
11. Luo, L.; Zhang, Z.h. Sensors Based On Galvanic Cell Generated Electrochemiluminescence And Its Application. *Anal. Chim Acta.* **2006,** *580,* 14–17.
12. Jiang, H.; Ju, H. Enzyme–Quantum Dots Architecture for Highly Sensitive Electrochemiluminescence Biosensing of Oxidase Substrates. *Chem. Comm.* **2007,** *(4),* 404–406.
13. Chovin, A.; Garrigue P.; Vinatier P.; Sojic N. Development of an Ordered Array of Optoelectrochemical Individually Readable Sensors with Submicrometer Dimensions: Application to Remote Electrochemiluminescence Imaging. *Anal. Chem.* **2004,** *76,* 357–364.

14. Cui, H.; Wang, W.; Duan, Ch. F.; Dong, Y. P.; Guo, J. Z. Synthesis, Characterization, and Electrochemiluminescence of Luminol-Reduced Gold Nanoparticles and their Application in a Hydrogen Peroxide Sensor. *Chem. Eur. J.* **2007,** *13,* 6975–6984.
15. Yang, M.; Liu, C.; Qian, K.; He, P.; Fang, Y. Study on the Electrochemiluminescence Behavior of ABEI and its Application in DNA Hybridization Analysis. *Analyst.* **2002,** *127,* 1267–1271.
16. Li ,Y.; Qi, H.; Fang, F.; Zhang, Ch. Ultrasensitive Electrogenerated Chemiluminescense Detection of DNA Hybridization using Carbon Nanotubes Loaded with Tris(2,2–Bipyridyl)Ruthenium Derivative Tags. *Talanta.* **2007,** *72,* 1704–1709.
17. Bard,A. J. New Challenges in Electrochemistry and Electroanalysis. *Pure Appl. Chem.* **1992,** *64,* 185–192.
18. Zhang, C. X.; Qi, H.; Zhang, M. Homogeneous Electrogenerated Chemiluminescense Immunoassay for the Determination of Digoxin Employing $Ru(Bpy)_2$ (Dcbpy) NHS and *Carrier Protein. Luminescence.* **2007,** *22,* 53–59.
19. Kankare, J.; Haapakka, K.; Kulmala, S.;Immunoassay by Time-Resolved Electrogenerated Luminescence. *Anal. Chim. Acta.* **1992,** *266,* 205–212.
20. Lee, W.–Y. Tris (2,2′–Bipyridyl)Ruthenium(II) Electrogenerated Chemiluminescence in Analytical Science. *Microchim. Acta.* **1997,** *127,* 19–39.
21. Richter, M. M. Electrochemiluminescence (ECL). *Chem. Rev.* **2004,** *104,* 3003–3036.
22. Mitschke ,U. Bäuerle P. The Electroluminescence of Organic Materials. *J. Mater. Chem.* **2000,** *10,* 1471–1507.
23. Reshetnyak, A. V.; Błażejowski, J. In *Electrochemical Generation Of Luminescence;* Gultiaj, V. P., Krivenko, A. G.,Tomiliv, A. P., Eds.; Company Sputnik+: Moscow, 2008; pp 6–55.
24. Reshetnyak, O. V.; Koval'chuk, E. P.; Błażejowski, J. Role of Molecular Oxygen and its Active Forms in Generation of Electrochemiluminescence. *Russ. J. Electrochem.* **2011,** *47,* 1111–1118.
25. Pei Q.; Yu, G.; Zhang, C.; Yang, Y.; Heeger, A. J. Polymer Light-Emitting Electrochemical Cells. *Science.* **1995,** *269,* 1086–1088.
26. Din,i D. Electrochemiluminescence from Organic Emitters. *J. Chem. Mater.* **2005,** *17,* 1933–1945.
27. Fu,Y.; Sun, M.; Wu, Y.; Bo, Z.; Ma, D. Conjugated Polymers Containing Electron-Transporting, Hole-Transporting, And Light-Emitting Units in the Polymer Main Chain. J. *Polym. Sci. Part A Polym. Chem.* **2008,** *46,* 1349–1356.
28. Daimon, T.; Nihei, E. Fabrication of Organic Electrochemiluminescence Devices with p-Conjugated Polymer Materials. *J. Mater. Chem. C.* **2013,** *1,* 2826–2833.
29. Kasahara, T.; Matsunami, S.; Edura, T.;Ishimatsu, R..; Edura,T.; Oshima , J .; Tsuwaki, M.; Imato, T.; Shoji, S.; Adachi, C.; Mizuno, J. Multi–Color Microfluidic Electrochemiluminescence Cells. *Sens. Actuators. A. Phys.* **2014,** *214,* 225–229.
30. Manzanares, J. A.; Reiss, H.; Heeger, A. J. Polymer Light–Emitting Electrochemical Cells: A Theoretical Study of Junction Formation under Steady-State Conditions. *J. Phys. Chem. B.* **1998,** *102,* 4327–4336.
31. Meier, S. B.; Tordera, D.; Pertegás, A.; Roldán-Carmona, C.; Ortí, E.; Bolink, H. J. Light–Emitting Electrochemical Cells: Recent Progress and Future Prospects. *Mater. Today.* **2014,** *17,* 217–223.
32. Yu, Zh. B.; Li, L.; Gao, H. E.; Pei, Q. B. Polymer Light–Emitting Electrochemical Cells: Recent Developments to Stabilize the P-I-N Junction and Explore Novel Device Applications. *Sci. China Chem.* **2013,** *56,* 1075–1086.

33. Yang, Y. Polymer Electroluminescent Devices. *MRS Bull.* **1997,** 22, 31–38.
34. Dennany, L.; Mohsan, Z.; Kanibolotsky, A. L.; Skabara, P. J. Novel Electrochemiluminescent Materials for Sensor Applications. *Faraday Discuss.* **2014,** *174,* 357–367.
35. Kapturkiewicz, A. Electrochemiluminescent Systems as Devices and Sensors. In *Electrochemistry Of Functional Supramolecular Systems;* Ceroni, P., Credi, A., Venturi, M., Eds.; John Wiley & Sons, Inc.: Hoboken, New Jersey, 2010; pp 477–522.
36. Shu, Q.; Adam, C.; Sojic, N.; Schmittel, V. Electrochemiluminescent Polymer Films with a Suitable Redox "Turn-Off" Absorbance Window for Remote Selective Sensing of Hg^{2+}. *Analyst.* **2013,** *138,* 4500–4504.
37. Szunerits, S.; Walt, D. R. The Use of Optical Fiber Bundles Combined with Electrochemistry for Chemical Imaging. *Chem. Phys. Chem.* **2003,** *4,* 186–192.
38. Koval'chuk, E. P.; Reshetnyak, O. V.; Chernyak A. O.; Kovalyshyn, Ya. S. Electrochemiluminescence on np^1–Metals. 1. The Analysis of Chemiluminescencent Reaction. *Electrochim. Acta.* **1999,** *44,* 4079–4086.
39. Lur'ie, Yu. Yu. *Handbook of Analytical Chemistry.* Khimia: Moscow,Russia, 1979 ; pp 267.
40. Barta P., Kugler Th., Salanek W. R., Monkman A. P. Libert, J.;Lazzaroni,R.;Bredas,J. L. Electronic Structure of Emeraldine and Pernigraniline Base: A Joint Theoretical and Experimental Study. *Synth. Met.* **1998,** *93,* 83–87.
41. Koval'chuk, E. P.; Whittingham, M. S.; Skolozdra, O. M.; Zavalij, P. Y.; Zavaliy, I. Yu.; Reshetnyak, O. V.;Seledets, M. Copolymers of Aniline and Nitroanilines. Part I. Mechanism of Aniline Oxidation Polycondensation. *Mater. Chem. Phys.* **2001,** *69,* 154–162.
42. Skolozdra, O.; Reshetnyak, O.; Koval'chuk, E. *The Oxidative Destruction as a Source of the Luminescence during Electrochemical Synthesis of Polyaniline*; *Visnyk Lviv Univ: Ser. Khim,* **2002**; *41,* 249–255 (in Ukrainian).
43. Batich, C. D.; Laitinen, H. A.; Zhou, H. C. Chromatic Changes in Polyaniline Films. *J. Electrochem. Soc.* **1990,** *137,* 883- 885.
44. Mortimer, R .J.; Dyer, A. L.; Reynolds, J. R. Electrochromic Organic and Polymeric Materials for Display Applications. *Displays.* **2006,** *27,* 2–18.
45. Kobayashi, T.; Yoneyama, H.; Tamura, H. Oxidative Degradation Pathway of Polyaniline Film Electrodes. *J. Electroanal.Chem.* **1984,** *177,* 293–297.
46. Koyanagi,M.; Kogo, Y.; Kanda, Y. N. p*–Electronic States of *P*–Benzoquinone; The T–S Emission Spectrum in Mixed Crystals. *J. Mol. Spectroscopy.* **1070,** *34,* 450–467.
47. Itoh, T. Emission Spectrum of *P*–Benzoquinone in a Fluid Solution. *Spectrochim. Acta. Part A.Mol. Spectrosc.* **1984,** *40,* 387–389.
48. Stiwell, D. E.; Park, S.–M. Electrochemistry of Conductive Polymers. *J. Electrochem. Soc.* 1988, 135, 2491–2496.
49. Koval'chuk, E .P.; Whittingham, M. S.; Skolozdra, O. M.; Zavalij, P. Y.; Zavaliy, I. Yu; Reshetnyak, O. V.; Błazejowski, J. Polyaniline and Copolymer Aniline and Ortho–Nitroaniline. Part II. Physicochemical Properties. *Mater. Chem. Phys.* **2001,** *70,* 38–48.
50. Pud, A. Stability and Degradation of Conducting Polymers in Electrochemical Systems. *Synth. Met.* **1994,** *66,* 1–18.
51. Albrecht, H. O. Über Die Chemiluminescenz Des Amonophthalsäurehydrazids. *Z. Phys. Chem.* **1928,** *136,* 321–330.
52. Harvey, N. Luminescence during Electrolysis. *J. Phys. Chem.* **1929,** *33,* 1456–1459.
53. Ramsthaler, F.; Ebach, S. C.; Birngruber, C. G.; Verhoff, M. A. Postmortem Interval of Skeletal Remains through the Detection of Intraosseal Hemin Traces. A Comparison of

UV-Fluorescence, Luminol, Hexagon-OBTI®, and Combur® Tests. *Forensic Sci. Int.* **2011,** *209,* 59–63.
54. Tan, X.; Song, Z.; Chen, D.; Wang, Z. Study on the Chemiluminescence Behavior of Bovine Serum Albumin with Luminol and its Analytical Application. *Spectrochim. Acta A.* **2011,** *79,* 232–235.
55. Liu, Q.; Wu, J.; Tian, J.; Zhang , C.; Gao, J.; Latep, N.; Ge, Y.; Qin, W. Sensitive and Selective Capillary Electrophoretic Analysis of Proteins by Zirconia Nanoparticle-Enhanced Copper (II)–Catalyzed Luminol–Hydrogen Peroxide Chemiluminescence. *Talanta.* **2012,** *97,* 193–198.
56. Li, M.; Li, J.; Sun, L.; Zhang, X.; Jin,W. Measuring Interactions And Conformational Changes of DNA Molecules using Electrochemiluminescence Resonance Energy Transfer in the Conjugates Consisting of Luminol, DNA and Quantum Dot. *Electrochim. Acta.* **2012,** *80,* 171–179.
57. Li, F.; Cui, H. A Label–Free Electrochemiluminescence Aptasensor for Thrombin Based on Novel Assembly Strategy of Oligonucleotide and Luminol Functionalized Gold Nanoparticles. *Biosens. Bioelectron.* **2013,** *39,* 261–267.
58. Dai, H.; Wu, X.; Wang, Y.; Zhou, W.; Chen, G. An Electrochemiluminescent Biosensor For Vitamin C Based On Inhibition Of Luminol Electrochemiluminescence On Graphite/Poly(Methylmethacrylate) Composite Electrode. *Electrochim. Acta.* **2008,** *53,* 5113–5117.
59. Tian, D.; Duan, C.; Wang, W.; Cui, H. Ultrasensitive Electrochemiluminescence Immunosensor Based On Luminol Functionalized Gold Nanoparticle Labeling. *Biosens. Bioelectron.* **2010,** *25,* 2290–2295.
60. Liu, X.; Niu, W.; Li, H.; Han, S.; Hu, L.; Xu, G. Glucose Biosensor Based on Gold Nanoparticle-Catalyzed Luminol Electrochemiluminescence on a Three-Dimensional Sol–Gel Network. *Electrochem. Commun.* **2008,** *10,* 1250–1253.
61. Chen, X. M.; Su, B. Y.; Song, X. H.; Chen, Q.– A.; Chen, X.; Wang, X.– R.. Recent Advances in Electrochemiluminescent Enzyme Biosensors. *Trac. Trends Anal. Chem.* **2011,** *30,* 665–676.
62. Zhang, G.–F.; Chen, H.–Y. Studies of Polyluminol Modified Electrode and its Application in Electrochemiluminescence Analysis with Flow System. *Analyt. Chim. Acta.* **2000,** *419,* 25–31.
63. Sassolas, A.; Blum, L. J.; Leca-Bouvier, B. D. New Electrochemiluminescent Biosensors Combining Polyluminol and an Enzymatic Matrix. *Anal. Bioanal. Chem.* **2009,** *394,* 971–980.
64. Li, G.; Zheng, X.; Zhang, Z. Electrogenerated Chemiluminescence (ECL) Determination of Ephedrine with a Self–Assembly Multilayer Ni(II)–Polyluminol Modified Electrode. *Microchim. Acta.* **2006,** *154,* 153- 161.
65. Wang, C. H.; Chen, S. M.; Wang, C. M. Co-Immobilization of Polymeric Luminol, Iron(II) Tris(5–Aminophenanthroline) and Glucose Oxidised at an Electrode Surface, and its Application as a Glucose Optrode. *Analyst.* **2002,** *127,* 1507–1511.
66. Koval'chuk, E. P.; Grynchyshyn, I. V.; Reshetnyak, O. V.; Błażejowski, J. Oxidative Condensation and Chemiluminescence of 5–Amino–2,3–Dihydro–1,4–Phtalazinedione. *Euro. Polym. J.* **2005,** *41,* 1315–1325.
67. Ferreira, V. C.; Cascalheira, A. C.; Abrantes, L. M. Electrochemical Copolymerisation of Luminol with Aniline: A New Route for the Preparation of Self–Doped Polyanilines. *Electrochim. Acta.* **2008,** *53,* 3803–3811.

68. Ferreira, V. C.; Cascalheira, A. C.; Abrantes, L. M. Electrochemical Preparation and Characterisation of Poly(Luminol–Aniline) Films. *Thin Solid Films.* **2008,** *516,* 3996–4001.
69. Ballesta–Claver, J.; Díaz Ortega, I. F.; Valencia–Mirón, M. C.; Capitán–Vallvey, L. F. Disposable Luminol Copolymer–Based Biosensor For Uric Acid In Urine. *Anal. Chim. Acta.* **2011,** *702,* 254–261.
70. Sassolas, A.; Blum, L. J.; Leca-Bouvier, B. D. Polymeric Luminol on Pre-Treated Screen–Printed Electrodes for the Design of Performant Reagentless (Bio)Sensors. *Sensor. Actuat. B – Chem.* **2009,** *139,* 214–221.
71. Li, G.; Lian, J.; Zheng, X.; Cao, J. Electrogenerated Chemiluminescence Biosensor for Glucose Based on Poly(Luminol–Aniline) Nanowires Composite Modified Electrode. *Biosens. Bioelectron.* **2010,** *26,* 643–648.
72. Ballesta–Claver, J.; Velázquez, P. S.; Valencia–Mirón, M. C.; Capitán-Vallvey, L. F. SPE Biosensor for Cholesterol in Serum Samples Based on Electrochemiluminescent Luminol Copolymer. *Talanta.* **2011,** *86,* 178–185.
73. Ballesta–Claver, J.; Ametis–Cabello, J.; Morales–Sanfrutos, J.; Megía-Fernández, A.; Valencia-Mirón , M. C.; Santoyo-González, F.; Capitán-Vallvey, L. F. Electrochemiluminescent Disposable Cholesterol Biosensor Based on Avidin–Biotin Assembling with the Electroformed Luminescent Conducting Polymer Poly(Luminol–Biotinylated Pyrrole). *Anal. Chim. Acta.* **2012,** *754,* 91–98.
74. Sassolas, A.; Blum, L. J.; Leca-Bouvier, B. D. Electrogeneration of Polyluminol and Chemilumine-Scence for New Disposable Reagentless Optical Sensors. *Anal. Bioanal. Chem.* **2008,** *390,* 865–871.
75. Reshetnyak , O.; Błażejowski, J.; Protsyshyn, K.h.; Pavlyuk, V. *Electrochemiluminescent Activity of The Co-Polymers of Aniline And Luminal; Visnyk Lviv. Univ: Ser. Khim,* **2011**; *52,* 249–260 (in Ukrainian).
76. Geniès E. M., Łapkowski M., Penneau J. F. Cyclic Voltammetry of Polyaniline: Interpretation of the Middle Peak. J. *Electroanal. Chem.* **1988,** *249,* 97–107.
77. Wroblewska, A.; Reshetnyak, O. V.; Koval'chuk , E. P.; Pasichnyuk, R. I.; Błażejowski, J.; Origin and Features of the Electrochemiluminescence of Luminol – Experimental and Theoretical Investigations. *J. Electroanal. Chem.* 2005, *580,* 41–49.
78. Koval'chuk, E. P.; Stratan, N. V.; Reshetnyak, O. V.; Błażejowski, J.; Whittingham, M. S.Synthesis and Properties of the Polyanisidines. *Solid State Ionics.* 2001, *142,* 217–224.

CHAPTER 11

SYNTHESIS AND PROPERTIES OF COMPOSITES BASED ON POLYANILINE AND POLY(ANILINE-*CO*-NITROANILINE) WITH XEROGEL OF VANADIUM (V) OXIDE

B. B. OSTAPOVYCH[1], O. V. RESHETNYAK[1], and M. V. BUZHANS'KA[2]

[1]*Department of Physical and Colloid Chemistry, Faculty of Chemistry, Ivan Franko National University of L'viv, 6 Kyryla & Mefodia Str., L'viv 79005, Ukraine*

[2]*Lviv Academy of Commerce, 9 Samchouk Str., L'viv 79011, Ukraine*

CONTENTS

ABSTRACT

The structure and physical chemical properties of the synthesized polymer–inorganic composites by intercalation type based on polyaniline (**PAn**)/copolymers of aniline and nitroaniline with xerogel of vanadium pentaoxide have been investigated. The methods of IR-spectroscopy, X-rays diffraction, and the elementary analysis were used to elucidate the structure and the composition of intercalates. The kinetics of polycondensation of anilines by oxidation of their gels was investigated by the method of differential scanning calorimetry. It was found that the rate of the oxidative copolycondensation in the presence of xerogel depends on the composition of the reacting mix due to the different reactivity of the nitroaniline isomers to copolycondensation reaction with aniline. It is shown that the thermolysis of synthesized polymer–inorganic composites at temperatures above 300 °C has a character of the oxidative degradation as a result of the vanadium (V) oxide composite component action.

The electrical conductivity of synthesized composites was studied and they were tested as cathode materials of lithium chemical power sources. It was shown that the use of copolymers of aniline with its nitro-derivatives improves the electrical characteristics of lithium chemical power sources based on hybrid polymer–xerogel composites by introducing of the electroactive nitro-groups into the structure of the polymer.

11.1 INTRODUCTION

Hybrid organic–mineral structures, the production, and using of which already has a long history, is the subject of intense investigation.[1–3] Use of the polymers having own electroconductivity[4–6] as the organic components of the composites gave a new impetus to the development of researches in this area. Organic–inorganic composites based on polymers with a system of conjugated p–bonds combine a high ability for intercalation and sorption properties of inorganic filler with the conductivity, flexibility, ability to doping, etc., which are typical for electroconductive polymers.

The oxides of the transition metals used as inorganic component in such materials attract an increasing attention of scientists. Sols of oxides of some 3*d*-elements form under the drying, the so-called xerogels, characterized by ordered quasi-one dimensional layered structure, and they are capable of intercalation not only with different ions but also with organic compounds of low molecular weight. Intercalation may be due to adsorption, ion exchange or the redox processes. In all cases, it leads to a change of the distance corresponding to wedging of one or more layers of intercalate in the interlayer without a change in the structure of xerogel. Development of new methods for preparing of the compounds of this class and the study of their characteristics remains an actual problem at the present moment, because today these hybrid organic–inorganic materials are widely used as catalytic, sensoric, and electrochemical systems.[7–10]

V_2O_5 is very attractive for electrochemical energy storage (**EES**) because the high oxidation state of vanadium leads to the possibility of storing more than one electron per formula unit (+5, +4, +3, and +2 are electrochemically accessible) and the ability to form layered compounds.[11] However, xerogel of vanadium (V) can be considered as inorganic graphene analog.[12] As a result, not only the ions of metals, but also the molecules of organic compounds can be intercalated in its lattice; in this way, it significantly differs[13] from the crystalline (orthorhombic form) V_2O_5. Moreover, in the case of using monomer molecules as intercalate, a result of the so-called intercalative polymerization (polycondensation) is the formation of composite materials, where the inorganic material is a matrix and the formed polymer is organic filler. In particular, the redox interaction between easy oxidized organic molecules such as aniline (**An**)[14–17] and xerogel of vanadium pentaoxide,[18–22] leads to the formation of the composite by total composition $(C_6H_4NH)_{0.44}{\cdot}V_2O_5{\cdot}0.5H_2O$. It is characterized by dark blue color with a characteristic metallic luster and it keeps the layered structure of the xerogel with the interplane distances ≈14 Å.[22] The conductivity of the obtained compound of **PAn** is four orders higher compared to the conductivity of original xerogel $V_2O_5{\cdot}nH_2O$ and is ~0.5 Ohm^{-1} $S{\cdot}cm^{-1}$ at room temperature. This material along with those nanocomposites by composition $(polypyrrol)_{0.4\text{-}0.9}{\cdot}V_2O_5$, $\mathbf{PAn}_{0.4}{\cdot}V_2O_5$ and $(polythiophene)_{0.3\text{-}0.8}{\cdot}V_2O_5$ been successfully used as a cathode in lithium electrochemical energy storage (**Li-EES**).[23,24] The best characteristics were obtained for **PAn**-containing composite cathode with the specific capacity 300 A hkg^{-1}.

The peculiarity of the oxide–polymer composite materials is that they can have a dual structure. Along with the formation of composites by intercalation type, the adsorption and, as a result the oxidative polycondensation of monomer can occur on the surface of particles of highly disperse inorganic oxide. In this case, already the polymer serves as a matrix; thereby the specific surface of such material significantly increases that is quite important at its using in **EES**. On the other hand, due to the drying of sol, including $V_2O_5{\cdot}nH_2O$, the fragile gel is derived, on the body of which a large number of capillaries are formed. As a result of the infiltration of monomer molecules, their subsequent polymerization leads to the formation of filled polymer composites by another type.

However, as the results presented in *ref.*[25,26] show, the specific electrical characteristics of **EES** are significantly increased at the replacing of **PAn** on the copolymeric poly(aniline–*co*–nitroaniline) cathode, due to the presence of electroactive nitro-groups in the polymer structure.

That is why the purpose of this work was the synthesis and comparison of physical and chemical properties of hybrid composites based on xerogel of vanadium (V) oxide and copolymers of aniline and nitroaniline (**NAn**), as well as the testing of the synthesized composites as cathodes in the models **Li-EES**.

11.2 EXPERIMENTAL PART

The aniline (**An**) and its *ortho*-(*o*-**NAn**), *metha*-(*m*-**NAn**) and *para*-nitroderivatives (*p*-**NAn**) by trademark "SIGMA-ALDRICH" (purity of 95% or higher), acetonitrile by trademark "KgaA" (Chromasolv™, purity $\geq$ 99.9%, water content $<$ 0.02%) and $LiClO_4$ by trademark "SIGMA-ALDRICH" (purity $\geq$ 98.0% (calculated per dry product), water content $\leq$ 3%) were used without further purification.

Synthesis of $V_2O_5{\cdot}nH_2O$ *xerogel* was carried out as described by *W. Biltz*[27]:

$$2\,NH_4VO_3 + 2\,HCl + nH_2O = V_2O_5{\cdot}nH_2O^{-} + 2\,NH_4Cl. \qquad (11.1)$$

For this purpose, the sample of 2.0 g of metavanadate ammonium (analytical grade) moistened with water was triturated in a mortar adding *a la carte* 20 ml of 2 N hydrochloric acid. The obtained red precipitate with liquid was put off a filter and was washed with distilled water to remove the yellow filtrate. At the beginning of the pass of reddish filtrate through the filter, the residue was transferred into an *Erlenmeyer's* flask and 100 ml of was added. Pinkish red sol of V_2O_5 was formed for some time, which matured for 14 days. The dialysis cleaning of sol was conducted in medium of bi-distilled water using the acetate cellulose membrane. The result of sol $V_2O_5{\cdot}nH_2O$ drying at air is brittle gel, which subsequently was dried under vacuum at 40 °C to the moment of xerogel $V_2O_5{\cdot}nH_2O$ (n = 0.5-1.38) formation.

Synthesis of polymer–inorganic composites based on vanadium oxide xerogel was carried out by dispersing of 1 g of $V_2O_5{\cdot}nH_2O$ powder xerogel in 1 g of aniline or in mixtures of monomers (aniline and nitroaniline isomers). In the latter case, the mass ratio of **An**:**NAn** was 9.5:0.5; 9:1, and 7:3. The reactive mixture was prepared in a glass ampoule, which was soldered in argon and was placed in a thermostat with mixing ($T = 60°C$) for 4 h. During the synthesis, an increase of viscosity of the mixture and the gradual change in color of the reaction mixture from dark red to dark green was observed; it was indication of the participation of V_2O_5 in the **An** oxidation. The next day, an ampoule was opened and the solid phase (synthesized composite by intercalation type) was separated from a viscous liquid phase with the use of a centrifuge (mode: 5000 r/min for 5 min).

Kinetics of oxidative polycondensation of **An** as same as its co-condensation with isomers of **NAn** in the presence of $V_2O_5 \cdot nH_2O$ xerogel was studied by differential calorimetry method.[28] An intensity of heat evolution of reaction system was investigated using the differential microcalorimeter *DAK–1A* by *Calvet type*. For this, a dosage amount of **An** or its derivatives was placed into a working cell of the calorimeter, and after that the xerogel sample (0.056 g per 1 ml of a monomers' mix) was injected after an ascertainment of the thermal equilibrium between the working and comparative (reference) cells of device. Weight ratio of **An** and its derivatives (*m*-, *o*- and *p*-nitroaniline) in the working cell was 7:3, when the total volume of the mixture of monomers was ~1 cm^3.

The relative depth of the monomers conversion was evaluated from the ratio:

$$\eta = \frac{Q_\tau}{Q_\infty} 100\%, \qquad (11.2)$$

where $Q\tau$ is amount of heat that is released at a time τ; Q_∞ is heat amount at full finish of process ($t \to \infty$). Rates of the heat evolution and the constant rate (k) of process (in $J \cdot s^{-1}$) were determined graphically from the dependence $Q\tau = f(\tau)$.

To study the structure of the formed xerogel and composites on its basis, the IR-spectroscopy (spectrophotometer *Specord®* M80, *Carl Zeiss Jena*, resolution ± 2 cm^{-1}) has been used. The synthesized samples were dispersed in KBr and pressed into pellets. The spectra were recorded in the range 4000–400 cm^{-1} at the speed scanning 8 cm^{-1}.

For the thermogravimetric analysis of the synthesized xerogel samples, the derivatograph *Q1500D* (*Paulik-Paulik-Erdei, Hungary*) was used. Scan speed of temperature in the investigated temperature range 20-480°C was 7 $°C \cdot min^{-1}$. The investigations were carried out in air. Thermogravimetric and differential thermogravimetric studies of synthesized samples of copolymers and composites V_2O_5/**PAn** and V_2O_5/P(**An**–*co*–**NAn**) were performed under argon at a volume rate of gas flow $s_V = 30$ cm^3·min^{-1} using a microbalance *NETZSCH TG*209. Scan speed of temperature in the investigated temperature range 20-720 °C was 2 °C·min^{-1}. Evolution of IR spectra of gaseous products of degradation of synthesized samples were studied with a time interval of 23 s, using a spectrophotometer *BRUKER IFS*66 *FTIR* (resolution ± 1 cm^{-1}; spectra were recorded in the range 4400–400 cm^{-1} at the speed scanning 30 sm^{-1}).

For elemental analysis of synthesized samples of polymer–inorganic composites, an analyzer *GA* 1108 (*Carlo Erba, Italy*) has been used.

For X-ray diffraction studies of $V_2O_5 \cdot nH_2O$ xerogel and its composites with polyanilines, the automatic powder diffractometer *STOE STADI P* (*Cu Kα* radiation, $\lambda = 0.15406$ nm) was used. Synthesis of samples for confirmation of the intercalation of **An** (**PAn**) in the lattice of xerogel by diffraction method was carried out with adding at the stirring of 0.5% mass of aqueous solution of **An** in undried hydrogel $V_2O_5 \cdot nH_2O$ (a content of dry residue of V_2O_5 was 0.12 mol/l) to achieve a certain molar ratio between the initial components (V_2O_5:**An**). The final surface texture of sediment in each case depended on the amount of **An**, which was added. After completion of the reaction (24 h), the product was filtered, washed with water, and dried under vacuum at room temperature.

The electroconductivity of the composites was determined by measuring the resistance using the sensitive integrating ohmmeter (*Щ* 306–1). For this, the powder-like samples of composite were compressed into the tablets by thickness of ~2 mm and by diameter of 10 mm in a metal mold with an effort of 15 MPa for 5 min at room temperature. The obtained tablets were placed in a Teflon cell between two plane–parallel copper contacts, which clamped the sample with a force of 0.05 MPa; and the readings of ohmmeter were recorded on a basis of which the expected value of conductivity was calculated.

The synthesized materials were used under constructing of the models **Li-EES** by prismatic type, produced in accordance with the trielectrode scheme, namely two lithium electrodes by a size 2 × 4 cm and one electrode based on a composite. To produce a positive electrode based on composites, 3 g of electroactive mixture with 20% (by mass) of *expanded graphite ABG*-81 (*Superior Graphite Co., USA*) was used, which was pressed on the nickel net by area 1 cm^2 with an effort of 15 MPa. 1 M of $LiClO_4$ in acetonitrile (**AcN**) was used as the electrolyte solution, and the porous polypropylene serves as the separator. The discharge of constructed models were carried out at constant resistance of the external circuit (2 kΩ), the charge of element occurred in galvanostatic mode at the charging current density 0.025 mA·cm^{-2}.

11.3 RESULTS & DISCUSSION

In order to determine the composition of the synthesized samples of xerogel $V_2O_5 \cdot nH_2O$, their thermal analysis was carried out. An analysis of integral and differential (**TG** and **DTG**, respectively) curves showed that the decrease in mass of samples in the temperature range of 60-120°C is connected with losing by them of physically bound water; at the same time, within the temperature range of 210-300°C, the loss of chemically bound crystallization water is observed (Fig. 11.1). Quantitative analysis of the obtained results showed that depending on the conditions of drying the amount of crystallization water may be differed significantly. In particular, the composition of the obtained samples was described by the formulas $V_2O_5 \cdot 0.5H_2O$, $V_2O_5 \cdot 1.1H_2O$, and $V_2O_5 \cdot 1.38H_2O$.

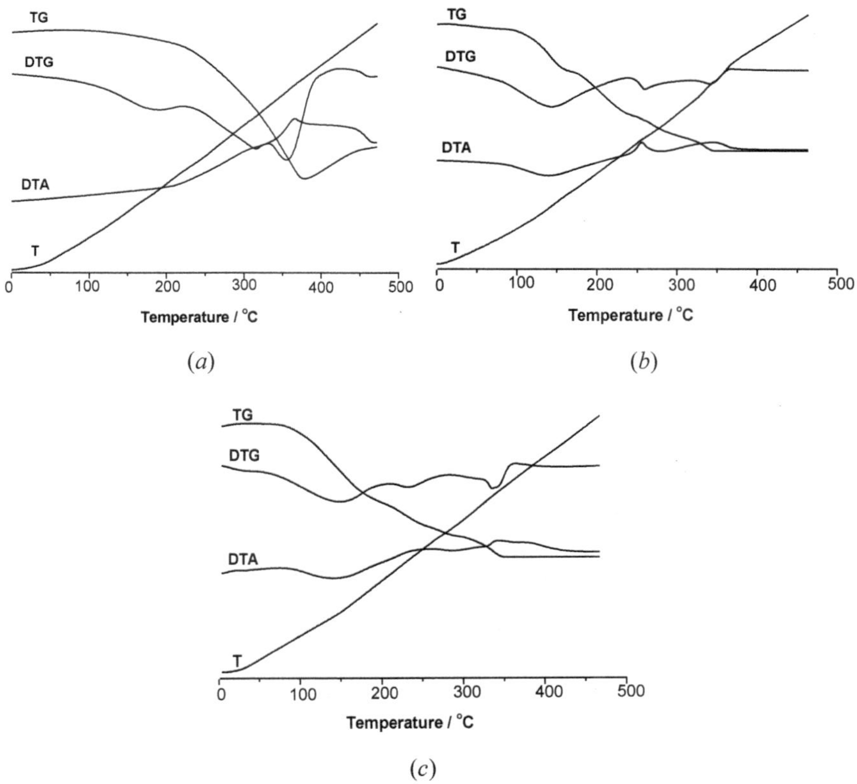

FIGURE 11.1 Derivatograms of the synthesized xerogel samples: (*a*) $V_2O_5 \cdot 0.5H_2O$; (*b*) $V_2O_5 \cdot 1.1H_2O$; and (*c*) $V_2O_5 \cdot 1.38H_2O$ (**T** is the curve of temperature change; **TG** is the curve of the mass loss; **DTG** is the differential curve (rate) of the mass loss; **DTA** is the curve of the rate of heat evolution/heat absorption).

The rate of intercalation of **An** into the interplanar space (interlamellar spacing) determines the distance between the vanadium-oxide layers, which, in turn, is determined by the number of intercalated water. Therefore, a sample of xerogel composition $V_2O_5 \cdot 1.38H_2O$ has been used for further researches. According to the sedimentation analysis (dispersion medium was undecane), the number average and weight average radius of the xerogel particles were 0.20 and 0.60 mm, respectively, at the values of the coefficient of polydispersity 3.1 and 3.01. The specific surface area of the particles of xerogel of $V_2O_5 \cdot 1.38H_2O$ and V_2O_5, obtained via dehydration, was 2.78 and 0.29 $m^2 \cdot g^{-1}$, respectively. Xerogel $V_2O_5 \cdot 1.38H_2O$ is characterized by a broader size distribution of particles compared to V_2O_5, and the largest fraction consists of the particles with a radius of 0.18 mm (38% mass). Results of the sedimentation analysis confirmed the possibility of gels $V_2O_5 \cdot nH_2O$ using as filler with advanced surface for the synthesis of polymeric composites.

At making of xerogel in **An** (in mixture of **An** and **NAn**), the redox reaction between the components of the initial reaction mixture takes place:

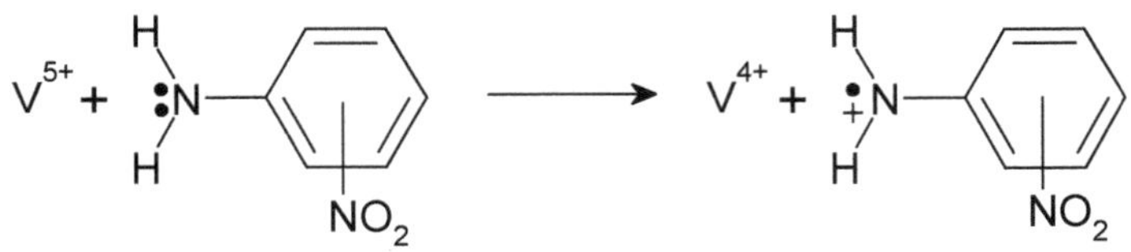

(11.3)

Ions of V^{+5} are reduced to V^{+4}, while being included in the lattice of xerogel.[13] At the same time, the molecules of anilines are oxidized to the primary cation-radicals, which, according to the results of *ref.*[26] are participating in the further growth of polymer chains up to the sizes (**Chapter 1**).

The results of X-ray diffraction studies are by proof of the polymer/copolymer inclusion in the lattice of xerogel. As can be seen from these data (Fig. 11.2, the diffractogram **2**), unmodified xerogel is characterized by a peak position (001) at $2q \approx 8°$ that corresponds to the interplanar distance ≈11.55 Å. Under obtaining of a hybrid composite **PAn**/$V_2O_5\ nH_2O$ (at different initial molar ratios of the components, namely 0.5:1 and 1:1, respectively, *see* Fig. 11.2, diffractograms **3** and **4**), the layered vanadium-oxide matrix $V_2O_5 \cdot nH_2O$ in the sample is kept, but there is an increase in half-width and mainly in the height of the small-angle (at 8°) reflex. At the same time, the result of the xerogel heat treatment, which leads to the formation of anhydrous V_2O_5, is removal of strongly bonded water molecules from the interspace that is involved in the formation of the polyoxovanadate network.

At this, an interlayer distance is decreased,[29] and the diffraction peak almost is disappeared (Fig. 11.2, the diffractogram **1**). So one could argue that a new calculated value of the interplanar distance (*see* diffractograms **3** and **4**), namely ~14 Å is the result of an intercalation of **PAn** in the interlayer space of xerogel.[22] The change of an interlayer distance is 2.45 Å. Within the mass ratios **An**/$V_2O_5 \cdot nH_2O$ in the initial reaction mixture investigated by us, more significant changes of interplanar distances were not observed. Comparing this value with the thickness of the layer, which is formed by the water molecules in the crystal structure of the xerogel (2.8 Å), and the thickness of **PAn**'s chain (5.3 Å), it can be argued that the intercalation of polymer into the interlayer space of the lattice is due to the displacement of water molecules of xerogel (Fig. 11.2). At this, synthesized composites represent by themselves the composites of intercalation type with the electron transfer from the unshared pair of the nitrogen atom of the **PAn**'s chain on the vanadium ion, which is included in a spatial crystalline lattice of vanadium (V) oxide.[22] The Figure 11.3 schematically represents the mechanism of the composite **PAn**/$V_2O_5 \cdot nH_2O$ formation. Therefore, an analysis of X-ray diffractograms for similar products allows us to predict the formation of composite by intercalation type, since a change of position and half-width of diffraction peak (001) at 8° is the result of modification of packaging layers caused by intercalation of **PAn** macromolecules.[30]

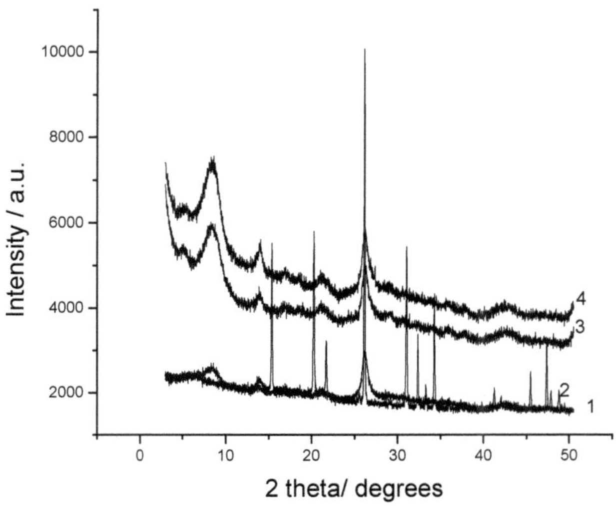

FIGURE 11.2 X-ray diffraction (*XRD*) patterns for initial components (vanadium (V) oxide (**1**) and xerogel $V_2O_5 \cdot 1.38H_2O$ (**2**)) and for the composites **PAn**/$V_2O_5 \cdot nH_2O$ at molar ratio of the components in the starting mix **An**:$V_2O_5 \cdot nH_2O$ 0.5:1 (**3**) та 1:1 (**4**).

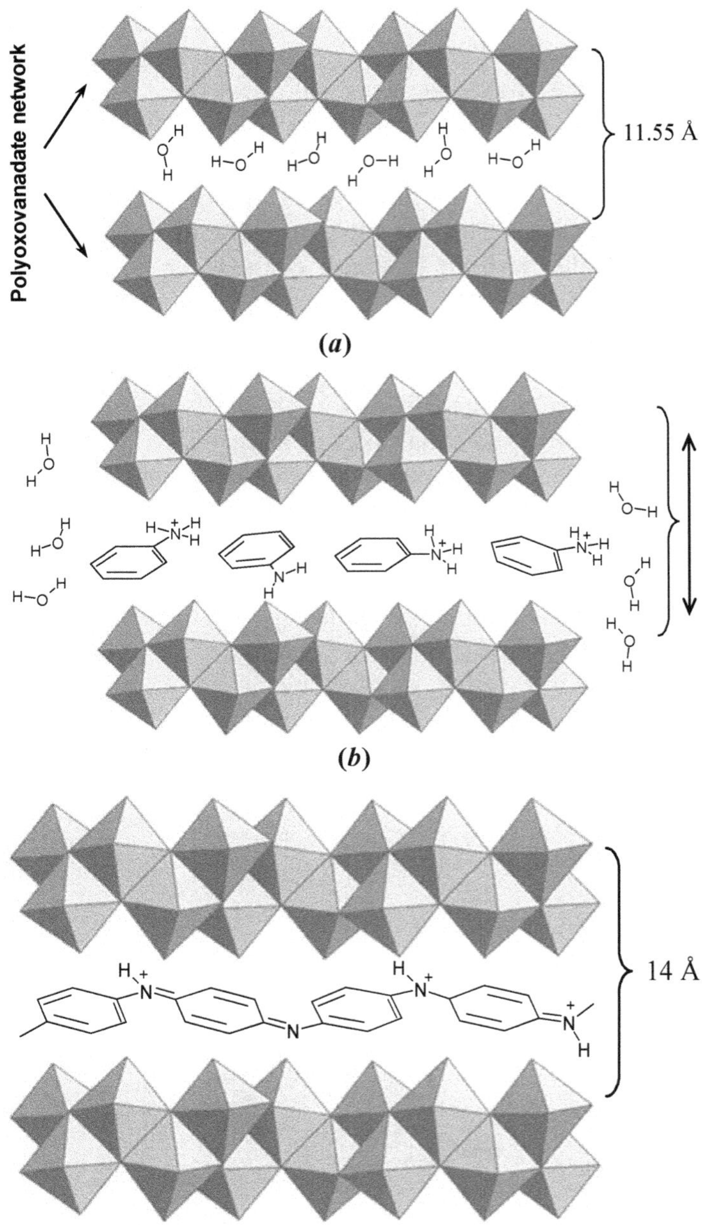

FIGURE 11.3 Scheme of the formation of **PAn**/V_2O_5 composite: (*a*) layered (lamellar) structure of starting xerogel $V_2O_5 \cdot nH_2O$; (*b*) an increase of the interlayer distance as a result of the extrusion of water molecules and an intercalation of **PAn**'s molecules; (*c*) synthesized composite with intercalated chains of **PAn**.

If a chemical reaction is accompanied by a change of enthalpy of the reacting system, the calorimetry can be a versatile method for the estimation of its rate. In some cases, it is the only possible method for studying the kinetics of chemical reactions, especially in dispersed environments that are opaque and viscous. In accordance with *ref.*,[31,32] the polycondensation reaction of **An** is the exothermic process ($\Delta H = -25$ kJ/mol), so the features of the kinetics of the initial stage of polycondensation can be explored by differential calorimetry. According to the obtained calorimetric data (Fig. 11.4), the differential heat curves of heat evolution in the reaction mixture during the chemical reactions proceeding in the studied systems (Fig. 11.4) were constructed. The most intense heat evolution, as can be seen from Figure 11.4, is observed in the initial stage of the process for 1500-2000 s from the time of reagents mixing. At this, the maximum of heat evolution is manifested through 300-500 s from the beginning of the reaction (Fig. 11.5). View of the integral heat curves shows that the reaction between anilines and xerogels is started without an induction period.

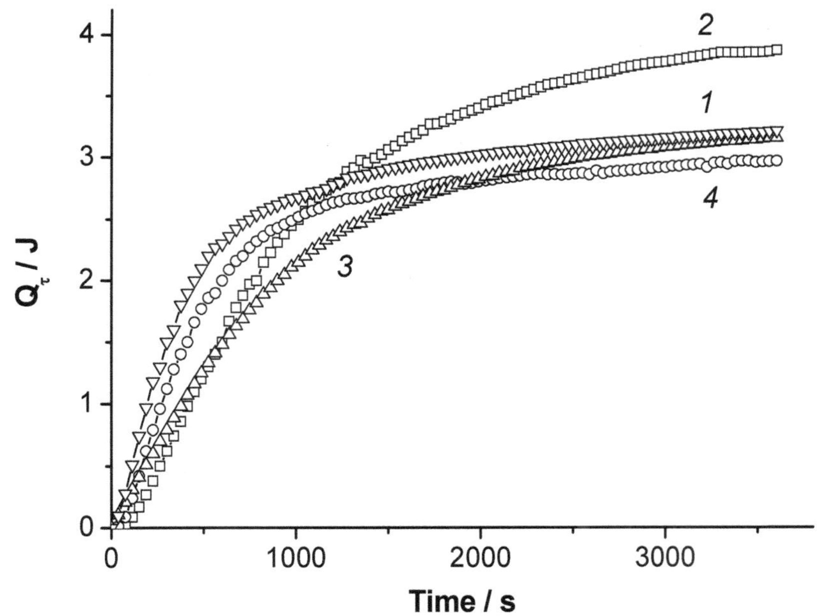

FIGURE 11.4 Typical integral curves of heat evolution during the oxidized polycondensation of **An** (**1**) and copolycondensation of **An** with isomeric nitroanilines (+ *o*-**NAn** (**2**), + *m*-**NAn** (**3**), + *p*-**NAn** (**4**); mass ratio of **An**:**NAn** in starting polymer mix was 9.5:0.5) by xerogel $V_2O_5 \cdot 1.38H_2O$. $T = 30°C$. The composition of reactive mix: 0.056 g of xerogel per 1 ml of monomer/mix of the monomers.

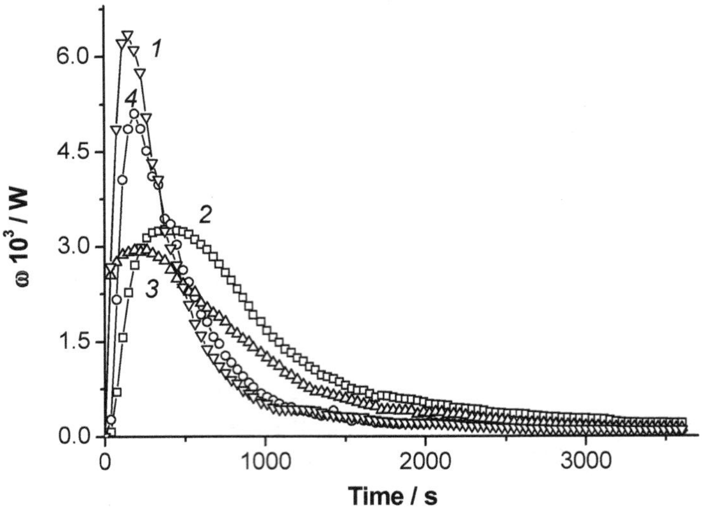

FIGURE 11.5 Typical kinetic curves of heat evolution (change of the heat capacity via time) during the oxidized polycondensation of **An** (**1**) and copolycondensation of **An** with isomeric nitroanilines (+ *o*-**NAn** (**2**), + *m*-**NAn** (**3**), + *p*-**NAn** (**4**); mass ratio of **An**: **NAn** = 9.5:0.5) by xerogel $V_2O_5 \cdot 1.38H_2O$. T = 30°C. The composition of reactive mix: 0.056 of xerogel per 1 ml of monomer/mix of the monomers.

As shown in Figure 11.5 and by numerical results presented in Table 11.1, the rate of heat evolution of oxidative polycondensation of **An** under the action of xerogel $V_2O_5 \cdot 1.38H_2O$ at an initial stage, which is defined as the tangent of the initial straight section of dependence Q = f (τ), is higher compared to the flow process involving two monomers (**An** + isomer **NAn**), which well correlates with the specific heats of reaction in the studied reaction media. It should be noted that the obtained thermal effects are not "pure" polycondensation reaction enthalpies, but some effective values. The components of these values are in particular the enthalpy of intercalation of monomers in the crystal lattice of xerogel, the enthalpy of interaction of the intercalated monomers (copolymers) with vanadium (V) oxide due to the possible formation of hydrogen bonds between the atoms of hydrogen of amine or imine groups of polymer with the oxygen of crystal component (V_2O_5). Analysis of the results shows that the copolymerization process proceeds much slower compared to pure polycondensation of **An**. The slowing effect of **NAn** on polycondensation of **An** appears in the increase of the rate constant of heat evaluation and in increase of total heat of reaction per unit of time with increasing of the relative content of **An** in the staring mixture (for some relationships **An**:**NAn** from 7:3 to 9:1, and to

9.5:0.5). In particular, the specific amount of heat that is released during the polycondensation oxidizing of **An** per unit of mass of xerogel accepts the most numeric value (69.46 $J \cdot g^{-1}$) for composite **An**:*o*-**NAn** = 9.5:0.5, which apparently could have been the greatest degree of conversion mixtures of these monomers at copolycondensation.

TABLE 11.1 Kinetic Parameters of Oxidizing Polycondensation of **An** and Copolycondensation of **An** with Isomeric Nitroanilines by Xerogel $V_2O_5 \cdot 1.38H_2O$ Based on Data of Differential Scanning Calorimetry. The Duration of the Experiment is 1 h.

Composition of reactive mix	Mass ratio An: NAn	$k \cdot 10^3/J \cdot s^{-1}$	Q^*/J	**Specific heat	
				$J \cdot mol^{-1}$	$J \cdot g^{-1}$
$V_2O_5 \cdot nH_2O$ + **An**	-	3.78	3.30	300.00	56.96
$V_2O_5 \cdot nH_2O$ + (**An**:*o*-**NAn**)	9.5:0.5	3.00	3.89	356.88	69.46
$V_2O_5 \cdot nH_2O$ + (**An**:*m*-**NAn**)		2.25	3.22	295.41	57.50
$V_2O_5 \cdot nH_2O$ + (**An**:*p*-**NAn**)		3.50	2.98	273.39	53.21
$V_2O_5 \cdot nH_2O$ + (**An**:*o*-**NAn**)	9.0:1.0	3.13	2.50	44.64	44.64
$V_2O_5 \cdot nH_2O$ + (**An**:*m*-**NAn**)		2.64	2.27	40.54	40.54
$V_2O_5 \cdot nH_2O$ + (**An**:*p*-**NAn**)		2.00	2.62	46.77	46.77
$V_2O_5 \cdot nH_2O$ + (**An**:*o*-**NAn**)	7.0:3.0	3.60	3.41	60.89	60.89
$V_2O_5 \cdot nH_2O$ + (**An**:*m*-**NAn**)		3.20	1.66	29.64	29.64
$V_2O_5 \cdot nH_2O$ + (**An**:*p*-**NAn**)		2.82	1.68	30.00	30.00

***Notes:** *Q* is an integral heat evaluation during the experiment (composition of reactive max: 0.056 g of xerogel on 1 ml of monomer/mix of monomers; **Specific heats represented from the calculations on 1 mol of monomer/mix of monomers ($J \cdot mol^{-1}$) and on 1 g of xerogel ($J \cdot g^{-1}$)).

As can be seen from Table 11.2, the rate of oxidizing copolycondensation of **An** + isomer of **NAn** in the presence of xerogel depends on the composition of the reaction mixture, confirming the different reactivity of **NAn** isomers to the copolycondensation reaction with **An**. For the reactive mixture of **An** + $V_2O_5 \cdot nH_2O$, the rate constant[33] at the initial stage of the reaction is $3.78 \cdot 10^{-3}$ $J \cdot s^{-1}$. The presence of *m*-, *p*- and *o*- isomers of **NAn** causes a slight slowdown of the initial stage of the polycondensation reaction, which is reflected on a decrease of the numerical value of the heat evaluation for the polycondensation process versus the polycondensation of pure **An**. Reaction rate constant of the heat evaluation of the copolycondensation process becomes the largest numerical value for *o*-derivative of **NAn** for the all investigated ratios **An**:**NAn**. Therefore, the investigated nitro-derivatives of **An** can be arranged in the following series as to reducing of their reactivity in the reaction of the copolymer formation:

$$o\text{-}\mathbf{NAn} \gg p\text{-}\mathbf{NAn} > m\text{-}\mathbf{NAn}. \quad (11.4)$$

However, this relationship is not entirely unambiguous. In particular, the decrease of **An** in a mixture of monomers from 9.5:0.5 to 7:3 leads to an increase of the rate of heat evaluation in the case of copolycondensation of **An** with *o*- and *m*- isomers of **NAn**, while the rate of heat evaluation in the system **An** + *m*-**NAn** begins to exceed this option for a mixture of **An** + *p*-**NAn**. Such ambiguous dependencies were observed at dilatometer study of copolycondensation of anilines in the presence of xerogel. The observed contraction of the polymerization mixture is the confirmation of anilines oxidation by xerogel with the formation of the polymer product. As shown in Figure 11.6, the contraction of volume of the reaction mixture, and thus the polycondensation reaction rate is the highest in a case of *o*-derivative of **NAn**. However, the rate of a process with the participation of *m*-derivative (which is the smallest) in the first 15 min of the beginning of the introduction of xerogels into a mixture of monomers next exceeds the rate of the process with the participation of *p*-derivative of **NAn**, and after 40 min even of *m*-derivative of **NAn**. At this, it should be noted that the analysis of kinetic data obtained by dilatometer method is complicated by the fact that the specific volume of polycondensation product is a function of not only the composition of the initial mixture, but the ratio of rates of polycondensation with the participation of cation radicals of **An** and its nitro-derivatives, respectively.

Obviously, in the presence of **Nan,** the labile complex by type ($V_2O_5 \cdot nH_2O$-nitroaniline–aniline-$V_2O_5 \cdot nH_2O$) in the reactive system is formed, the decay rate of which will be determined by the stabilizing action of tricenter hydrogen bonds N···H···O, which appears between atoms of hydrogen of amino groups of **An** and **NAn** with oxygen atoms of polyvanadate layers and nitro-groups of **NAn**. In addition, in the complexes of the type [metal ion-ligands] the electrons are distributed on energy levels differently than the electrons of ligands taken separately, leading to the "inflexible" spatial distribution of ligands around the central ion in a relatively small space. This brings the ligands (in our case, the molecules of **An** and **NAn**s) and opens the possibility for the occurrence of such reactions which could met the difficulties in a medium containing the randomly distributed particles. At this, it should be mentioned that the nitroanilines (in the absence of **An**) do not give the polymer products of polycondensation.

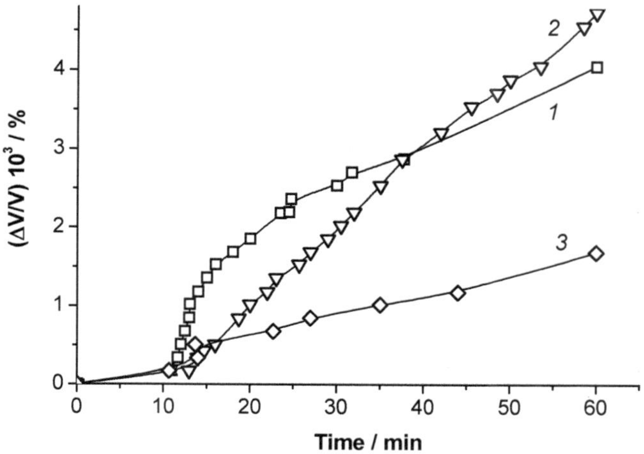

FIGURE 11.6 Dependence of relative change of the volume of reactive mix via time during the oxidizing copolycondensation of **An** with *o*-**NAn** (**1**), *m*-**NAn** (**2**) and *p*-**NAn** (**3**) in the presence of xerogel $V_2O_5 \cdot 1.38H_2O$. Molar ratio **An** : **NAn** = 7:3; mass ratio xerogel:monomers = 1:10; T = 293°C.

On a basis of elementary analysis, it was calculated that the composition of the obtained composite based on **An** corresponds to the gross formula $(C_6H_4\text{-}NH)_{0.72} \cdot V_2O_5 \cdot yH_2O$ (where $y < 1.38$, since the **An** as a result of its intercalation into interlamellar structure of xerogel replaces water molecules from it (Fig. 11.3)). So, next at recording of the structure of composites, an index 1.38 shows not actual content of water into the structure of the inorganic component, but the composition of the original xerogel, which was used for its synthesis.

In the case of copolymers of **An** with **NAns**, a composition of the resulting composite cannot be determined accurately due to the presence in the structure of copolymer of two functional groups containing the nitrogen. The most content of nitrogen in the composite is typical for the copolymer of **An** with *o*-**NAn** (7.05% mass compared with 5.85 and 5.10% mass for *p*- and for *m*-**NAn**, respectively) (Table 11.2). This fact, which is connected with an enrichment of copolymer composition by links, containing the nitrogroup, again confirms the most reactivity toward this process of *o*-derivative of **NAn**.

Typical integral and differential derivatographic curves of synthesized samples (on example of the composite of poly(**An**–*co*–*o*–**NAn**)/ $V_2O_5 \cdot 1.38H_2O$) are shown in Figure 11.7 (*a*). Comparison of these relationships with the derivatograms of poly(**An**–*co*–*o*–**NAn**) copolymer obtained under the same conditions, shows that the main difference in the thermal properties of the composite and of the polymer is observed at the temperatures in the range of 300-500 °C. However, even the first maximum on **DTG** curve of the composite (in the range 20-300 °C) is significantly wider compared to the maximum, which is observed for the samples of copolymer. This difference in thermal properties is primarily concerned with the fact that the removal processes of the remaining water after the synthesis of composite in the structure of xerogel (in the range 150-250 °C) are superimposed on the processes of thermal desorption of water physically bounded to the amino- and imino- groups of the polymer (in the range 30-170 °C).[34,35] It is bounded much more strongly due to the fact that its molecules are involved in the formation of the polyoxovanadate network.

TABLE 11.2 Data of Elementary Analysis of the Synthesized Hybrid Polymer–Inorganic Composites Based on **PAns** and Xerogel $V_2O_5\cdot 1.38H_2O$: **I** – **An** + $V_2O_5\cdot 1.38H_2O$; **II** – **An**:*o*-**NAn** (9.5:0.5) + $V_2O_5\cdot 1.38H_2O$; **III** – **An**:*m*-**NAn** (9.5:0.5) + $V_2O_5\cdot 1.38H_2O$; **IV** – **An**:*p*-**NAn** (9.5:0.5) + $V_2O_5\cdot 1.38H_2O$.

Sample	Content, % mass			
	C	N	H	V_2O_5
I	18.40	5.19	3.26	73.15
II	20.41	7.05	3.04	72.54
III	15.07	5.10	2.19	77.64
IV	21.59	5.85	3.11	69.45

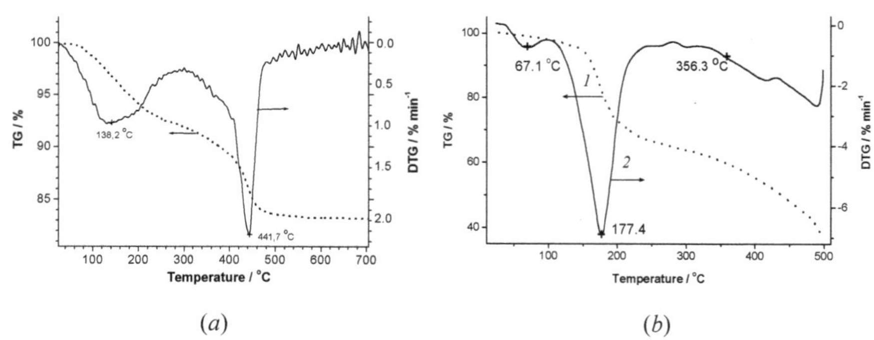

FIGURE 11.7 Integral (**1**) and differential (**2**) derivatograms of composite samples of poly(**An**–*co*–*o*–**NAn**)/$V_2O_5\cdot 1.38H_2O$ (*a*) and copolymer of poly(**An**–*co*–*o*–**NAn**) (*b*). Mass ratio of monomers in starting polymerizing mix was 7:3.

This fact is confirmed by typical IR-spectrum of the obtained composite. In particular, the band at ~925 cm^{-1} is available in spectra both of xerogel (Fig. 11.8, spectrum 2) and composite (Fig. 11.8, spectrum 3), and can be assigned to the V–OH$_2$ valence vibrational mode indicating the formation of coordination bonds of water molecules with vanadium atoms of vanadyl groups in the interlamellar domain.[36] In both spectra, the absorption bands at 1605, 1004, 730, and 537 cm^{-1} are also appeared, which correspond to the δ(H-O-H), ν(V=O), δ(V-O-V), and δ(V-O) fluctuations of the "constituents" of xerogel.[37] In accordance with *ref.*,[29] even following a xerogel annealing for 15 min at a temperature 270 °C, the composition of the annealed sample of xerogel is described by the formula $V_2O_5 \cdot 0.32H_2O$, and anhydrous V_2O_5 was obtained by authors at the annealing temperature 600 °C. So, we can conclude that the second maximum on **DTG** curve of the composite is formed by the superposition of two processes, namely thermal degradation of the polymer (at temperatures over 300 °C, just as it is observed for copolymer, Fig. 11.7 (*b*)) and removing of the strongest bounded water from the crystal lattice of xerogel (at the temperatures over 400 °C). This is also confirmed by IR-spectra of gaseous products of synthesized composite thermolysis (Fig. 11.9 (*b*)), in which the lines in the range of 1250–1650 cm^{-1} (with a maximum at ~1530 cm^{-1}) and 1650–2100 cm^{-1} (with a maximum at ~1700 cm^{-1}) are traced, which correspond to the absorption of gaseous water.[38]

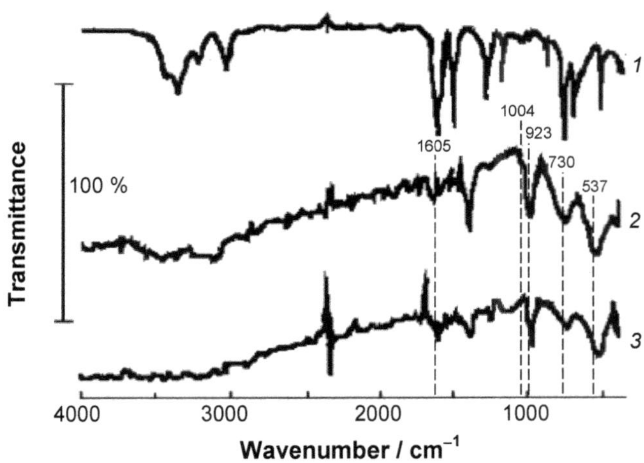

FIGURE 11.8 IR-spectra of **An** (**1**), synthesized xerogel $V_2O_5 \cdot 1.38H_2O$ (**2**) and poly(**An**–*co*–*o*–**NAn**)/$V_2O_5 \cdot 1.38H_2O$ composite (**3**) (mass ratio of monomers in starting polymerizing mix was 7:3).

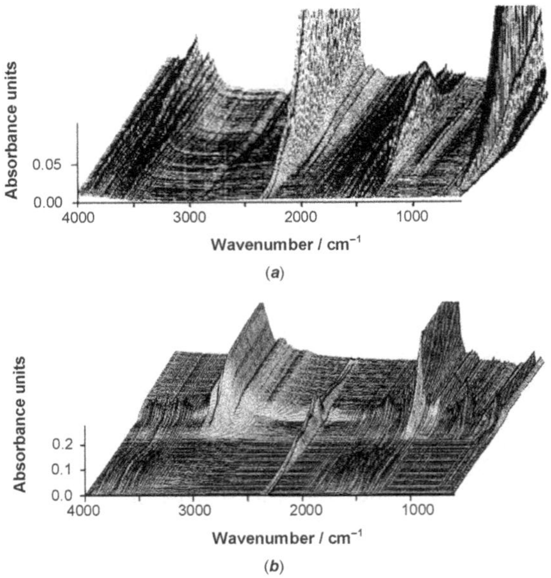

FIGURE 11.9 Change of the infrared absorption spectrum of the thermodestruction products of the aniline copolymer with the *o*-nitroaniline (*a*) and poly(**An**–*co*–*o*–**NAn**)/$V_2O_5 \cdot 1.38H_2O$ composite (*b*) with time (mass ratio of **An**:*o*-**NAn** 7:3).

Generally, the characteristic bands corresponding to the formation of polymer in the IR-spectra of the synthesized composites are less intense and more blurred compared with the spectra of samples of identical copolymers (Fig. 11.10), due to apparently the influence of xerogel matrix. At the same time, the available bands in spectrum in the range 1500–1570 cm^{-1} and 1330–1370 cm^{-1} corresponding to asymmetric and symmetric valence vibrations of $-NO_2$ group[39] indicate the formation of copolymer in xerogel matrix.

Returning to the results of thermogravimetric analysis of the composite, it can be assumed that at temperatures above 300 °C, not only the thermolysis of intercalated copolymer (including argon) will proceed, but also its oxidative destruction with the participation of ions vanadium (V) and nitro-groups of copolymer that causes much faster decrease in mass of the sample. Significant differences in the IR-spectra of the gaseous products of thermolysis of composite (Fig. 11.9 (*a*)) and of the corresponding copolymer (Fig. 11.9 (*b*)) are the evidences of this fact. Spectral analysis of gaseous products degradation of the polymer and of the composite samples showed that in both cases, the absorption bands with maxima at 2359 and 2344 cm^{-1} arising at the initial stages of thermolysis are typical for the main products. Somewhat less intense bands are observed at 668 and at 1370–1320 cm^{-1}.

If the first two bands can be attributed to the valency vibrations of $-NH_2^+$ and $C{=}NH^+$ groups, then the next two bands correspond to asymmetric valency vibrations of NO_2-group, amino-group in aromatic amines, and to OH-groups of phenols.[39,40] A set of absorption bands in the range 3500-3800 cm^{-1} confirms the presence of OH-group in the products of thermolysis.[39] Besides, the intense band at 668 cm^{-1} may be also connected with the deformation vibrations of N-H-group.

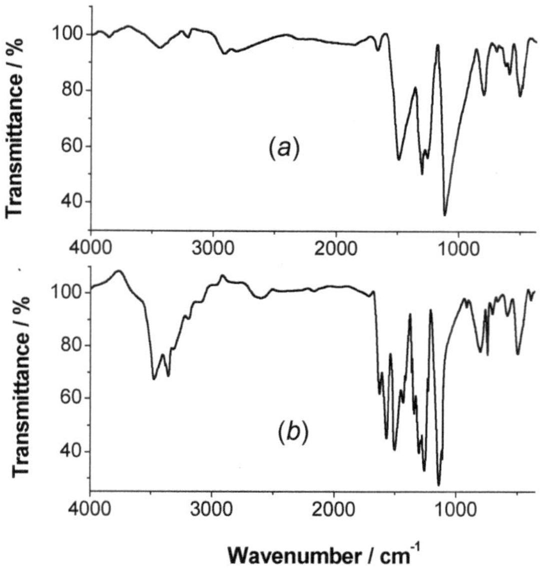

FIGURE 11.10 *FTIR* spectra of copolymers of **An** and *o*-**NAn** at molar ratio of monomers in starting polymerizing mix 9:1 (*a*) and 7:3 (*b*).

The main peculiarity of the spectrum of the gaseous products of thermolysis of composite is that the intensity of the above bands is essentially decreased. Instead, in 1000 s after the start of thermolysis, which corresponds to the temperature ~300 °C, there is a set of the intense bands in the range of 2500-3500 cm^{-1} (Fig. 11.9 (*b*)). The appearance of this set of bands is in good agreement with the beginning of the second maximum on the **DTG** curve of the composite sample (Fig. 11.7 (*b*)). The valency vibrations of C-N aromatic ring (3030-3080 cm^{-1}) and of C=O group of benzoquinones (3200-3550 cm^{-1}) can be associated with this wavenumber interval band.[39] Moreover, the absorption of gaseous nitrogen is observed in the same range.[38] These results confirm the above-done assumption about the oxidative nature of the copolymer degradation at temperatures above 300 °C. If the thermolysis of copolymers of **An** and **NAn** mainly proceeds with the formation of aniline or its protonated derivatives as well as hydroquinones, then the xerogels at higher temperatures oxidizes the copolymer, resulting in the formation of the main degradation products such as benzene, benzoquinones, and ammonia.

According to the data given in Table 11.3, the electrical conductivity of synthesized composites is in the range (1.77–2.49)$\cdot 10^{-4}$ S$\cdot$cm^{-1} that satisfies the requirements[10] on the use of synthesized composites based on **PAn** intercalated in the structure of xerogel $V_2O_5 \cdot nH_2O$ or its copolymer with nitroanilines as cathode materials of Li-EES. Designed on their basis models of lithium power sources were characterized by open circles 3.20-3.62 V, which is in good agreement with the results presented in *ref.*[13]. The discharge of designed models of Li-EES was carried out by an external constant resistance circle. Typical discharge curves of elements constructed based on composites **PAn**$\cdot V_2O_5 \cdot 1.38H_2O$ (the element **I**) and poly(**An**–*co*–*o*–**NAn**)$\cdot V_2O_5 \cdot$1.38H2O (the element **II**) are shown in Figure 11.11 and in Figure 11.12, respectively. Specific characteristics of the element, namely specific capacity and specific energy calculated by integrating of the discharge curves (without taking into account of electrolyte mass and of constructive materials of **EES** models[41]), were within 248–298 A$\cdot$h$\cdot$kg^{-1} and 158–394 W$\cdot$h$\cdot$kg^{-1}, respectively.[42]

TABLE 11.3 Conductivity of Synthesized Samples of Composites. Ratio (**An**:isomer of **NAn**) = 7:3 (mass. parts) Into the Starting Reactive Mix.

The composition of composite	**Conductivity $\chi \cdot 10^4$, S$\cdot$cm^{-1}**
Copolymer (**An** + *m*-**NAn**) + $V_2O_5 \cdot 1.38H_2O$	1.77
Copolymer (**An** + *p*-**NAn**) + $V_2O_5 \cdot 1.38H_2O$	2.98
Copolymer (**An** + *o*-**NAn**) + $V_2O_5 \cdot 1.38H_2O$	2.49
PAn + $V_2O_5 \cdot 1.38H_2O$	2.42

According to the presented results, synthesized materials can be used not only in primary but also in secondary electrochemical systems, namely storages. At this, the calculations based on the discharge capacity of the galvanic cells show that both the redox processes of the polymer component and the reverse transition $V^{5+} \leftrightarrow V^{4+}$ are current-forming processes. The results of the designed models cycling are shown in Table 11.4. As it was evidenced by the results, the specific characteristics of models are improved during the cycling first of all at the expense of the internal resistance of elements reducing. Somewhat worse characteristics (in the case of element **I** based on **PAn**) or some their deterioration (in the case of element **II** based on composite of copolymer of **An** with *o*-**NAn**) on the initial charge–discharge cycles, first of all are due to the formation of the oxide film on the surface of lithium and its destruction during the cycling. The reason of this phenomenon is the residual moisture containing in the electrolyte or in the electrode materials, or it can be connected with the displacement of water molecules by electrolyte as a result of its penetration into the interlayer galleries of xerogel. Another reason could also be the better penetration of the electrolyte into the structure of active mass of pressed cathode and its swelling that slightly reduces the polarization loss (the overvoltage during the polarization of electrodes).

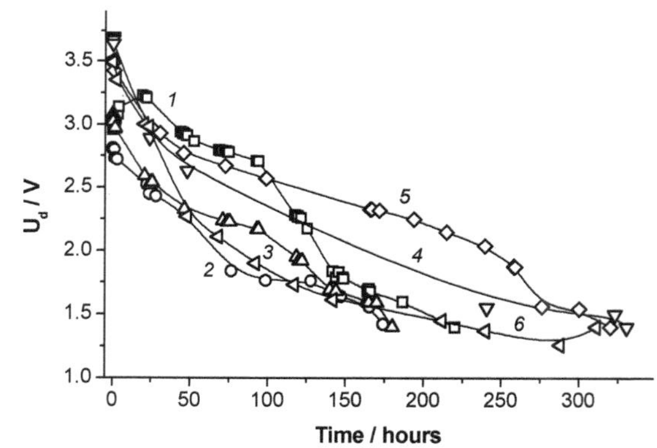

FIGURE 11.11 Discharge curves of Li-**EES** prototype by structure Li | 1 M $LiClO_4$ in **AcN** | **PAn**·V_2O_5·$1.38H_2O$ + ABG-81 (the ciphers indicate on number of the discharge cycle).

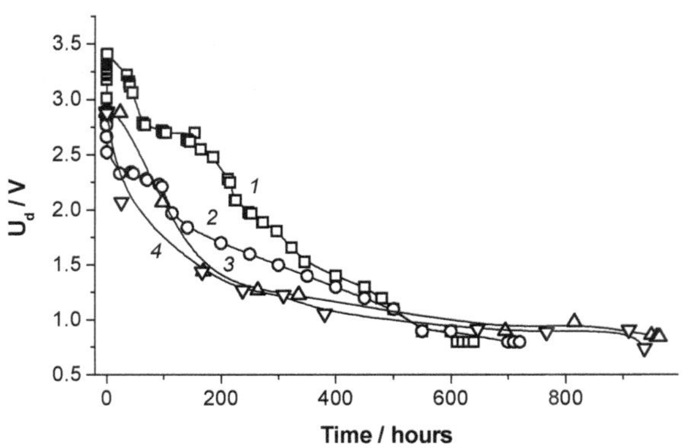

FIGURE 11.12 Discharge curves of Li-**EES** prototype by structure Li | 1 M $LiClO_4$ in **AcN** | poly(**An**–*co*–*o*–**NAn**)·V_2O_5·$1.38H_2O$ + ABG-81 (the ciphers indicate on number of the discharge cycle).

TABLE 11.4 Discharge Characteristics* of the Designated Lithium Electrochemical Energy Storage (**EES**): Li | PAn·V_2O_5·1.38H_2O (**I**); Li | poly(**An**–*co*–*o*–**NAn**)·V_2O_5·1.38H_2O (**II**). Mass Ratio of Monomers into the Starting Polymerizing Mix Was 7:3.

Element	Cycle	U_{OC}/V	$\bar{U}_d/V$	j_d/mA·cm^{-2}	R_C/Ohm	Q_g/A h·kg^{-1}	W_g/W·h·kg^{-1}
I	1	3.62	2.51	0.09	123	248	158
	2	3.19	2.45	0.09	271	165	338
	3	3.25	2.11	0.09	112	169	379
	4	3.82	2.23	0.10	65	266	323
	5	3.61	2.39	0.11	64	298	394
	6	3.62	2.41	0.11	64	296	394
Average value		3.52	2.35	0.10	117	240	331
II	1	3.67	2.91	0.09	44	418	550
	2	3.34	2.50	0.06	155	205	446
	3	3.18	2.35	0.08	201	242	339
	4	3.20	2.68	0.08	90	300	460
Average value		3.35	2.61	0.08	123	292	449

***Note:** U_{OC} is the voltage of the open circuit; $\bar{U}_d$ is an average voltage of the discharge, j_d is the density of current discharge at the initial moment of time; R_C is the internal resistance; Q_g is the specific capacity; W_g is the specific energy.

Comparison of the results obtained for two types of the constructed models of elements (Table 11.4) shows that the introduction of the *o*-**NAn** links into the structure of polymer leads to some improvement (on ~20-25%) of the specific characteristics of Li-**EES**. First of all, this effect is due to the presence of the electroactive nitro-groups in the structure of the copolymer and is another proof that the copolymer but not the mechanical mixture of **PAn** and nitro-derivative is formed during the oxidation of a mixture of **An** with *o*-**NAn.**[25]

11.4 CONCLUSIONS

So, the presented results confirm the possibility of obtaining the nanocomposites by "guest-host" type based on vanadium (V) oxide sol, and conducting polymer by intercalation of monomer molecules into the interlayer galleries of sol nanoparticles. Further oxidative polycondensation of anilines under the action of host material (V_2O_5) leads to the formation of hybrid organic–mineral composite. The monomer displaces water molecules from the interlamellar space of xerogel during the intercalation. At this, further interaction of the formed polymer with polyoxovanadate network leads to significant changes of physical and chemical properties both of a xerogel and of a conducting polymer/copolymer.

Analyzing the research results, we can conclude that the polymers based on derivatives of aniline intercalated into the crystalline lattice of xerogel $V_2O_5 \cdot nH_2O$ are promising electroactive materials for the manufacture of cathodes for secondary Li-**EES**. Cations of Li^+ are capable reversibly to intercalate into cathode material during discharge of chemical power source of this kind just as in the case of inorganic layered cathode materials. The presence of conductive polymer layer into the interlayer space of xerogel permits to increase the lifetime of the cathode materials, because of their ability to withstand a large number of charge–discharge cycles is increased.

KEYWORDS

- **polyaniline**
- **vanadium (V) oxide**
- **xerogel**
- **hybrid organic–inorganic composites**
- **thermogravimetric analysis**
- **IR-spectroscopy**
- **lithium electrochemical energy storage**

REFERENCES

1. Posudievsky O. Yu.; Kozarenko, O. A.; Dyadyun, V. S.; Jorgensen, S. W.; Spearot, J. A.; Koshechko, V. G.; Pokhodenko, V. D. Mechanochemically Prepared Ternary Hybrid Cathode Material for Lithium Batteries. *Electrochim. Acta.* **2013,** *109,* 866–873.
2. Boyano, I.; Bengoechea, M.; Meatza, I.; Miguel, O.; Cantero, I.; Ochoteco, E.; Rodríguez, J.; Lira-Cantú, M.; Gómez-Romero, P. Improvement in the Ppy/V_2O_5 Hybrid as a Cathode Material for Li- Ion Batteries Using PSA as an Organic Additive. *J. Power Sources.* **2007,** *166,* 471–477.
3. Lira-Cantu, M.; Gomez-Romero, P. Synthesis and Characterization of Intercalate Phases in the Organic-Inorganic Polyaniline/ V_2O_5 System. *J. Solid State Chem.* **1999,** *147,* 601–608.

4. Macdiarmid, Alan G. "Synthetic Metals": A Novel Role for Organic Polymers. *Curr. Appl. Phys.* **2000,** *1,* 269–279.
5. Skotheim, T. A. Ed. *Handbook of Conducting Polymers*. Marcel Dekker Inc.: New York, 1986; Vol. 1, pp 2.
6. Posudievsky O. Yu.; Kozarenko, O. A.; Dyadyun, V. S.; Jorgensen, S. W.; Spearot, J. A.; Koshechko, V. G.; Pokhodenko, V. D. Characteristics of Mechanochemically Prepared Host–Guest Hybrid Nanocomposites of Vanadium Oxide and Conducting Polymers. *J. Power Sources.* **2011,** *196,* 3331–3341.
7. Sun, W.; Qin, P.; Gao, H.; Li, G.; Jiao, K. Electrochemical DNA Biosensor Based on Chitosan/Nano- V_2O_5/Mwcnts Composite Film Modified Carbon Ionic Liquid Electrode and its Application to the Lamp Product of *Yersinia Enterocolitica* Gene Sequence. *Biosens. Bioelectron.* **2010,** *25,* 1264–1270.
8. Anaissi, F. J.; Demets, G. J. F.; Timm, R. A.; Toma, H. E. Hybrid Polyaniline/Bentonite - Vanadium(V) Oxide Nanocomposites. *Mat. Sci. Eng. A.* **2003,** *347,* 374–381.
9. Ragupathy, P.; Shivakumara, S.; Vasan, H. N.; Munichandraiah, N. Preparation of Nanostrip V_2O_5 by the Polyol Method and its Electrochemical Characterization as Cathode Material for Rechargeable Lithium Batteries. *J. Phys. Chem. C.* **2008,** *112,* 16700–16707.
10. Koval'chuk, E. P.; Ostapovych, B. B.; Seledets, M. V.; Turyk, Z. L. Chemical Power Sources with Hybrid Cathodes on the Base of Polyaniline and $V_2O_5 \cdot nH_2O$ Xerogel. *Ukr. Khim. Zh.+* (Ukrainian Chemical Journal) **2005,** *71,* 52–55 (In Ukrainian).
11. Augustyn, V.; Simon, P.; Dunn, B. Pseudocapacitive Oxide Materials for High-Rate Electrochemical Energy Storage. *Energy Environ. Sci.* **2014,** *7,* 1597-1614.
12. Jing, Y.; Zhou, Z.; Cabrera, C. R.; Chen, Z. F. Graphene, Inorganic Graphene Analogs and their Composites for Lithium Ion Batteries. *J. Mater. Chem. A.* (2014), *2,* 12104-12122.
13. Zakharova, G. S.; Volkov, V. L. Intercalation Compounds Based on Vanadium(V) Oxide Xerogel. *Russ. Chem. Rev.* **2003,** *72,* 311–325.
14. Ferreira, M.; Zucolotto, V.; Constantino, C. J. L.; Temperini, M. L. A.; Torresi, R. M.; Oliveira, O. N. Layer-By-Layer Hybrid Films of Polyaniline and Vanadium Oxide. *Synth. Met.* **2003,** *137,* 969-972.
15. Ferreira M., Zucolotto, V., Huguenin, F., Torresi, R. M., Oliveira, O. N. Layer-By-Layer Nanostructures Hybrid Films of Polyaniline and Vanadium Oxide (V_2O_5). *J. Nanosci. Nanotechno.* **2002,** *2,* 29–33.
16. Huguenin, F.; Ticianelli, E. A.; Torresi, R. M. Xanes Study of Polyaniline – V_2O_5 and Sulfonated Polyaniline – V_2O_5 Nanocomposites. *Electrochim. Acta.* **2002,** *47,* 3179–3186.
17. Anaissi, F. J.; Demets, G. J. F.; Timm, R. A.; Toma, H. E. Hybrid Polyaniline/Bentonite - Vanadium (V) Oxide Nanocomposites. *Mat. Sci. Eng. A.* **2003,** *347,* 374–381.
18. Huguenin, F.; Torresi, R. M. Electrochemical Behavior and Structural Changes of V_2o_5 Xerogel. *J. Braz. Chem. Soc.* **2003,** *14,* 536–540.
19. Guerra, E. M.; Brunello, C. A.; Graeff, C. F. O., Herenilton P. Synthesis, Characterization, and Conductivity Studies of Poly- O-Methoxyaniline Intercalated into V_2O_5 Xerogel. *J. Solid State Chem.* **2002,** *168,* 134–139.
20. Harreld, J.; Wong, H. P.; Dave, B. C.; Dunn, B.; Nazar, L. F. Synthesis and Properties of Polypyrrole–Vanadium Oxide Hybrid Aerogels. *J. Non-Cryst. Solids.* **1998,** *225,* 319–324.
21. Demets, G.; Toma, H. Strong Electric Promote Oriented Intercalative Polymerization of Pyrrole Inside the Lamellar Matrices of Vanadium Pentoxide. *Electrochem. Commun.* **2003,** *5,* 73–77.

22. Kanatzidis, M. G.; Wu, C. G.; Macy, H. O.; Kannewurf, C. R. Conductive-Polymer Bronzes. Intercalated Polyaniline in Vanadium Oxide Xerogels. *J. Am. Chem. Soc.* **1989,** *111,* 4139–4141.
23. Novak, P.; Muller, K.; Santhanam, K. S. V.; Haas, O. Electrochemically Active Polymers for Rechargeable Batteries. *Chem. Rev.* **1997,** *97,* 207-281.
24. Govard, G. R.; Leroux, F.; Nazar, L. F. Poly(Pyrole) and (Thiophene)/Vanadium Oxide Interleaved Nanocomposites: Positive Electrodes for Lithium Batteries. *Electrochim. Acta.* **1998,** *43,* 1307–1313.
25. Koval'chuk, E. P.; Skolozdra, O. M.; Reshetnyak, O. V.; Błażejowski, J. Use of Polyanilines as Cathodic Materials in Magnesium Power Sources. *Ukr. Khim. Zh.+* (Ukrainian Chemical Journal) **2001,** *67* (Is. 4), 99–103 (In Ukrainian).
26. Koval¢Chuk, E. P.; Whittingham, S.; Skolozdra, O. M.; Ather. Co-Polymers of Aniline and Nitroanilines. Part I. Mechanism of Aniline Oxidation Polycondensation. *Mater. Chem. Phys.* **2001,** *69,* 154–162.
27. Brauer, G. M., Ed. Guidelines for the Inorganic Synthesis; Mir: Moscow, Russia, 1985; Vol. 5, pp 1525.
28. Kalve, E.; Prat, A. *Microcalorimetry*. Publ Foreign Lit.: Moscow, Russia, 1963.
29. Barbosa, G. N.; Graeff, C. F. O.; Oliveira, H. P. Thermal Annealing Effects On Vanadium Pentoxide Xerogel Films. *Ecl. Quím.*, São Paulo, **2005,** *30,* 7-15.
30. Pokhodenko, V. D.; Krylov, V. A.; Kurys', Ya. I. Spectral and Electrochemical Properties of Polyaniline-V_2O_5 Composites of the Intercalation Type. *Theor. Exp. Chem.* **1995,** *31,* 301– 304.
31. Gupta, V. K. Kinetics and Mechanism of Oxidation of p-Substituted Anilines by Peroxydisulfate Ion in Acetic Acid-Water Medium. *React. Kinet. Catal. Lett.* **1985,** *27,* 207–211.
32. Gupta, V. K. Thermodynamic and LFER Studies for the Oxidation of Anilines by Peroxydisulphate Ion in Acetic Acid-Water Medium, Z. *Physik. Chemie. Leipzig.* **1986,** *267,* 204-210.
33. Seledets, M. V.; Koval'chuk, E. P.; Ostapovych, B. B.; Turyk, Z. L. *Kinetics of Oxidative Polycondensation of Polyaniline Derivatives on the Xerogel $V_2O_5{\cdot}nH_2O$ Surface,* VII Polish-Ukrainian Symposium on Theoretical and Experimental Studies of Interfacial Phenomena and their Technological Applications (Book of Abstracts), Lublin, Poland, September 15–18, 2003; pp 215.
34. Koval'chuk, E. P.; Whittingham, S.; Skolozdra, O. M., Co-Polymers of Aniline and Nitroanilines. Part II. Physicochemical Properties. *Mater. Chem. Phys.* **2001,** *70,* 38–48.
35. Anand, J.; Palaniappan, S.; Sathyanarayana, D. N. Spectral, Thermal, and Electrical Properties of Poly(o- and m-Toluidine)-Polystyrene Blends Prepared By Emulsion Pathway. *J. Polym. Sci. Pol. Chem.* **1998,** *36,* 2291–2299.
36. Petkov, V.; Trikalitis, P. N.; Bozin, E. S.; Bollinge, S. J.; Vogtand T., Kanatzidis, M. G. Structure of $V_2O_5{\cdot}nH_2o$ Xerogel Solved by the Atomic Pair Distribution Function Technique. *J. Am. Chem. Soc.* **2002,** *124,* 10157- 10162.
37. Parakash, R.; Marimuthu, R.; Mandale, A. B. Structural and Electrochemical Properties of Nanocomposites Formed by V_2o_5 and Polyaniline. *Polymer*. **2001,** *42,* 2991-3001.
38. NIST (The National Institute of Standards and Technology) Standard Reference Database 69: *NIST Chemistry Webbook.* http://webbook.nist.gov/chemistry
39. Kazitsina, L. A.; Kupletskaya, N. B. *Application of IR, UV and NMR Spectroscopy in Organic Chemistry*. Vysshaya Shkola: Moscow, 1971 (In Russian).

40. Nakanishi, K. *Infrared Absorption Spectroscopy, Practical*. Mir: Moscow, 1965; pp 216
41. Kedrinski, I. A.; Dmitrenko, V. E.; Hrudianov, I. I. *Lithium Current Sources*. Energoizdat: Moscow, Russia, 1992.
42. Kovalchuk, E. P.; Ostapovich, B. B.; Seledets, M. V.; Kovalyshyn, Ya. S. In *Lithium Current Sources Using Composite V_2O_5 – Polyaniline*. Reports VII International Conference. Fundamental Problems of Energy Conversion in Lithium Electrochemical Systems, Saratov, Russia, June 24–28, 2002; pp 85–86.

CHAPTER 12

SURFACE MODIFICATION OF POLYMERIC MATERIALS BY POLYANILINE AND APPLICATION OF POLYANILINE/POLYMERIC COMPOSITES

YU. A. HNIZDIUKH, M. M. YATSYSHYN, and O. V. RESHETNYAK

Department of Physical and Colloid Chemistry, Faculty of Chemistry, Ivan Franko National University of Lviv, 6 Kyryla & Mefodia Str., Lviv 79005, Ukraine. E-mail: yulya_hnisdyuch@ukr.net

CONTENTS

ABSTRACT

The subject of a brief review is the modification of large-scale (films, sheets, tapes, threads, fibers, etc.) non-conductive material substrates of different polymeric nature by polyaniline layer and also obtaining of homogeneous polyaniline/polymeric composites. The methods of preparation of non-conductive polymeric substrates and also the basic techniques of applying of the polyaniline layers or films on such substrates, namely the technique from the solution *in situ*, from the gaseous phase, by polymerization method in plasma, by forming polyaniline films on the surface of polymeric substrates from its solutions and suspensions, by preparation of polyaniline/polymeric composites using the methods of co-dissolving, and by making melts as well as by combined methods were discussed. The basic stages of the processes have been analyzed, the physico-chemical properties and morphology of the obtained composites in each case were considered, the examples of composite materials based on polyaniline and non-conducting massive polymers of natural, artificial, and synthetic origin have been demonstrated. It is shown that due to their physical and chemical properties, such materials are promising for applications as sensitive sensory materials, anticorrosive, antistatic, electrostatic coating materials for shielding of electromagnetic waves, elements of organic optoelectronic devices, diaphragms, artificial muscles, electroconductive fabrics, biologically active substrates, etc.

12.1 INTRODUCTION

Polyaniline (**PAn**) is an important representative of a relatively new class of polymers such as electroconducting polymers (**ECP**) or as they are called, organic metals. Due to the simplicity of the synthesis techniques, unique physical and chemical properties (among which the important are the high electrical conductivity, multicolored electrochromism, chemical sensitivity, limited solubility, chemical and thermal resistance, and high adhesiveness to the surfaces of different nature), the possibilities of reversible transitions between its various forms with characteristic properties (through the redox processes and mechanism of doping–dedoping), this substance and composites on its basis are important materials in modern technologies.[1-3] In recent years, **PAn** was tested in the various technologies and thus more than encouraging results were received. In particular, **PAn** is the first among the electroconductive polymers that has been used for industrial use in the current sources[4,5] and biosensors.[6] Methods of **PAn** synthesis and the mechanism of oxidative polycondensation of aniline (**An**) as well as the possibility of **PAn** application are described in detail in *ref.*[2,7–19]

Currently, the investigations concerning to the improvement of existing devices based on **PAn** and expand of the scope of its application actively are conducted. Particularly, actual for today is the problem of electroactive polymers applying in general and of polyaniline in particular both on efficiently large and on small-scale materials by polymeric nature, namely synthetic, artificial, and natural macromolecular compounds in the form of thin films or nanoparticles, or microdispersible sediments. For this purpose, different methods are used, such as an application of **PAn** films *in situ*, from steam or gaseous phase, by the method of the solvent evaporation, etc. This approach allows obtaining the polymer–polymeric composites that combine the properties of individual components, including the electroconductivity of **PAn**, with optical transparency and plasticity of the polymer undercoat (substrate). This allows the use of such composites as adsorption or antistatic materials, electroconductive fabrics, or as sensitive and selective layers for separation of gases in chemo- and biosensors, protective layers from UV-radiation, electromagnetic shielding, etc.

12.2 PREPARATION OF THE SUBSTRATES FOR THE MODIFICATION BY POLYANILINE

For creating **PAn**-polymeric composites different substrates of natural, artificial, or synthetic origin are used, namely different high-molecular compounds, for example cotton, silk, cellulose, polyethylene (**PE**), polyethylene terephthalate (**PET**) and many others. These substrates come in different forms, in particular in the form of threads, fabrics, films, granules, pipes, etc. One of the factors that determines the quality of **PAn**-polymeric composites, including their durability from a mechanical point of view, is the adhesion of polyaniline to the surface of the substrate. However, the surfaces of polymeric materials, which are often used as the substrates, are mostly hydrophobic, and that is unfavorable moment for applying films or layers of electroconductive polymers on them.

Depending on the material of substrate, its preparation can be relatively simple (washing by the solvent) and quite labor consuming. In general, the methods of surface preparation can be divided into three groups: *physical* and *chemical* (washing by the solvent), *chemical* (processing by oxidants, the grafting of functional groups), and *physical* (the treatment by plasma, by ultrasound, by high-energy irradiation, laser irradiation, etc.), although we note at once that this division is rather conventional. However, the combined methods of the surface pretreatment of polymer substrate are the most commonly used; such methods combine previously listed three groups of methods. The approaches concerning to the preparation of substrates for different methods of polyaniline films applying, which allow to obtain the films of electroconductive polymer with different properties and for different applications have been developed. Let us consider the above-mentioned groups of the methods in detail.

12.2.1 PHYSICAL–CHEMICAL METHODS OF POLYMERIC SUBSTRATES SURFACES PREPARATION

The simplest method of the pre-treatment of polymeric substrates is their treatment by the solvent. In particular, during the preparation of non-woven material made from the polypropylene fibers[20] or the material made from the nylon-6,[21] the preparation of samples was limited by their washing with distilled or deionized water.[22,23] But more often, polymeric substrates for degreasing of their surface successively are washed with organic solvents and water. For this purpose, primarily the aliphatic alcohols, such as metanol[22,24,25] and ethanol[26–29] are used. Other solvents are used much less frequently. As an example, the results of the *ref.*[30] can be presented, when the fibers of nylon-6,6 before the modification were kept in formic acid.[30]

In some cases, the substrates before modification with polyaniline were boiled for a long time. In particular, the fabrics of wool, **PET**, nylon-6, acrylic, and cotton before the modification by polyaniline were boiled in distilled water.[25] The films of low-density polyethylene (**LDPE**), high-density polyethylene (**HDPE**), polypropylene (**PP**), **PET**, poly(tetrafluoroethylene) (**PTFE**) before modification by **PAn** were washed by extraction in a *Soxhlet* apparatus with methanol for ~6 h.[31,32] Against this background, stands out the approach of the authors[33] when *curauá* fibers were extracted with acetone in a *Soxhlet* apparatus for 48 h. The aim of such treatment was to eliminate part of the lignin and promote fibrillation of the microfibril bundles.[33]

In the case of possible more intensive contaminations, particularly when the **PAn** is applied on natural fibers (e.g., jute fibers,[34] wool-nylon-Lycra fabric (*CSIRO Textile and Fiber Technology*),[35] etc.) such substrates, for a long time, were kept in hot detergent solutions following by washing with water. Even tougher treatment has been used by the authors of *ref.*[23] in which the substrate (microslides) was placed firstly into a H_2SO_4/H_2O_2 (7:3 by volume) solution for about 5 min and then repeatedly rinsing with deionized water.

Compulsory stage of the substrate preparation after washing with the solvents is their drying. Depending on which method of application the **PAn** further is used, drying is carried out both in air and under reduced pressure (vacuum box)[22,23] or under infrared lamps for 4-24 h. The temperature at which drying is done is determined by the thermal properties of substrates, but often is limited by the temperature range 60-120 °C.

12.2.2 PHYSICAL METHODS OF THE POLYMERIC SUBSTRATES SURFACES PREPARATION

Among physical methods of pre-treatment of polymeric substrates, three approaches can be distinguished. *The first approach* is echoed with the methods described in **Section 12.2.1.**, but for the better degreasing of the samples, their washing by organic solvents is also accompanied by ultrasonic treatment. For example, the samples of polyester (**PES**)[36] or **LDPE**[32,37] were treated into ultrasonic bath in the medium of acetone whereas, the substrates with **PET** in the ethanol for 1 h.[29] *The second approach* is used to change the hydrophobic surface of the substrate on a hydrophilic for selective deposition of **PAn** on certain parts of the surface. However, such treatment also has other purposes. Particularly, an important role in the formation of adhesive films of **PAn** on polymeric substrates plays the hydrogen bond[27], which is formed between the macromolecules of the substrate and **PAn**. Exactly in the absence of the necessary functional groups for this purpose (–COOH, –CONH-, –OH, etc.), the inert substrates are subjected to an appropriate treatment (modification). In particular, for this purpose, the samples of nylon-6 fabric[38] before the applying of **PAn** film were subjected to low-temperature plasma treatment. For this, oxygen, ammonia, and argon gases were used for the plasma treatment, which was performed in vacuo at 100 Torr for 10 min with radio frequency (**RF**) power at 60 W. On the surface of fibers, such functional groups as COOH or C=O are formed after oxygen treatment. Flexible glossy photo paper (*Hansol Co. Ltd.*) and the **PET** (3 M) film was treated with oxygen plasma (power is 100 W), only for 30 s.[39] The films of **LDPE** were treated in similar way (only 10 s).[32,37] The purpose of such treatment is slightly different. In the presence of molecular oxygen, the peroxide groups are formed on the surface of polymers and continue they are used to initiate free-radical polymerization (grafting to the surface) of various monomers having the hydrophilic functional groups. The treatment by Ar plasma after prior exposure of the substrate surface in the atmosphere of the ozone[40,41] gives the identical result.

However, often, physical methods of the surface treatment are used for obtaining of porous substrates, including membranes. For example, **PET** with the thickness of 10.0 ± 0.1 μm and with the effective diameter of pores equal to 0.215 μm were obtained by the radiation of initial **PET** film with the ions of krypton with an energy of 3 MeV/nucleon, accelerated in U-400 cyclotron.[42,43] Microporous substrates, which were subsequently used to obtain **PAn–PE** composites, were prepared from **HDPE** in the process, based on the melt extrusion technique with subsequent annealing, uni-axial extension, and thermofixation. These **PE** films contain a large number of through-flow pores with 0.05–0.45 mm in size.[44]

Speaking about porous polymeric films, it should also be mentioned that such substrates can be also obtained by evaporation of the solvent from a solution of the corresponding polymer. The obtaining of sulfonated polysulfones **(SPSF)**[45,46] membrane as a result of the polymeric solution drying on a glass substrate at 25 °C and at relative humidity of 75% can be viewed as an example.

12.2.3 CHEMICAL MODIFICATION OF THE SURFACES OF POLYMERIC SUBSTRATES

The most radical are the chemical techniques of the substrate surface preparation since they provide the surface functionalization. The result is that the hydrophilizing of initially hydrophobic surface of the substrates than the better adhesion of **PAn** to the surface is provided. The simplest method is the oxidation of the surface of polymeric substrates (**LDPE**, **HDPE**, **PP,** and **PET**), which can be made by the treatment with ozone,[40] or the grafting of sulfo-groups as a result of the action of steaming sulfuric acid from 15 min to 15 h.[41] In addition, the surface functionalization of polymeric substrates can be realized as a result of near-UV light-induced graft copolymerization of acrylic acid (**AAc**), acrylamide, or Na salt of styrene sulfonic acid (**NaSS**)[47,48] to the substrates with **LDPE**, **HDPE**, **PP**, **PET**, **PTFE,** etc. pretreated with Ar plasma (to form the peroxide groups on the surface). A similar result can also be obtained if the polymeric chain of the material-substrate contains relatively reactive functional groups. So, to improve the adhesive interaction of the film of **PAn** with the surface of polyacrylonitrile fibers, the last were modified with carboxyl group.[49]

12.3 CHEMICAL POLYMERIZATION OF ANILINE ON POLYMERIC SUBSTRATES

12.3.1 MODIFICATION OF POLYMERIC SUBSTRATES BY POLYANILINE IN SITU

Chemical oxidation of aniline by various oxidants in aqueous solutions of mineral acids permits to obtain on the surface of the materials of different nature (glass, **PES**, **PET**, wool fibers, cellulose, etc.), the electroactive coatings of different functional applications. An application of electroactive polymers and **PAn**, in particular, on the surface of polymeric materials of different nature and forms during chemical polymerization is considered in *ref.*[7,12,50,51] The essence of this the simplest and the most commonly used method of the **PAn**-polymeric composites obtaining is that at the chemical oxidation of aniline, the part of **PAn**, which is formed in the volume of the reaction medium, illegally deposits on the surface of the material-substrates immersed in the polymerization solution. The ratio between the amounts of **PAn**, which covers the surface of the material, is deposited on the bottom of the reactor and remains in the volume of the reaction dishes in the form of the suspension which is varied widely. To effectively cover a material by the layer of **PAn**, it is necessary to change this distribution in favor of the formation of a polymeric layer on the substrate.[7] This can be achieved by choosing the reaction conditions, appropriate surface pretreatment of the material which is covered, the ratio of the concentrations of oxidant/monomer, the reaction temperature, and the presence of the dopant substances in polymerization solution. The polymerization in the volume cannot be fully depressed, but acceptable high contents of **ECP** deposited on the surface, can be achieved properly through the selection of the reaction conditions.[7] In particular, for this type of **PAn** application typical is the process proceeding at low (0-5 °C) temperatures. Overall, the sequence of operations during the five (*a-e*) kinds of the process of polymeric substrates modification by **PAn** *in situ* is shown in Figure 12.1 (the stage of the substrate preparation, their washing, and the final drying are not shown in the figure). The procedure consists in the substrate holding during the certain time intervals in appropriate reaction solutions. In the case of fibrous substrates or tissues at the first step (Fig. 12.1), the using of ultrasound (for the better dispersion of fibers) and/or the stirring of the solution for some time is possible.

From the point of view of the possibility of further use of **PAn**/polymeric composites, which are obtained exactly by chemical polymerization of aniline, the important are the conclusions made by the authors of *ref.*,[52] which compared the properties of chemically and electrochemically deposited **PAn**, namely on the surface of platinum. The detailed comparison between both types of **PAn** coatings, produced under the identical conditions, shows that the chemical coatings, if obtained under suitable synthesis conditions, can compete with the electrochemical ones in what concerns their surface homogeneity and stability. This result is important from a practical point of view, having in mind that chemical synthesis is cheaper and more appropriate for large-scale polymerization than the electrochemical polymerization.

Since during the synthesis of **PAn**/polymeric composites, a great number of factors (namely, the nature of oxidant, its concentration, duration of deposition, temperature, pretreatment methods of the initial substrate and the obtaining composite, etc.) can be varied, then further let us consider the results of the studies of **PAn**/polymeric composites depending only on the nature of used substrates.

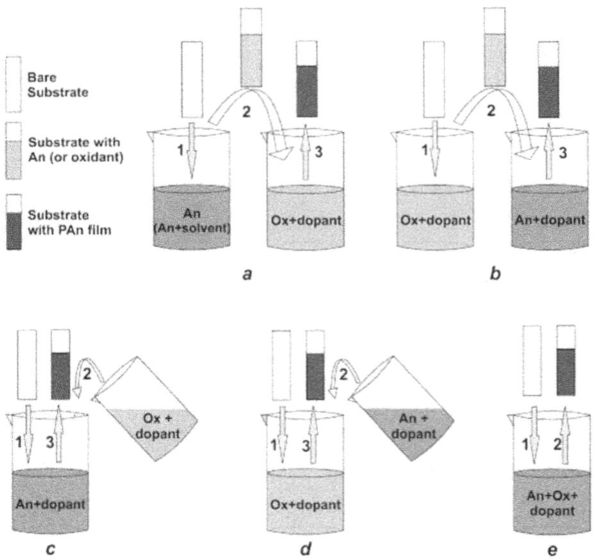

FIGURE 12.1 Schematic representation of the modification of polymeric substrates by **PAn** *in situ*: (*a*) is initial holding of the substrate in a solution of aniline (or in pure aniline) with following holding in a solution of oxidant; (*b*) is the procedure (*a*) carried out in reverse order; (*c*) is holding of the substrate in a solution of aniline with the following addition of oxidant solution; (*d*) is the procedure (*c*) performed in reverse order; (*e*) is holding of the substrate in a solution of aniline and oxidant. Stages: **1** is dipping of the substrate into the first solution and holding therein; **2** is transferring of the substrate from the first into the second solution (*a*, *b*) and holding therein; an addition of the solution of oxidant or aniline (*c*, *d*); **3** is the removing of a modified substrate (*a-d*) and **2** (*e*).

POLYACRYLONITRILE (PAN)[53,54]

Production of flexible and conductive composite films of **PAn** on the **PAN** substrates is described in *ref.*[53] Chemical oxidative polymerization of **An** in 1 M aqueous solution of HCl was initiated by adding of an ammonium persulfate (**APS**) solution in 1 M HCl, the reaction time was 1-5 h. For the obtained composite film, the content of **PAn** in which was about ~1 mass.%, the minimal surface resistance at room temperature was 7.4 kOhm/□.

It was established, that the thermal stability of **PAn/PAN** composite films is significantly improved compared with pure **PAN** at the retaining of good mechanical properties. However, significantly better results were obtained using the polyacrylonitrile fiber (**CPAN**),[54] which has been chemically modified with carboxyl groups in accordance with the procedure described in *ref.*[49] **CPAN** fibers were dispersed in 1 M aqueous solution of HCl containing the aniline, a solution of **APS** was added, and the mixture was treated with ultrasound for 30 min. The synthesis lasted overnight at 5–10 °C. Results from batch adsorption tests showed that grafting of carboxylic group could substantially improve the adsorption capacity of **CPAN** for aniline monomer at optimal pH 3.0–6.0; however, in strong acidic solutions, both the raw poly(acrylonitrile) fiber and the as synthesized **CPAN** had no obvious adsorption for aniline. Both **PAN** and **CPAN** could obviously accelerate the polymerization of aniline; however, the deposited **PAn** was more uniform on **CPAN** as compared with **PAN**. Besides, the produced **PAn/CPAN** composite showed higher **PAn** content and conductivity than the **PAn/PAN** composite. **PAn/CPAN** composite has potential applications for the adsorption of heavy metals and the manufacture of protective clothings.

POLYCAPROLACTAM (POLYAMIDE 6, PA6, NYLON 6)[21,22,38,55]

Just as in the case of **PET** substrates, **PAn** coatings on **PA6** substrates often are obtained standing the initial substrates in aniline after which they are dipped into the solution of oxidizing agent.[12,56,57] By applying this method, as was shown by the authors of *ref.*,[55] the speed polymerization of aniline affects the nature of the polymeric matrix, which also determines the distribution of **PAn** in a polymeric matrix (which has mainly laminate and cluster character) and depth of the aniline penetration. The difference of the used method of synthesis of **PAn/PA6** composites compared with obtaining **PAn/PET** is much less (only 1-3 h) duration of the substrate holding in pure aniline. In particular, the synthesis of electroactive **PAn/PA6** composite films was made by authors of *ref.*[22] who have shown that the increasing of the duration of the film substrate immersion into solution of oxidant (0.25 M **APS** solution in 2 M HCl) can lead to excessive oxidation of **PAn**, which reduces its yield. At the same time, the authors of *ref.*[21] showed that at the same method of **PAn** application on **PA6** fabric the maximum conductivity of the samples is achieved under fabric systems holding in pure aniline at 40 °C for 3 h and the duration of polymerization in 0.25 M **APS** at 5 °C for 1 h. The same authors studied the applying of **PAn** at the holding of fabric in aqueous solutions of aniline. Under these conditions, the maximum conductivity ($0.6 \cdot 10^{-1}$ $S \cdot sm^{-1}$) has been reached when the fabrics were previously kept in an aqueous solution of aniline in 0.5 M HCl at 40 °C for 1 h, and the polymerization was performed by adding of the oxidant solution at stirring for 30 min at 5 °C. It was established that with multiple acidic and alkaline treatments, no significant changes in the conductivity of the composite fabrics were made. It was shown that under the action of light for 100 h, the conductivity was decreased by less than one order. The effect of plasma gas (oxygen, ammonia, and argon) treatment on electroconductivity and productivity of **PAn/PA6** composites was studied in *ref.*[38] The samples of nylon fabrics (*Korea Apparel Testing Research Institute, Korea*) was kept in an aqueous solution of aniline in 0.35 M HCl at 40 °C for 2 h, the polymerization was initiated by **APS** solution in HCl (molar ratio **An**:**APS**:HCl was 1:0.7:1) and was carried out at 5 °C for 30 min. It was found, that the oxygen is the most aggressive and efficient gas for plasma treatment compared with ammonia and argon. It is shown that the conductivity of the treated by oxygen plasma **PAn/PA6** fabrics is more stable for repeated washing cycles than of the same fabrics without prior plasma treatment. It was fixed, the increasing of the conductivity of fabrics with increasing of the concentration of **An** to 0.5 M. At this, the quantity of deposited layers practically does not change the value of their electrical conductivity: the conductivity of fabrics with mono-, bi-, and triple layers of deposited **PAn** was 22.2-21.6; 21.9-21.4, and 21.6-21.1 $S \cdot sm^{-1}$, respectively.

POLYESTER (PES)[36,58,59]

Polyaniline layers on the surface of **PES**' films (75 mm thick, *Dupont de Nemours & Co*) were obtained by chemical oxidation of aniline by Fe (III) chloride for 30 min in aqueous solutions of HCl.[58] It was shown, that such films can be used for about ~10 measurements for low concentrations of the sulfite-ion (up to 0.5 $mg \cdot l^{-1}$) and ~5 measurements for higher concentrations of analyte. **PES**' samples modified by **PAn** were used to determine the sulfite-ions and sulfur dioxide in wine samples and the results were consistent with those obtained by iodometric titration. Reproducibility, ease of preparation and low cost of films allow using them as a single system for such measurements.

Materials obtained via polymerization of **An** on the non-conducting polyester fabrics deserve an especial attention. **PES**' fabrics were covered with **PAn** during chemical synthesis[36,59] when the substrates were previously kept in acid (*p*-toluene sulfonic acid, HCl, H_2SO_4; pH 1–2) aqueous solutions at 2–5 °C. After adsorption of **An** and doping acid on the samples of fabric the solution of oxidant (ammonium[59] or potassium[36] peroxydisulfate) was added into a solution. In the first case, the substrates were continuously (within 4 h) rotated to provide uniform reaction on the substrate, while in the second case, the polymerization was carried out for 180 min without stirring. An increase of the weight for the fabrics covered with **PAn**–HSO_4^-, and **PAn**–Cl^- was 6% and 2%, respectively. It was established, that the higher degree of the doping under the same conditions is achieved for **PAn**–HSO_4^-/**PES** (0.66 compared to 0.28 for samples of **PAn**–Cl^-/**PES**), indicating a higher degree of oxidation of PAn chains in such coverage and higher electrical conductivity. It was shown that such fabrics (**PAn**–HSO_4^-/**PES**, **PAn**–Cl^-/**PES**) can be used as conductive materials as well as materials for antistatic applications.[36] Also, it was found that these coatings can be effectively used for the screening of electromagnetic waves.[59] In the radio frequency range from 100 to 1000 MHz, conducting polyaniline-coated fabrics show a shielding effectiveness in the range 30–40 dB. Shielding effectiveness of the conducting fabrics in W-band region at 101 GHz shows an attenuation of 35.61 dB. The microwave reflectance studies of the coated fabrics in 8–12 GHz range show that conducting fabrics gives shielding effectiveness value of −3 to −11 dB. The reflectance studies of conducting polyaniline-coated fabric shows that 98% of the energy is absorbed in the UV–Vis–NIR range and 2% is reflected back.

POLYETHYLENE[32,37,44,60]

Electroconductive films of **PAn** on **LDPE** substrate were obtained by holding the substrate in the reaction mixture of 0.025 M **APS**, 0.05 M **An**, 0.5 M *p*-toluenesulfonic acid (**TSA**) or 0.5 M sulfosalicylic acid (**SSA**) as dopants.[32] It was established, that the exposure time of the substrate in the reaction solution practically does not effect on the thickness of the **PAn** coating. It was shown that the use of **SSA** instead of **TSA** leads to the formation of thicker, though less conductive films of **PAn**. Using of the **LDPE** surface, to which the **AAc** was grafted, significantly increases an adhesion of the electroactive polymeric coatings as well as accelerates the growth of coating in such way that the thicker films are formed in a short period of time. The authors attribute such effect to the effect of branching of the polymeric chain and electrostatic interaction between the links of the grafted polyacrylic acid and electroactive polymer.[32] The same substrates further were used for obtaining the polyaniline–palladium composites.[37] Deposition of Pd was performed on previously prepared film of **PAn** (in the form of leucoemeraldine) on the **LDPE** substrate. The result of this method is the formation of Pd clusters by size of 100 nm, which eventually are merged into a layer of Pd on the surface, which gives the conductivity to the film even after **PAn** is dedoped. Pd particles are actually distributed throughout the volume of the matrix of **PAn**, but not limited to the surface area, and their size depends on the reaction time and the starting ratio Pd/**PAn**. The high density of Pd particles in the **PAn**'s matrix provides the conductivity of the films, although **PAn** is in the dedoped form. However, it was shown, that the increase of the palladium content in **PAn** leads to the decrease of adhesion between **PAn**–Pd layer and the substrate **LDPE**.

Microporous films of **PE** were modified with **PAn** from an aqueous solution of aniline in the presence of HCl as dopant and **APS** as oxidizing agent.[44] The content of **PAn** in composites **PE/PAn** was no more than 12%. It was established that due to the formation of dense homogeneous electroconductive polymeric layer on the surface of porous substrate, the composite **PE/PAn** acquires selective properties concerning to the separation of gases O_2 and N_2 through interaction of the first with the electroconductive form of **PAn**. It is shown, that the maximum coefficient of the selectivity for the separation of mixture O_2/N_2 is 9.5 and depends on the degree of **PAn** doping.

PAn's film by the thickness of 100 nm was deposited by authors of the *ref.*[60] during chemical oxidative polymerization of **An** on polyethylene tube with an inner and an outer diameter of 3.5 and 4.2 mm, respectively. The obtained samples were tested with the optical sensor on gaseous ammonia. It was found that the dependence $lg(A/A_0)$ on $lgC(NH_3)$ is strictly linear within 180–18000 ppm, indicating the prospects of such composites using in sensorics.

POLY(ETHYLENE TEREPHTHALATE)[28,55,56,61,62–66]

PET substrates are the most popular for creating of **PAn**/polymeric composites. At the use of such substrates in the most cases during the synthesis, the preference has the method at which the substrates were previously maintained for some time in aniline. The process was continued until the weight of **PET** films do not increase on 10–15% (to 130 h).[28,64] Since the swelling of only a thin surface layer of the polymer matrix takes place, transparent composite films are obtained (under immersion of swollen film of **PET** into solution of oxidant) as the result of further polymerization of aniline. Thus, the authors of *ref.*[61] used double axially stretched films with the crystallinity degree 69% and the **APS** was used as the oxidizing agent. These authors found that the penetration depth of aniline in the film is ~4–5 microns, and the thickness of the formed surface layer of **PAn** consists of ~2–4 microns. Composite samples obtained in this way are also described in *ref.*[55,56,62] Another feature of this approach to the synthesis is the use of chlorine solution in water (actually, hypochlorous acid HOCl)[28,64] as oxidant. That is why composite films after polymerization contain the doped HCl in a surface layer of **PAn** (**PAn**–HCl/**PET**). The remainders of **An** and by-products were removed by their extraction with organic solvent for 5-8 h.[67]

Authors of *ref.*[62] investigated the doping–dedoping processes of such composites (by processing of 37% HCl solution or 20% $HClO_4$ solution and 5% NH_4OH solution, respectively). It was shown, that the presence of **PAn** significantly changes the dielectric properties of the polymeric matrix. It was established that the value of the dielectric losses in the range of low frequencies is increased with the size of doping anion, which likely causes the smaller size of the formed clusters and, therefore, creates a greater surface specific area. The presence of paramagnetic centers in doped and dedoped samples was also determined.

One of the main possible applications of composites obtained in this way is the chemosensorics, as sensitive element of the transparent sensors. It was established that the strength of the molecules binding for detectable substance of this sensitive material strongly depends on the state of its surface and on accessibility of the analyte's molecules to the clusters of **PAn**. It is shown, that the treatment of surface of **PAn–PET** composite with KOH solution for 12–48 h allows to obtain more specific surface area[56,64] with an average surface resistance ~10^5 Ohm/cm.[56,62] This facilitates an access of the analyte's molecules to the clusters of **PAn**, and correspondingly provides strong, fast, reversible, and linear responses of optical sensor of double character depending on the form of **PAn**: on formic acid (if **PAn** is in dedoping state) and on ammonia (after repeated doping of **PAn** in the atmosphere or in HCl solution). Based on the data of *ref.*,[64] the use of **PAn**–HCl/**PET** composite with dual channel optical scheme makes it possible to work in the range of concentrations 1–20,000 ppm at the determination of the ammonia.

Authors of the *ref.*[62] produced such transparent organic conductive films, previously, maintaining the **PET** film not in liquid aniline, but in its 0.2 M aqueous solution (at 5 °C for 2 h). The thickness of the surface film of **PAn** was regulated by changing of the polymerization time in 0.2 M $(NH_4)_2S_2O_8$ aqueous solution cooled to 5 °C. The obtained samples are characterized by significantly lower conductivity (~10 $S \cdot cm^{-1}$), which points to the possibility of application of such films as materials for chemical and biological sensors.

Electroconductive films of **PAn** also were obtained on the samples of non-woven **PET** fabric.[65,66] Polymerization of **An** on prepared samples was carried out in HCl aqueous solution using $K_2Cr_2O_7$[65] as oxidant or in the solutions of *p*-dodecylbenzylsulfonate, when $FeCl_3$ was used as the oxidant. It is shown, that the **PAn/PET** composite fabrics have good mechanical properties of **PET** and electrical conductivity of **PAn**. It was established, that the electroconductivity of fabrics is increased with the **PAn** content increasing. It was investigated, that the quantity of **PAn**, which is formed on the **PET** fabric, achieves the saturation for 2 h at 0 °C and for 1 h at 40 °C. It was shown, that the surface resistance of **PAn/PET** composite fabrics kept under vacuum is five times higher than the **PAn/PETs** kept under atmospheric conditions.[65] Also it was found that such composites with the electroconductivity over 0.06 $S \cdot cm^{-1}$ have the electrochromic properties.[66] The color change was observed from green (emeraldine salt) to blue/purple (*pernigraniline*) under applied potential ±3 V for 1 min. The change of the color was reverse for 10 cycles of imposition/removal of the potential inclusive.

POLY(METHYL METHACRYLATE) (PLEXIGLASS, PMMA)[26,68-70]

A feature of the **PMMA** substrates is the fact that at their use, the optically transparent and biocompatible **PAn**/polymeric composites are obtained. In particular, the authors of *ref.*[70] obtained the films via oxidation of aniline by $FeCl_3$ after exposure of respective substrates in pure aniline, and in *ref.*[28,68,69] via oxidation of aniline by **APS** after adsorption of aniline from the acidic (HCl) aqueous solutions. With the use of the second method[68-69] it were obtained optically transparent layers of **PAn** by the thickness of 21 μm after 5 min of the substrate holding in the reaction solution. All the samples have the surface conductivity, the mechanism of which corresponds to charge-energy-limited tunneling model. The films demonstrate the reversible changes of color in the field of pH 2-12[26] and therefore can be used for optical determination of pH or for the measurement of NH_3 (<10 ppm).[68,69] Since the films have $pK_a \sim 6.7$, the accurate measurement of pH from 5 to 8 can eventually carry out with the recall time less than 1 s; taking into account of non-toxic and biological compatibility, they are suitable for the use in optical sensors to measure the pH of blood *in vivo*.

POLYPROPYLENE[20]

Simple method of **PAn** obtaining on the non-woven materials made from the polypropylene fibers is described in *ref.*[20] The samples of fabrics were kept in a solution containing of 0.2 M aniline and 0.2 M H_2SO_4 for 10 min at room temperature and then were immersed in 0.1 M **APS**. **PAn** coatings of different thickness can be obtained by varying the number of procedures and their application duration. Gas sensors based on this material showed the high performance to identify the volatile organic compounds (the response time was about 10 s). The investigated determined substances as to reducing of the sensor to them were located in a row: ethanol > chloroform > toluene > acetone > ethyl acetate.

POLY(TETRAFLUOROETHYLENE)[27,71]

Modification of **PTFE** substrates by the **PAn** films in solutions of hydrochloric, perchloric, malic, and citrate acids is described in *ref.*[71] **PTFE** substrates were consistently maintained (ice bath) for 2 h in solutions of **An** (in methylene chloride) and **APS** (in 1 M solution of doping acid). After synthesis, the films were washed by 1 M solution of corresponding doping acid and were dried under infrared light. It was shown, that the process of doping–dedoping changes the electric and surface properties of the films. It was established the reduction of conductivity with the duration of exposure time in water and also achievement of virtually non-conductive (fully dedoped) state of the obtained **PAn** films after exposure for 1 h. After repeated doping, the electroconductivity of these samples was recovered. The authors of *ref.*[27] obtained the multilayer coatings of **PAn** on **PTFE** film (*Sinoma Science & Technolog Co. Ltd., Nanjing, China,* the thickness is 20 μm) substrates, previously kept in a pure **An** for 8 min. The polymerization of adsorbed aniline further was performed in a solution of **APS** in 1 M

HCl at 0 °C for 20 h. It was established that the composite film has a high conductivity, easy processability, good stretching.

The feature of **PAn/PTFE** composites is their adjustable hydrophilic surface, which can be provided by the control of dedoping degree of **PAn**'s layer by exposure of composites in water.[27,71] Moreover, the transition between hydrophobicity and hydrophilicity are completely reversible: initial hydrophilic films of **PAn** during dedoping become hydrophobic, and their hydrophilicity easily is restored by the repeated doping. These results indicate on the prospects of these materials using as neutral tube coatings,[71] they can also find a wide application in shielding of electromagnetic radiation, in other high-tech industries.

POLYURETHANE[72]

The surface of polyurethane (*Tecoflex 60* D, *Thermidic Inc., MA, USA*) has been modified by applying of a thin layer of **An**, maintaining the substrate in the polymerization solution containing an excess of **APS** as oxidant and salicylic acid as dosing agent) at room temperature overnight. It is shown, that for the deposited films of **PAn** the coefficient of the selectivity of absorption degree of *o*-aminobenzoic acid and acetyl salicylic acid (**ASA**) was 22 and 16.5, respectively, which allows to use of such composites for the separation or express analysis of these substances.

POLYVINYLCHLORIDE (PVC)[56,70]

Authors of *ref.*[70] obtained the electroconductive films **PAn/PVC** via polymerization of **An** after the holding of **PVC** films in solution of **An** with followed oxidation by $FeCl_3$. Surface resistance and transparency of films at l = 500 nm were 10^3–10^4 Ω/sq and 60–70% respectively. It was shown, that the mechanism of conduction of such coatings can be described by charge-energy-limited tunneling model. A feature of the methodology used in *ref.*[56] was that before the formation of electroconductive films of **PAn**, the **PVC**-films were kept not in pure aniline or its aqueous solution, but in 50% solution of **An** in organic solvent (*n*-hexane). This approach enables deeper insight of the monomer into the structure of substrate due to its swelling in organic solvent. It is shown, that in a case of this method applying as well as at the using of aqueous solutions of Cl_2, HOCl, **APS**, $K_2Cr_2O_7$, $KMnO_4$ as oxidant and HCl, HBr, HI, H_2SO_4, $HClO_4$, CH_3COOH, HNO_3 as dopant, the obtained **PAn** films are characterized with surface conductivity.

SULFONATED POLYSULFONES (SPSF)[46]

For obtaining of electroconductive film of **PAn** on the surface of the honeycomb ordered sulfonated polysulfone film as template, it was kept in 0.325 M hydrochloride solution of aniline. 0.125 M solution of sodium vanadate in 1 M aqueous solution of HCl was used as oxidant. The reaction was carried out for different time in an ice bath. It is shown, that the introduction of a sulfo-group into polysulfone matrix is useful to form the regular structure of deposited **PAn**. Composite films of **PAn** on the **SPSF** substrates may have potential use as electroconductive membrane materials as well as materials for the sensors.

CELLULOSE[73-76]

Composites "polyaniline–cellulose fibers (**PAn/CeFs**)" in the most cases were obtained in the "classic" way via *in situ* polymerization, when the cellulose fibers were kept in **An**[73] or in aqueous solution of **An**[76] (in an ice bath) and then the solutions of a dopant and an oxidant (**APS**) were added at stirring. The duration of the polymerization was 1-2 h. At this, inorganic[73] as well as organic[76] acids were used as a dopant. It is shown, that the conductivity, thermal stability and fire resistance of composite increases with the increase of **PAn** amount deposited on **CeFs**. It was established, that the properties of composites can be significantly improved by doping of **PAn** not individual acids, but their mixtures. Thus, the presence of H_2SO_4 and HCl in acidic mixture provides the higher conductivity of the composite and in the presence of H_3PO_4, the fire-resistance[73] is increased significantly. It was established, that the conductivity and fire-resistance of the composite materials remained stable after 30 days storage under natural conditions. At the same time, the replacement of inorganic dopants to organic acids changes mainly the morphology of **PAn** on the surface of the substrate. In particular, at the use of **TSA** on bactericidal **CeFs** it was obtained the **PAn** with spherical structure and an average grain size ranging from 100 to 200 nm. Such changes in the morphology of deposited **PAn** make a significant impact on the electrical conductivity and thermal stability of **PAn/CeFs** nanocomposite films.

Some separate are the results obtained by the authors of *ref.*[74,75] They synthesized the hybrid materials based on **PAn/CeFs** composites. Thus, the **PAn** application on the surface of **CeFs** (*Carter Holt Harvey, NZ*)[74] took place via their successive holding in an aqueous solution of **An** (0.5 M) in water under the stirring for 1 h, adding the oxidant solution (0.5 M **APS**) with followed polymerization for 3 h. The feature of this method was the presence in the reaction mixture of the surfactant, namely 0.05 M solution of *p*-dodecylbenzyl sulfonate. The obtained electroconductive **PAn/CeFs** were washed with water and then they were treated by ultrasound with ethanol and were dopped by 1.0 M solution of HCl. The electroconductivity of the obtained hybrid materials was $2.6 \cdot 10^{-1}$ $S \cdot cm^{-1}$. The authors of *ref.*[75] after deposition of **PAn**'s layer with the thickness of 0.1 μm on the surface of **CeFs** (*J. Rettenmaier & Söhne Gmbh–Co, Rosenberg, Germany*) under similar conditions of synthesis was increased the specific conductivity of the fibers from $4.0 \cdot 10^{-14}$ to 0.41 $S \cdot cm^{-1}$.

Silver particles were deposited on the surface of **PAn/CeFs** composites via chemical reduction of Ag^{+} ions by aniline in the form of emeraldine base (**EB**). At the addition of **PAn/CeFs,** directly in 0.1 M $AgNO_3$ solution the emeraldine base was oxidized to pernigraniline and Ag^{+} was reduced to Ag^{0} (in the form of nanoparticles) on the surface of the fibers. The obtained fibers further were washed with distilled water and were sonicated in ethanol. As a result, the conductivity of these samples become less concerning to composite **PAn/CeFs**, probably due to poor electrical contact between silver nanoparticles, and also lesser own conductivity of the oxidized polymer.[74] In particular, it was determined[75] that for the samples with content of silver up to 10.6 mass.%, their conductivity was $4.1 \cdot 10^{-4}$ $S \cdot cm^{-1}$. Thanks to nano-dispersed silver Ag-**PAn/CeFs** composites showed antimicrobial activity against *Staphylococcus Aureus*, and therefore can be used primarily in the production of packaging materials for food products, health products and medical devices.

COCONUT FIBER [77]

Electroconductive fibers based on coconut fibers (**CcFs**) (*EPAGRI Empresa de Pesquisa Agropecuária e Extensao Rural de Santa Catarina, Brazil*) and **PAn** were obtained *in situ* oxidative polymerization of **An** on **CcFs** with the use of $FeCl_3$ or **APS** as oxidant. The fibers were kept in aniline aqueous solution under the stirring at room temperature for 10 min, then an aqueous solution of oxidant was added. The molar ratios $FeCl_3$/**An** and **APS/An** were 3:1

and 1:1 respectively. Electrical conductivity of composite materials obtained after 6 h of synthesis was ~$1.5 \cdot 10^{-1}$ and $1.9 \cdot 10^{-2}$ $S \cdot cm^{-1}$ correspondingly for various oxidants. Composites **CsFs/PAn** were tested as electroconductive additives for the change of the structure and properties of polyurethane (**PU**). Reinforced by **PAn** modified with coconut fibers composites had a higher conductivity compared to **PAn/PU** composites and for them characteristic is a change in electrical resistance at the compression of the samples. For this reason, these materials can be used as sensitive materials in the load cells.

COTTON FABRICS[78]

Chemical oxidative polymerization of **An** on the samples of fabrics (*Ranama weave*, surface density 239 $g \cdot m^{-2}$) has been carried out as described in *ref.*[78] The samples of fabrics were kept for 3 h at 0–5 °C in 1 M solution of **An** in 1 M HCl and cold (0–5 °C) 1 M solution of **APS** in 1 M HCl was gradually added. The ratio oxidant/**An** was 1:1. The polymerization was carried out at constant moderate stirring for 1 h at the same temperature. It is reported that fabrics covered with **PAn** have excellent properties as for UV-protection; electric value of resistance for **PAn**-modified fabric was 350 W; average electromagnetic effectiveness of shielding and the average absorption of this fabric were 3.8 dB and 48%, respectively. These characteristics make the cotton fabric with coated polyaniline modifying layer by promising not only for making everyday summer clothes, but also the clothes of special purpose.

CURAUÁ FIBERS (CFs)[33,79]

Curauá (*Ananas erectifolius*) fiber refer to lignocellulosic fibers. The plant is native to the Amazon where the Indians of that region have known it since the pre-Columbian era for making hammocks. **CFs** with **PAn** coating has been obtained in *ref.*[33] via addition dropwise of the oxidant solution (0.3 M **APS** in aqueous solution 2 M **TSA**) under stirring at 5 °C to 0.1 M aqueous solution of **An** + 1.0 M **TSA**, in which the fibers previously were mixed for 1 h. These modified fibers **CFs–PAn** further were used as a reinforcing material of composite based on **PA6**. Such composites were obtained in the extruder at 200–220 °C at a screw rotation speed of 300 rpm. It was established that the percolation threshold is achieved at very low content of **PAn** on the fibers (1–2 wt.%). At the content of **PAn** equal to 12 wt.%

the conductivity of the sample is near to the conductivity of pure **PAn**. An applying of **PAn** on vegetal fiber, including the surauá fibers, and their inclusion into the polymeric matrix by **PA6** type enables to obtain the advanced antistatic reinforced materials.[79]

JUTE FIBERS (JFs)[34]

In order to obtain the electroconductive polymeric coating the authors of *ref.*[34] used a mixture of **An** and 1,4–phenylenediamine (molar ratio was 1:7). Jute fibers with the length of 0.5–1.0 cm were placed into 1 M HCl aqueous solution, which also contained the mixture of monomers and after that were cooled in an ice bath to 0–5 °C, and the polymerization of **An** was carried with the **APS** solution in 1 M HCl. It is shown, that the obtained composite is an effective adsorbent of the chromium Cr(VI) ions at pH = 3-4 due to electrostatic attraction of acid chromate ion with protonated **PAn/JFs**: adsorption equilibrium was achieved within 40–120 min for initial Cr(VI) at 50–500 mg/L; maximum monolayer capacity of composite was observed at 62.9 mg/g at temperature 20 °C. During desorption almost 83% efficiency was achieved within 10 min by 2 M NaOH. By ignition of chromium bounded **PAn** jute, 94% Cr(VI) were able to recover as Cr(III) along with reduction in weight by 95%. These characteristics show that **PAn/JFs** composite can be used effectively to remove the chromium from wastewater.

NYLON-6,6 (NY-6,6)[30]

Electroconductive and isothermally stable fibers of **PAn**:**Ny-6,6** were obtained by holding of the initial polymeric fibers in formic acid, to which a solution of **An** and through the night also 1 M **APS** solution dropwise were added. For the obtaining of more specific surface the fibers after washing with water and acetone were left overnight in 2% aqueous solution of ammonia. It is shown, that depending on the content of **An** in reactive mixture (5-20%), the electrical conductivity of **PAn**:**Ny-6,6** was from 10^{-5} to 10^{-1} S·cm^{-1}. This fibrous material may be promising for many applications that require a combination of electrical conductivity of isothermal stability.

PAPER[80]

Authors of *ref.*[80] precipitated of **PAn** on the commercial filter paper (*Whatman* 40) by impregnating it with an aqueous solution of **An** in 2 M HCl for 2–3 min, then added dropwise the 30% H_2O_2. The obtained product (emeraldine salt of **PAn**, green color) further was transformed into emeraldine base (blue color) by treatment with 0.1 M solution of NaOH. The next stage of processing by 0.01 M solution of $NaBH_4$ transfers the coating of **PAn** in colorless form of emeraldine base. The possibility of using of such filter paper with **PAn** coating as the acid–base sensor as well as for filtering of acids and bases was shown. The sensitivity of produced on its basis sensor on ammonia was 45 ppm in pairs and 14 ppm for its determination in solution.

SILK[81]

Electroconductive polyaniline/silk fibroin composite fibers **(PAn/SF)** were obtained by polymerization *in situ*. Silk fibers were dispersed in an aqueous solution of **An**, to which an **APS** aqueous solution in HCl solution was added for 30 min. The polymerization time of **An** was 24 h at room temperature under continuous stirring. The conductivity of composites **PAn/SF** was $0.9–1.2 \cdot 10^{-2}$ $S \cdot cm^{-1}$. These composite fibers still possessed former fibrillar morphology and strength properties. Structural analysis implied that conjugated polymers molecules have different interactions to certain extent with **SF** chains. Namely, the formation of hydrogen bonds (-C=O...H-N- between the peptide group of **SF** and non-protonated amide group of **PAn**), and also the electrostatic interaction (between the partial negative charge on the oxygen of the carbonyl group of peptide group and positive charged protonated by nitrogen atom of **PAn** $-C{=}O^{d-}...-^{+}NN-$) take place between the macromolecules of **SF** and **PAn**; the result of this are the significantly better electrical (conductive) and thermal properties of this type of composites that extends their potential applications in textiles, biological and other functional materials.

WOOL[25]

Taking into account the multicomponent composition of modern woolen fabrics, the authors of *ref.*[25] conducted a comparative analysis of **PAn** deposition on the substrates that can be components of woolen fabrics, such as wool, **PET**, **PA6**, acrylic and cotton (*Japanese Standards Association*). Prepared samples of fabrics were kept in 0.5 N HCl in aqueous solution containing of **An** and oxidant (KIO_3, **APS** or $K_2Cr_2O_7$). The results showed that the layer of **PAn** selectively is formed on the wool when the polymerization of **An** takes place with the use of KIO_3. Proof of this is the presence of cysteic acid units (Cy-SO_3H) resulted from the oxidation of cystine bonds (Cy-S-S-Cy) in wool. Selective formation of **PAn** on wool was considered to occur by a concentration of **An** on wool bearing cysteic acid units. It became apparent by *IR* analysis of oxidized wool that the oxidized layer of the wool textiles was thicker and cystine bonds in wool were converted into more cysteic acids by oxidation, using KIO_3 than using $(NH_4)_2S_2O_8$ and $K_2Cr_2O_7$. In addition, the surface resistance of deposited polyaniline' layers at the use of KIO_3 was on order smaller compared with coatings obtained under $(NH_4)_2S_2O_8$ and $K_2Cr_2O_7$. Therefore, formation of **PAn** on wool occurred remarkably in the case of polymerizing aniline, using KIO_3.

12.3.2 POLYMERIZATION OF ANILINE ON POLYMERIC SUBSTRATES FROM GASEOUS PHASE

The scheme of modification of polymeric substrates with **PAn** from vaporous phase is summarized by Figure 12.2. It includes the holding of the substrate in a solution of oxidant, which also contains the dopant (acid), it's drying from the solvent, and then transference and holding of the substrate in vapor of **An**. The temperature of the substrate holding in vapor of **An** could vary from room temperature to 80 °C depending on the motivation of the experimenter. The stages of the substrates preparation and also their extraction, washing and drying after polymerization of **An** are not shown in this figure. The principal difference of this composites obtaining method described in *Chapter* 12.3.1 is the use of solutions based on organic compounds, primarily aliphatic alcohols, which are then easily evaporated (Fig. 12.2) in the most cases instead of aqueous solution (oxidant and dopant).

An illustration of this method of **PAn**/polymeric composites synthesis can be the results of *ref.*[29,82,83] Such composites on **PET**[29,82,83] and also on polyimide (**PI**), **PVC** and polystyrene (**PS**)[83] substrates have been obtained by these authors. $FeCl_3 \cdot 6H_2O$ was used as oxidant, and camphorsulfonic acid was used as dopant; both oxidant and dopant were applied on the substrates from butanol and methanol solutions, respectively. After drying of the polymeric substrate from the organic solvent, it was subjected to the impact of

An during 5–120 min in the reaction chamber. According to the data of *ref.*[29] the pressure of vapor of **An** at 60 °C was about 10 mm *Hg*. Modified films after washing (usually with methanol) and drying were characterized by stable electrochemical activity in an acidic medium. According to the *ref.*[82] as well as *ref.*[83] the conductivity of such composite films ranged from 0.6 to 1.8 $S{\cdot}cm^{-1}$; the surface resistance of **PAn** film was ~104 $\Omega{\cdot}sq^{-1}$, a light transmission was 90% for a film with the thickness less than 30 nm, respectively. Due to the fact that modified thin polymeric films possess by high transparency and conductivity, it is proposed to use them as optical films, antistatic coatings in flat displays, as materials for organic optoelectronic devices. Besides, the authors of *ref.*[29] showed that the **PAn/PET** composites could be used as a stable electron field emitter of electrons (electron field emitter). In addition, the use of **PET** as a substrate for the deposition of **PAn** enhances the possibilities of their using, especially in displays of autoelectric emission. Also due to the large surface area of the **PAn**'s film, stipulated by their porous structure with elongated form of clusters, there is possible their use as materials for the sensors.[29]

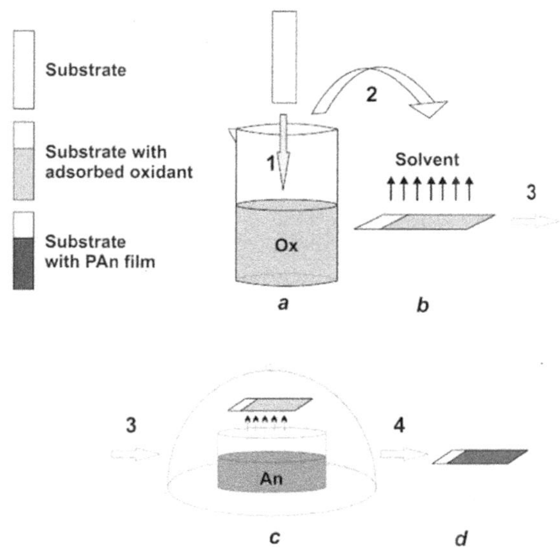

FIGURE 12.2 Schematic representation of the polymeric substrates modification with **PAn** from vaporous phase: *a* is primary holding of the substrate in a solution of the oxidant and dopant; *b* is holding of the substrate in vapor of **An**. Stages: **1** is dipping of the substrate in a solution and holding; **2** is drying of the substrate from the solvent; **3** is transferring of the substrate from a solution into vapor of **An**.

Somewhat different approach of the **PAn** deposition on flexible substrates, namely on glossy photopaper (*Hansol Co. Ltd.*) and films **PET** (3*M*) using a vapor–phase polymerization was proposed by the authors of *ref.*[39] Its peculiarity consists in fact that the solution of oxidant (aqueous solution of **APS**) was printed on a substrate using a modified inkjet printer. After drying from water the sample was placed in a reaction chamber, which contained with the exception of **An** also diluted solution of HCl, since the solution for print, in view of the corrosion activity, does not contain the solution of dopant, namely strong acid. The holding of the substrate in vapor of **An** was occurred with a gradual increase in temperature from room temperature to 75 °C. It was found that at the thickness of ~450 nm the surface resistance of such films was 3.8·103 Ω/sq.

Also, let us note that initially for the obtaining of **PAn/PET** composite films, also, the technique described in *ref.*[84] has been used, which was reverse to that shown in Figure 12.2. In particular, transparent and conductive **PAn/PET** composites were obtained by the original holding of dried **PET** films in pure **An** at room temperature for 0.5-8 h with their subsequent exposure into vapors of aqueous solution of **APS** and HCl. It is shown, that the conductivity of the films was 0.1–0.2 $S \cdot cm^{-1}$, and light transmission at l = 450–700 nm was ~70%. But, now this approach is used very rarely.

12.3.3 PLASMOUS POLYMERIZATION OF ANILINE ON POLYMERIC SUBSTRATES

With the use of plasmous polymerization the **PAn**'s layers were obtained, in particular, on fluorinated ethylene propylene copolymer (**FEP**),[85] **LDPE**,[86] **PET**[43], and other substrates. Thin films of **An** polymer are obtained using radio frequency (**RF**)–induced plasma, for example under variable frequency between 125 i 375 kHz.[785] The samples of substrates are placed on anode of plasma–chemical reactor. After the pumping of gases from the system (up to the pressures consisting of 0.1-5 Pa) the vapors of **An** are introduced into the reactor and initiate the polymerization process.

The authors of *ref.*[86] obtained the **PAn**'s film on the substrates made from **LDPE** (*Goodfellow, UK*) and **LDPE** copolymerized with **AAc**. The feature of the process was that before the deposition of **An** the argon was injected into the reactor for the pretreatment of the substrate with argon plasma for 10 s at a pressure of 100 Pa and *RF* power of 20 W. **PAn** obtained on such substrates by plasma polymerization for 5 min has significant differences in the properties compared with the polymer obtained via conventional chemical or electrochemical synthesis. This is explained mainly by the formation of oligomers and substitutions of benzene rings, which result in some $-NH_2$ groups remaining unreacted during the course of polymerization. Electrical conductivity can be induced in the plasma-polymerized **PAn** through protonation by acid solutions although the level of conductivity is significantly lower than that achievable with chemically synthesized **PAn**. Copolymerization of **LDPE** with acrylic acid significantly improves an adhesion between **PAn** and substrate of **LDPE** compared to the film of **LDPE**. This improvement of an adhesion prevents the loss of **PAn** during the holding in aqueous solutions. It is shown, that plasma-polymerized **PAn** on **LDPE** copolymerized with **AAc** can react with $AuCl_3$ and $Pd(NO_3)_2$ in acidic solutions, resulting to the deposition of Au (0) and Pd (II), respectively[86] on the surface of **PAn** indicating that the obtained polymer is in the reduced emeraldine or leucoemeraldine form.

An increase of *RF* power during plasma polymerization leads to an increase of the thickness of layer of the deposited **PAn.**[86] However, as it was shown by the authors of *ref.*,[85] for the polyaniline layers deposited on fluorinated ethylene propylene copolymer (*Du Pont* **FEP** 100 *Type A*), the contents of quinoid sequences and aliphatic crosslinking moieties to increase with increasing power input and/or discharge duration. By contrast, the number of free radicals trapped in the polyaniline films and their mobility were shown to increase with decreasing the power input and/or discharge duration within the plasma conditions covered in this study.

The authors of *ref.*[43] using the plasma polymerization of **An** have modified the surface of **PET** membrane. The deposition of **PAn** on the surface of the membrane was carried out at constant current of plasma of glow discharge under the pressure of aniline vapors of 26.6 Pa and discharge current density of 0.1 $mA \cdot cm^{-2}$ for 120 s. The pore walls and one of the surfaces of track membrane were modified by plasma of aniline, which resulted in formation of the polyelectrolyte transient cationite–anionite layers. It is shown that the membrane produced possesses asymmetry of conductivity, that is, the effect of the current rectification similar to *p–n* junction in semiconductors is observed.

12.3.4 COMBINED METHODS OF MODIFICATION OF POLYMERIC SUBSTRATES BY POLYANILINE

The development of combined methods of an applying of the films of electroconductive polymers on non-conductive polymeric substrates is caused by the aim to ensure high adhesion of **PAn**'s film and to achieve the optimal thickness, corresponding structure, and morphology. Combined methods typically combine two different ways of the applying of films of electro-conductive polymers and **PAn**, including widely practiced combination of chemical and electrochemical methods, or physical and chemical. The purpose of the initial physical and chemical treatment of the surface of non-conductive polymer is the formation of the electroconductive or other type of surface layer on which subsequently the precipitated **PAn** may be deposited. Such treatment can include the chemical transformation of the surface, for example as a result of photocatalytic oxidation due to the presence of oxidant,[87] annealing of the substrate[88], etc. Further chemical (electrochemical) deposition results in non-selective or selective formation of **PAn** films or thin layers on the substrate surface. Under non-selective formation of electroconductive layer of **PAn** understand its deposition on the entire surface of non-conductive polymer substrate, while during the selective polymerization of **PAn** the coating is formed only in certain areas, allowing just enough to get various images and micropatterns.

Schematically the modification of polymeric substrates by **PAn** via combined chemical and electrochemical method is shown in Figure 12.3 (the stages of preparation of the substrate, and also washing and drying of the obtained composite are not shown in this figure). The examples of this approach are the *ref.*[88–90] the authors of which have previously created the electroconductive layer on the polymeric substrate on which subsequently **PAn** was precipitated by electrochemical method. In particular, such conductive layer on the surface of the paper was obtained by carbonization of polyacrylonitrile aerogel (*PANCF*).[88] On this layer–electrode further the **PAn** was precipitated for different (2-50 s) time electrochemically in potentiostatic mode (at the potential of 1.0 V). It is shown, that these electrodes can be used as material for the supercapacitors. Specific capacity of the symmetrical ultracapacitor based on **PAn**-modified paper electrodes reaches of 230 $F \cdot g^{-1}$ under good cycling (over 3000 repeat cycles of charge discharge).

Cases and co-workers obtained the electroconductive **PAn** coatings on conducting textiles of polyester.[89,90] For this purpose, the substrate from the **PES** fabric (*Viatex SA*, surface density 140 $g \cdot m^{-2}$) initially has been modified by the polypyrrol doped with anthraquinone sulfonic acid, and further **PAn** was electrochemically deposited.[89] The electrode was kept for 10 min before electrochemical deposition of **PAn** for better infiltration and, as a result, for the adsorption of **An** on the surface of the fabric electrode. An oxidation of **An** was performed by potentiostatically or by cyclic voltammetry (potential was cycled between −0.2 V and 1.1 V) in aqueous medium with 0.5 M

H_2SO_4 and 0.5 M aniline in N_2 atmosphere. When potentiostatic synthesis was employed, globular morphology was obtained with a 2-*D* growth at the beginning of the synthesis. In the potentiodynamic synthesis, fibers' length was 10–50 mm and an average diameter of 1–2 mm was obtained in a first stage of electrodeposition. As the nucleation sites increased, the current density grew and polyaniline with globular morphology was obtained. A mass variation value of 508 mg/C was obtained. Surface resistivity of the textile after the electrochemical deposition of polyaniline was around 20 W/sq for both methods of synthesis.[89] Also, it is shown that a great influence on the morphology of **PAn** deposited under potentiodynamic conditions has the potential scanning speed.[90]

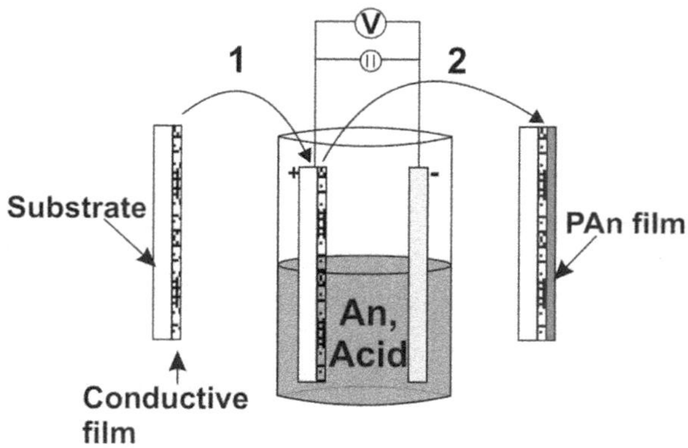

FIGURE 12.3 Schematic representation of the modification of polymeric substrates by polyaniline with combined chemical and electrochemical method: **1** is dipping of the modified with electroconductive film substrate into solution and electropolymerization; **2** is remove of the modified substrate by **PAn**.

As example of physical- and chemical-combined method of the **PAn**/polymer composite obtaining can serve the investigation described in *ref.*[87] Authors of this paper have proposed a simple and general (for various polymer surfaces including **PET**, **PI**, **PP** and so on) method for creating of electroconductive polymeric layer using the photocatalytic oxidation. For this purpose the layer of the aqueous solution (30 wt.%) of **APS** was pressed between two polymeric films (the bottom further served as a substrate for the application of **PAn**, the top was by photomask with the printed image) and was subjected to UV-radiation (the mercury lamp of high pressure—1000 W, the UV intensity was 8000 μm·sm^{-2}, the exposure time was 120 s). During this reaction, the sulfate anion groups (SO_4^-) are implanted into polymer substrates that gives to them a high hydrophilicity. With the use of photomask, the hydrophilic properties acquired only those surface areas, which are irradiated with light. Thereafter, such substrate was kept in a solution of **An** in 1 M HCl with the following immersion in an aqueous solution of the oxidant (**APS** in 1 M HCl solution). Furthermore, *Whitesides*, *MacDiarmid* and co-workers found that on a hydrophilic glass surface, hydrophobic **PAn** could be selectively deposited on a patterned hydrophobic region, as a result of the "like dissolves like" principle.[91,92] Therefore, at the use of **PET**' films as the substrate, the selective deposition of **PAn** took place and also the so-called negative image has been received, since the **An** was adsorbed and, as a result, it was polymerized on the unoxidized areas of the substrate. It was established that at the use of **PI** or **PP** as the substrates under identical procedure a non-selective formation of **PAn** on the entire surface of the substrates took place. To form the pattern on the deposited **PP** or **PI** films, a piece of 3 M Scotch® adhesive tape, Scotch® Crystal Clear Tape (*Cinta Cristal, CC*1920–*Bx*) was placed on the deposited film, pressed gently to achieve a homogeneous contact between the tape and the film, and then peeled off quickly. Due to electrostatic forces the **PAn**'s protonated positively charged layer stronger is held up by the oxidized surface, and therefore, at the separation of adhesive tape is easily removed from the hydrophobic areas and is remained on hydrophilic areas of the surface of the substrate. As a result, the positive **PAn**'s image is formed on the surface of the substrates. This method of the PAn obtaining on different organic polymeric substrates is simple and cheap, and therefore can be easily implemented in large-scale industrial production at the manufacture of flexible, portable, or disposable materials for electrical and optical devices. Easy, fast, inexpensive technology to produce high quality conductive polymeric templates using the ordinary printer *Cannon BCJ*-4000 *DeskJet* is described in *ref.*[93] Instead of the ink of cartridges into inkjet printers, the solution of AgNO3 was used. The image was deposited as a result of printing on the substrates (paper, glossy paper or transparence sheet), which were kept previously in an aqueous solution of **An**. For display of the images into result of chemical reaction (the oxidation of **An** to conductive form of **PAn** and recovery of Ag^{+} ions to metallic silver) the germicidal lamp 20 W was used. The conductivity of the samples printed on glossy films was 2×10^{-2} S·cm^{-1}. This technique can be used in the development of conducting polymer patterns in standard office paper using a DeskJet printer, and could be used for the development of low prices electronic devices.

12.4 THE FORMATION OF POLYANILINE FILMS ON THE SURFACE OF POLYMERIC SUBSTRATES FROM ITS SOLUTIONS AND SUSPENSIONS

The easiest method to obtain the **PAn**/polymeric composites is the formation of **PAn** films on the surface of polymeric substrates from its solutions and suspensions. This method involves two stages: an immersion of the pre-prepared substrate into a solution of **PAn**, holding in it, washing, and drying of the obtained modified substrate by electroconductive polyaniline layer (Fig. 12.4). However, the use of this method is limited by the fact that **PAn**, especially in the form of doped form of emeraldine is very poorly dissolved in the all solvents. The exception may be the *N*-methylpyrrolidone. An example of such approach to the formation of polyaniline layers on polymeric substrates can be the result of the *ref.*[31] The authors covered with a thin layer of **PAn** in the form of neutral emeraldine base the substrates **LDPE**, **HDPE**, **PP**, **PET**, **PTFE** (*Goodfellow Inc., UK,* 100-130 μm *thickness*) through their holding in 0.1 wt.% solution of **PAn** in *N*-methylpyrrolidinone (**NMP**) with the following drying of organic solvent. It was established that the charge transfer between **PAn** coating and surface functional groups of polymeric substrates leads to the improvement of adhesion of electroactive polymer.

The most widely used is the method of PAn deposition from it suspensions. The modification of cellulose and paper (20 × 5 cm^2, *Aracruz Cellulose, Brazil*) by **PAn** was made in *ref.*[94] The samples of paper were kept in the dispersion of **PAn** in chloroform for 12 h and were dried under vacuum for 2 h. It was shown that the cellulose paper with **PAn** coating can be used for the determination of low concentrations of acids (below 500 ppm), that is essential for an environmental control of liquid and gaseous fluids. A similar dispersion of **PAn** also was used for the production of polymeric electrodes based on **PET** substrates.[95] High-conductive **PAn** was prepared in accordance with the method described in *ref.*[96,97]: aqueous solution of **An** in HCl was added to a cooled water–chloroform mixture and under the stirring, an aqueous solution of **APS** in HCl was added dropwise. Further **PAn** was doped by camphorosulfonic acid (**CSA**), and the **CSA–PAn** solution was spin-cast on plastic substrates to give the conductive **CSA–PAn/PET** substrate electrode. It was determined, a surface-interpenetrated conducting polymer, **CSA–PAn** on plastic substrate, especially cheap **PET** significantly boosts mechanical properties for outstanding performance as an ITO-free organic electrode in flexible organic solar cells and exhibits best performance of an average power conversion efficiency of 3.5 ± 0.2% under 100 mW/cm^2 illumination, with a short-circuit current density of 8.7 mA/cm^2, an open-circuit voltage of 0.83 V, and a fill factor of 0.49 comparable to conventional organic solar cells made with ITO electrodes on glass or **PET**.

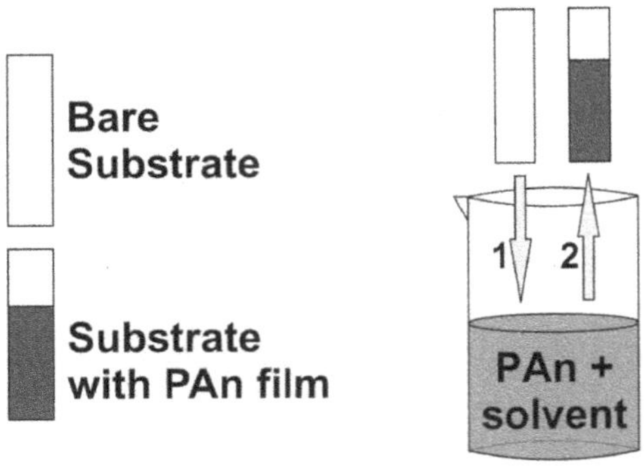

FIGURE 12.4 Schematic representation of the formation of polyaniline films from its solutions and suspensions: **1** is the dipping of the substrate into a solution (suspension) of **PAn** and holding; **2** is the removing and drying of the modified by polyaniline substrate (the stages of the previous preparation of the substrate, its washing and drying after deposition of **PAn** are not shown).

The method of formation of polyaniline films from it solutions was also successfully used to form the multilayer **PAn**/polymeric layers. In particular, at the manufacture of poly(sodium styrenesulfonate)/**PAn** (**PSS**/**PAn**) self-organized multilayer film the technology layer-by-layer was used, the driving force of which was the electrostatic interaction between the positively charged chains of **PAn** (at the expense of the presence of protonated imino-groups $-NH^{+}-$) and negatively charged macromolecules of **PSS** (at the expense of ionized sulfo-group $-SO_3^{-}$).[98] For the self-assembly process a glass substrate was kept in 0.5 M aqueous suspension of **PAn** for 15 min with the following washing by deionized water and drying in a stream of nitrogen. Further, the substrate with **PAn** coating was kept in solution of 5 mM **PSS** for 5 min, was washed, and dried forming a monolayer film. By the repetition of these stages, the multilayer self-organized films of **PSS**/**PAn** were obtained. The conductivity of these multilayer films was about 1 $S \cdot sm^{-1}$. It was shown, that by changing of the number of bilayers into assembly, the conductivity of multilayer films can be adjusted. Cyclic voltammetry studies have shown that such multilayer films with more quantity of deposited layers are characterized by low charge transfer resistance and high current density, particularly in the case of I_2/I^- redox reaction.[98] It was proposed that such multilayer films can be used as the components for the design of optical and electronic devices.

12.5 THE FORMATION OF COMPOSITE FILMS OF NON-CONDUCTIVE POLYMERS AND POLYANILINE BY SPREADING METHOD

The formation of composite films of **PAn** and non-conductive polymers by spreading method is one of the most common used and descr ibed.[3,7,11,12,15–19,50,99,100] At this, there are two types of the implementation of process: 1) the first is co-dissolving of non-conductive polymer and **PAn** for the formation of films by spreading method with the following evaporation of the solvent; and 2) the second is obtaining of the melts of non-conductive polymers with dispersed **PAn** in them with the following spreading and hardening of the composite or formation of granules, tapes or fibers using the various extruders. The first method schematically is shown in Figure 12.5. Its advantage is that, because of poor solubility of **PAn**, it can be introduced into solution of non-conductive polymer also in the form of dispersed powder. This requires as same as for method to produce the melts, additional processing of the suspensions, for example by ultrasound to homogenize the solution/melt (**PAn** distribution evenly throughout the volume). In addition, an important factor is that the range of the solvents applying in this case is expanded, since their choice is determined by the nature of the non-conductive polymer. The formation of films mainly is engaged on glass substrates including, in view of the possibility of further use, on ITO substrates.

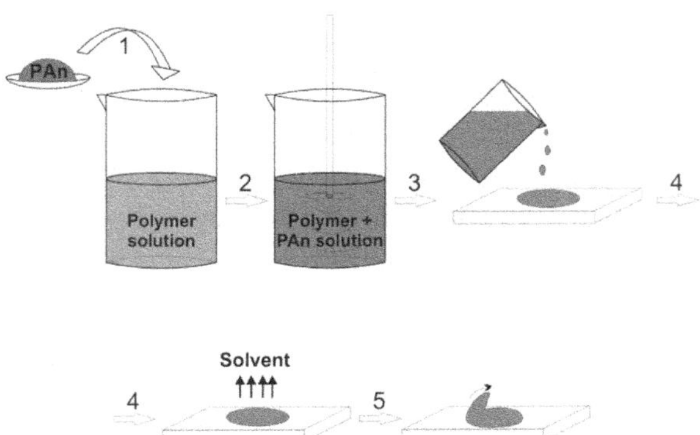

FIGURE 12.5 Schematic representation of the formation of composite films of non-conductive polymers and **PAn** by co-dissolving method, preparation of the solutions and evaporation of the solvent. *Stages*: **1** is an introduction of dispersed **PAn** into solution of non-conductive polymer; **2** is the homogenization of the solution by stirring or by ultrasound); **3** is transfer of the solution on the surface of the substrate; **4** is the evaporation of solvent; **5** is obtaining of the film. (The stages of previous preparation of the substrate, its washing and drying after precipitation of **PAn** are not shown).

12.5.1 THE FORMATION OF COMPOSITE FILMS OF THE NON-CONDUCTIVE POLYMERS AND POLYANILINE BY CO-DISSOLVING METHOD

The results described in *ref.*[101] as to obtaining of the electroconductive **PAn**/polysulfone (**PAn/PSF**) films can serve by the illustration of the true case of non-conductive polymer and **PAn** co-disolving. For this purpose powder of **PAn** in the form of **EB** was dissolved in a solution of *N*–methylpyrrolidone (**NMP**) and 1,3-dimethyl-2-imidazolidinone (**DMI**) (10%). Polysulfone, dissolved in **NMP** was added to a solution of **PAn** under stirring to form a polymeric mixture; the films of **PAn/PSF** were obtained in the *Petri* dish through the solvent evaporation in a vacuum for 18 h at room temperature. The thickness of film-membranes of **PAn** was regulated by the quantity of **PAn/PSF**, which was used. The obtained films were doped by holding of them in a solution of 1 M HCl for 4 h, which is optimal for achieving of the maximal conductivity.[102] It was found, that the optimal conductivity and shape stability have the films made in accordance with this method from the 3% solution of **PAn** and 40% **PSU**.

But often the **PAn** is in solution of non-conductive polymer as suspension. For example composite films based on **PAn** were prepared by dissolving of the polymer powder (M_η = 7.8 × 104 g/mol, *Shanghai Institute of Synthetic Fiber*) in a mixture of **DMSO** and **An.**[53] The obtained suspension was left at 50 °C for 2 h, then stirred for 2 h at 60 °C, then at 70 °C for another 6 h. After the transferring of **PAn** solution on a glass substrate and evaporation of the solvent, the films with the thickness of 50 μm were obtained. The same method has been used to obtain the electroconductive composite **PAn**-(acrylonitrile–butadiene–styrene) (**PAn–ABS**).[103] For this, **PAn** and **ABS** at a ratio (5:5) were dissolved in chloroform. The surface resistance of composite films obtained after the solvent evaporation was between 1302 kΩ and 10^{11} Ω. It was shown the possibility to use of these composites for the manufacture of multiple conductometric sensors for the ammonium ions in aqueous solutions.

Electroconductive polyurethane/polyaniline (**PU/PAn**) films were obtained in *ref.*[104] The different amounts of **PAn** doped with dodecyl benzene sulfonic acid were added under the stirring to **PU** aqueous dispersion. Next, consistently the curing agent was added to the mixture under mechanical stirring, the mixture was decontaminated with the use of the vacuum pump and was poured on the **PET** substrate, where was heated at 120 °C for 2 min, and then (after finish of the **PU** curing reaction) at 150 °C for another 2 min. It is established, that the electrical conductivity and thermal stability of

PU/PAn films is increased with the content of **PAn** raising. It is shown, that the electroconductive **PU/PAn** films are promising for the use as anticorrosion, antistatic, electrostatic coatings and for shielding of electromagnetic interferences.

Composite films **PAn–PVDF** were obtained from a mixture of powders of polyvinylidenefluoride (**PVDF**) (*Mr* = 534,000) and **PAn** (*PANIPLAST, France*) at weight ratio of 5:95.[105] **PAn** was dissolved in dichloric-acetic acid (**DCA**) for 1 day at 90 °C, then the solution was cooled to 30 °C. **PVDF** was dissolved in dimethylformamide (**DMFA**) for 60 min at 30 °C. Dissolved **PAn** was added dropwise to a solution of **PVDF**, was stirred for 48 h at 30 °C, and then the solution was applied on the ITO glass substrate at 800 rev/min for 60 sec. The authors showed that the highest conductivity and conductivity of **PAn–PVDF** film are connected with α- to β-phase transitions in **PVDF**, depending on the drying temperature, and are achieved at 90 °C. It was shown, that the use of **PAn–PVDF** (90 °C) in organic solar cells significantly increases the photovoltaic characteristic of the latter. In particular, short-circuit current density are increased from 3.81 to 7.96 $mA \cdot cm^{-2}$, and power conversion efficiency—from 0.96 to 3.06%.

One of the most popular polymers which are used for obtaining composites with **PAn** by the spreading method is cellulose, namely ethyl cellulose (**ECe**),[106] cellulose acetate nanofibers (**CeAFs**),[107] nanocellulose[108], etc., since the flexible and conductive materials are obtained on their basis. In particular, for the manufacture of composite the powders of **ECe** and **PAn** was dissolved in ethanol (the latter additionally was treated with ultrasound for 5 min) were mixed with stirring for 3 h and were poured onto Teflon sheet, evaporating the solvent at room temperature. The investigations have shown that the obtained **PAn/ECe** films are characterized by significant peroxide free radical activity. Their antioxidant effect is increased by increasing the area of film when it contains a higher percentage of **PAn** (17%) and when **PAn** chemically is reduced using the hydrazine. It was proposed that such **PAn/ECe** antioxidant films could be used as high-effective packaging materials.

Biocomposite drive based on composites **PAn/CeAFs** was developed in *ref.*[107] Electrically responsive polymer actuators showing similar responses to natural muscles have received great attention as an important topic in biomimetic engineering, because of their potential applications to biomimetic robots and biomedical devices. During the synthesis cellulose acetate (M_n 30,000, *Sigma-Aldrich*) was dissolved into solution of N,N-dimethylacetamide/acetone (volume ratio 2:1), stirred vigorously at room temperature, 0.1 or 0.5 mass.% of **PAn** was added, and then the solution was homogenized in the ultrasonic bath. Next, using the electrospinning, the fibers **PAn/CeAFs** were obtained, from which the membrane covered with gold was formed for obtaining of the electroactive biocomposite drive. In the electrospun **PAn/CeAFs** biocomposite, the hydroxyl groups of cellulose acetate could interact physically with the secondary amine groups of polyaniline to form weak hydrogen bonds that ensure the well invasion of chopped polyaniline nanoparticles into cellulose acetate matrix. The morphology of the membranes showed nanoporous structures mimicking the extracellular form of natural muscles. The tests of the drives showed an increase in bending deformation even at the lowest concentrations **PAn**, indicating the prospects of practical use in the future such materials.

Methods of synthesis of composites based on nanocellulose and **PAn** is slightly different from those described above, since they were obtained *in situ* by oxidative polymerization of hydrochloride of **An** in aqueous 0.5 wt% suspension of nanocellulose (filamentary nanocrystals-bleached flax, *Jayashree Textiles, India*) using of **APS** as an oxidant.[108] The mixture was stirred for 24 h at room temperature, then was diluted with water and washed by deionized water during the centrifugation. The films were formed by spreading method and by evaporation of the solvent on the *Petri* dishes. It was established that thin film of the composite has a good flexibility, namely the film with the thickness of 50 μm can be bent at 180° without breaking. Electroconductivity of the obtained composite films increases from 10^{-6} to 10^{-2} $S{\cdot}cm^{-1}$ at the **PAn** content increasing from 10 to 20 mass.% and remains at the same level under further increase of **PAn** till 30 mass.%. In addition, the composite film has improved thermal stability above 300 °C. It was proposed, that such flexible, electrically conductive and, the most importantly, biodegradable composites as nanocellulose/**PAn** are promising materials for the use as sensors, elements of batteries and conductive glues.[108]

Polyimide–polyaniline hollow fibers were produced by dissolution of the polymers in N-methylpyrrolidone (**NMP**) and dry/wet spinning of the resulting solution in a non-solvent (H_2O).[109] The polymer were dried overnight at 120 °C under vacuum prior to dissolution to **NMP**. This composite was used to create the membranes, the structural elements, which were well-structured hollow fibers thermally stable up to 500 °C. It was established, that the introduction of **PAn** into **PI** matrix results in a significant increase in the permeability of gases (in 60–600 times) into the fibers. Permanent ratio of the selectivity H_2/CH_4, He/N_2, H_2/N_2 and H_2/CO_2 for composite membranes compared to membranes obtained using only the **PI** was less than 6.4; 8.9; 7.7 and 1.47, respectively, and therefore such materials can be used in the manufacture of membranes for industrial separation of the gases mixtures.

The obtaining of composite films of **PAn**/polymer acid calls an especial interest, since the latter acts as a dopant of electroconductive polymer, and also as a result of the transfer of a proton of functional group on the macromolecule of **PAn** forms the electrostatically bound complex (between protonated imino-groups -NH^+- of **PAn** and, for example, ionized groups -COO^-). It was established, that obtained films have the properties similar to those obtained by electrochemical method.[110] Moreover, the inclusion of polymeric acid (namely, poly(2–acrylamido-2-methyl-1-propanosulfonic acid) into the **PAn**'s matrix, reduces the values of the potentials corresponding to the transitions of **PAn**, causing this lower optical voltage of switch of the electrochromic device based on this composite. Similar properties are also the polyanions of poly(acrylic acid) (**PAA**), **PSS**, poly(methylvinylether-*alt*-maleic acid), poly(methylacrylate-*co*-acrylic acid), poly(acrylamide-*co*-acrylic acid), etc.

12.5.2 FORMATION OF FILMS OF NON-CONDUCTIVE POLYMERS AND POLYANILINE BY THE METHOD OF MELTS PREPARATION

The formation of films of non-conductive polymers and **PAn** by the method of melts preparation has the purpose to combine the technological properties of thermoplastic polymers and conductivity of conductive polymers. It is well known, that for the **PAn** typical is high (over 200 °C) thermal stability, which allows to process it at the melting point of many thermoplastics, among which the polystyrene (**PS**) and **LDPE** are the most common applied. This method of **PAn**/polymeric composites obtaining is one among the most technologically challenging because it requires the use of special equipment such as extruders of various designs, the precision of temperature conditions and lack of contact with air melts in order to avoid the thermolysis and thermal-oxidative destruction of the components of composite, respectively.

Thus, blending of **PS** with **PAn** doped with dodecylbenzeneosulfonic acid was carried out in double-auger extruder at a temperature 160-190 °C[111] after pre-mixing of powder ingredients at room temperature. The conductivity of formed composite in the form of a tape by a size 16 × 1.5 mm was 10^{-6}–10^{-2} S·cm^{-1}. It was shown, that due to mechanical and electrical properties of such thermoplastic composite it can be used as antistatic additives.

In *ref.*[112] and *ref.*,[113] the methods of producing of nanocomposites **PAn**/**LDPE** are described; in these cases the **PAn** was injected into the matrix of non-conductive polymer in the form of nanofibers and nanorods, respectively.

In the first case, the content of **PAn** (nanorods were synthesized in accordance with the technique described in *ref.*[113]) in composites was 5-20%. Before processing, the granules of **LDPE** (*TCL Hunt Ltd, New Zealand*) were dried at 80 °C overnight and nanorods of **PAn** were dried at 60 °C in a vacuum oven and before loading them into the extruder were mixed. An extrusion was carried out at 150 °C, and cooled extruded mixture in the form of rods was granulated. The granules further were pressed between Teflon plates at 150 °C for obtaining of films with the thickness of 0.35–0.05 mm. It was shown, that the electroconductivity of the blends is increased with the content of **PAn** increasing, but it also leads to the coalescence of **PAn**'s particles dispersed in a matrix of **LDPE**. In addition, the strength on tensile and elongation is reduced at a break of **LDPE** due to insufficient interfacial adhesion between **PAn** and **LDPE**. Nevertheless, these composites have sufficient mechanical properties and absorption capacity, as well as antioxidant activity. Likewise, biocompatible composite based on **PAn**'s nanofibers and **LDPE** matrix was obtained.[113] **LDPE** and **PAn**'s nanofibers were dried at 60 °C, were crushed, and mixed with an electric mixer. The temperature in the extruder was 130 °C. The melted composite was squeezed through the filters with the size 2 mm in the form of rods was cooled. It was shown that the obtained composite may prove beneficial as a low cost biocompatible and electrically simulated nerve tissue repair scaffolds.

12.6 MORPHOLOGY OF FILMS OF PAn ON THE SURFACE OF POLYMERIC SUBSTRATES

Artificial or synthetic polymeric substrates that are used as the platforms for the deposition of **PAn** are not perfectly smooth. They usually have a surface that reflects the working surface of the devices (rollers, spinnerets, etc.) that form the films, fabrics, thread and more. Natural polymeric substrates such as silk, wool have enough highly specific surface that is not smooth and contains many branches in the form of villus. From this review, it is expected that in this case formed **PAn** layers will reflect the surface of the substrate. Obviously, the defects of the surface (roughness, inhomogeneities) will serve as nucleation centers of **PAn** macromolecules. The samples obtained on the surface of **PAn** (Fig. 12.6),[54] **PET** and viscose fibers (Fig. 12.7)[66] can by an illustration of the morphology of **PAn** films on synthetic fibers. According to the presented images, the result of increasing the duration of the polymerization and consequently of the amount of deposited **PAn**, is the formation of a specific surface of **PAn**'s surface layer.

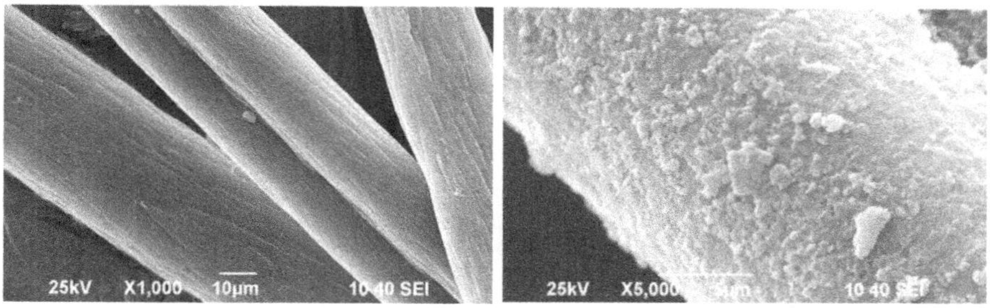

FIGURE 12.6 *SEM* pictures of the initial (*left*) as synthesized **CPAN** fiber and (*right*) the **PAn/CPAN** composite with 1000– and 5000–fold magnification, respectively (From: Wang, J., Zhang, K., Zhao, L., Ma, W., and Liu, T. "Adsorption and Polymerization of Aniline on a Carboxylic Group-Modified Fibrous Substrate," in *Synth. Met.* **2014**, *188*, 6–12. Used with permission from Elsevier.)

FIGURE 12.7 *SEM* images of the fibers of **PET** (*left*) and viscose (*right*) non-wovens before (*a*) and after deposition of **PAn** on its surface under the duration of polymerization 1 h (*b*) and 24 h (*c*) (From: Kelly, F. M. and Meunier, L., in "Cochrane, C.; Koncar, V. Polyaniline: Application as Solid State Electrochromic in a Flexible Textile Display. *Displays*," **2013**, *34*, 1–7. Used with permission from Elsevier.)

However, the modification of porous substrates by **PAn**, such as filter paper, leads mainly to the filling of the substrates pores with the formation of dense surface (Fig. 12.8)[80] or to the formation of new porous microstructure, but in this case already **PAn**' surface layer (Fig. 12.9).[73]

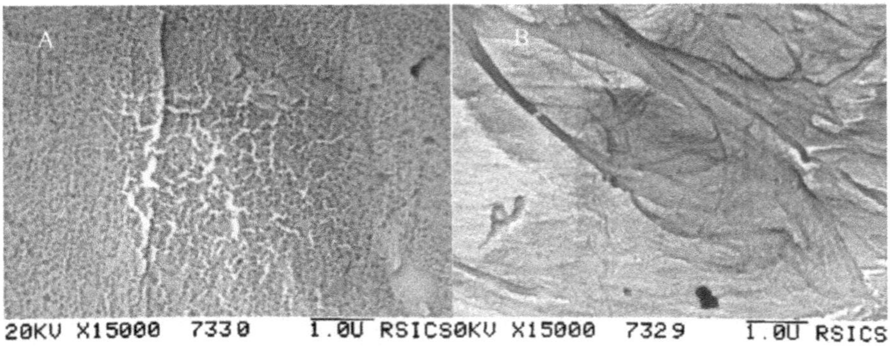

FIGURE 12.8 Scanning electron micrographs of a filter paper (**A**) and the paper coated with **PAn** (**B**). Bar is 1 μm (From: Dutta, D., Sarma, T. K., Chowdhury, D., Chattopadhyay, A. A, "Polyaniline-Containing Filter Paper that Acts as a Sensor, Acid, Base, and Endpoint Indicator and also Filters Acids and Bases," in *J. Colloid Interf. Sci.* **2005**, *283*, 153–159. Used with permission from Elsevier.)

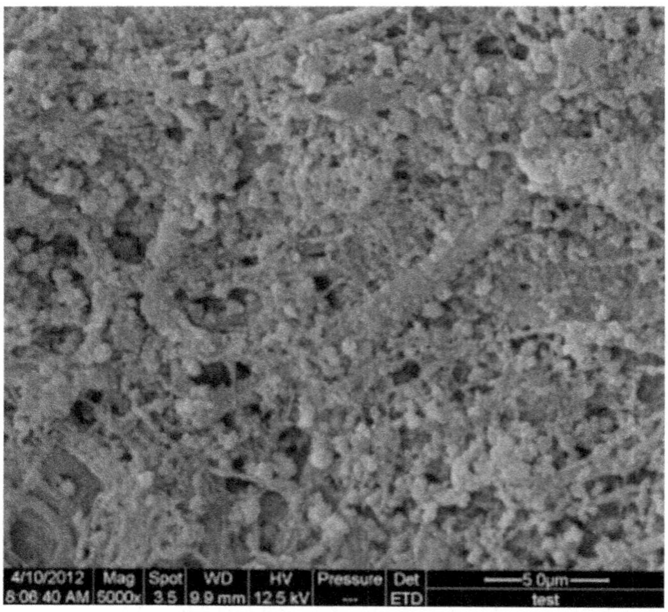

FIGURE 12.9 *SEM* images of polyaniline-deposited paper (sample was prepared in 1 mol/L phosphoric acid solution) (From: Wu, X., Qian, X., An, X., "Flame Retardancy of Polyaniline-Deposited Paper Composites Prepared Via in Situ Polymerization.," in *Carbohyd. Polym.* **2013**, *92*, 435–440. Used with permission from Elsevier.)

Particular attention has been paid into the treatment by alkalis of the **PAn**'s coatings, which were obtained after swelling of the polymeric substrate in pure aniline with the following initiation of the polymerization by solution of the strong oxidant. As it was shown by the authors of *ref.*,[28] the alkaline etching/dedoping procedure accelerates redoping the **PAn**/polymer composite, namely **PAn/PET**. This fact indicates that the etching facilitates an access of the dopant to **PAn** clusters distributed in the **PET** matrix. Therefore, it was assumed that a part of the **PET** matrix is etched out of the composite surface that leads to baring a part of surface of some **PAn** clusters. Indeed, the SEM image shows the relatively homogenous and smooth surface of the initial **PAn/PET** composite surface before the alkaline treatment (Fig. 12.10, *a*). However, after this treatment the surface morphology is completely changed to the highly increased and porous surface area of the **PAn/PET** composite film (Fig.12.10, *b*). Specifically, one can see sticking out flakes and globules of irregular shape and pores on the composite surface. Their dimensions strongly vary in the range 1–10 μm. Obviously, these areas can include both **PET** matrix pieces, which were not destroyed completely during the alkali etching, as well as partially bare **PAn** clusters appeared due to a removal of a part of the **PET** matrix, which screened **PAn** clusters. The same result can be obtained also after processing by alkalis of other substrates coated with **PAn**'s layers.

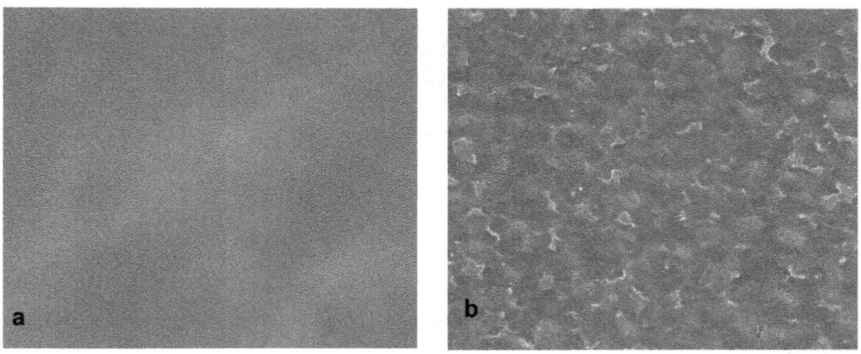

FIGURE 12.10 *SEM* images of the **PAn/PET** composite surface before (*a*) and after (*b*) the alkaline treatment (From: Duboriz, I. and Pud, A., "Polyaniline/Poly(Ethylene Terephthalate) Film as a New Optical Sensing Material," in *Sensor. Actuat. B-Chem.* **2014**, *190*, 398–407. Used with permission from Elsevier.)

Generally, an analysis of the morphology of the **PAn**'s films, deposited *in situ* on the non-conducting polymeric substrates, shows that their surface mainly has specific and nanostructured character. For **PAn**'s layers the presence of nano- and microformations from the aggregates of macromolecules of **PAn** is typical. The films of **PAn** deposited on polymeric substrates are formed by merging of assemblies of macromolecules of different shapes,[20] the nucleation of which took place on active centers of the polymeric matrix. As a result, under these conditions of synthesis in the most cases the deposited layer of electroconductive polymer has a granular morphology, typical for **PAn** coatings.75 The change of **PAn**'s films deposition conditions, namely the concentrations of **An**, oxidant, deposition time, special surface preparation leads to increase of the sizes and number of microaggregated entities, and thus to the formation of **PAn** layers with more specific surface.[27,29,38,54,66,77,81] However, the main factor, of course, is the method of the composite obtaining. In particular, at the use of aniline polymerization from gaseous phase the obtained polymer mainly has a very specific microporous structure.[29] As it is illustrated by Figure 12.11, analogous results can be obtained also during **PAn** deposition from solution.

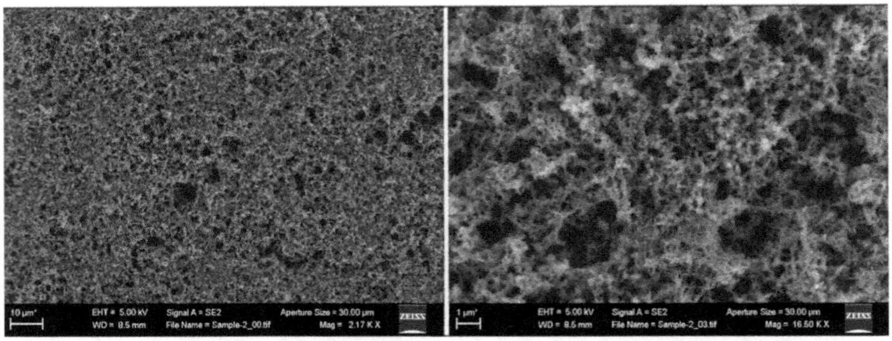

FIGURE 12.11 Detail of the surface morphology of cellulose film coated with **PAn**.

Separately let us dwell on the morphology of **PAn** obtained by combined methods of modification of polymeric substrates. In this aspect there are interesting results obtained by electrochemical deposition of **PAn** in potentiodynamic mode on **PES** fabrics modified by polypyrrol doped by anthraquinone sulfonic acid (**PES/PPy-AQSA** substrates), since the authors studied in detail the effect on the morphology of deposited **PAn** both of the polyaniline deposition duration (the number of cycles scanning of the potential) and the deposition conditions, namely potential speed scanning.[90] It was revealed that fibrillar and globular morphology are the most observed **PAn** forms. In particular, **PAn** with globular morphology has been observed in all micrographs for all scan rates and numbers of scans.

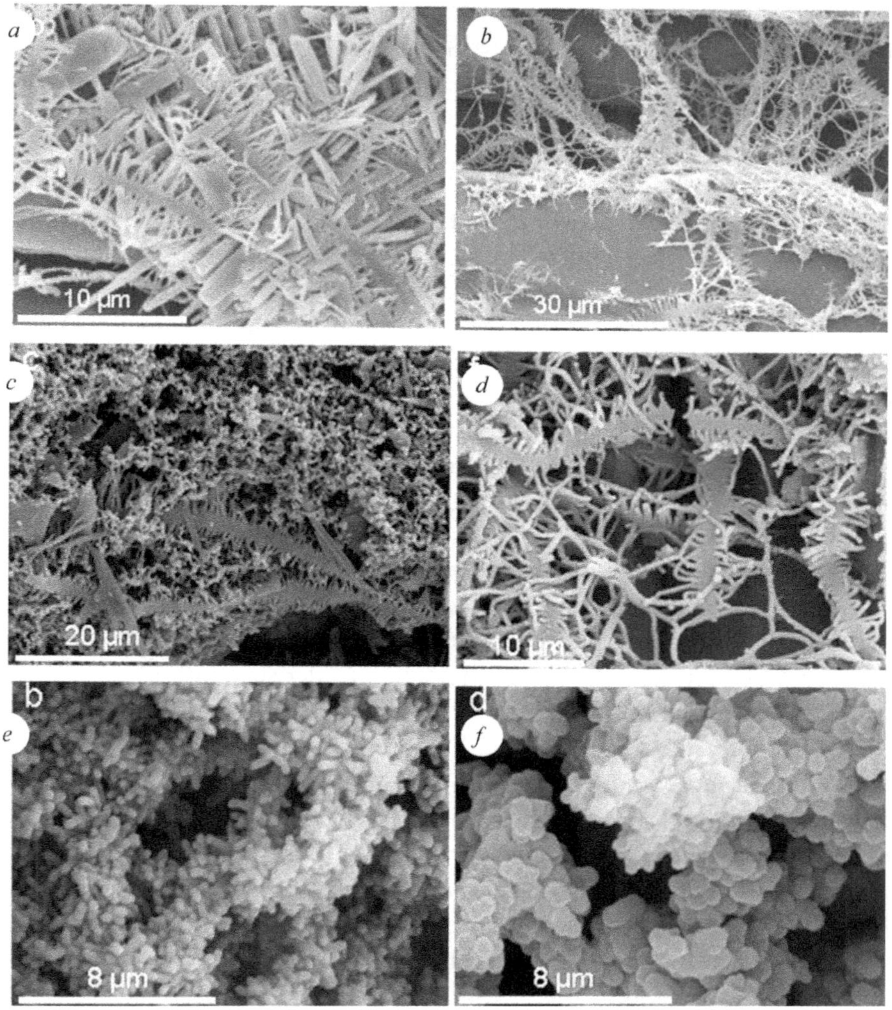

FIGURE 12.12 Micrographs of **PAn** synthesized by cyclic voltammetry method on **PES/PPy-AQSA** substrates in the 0.5 M H_2SO_4 + 0.5 M aniline solutions under the (-0.2)-(+1.1) V potential scanning interval under the scan rate 50 (*a*, *b*), 5 (*c*, *d*) and 1 (*e*, *f*) $mV{\cdot}s^{-1}$ after the 2 (*c*, *e*), 3 (*d*, *f*), 5 (*a*), and 10 (*b*) scans of potential (From: Molina, J., Río, A. I., Bonastre, J., "Cases, F. Influence of the Scan Rate on the Morphology of Polyaniline Grown on Conducting Fabrics. Centipede-Like Morphology," in *Synth. Met.* **2010**, *160*, 99–107. Used with permission from Elsevier.)

Under scanning speed of 50 $mV{\cdot}s^{-1}$ after initial five cycles of scan the development of **PAn** with different morphologies is obvious, namely there was fixed formation of **PAn** fibrils, and **PAn** with stalagmite-like and with centipede-like morphology (Fig. 12.12, *a*). The structures with stalagmite-like morphology grow with different orientations and their length is lower than 10 mm. The average diameter of the branched nanofibers is around 200 nm and their length is 1-3 mm approximately. It indicates that cyclic voltammetry can be used as an alternative method to produce **PAn** nanostructures on conducting fabrics. Globular morphology has not been observed during the first scans. After the 10 scans, a lot of centipede-like structures can be observed as well as **PAn** fibers (Fig. 12.12, *b*). With higher number of scans, the globular morphology gains importance. The result of the scan rate reducing to 5 $mV \cdot s^{-1}$ is fact that only in some areas the growth of **PAn** with fibrillar morphology was observed, but the actual morphology in some parts of the micrograph is a centipede-like morphology (Fig. 12.12, *c*). It can be seen that the two centipede-like structures present in this micrograph are longer than 60 mm. The diameter of the branched nanofibers was the same that for 50 $mV \cdot s^{-1}$, 200 nm approximately. So each of the structures in Figure 12.12, *c* had nearly 600-branched microfibers (300 on each side of the central fiber). The length of the branched nanofibers was approximately 3 mm. Figure 12.12, *d* shows the morphology of the sample after three scans. This micrograph several centipede-like structures and several isolated fibers can be observed. After seven scans the surface of the sample was covered by **PAn** with globular and fibrillar morphology. Under scan rate 1 $mV \cdot s^{-1}$ after two scans the fabric was entirely covered with fibrillar and globular **PAn** and the textile substratum could be hardly observed. The development of **PAn** with globular morphology is general on the fabric surface. **PAn** with coral-like morphology has also been observed as it can be seen in Figure12.12, *e*. The fibers of the fabric cannot be observed after three scans. **PAn** with globular morphology (Fig. 12.12, *f*) has covered all the surface of the fabric. Higher number of scans does not produce any changes in the samples and only globular morphology is observed.

12.7 CONCLUSIONS

So, as follows from the analysis of the references, the choice of synthesis conditions and appropriate substrate enables to obtain on different on nature polymeric surfaces the high adhesive, electroactive, selective and stable films of **PAn** that successfully can be used for various technological purposes. At the chemical oxidative polymerization of **An** on different polymeric surfaces the ammonium peroxydisulfate or $FeCl_3 \cdot 6H_2O$ mainly are used as an oxidizing agent. Transparent and flexible polymeric films or woven and non-woven fabrics made from synthetic, artificial or natural fibers mostly are used as the substrates.

The choice of the method of **PAn**/polymer composite obtaining is determined mainly by two factors. The first is the field of the following application of synthesized composite. **PAn**/polymeric composites today successfully are tested as touch sensitive materials, anticorrosive, antistatic, electrostatic coating materials for shielding of electromagnetic waves, elements of organic optoelectronic devices, diaphragms, artificial muscles, electroconducting fabrics, biologically active substrates, etc.

Another factor is the nature and form of the substrate. So, for the film substrates the most commonly used is the synthesis method *in situ*. In this case, the best results can be obtained when such substrate before the polymerization previously was withstood in pure aniline: the swelling of substrate into aniline provides a layered cluster structure of polyaniline component in the composite. Due to specific surface of porous and fibrous substrates, good results are obtained under the synthesis of **PAn** *in situ* from the solutions. However, the uniform composites of non-conductive polymers and **PAn** usually are obtained via spreading method and via the method of melts preparation. At the same time, from a technological point of view, the most productive and, as a result, the most attractive in terms of industry, is the last method.

One can predict that further development of physical chemistry of **PAn**/polymeric composites will enter into the plane of nanoscience and nanotechnologies. This is applied to both methods and technologies of nanoscale (as to thickness) composite coatings electroconductive/non-conducting polymer preparation for various applications, as well as the use of nanopowder or nanostructured components at their obtaining, etc. The separate references above-mentioned in this review are the proof of this fact.

KEYWORDS

- **polyaniline**
- **non-conductive polymers**
- **modification of surface**
- **polyaniline/polymeric composites**

REFERENCES

1. Mortimer, R. J. Electrochromic Materials. *Chem. Soc. Rev.* **1997,** *26,* 147–156.
2. Gurunathan, K.; Murugan, A. V.; Marimuthu, R.; Mulik, U. P.; Amalnerkar, D. P. Electrochemi-Cally Synthesised Conducting Polymeric Materials for Applications towards Technology in Electronics, Optoelectronics and Energy Storage Devices. *Mater. Chem. Phys.* **1999,** *61,* 173-191.
3. Hatchett, D. W.; Josowicz, M. Composites of Intrinsically Conducting Polymers as Sensing Nanomaterials. *Chem. Rev.* **2008,** *108,* 746-769.
4. Osama, O.; Kimura, O.; Kabata, T. A Solid Electrolytic Paper Battery Containing Electroconductive Polymers. *Electron. Commun. Jpn.* **1992,** *75,* 1123-1129.
5. Novak, P.; Müller, K.; Santhanam, K. S. V.; Haas, O. Electrochemically Active Polymers for Rechargeable Batteries. *Chem. Rev.* **1997,** *97,* 207-281.
6. Kim, J. H.; Cho, J. H.; Cha, G. S. Conductimetric Membrane Strip Immunosensor with Polyaniline-Bound Gold Colloids as Signal Generator. *Biosens. Bioelectron.* **2000,** *14,* 907-915.
7. Malinauskas, A. Chemical Deposition of Conducting Polymers. *Polymer.* 2001, *42,* 3957-3972.
8. Toshima, N.; Hara, S. Direct Synthesis of Conducting Polymers from Simple Monomers. *Prog. Polym. Sci.* **1995,** *20,* 155-183.
9. Gospodinova, N.; Terlemezyan, L. Conducting Polymers Prepared by Oxidative Polymerization: Polyaniline. *Prog. Polym. Sci.* **1998,** *23,* 1443-1484.
10. Koval'chuk, E. P.; Whittingham, S.; Skolozdra, O. M.; Zavaliy, P. Y.; Zavaliy, I. Yu.; Reshetnyak, O. V.; Seledets, M. Co-Polymers of Aniline and Nitroanilines. Part I. Mechanism of Aniline Oxidation Polycondensation. *Mater. Chem. Phys.* **2001,** *69,* 154-162.
11. Steskal, J.; Gilbert, R. G. Polyaniline. Preparation of a Conducting Polymer. IUPAC Technical Report. *Pure Appl. Chem.* **2002,** *74,* 857–867.
12. Pud, A.; Ogurtsov, N.; Korzhenko, A.; Shapoval, G. Some Aspects of Preparation Methods and Properties of Polyaniline Blends and Composites with Organic Polymers. *Prog. Polym. Sci.* **2003,** *28,* 1701–1753.
13. Biallozor, S.; Kupniewska, A. Conducting Polymers Electrodeposited on Active Metals. *Synth. Met..* **2005,** *155,* 443–449.
14. Yatsyshyn, M.; Koval'chuk, Ye. Polyaniline: Chemical Synthesis, Synthesis Mechanism, Structure and Properties, Doping. Proc. Shevchenko Sci. Soc. *Chem. Biochem.* **2008,** *21,* 87–102 (In Ukrainian).
15. Koval'chuk, E. P.; Tomilov, A. P.; Ostapovich, B. B.; Yatsyshyn, M. N. Electroconducting Polymers. In *Electrochemistry of Organic Compounds at the Beginning of XXI Century*; Moscow: Kompania Sputnik+, Russia, **2008**; pp 496–537.
16. Bhadra, S.; Khastgir, D.; Singh, N. K.; Lee, J. H. Progress in Preparation, Processing and Applications of Polyaniline. *Prog. Polymer Sci.* **2009,** *34,* 783–810.
17. Shirota, Y. Organic Materials For Electronic and Optoelectronic Devices. *J. Mater. Chem.* **2000,** *10,* 1–25.
18. Sapurina, I.; Stejskal, J. The Mechanism of the Oxidative Polymerization of Aniline and the Formation of Supramolecular Polyaniline Structures. *Polym. Int.* **2008,** *57,* 1295–1325.
19. Ciric-Marjanovic, G. Recent Advances in Polyaniline Research: Polymerization Mechanisms, Structural Aspects, Properties and Applications. *Synth. Met.* **2013,** *177,* 1–47.

20. Qi, J.; Xu, X.; Liu, X. X.; Lau, K. T. Fabrication of Textile Based Conductometric Polyaniline Gas Sensor. *Sensor. Actuat. B-Chem.* **2014,** *202,* 732–740.
21. Oh, K. W.; Hong, K. H.; Kim, S. H. Electrically Conductive Textiles by *in Situ* Polymerization of Aniline. *J. Appl. Polym. Sci.* **1999,** *74,* 2094–2101.
22. Neoh, K. G.; Tay, B. K.; Kang, E. T. Oxidation and Ion Migration during Synthesis and Degradation of Electroactive Polymer–Nylon 6 Composite Films. *Polymer.* **2000,** *41,* 9–15.
23. Xing, S.; Zhao, C.; Jing, S.; Wu, Y.; Wang, Z. Morphology and Gas-Sensing Behavior of *in Situ* Polymerized Nanostructured Polyaniline Films. *Eur. Polym. J.* **2006,** *42,* 2730–2735.
24. Ge, Z.; Brown, C. W.; Sun, L.; Yang, S. C. Fiberoptic Ph Sensor-Based on Evanescent-Wave Absorption-Spectroscopy. *Anal. Chem.* **1993,** *65,* 2335–2338.
25. Hirase, R.; Shikata, T.; Shirai, M. Selective Formation of Polyaniline on Wool by Chemical Polymerization, Using Potassium Iodate. *Synth. Met.* **2004,** *146,* 73–77.
26. Jin, Z.; Su, Y.; Duan, Y. An Improved Optical Ph Sensor Based on Polyaniline. *Sensor. Actuat. B.Chem.* **2000,** 71, 118–122.
27. Zhou, H.; Shi, Z.; Lu, Y. Conducting Polyaniline/Poly (Tetrafluoroethylene) Composite Films with Tunable Surface Morphology and Hydrophilicity. *Synth. Met.* **2010,** 160, 1925–1930.
28. Duboriz, I.; Pud, A. Polyaniline/Poly(Ethylene Terephthalate) Film as a New Optical Sensing Material. *Sensor. Actuat. B-Chem.* **2014,** *190,* 398–407.
29. Goswami, S.; Mitra, M. K.; Chattopadhyay, K. K. Enhanced Field Emission from Polyaniline Nano-Porous Thin Films on PET Substrate. *Synth. Met.* **2009,** *159,* 2430–2436.
30. Khalid, M.; Mohammad, F. Preparation, FTIR Spectroscopic Characterization and Isothermal Stability of Differently Doped Fibrous Conducting Polymers Based on Polyaniline and Nylon-6,6. *Synth. Met.* **2009,** *159,* 119–122.
31. Kang, E. T.; Neoh, K. G.; Pun, M. Y. Charge Transfer Interactions between Polyaniline and Surface Functionalized Polymer Substrates. *Synth. Met.* **1995,** *69,* 105–108.
32. Neoh, K. G.; Teo, H. W.; Kang, E. T. Enhancement of Growth and Adhesion of Electroactive Polymer Coatings on Polyolefin Substrates. Langmuir. **1998,** *14,* 2820–2826.
33. Araujo, J. R.; Adamo, C. B.; De Robertis, E.; Kuznetsov, A. Yu.; Archanjo, B. S.; Fragneaud, B.; Achete, C. A.; De Paol, M. A. Crystallinity, Oxidation States and Morphology of Polyaniline Coated Curauá Fibers in Polyamide-6 Composites. *Compos. Sci. Technol.* **2013,** *88,* 106–112.
34. Kumar, P. A.; Chakraborty, S.; Ray, M. Removal and Recovery of Chromium from Wastewater Using Short Chain Polyaniline Synthesized On Jute Fiber. *Chem. Eng. J.* **2008,** *141,* 130–140.
35. Wu, J.; Zhou, D.; Looney, M. G.; Waters, P. J.; Wallace, G. G.; Too, C. O. A Molecular Template Approach to Integration of Polyaniline into Textiles. *Synth. Met.* **2009,** *159,* 1135–1140.
36. Molina, J.; Esteves, M. F.; Fernández, J.; Bonastre, J.; Cases, F. Polyaniline Coated Conducting Fabrics. Chemical and Electrochemical Characterization. *Eur. Polym. J.* **2011,** *47,* 2003–2015.
37. Wang, J. G.; Neoh, K. G.; Kang, E. T. Polyaniline–Palladium Composite Coatings for Metallization of Polyethylene Substrate. *Appl. Surf. Sci.* **2003,** *218,* 231–244.
38. Oh, K. W.; Kim, S. H.; Kim, E. A. Improved Surface Characteristics and the Conductivity of Polyaniline–Nylon 6 Fabrics By Plasma Treatment. *J. Appl. Polym. Sci.* **2001,** *81,* 684–694.

39. Cho, J.; Shin, K. H.; Jang, J. Polyaniline Micropattern onto Flexible Substrate by Vapor Deposition Polymerization-Mediated Inkjet Printing. *Thin Solid Films.* **2010,** *518,* 5066–5070.
40. Kang, E. T.; Neoh, K. G.; Tan, K. L. Surface Modifications of Poly(3-Alkylthiophene) Films by Graft Copolymerization. *Macromolecules.* **1992,** *25,* 6842–6848.
41. Tan, K. L.; Woon, L. L.; Wong, H. K.; Kang, E. T.; Neoh, K. G. Surface Modification of Plasma-Pretreated Poly(Tetrafluoroethylene) Films by Graft Copolymerization. *Macromolecules.* **1993,** *26,* 2832.
42. Flerov, G. N. Synthesis of Superheavy Elements and Application of the Nuclear Physics Methods in Adjacement Fields. *Vestn. Akad.* Nauk SSSR, **1984,** *4,* 35–48.
43. Demidovaz, E. N.; Drachev, A. I.; Grigor'eva, G. A. Investigation of Electrotransport Properties of Poly(Ethylene Terephtalate) Track Membranes Modified by Plasma of Aniline. *Russ. J. Electrochem.* **2009,** *45,* 533–537.
44. Elyashevich, G. K.; Smirnov, M. A.; Kuryndin, I. S.; Bukošek, V. Electroactive Composite Systems Containing High Conductive Polymer Layers on Poly(Ethylene) Porous Films. *Polym. Adv. Technol.* **2006,** *17,* 700–704.
45. Chen, M. H.; Chiao, T. C.; Tseng, T. W. Preparation of Sulfonated Polysulfone/Polysulfone and Aminated Polysulfone/Polysulfone Blend Membranes. *J. Appl. Polym. Sci.* **1996,** *61,* 1205–1209.
46. Lu, Y.; Wang, L.; Zhao, B.; Xiao, G.; Ren, Y.; Wang, X.; Li, C. Fabrication of Conducting Polyaniline Composite Film Using Honeycomb Ordered Sulfonated Polysulfone Film as Template. *Thin Solid Films.* **2008,** *516,* 6365–6370.
47. Kang, E. T.; Neoh, K. G., Tan, K. L.; Uyama, Y.; Morikawa, N.; Ikada, Y. Surface Modifications of Polyaniline Films b y Graft Copolymerization. *Macromolecules.* **1992,** *25,* 1959–1965.
48. Wang, T.; Kang, E. T.; Neoh, K. G.; Tan, K. L.; Liaw, D. J. Surface Modification of Low-Density Polyethylene Films by UV–Induced Graft Copolymerization and its Relevance to Photolamination. *Langmuir.* **1998,** *14,* 921–927.
49. Sha, B.; Wang, J.; Zhou, L.; Zhang, X.; Han, L.; Zhao, L. Adsorption of Organic Amines from Wastewater by Carboxyl Group-Modified Polyacrylonitrile Fibers. *J. Appl. Polym. Sci.* **2013,** *128,* 4124–4129.
50. Anand, J.; Palaniappan, S.; Sathyanarayana, D. N. Condacting Polyaniline Blends and Composites. *Prog. Polym. Sci.* **1998,** *23,* 993–1018.
51. Riede, A.; Stejskal, J.; Helmstedt, M. *In-Situ* Prepared Composite Polyaniline Films. *Synth. Met.* **2001,** *121,* 1365–1366.
52. Ivanov, S.; Mokreva, P.; Tsakova, V.; Terlemezyan, L. Electrochemical and Surface Structural Characterization of Chemically and Electrochemically Synthesized Polyaniline Coatings. Thin Solid Films. **2003,** 441, 44–49.
53. Zhai, G.; Fan, Q.; Tang, Y.; Zhang, Y.; Pan, D.; Qin, Z. Conductive Compositefilms Composed Of Polyaniline Thin Layers On Microporous Polyacrylonitrile Surfaces. *Thin Solid Films.* **2010,** *519,* 169–173.
54. Wang, J.; Zhang, K.; Zhao, L.; Ma, W.; Liu, T. Adsorption and Polymerization of Aniline on a Carboxylic Group-Modified Fibrous Substrate. *Synth. Met.* **2014,** *188,* 6–12.
55. Tabellouta, M.; Fatyeyeva, K.; Baillif, P. Y.; Bardeau, J. F.; Pud, A. A. The Influence of the Polymer Matrix on the Dielectric and Electrical Properties of Conductive Polymer Composites Based on Polyaniline. *J. Non-Cryst. Solids.* **2005,** *351,* 2835–2841.

56. Pud, A. A.; Rogalsky, S. P.; Shapoval, G. S.; Korzhenko, A. A. The Polyaniline/Poly(Ethylene Terephthalate) Composite 1. Peculiarities of the Matrix Aniline Redox Polymerization. *Synth. Met.* **1999,** *99,* 175–179.
57. Pud, A. A.; Shapoval, G. S.; Kukhar, V. P. The Method of Manufacture of the Sensor Conducting Polymer Composite Material. Ukrainian Claim for Invention Rights N940**10153,** May 11, 1993.
58. Demarcos, S.; Alcubierre, N.; Galban, J.; Castillo, J. R. Reagentless System for Sulphite Determination Based on Polyaniline. *Anal. Chem. Acta.* **2004,** *502,* 7–13.
59. Dhawan, S. K.; Singh, N.; Venkatachalam, S. Shielding Behaviour of Conducting Polymer–Coated Fabrics in X-Band, W-Band and Radio Frequency Range. *Synth. Met.* **2002,** *129,* 261–267.
60. Jin, Z.; Su, Y.; Duan, Y. Development of a Polyaniline-Based Optical Ammonia Sensor. *Sens. Actuators. B-Chem.* 2001. *72,* 75–79.
61. Korzhenko, A. A.; Tabellout, M.; Emery, J. R.; Pud, A. A.; Rogalsky, S.; Shapoval, G. S. Dielectric Relaxation Properties of Poly(Ethylene-Terephthalate)–Polyaniline Composite Films. *Synth. Met.* **1998,** *98,* 157–160.
62. Pud, A. A.; Tabellout, M.; Kassiba, A.; Korzhenko, A. A.; Rogalsky, S. P.; Shapoval, G. S.; Houzé, F.; Schneegans, O.; Emery, J. R. The Poly(Ethylene Terephthalate) / Polyaniline Composite: AFM, DRS and EPR Investigations of Some Doping Effects. *J. Mater. Sci.* **2001,** *36,* 3355–3363.
63. Liu, C. D.; Wu, S. Y.; Han, J. L.; Hsieh, K. H. Patterned Conductive Polyaniline Films Fabricated Using Lithography and *in Situ* Polymerization. *J. Appl. Polym. Sci.* 2010, *115,* 2271–2276.
64. Duboriz, Ye. P.; Fateeva, K. Yu.; Pud, O. A.; Shapoval, G. S. Device for Registration of the Optical Response of Sensoric Film Polymeric Materials. Ukraine Patent 30781, Aprile 11. 2008.
65. Kutanis, S.; Karakisla, M.; Akbulut, U.; Sacëak, M. The Conductive Polyaniline/Poly(Ethylene Terephthalate) Composite Fabrics. *Composites. A.* **2007,** *38,* 609–614.
66. Kelly, F. M.; Meunier, L.; Cochrane, C.; Koncar, V. Polyaniline: Application as Solid State Electrochromic in a Flexible Textile Display. *Displays.* **2013,** *34,* 1–7.
67. Pud, O. A.; Rogalskiy, S. P.; Fateeva, K. Yu.; Shapoval, G. S. The Method of the Production of Sensoric Electroconductive Polymeric Material. Ukraine Patent 75761, May 15, 2006.
68. Nicho, M. E.; Trejo, M.; Garcia–Valenzuela, A.; Saniger, J. M.; Palacios, J.; Hu, H. Polyaniline Composite Coating Interrogated by a Nulling Optical-Transmittance Bridge for Sensing Low Concentrations of Ammonia Gas. *Sens. Actuators. B-Chem.* **2001,** *76,* 18–24.
69. Hu, H.; Trejo, M.; Nicho, M. E.; Saniger, J. M.; Garcia-Valenzuela, A. Adsorption Kinetics of Optochemical NH_3 Gas Sensing with Semiconductor Polyaniline Films. *Sens. Actuators. B-Chem.* **2002,** *82,* 14–23.
70. Wan, M.; Li, M.; Li, J.; Liu, Z. Transpared and Conducting Coatings of Polyaniline Composites. *Thin Solid Films.* **1995,** *259,* 188–193.
71. Li, D. F.; Wang, W.; Wang, H. J.; Jia, X. S.; Wang, J. Y. Polyaniline Films with Nanostructure Used as Neural Probe Coating Surfaces. *Appl. Surf. Sci.* **2008,** *255,* 581–584.
72. Sreenivasan, K. Identification of Salicylic Acid Using Surface Modified Polyurethane Film Using an Imprinted Layer of Polyaniline. *Anal. Chim. Acta.* **2007,** *583,* 284–288.
73. Wu, X.; Qian, X.; An, X. Flame Retardancy of Polyaniline-Deposited Paper Composites Prepared Via *in Situ* Polymerization. *Carbohyd. Polym.* **2013,** *92,* 435–440.

74. Kelly, F. M.; Johnston, J. H.; Borrmann, T.; Richardson, M. J. Functionalised Hybrid Materials of Conducting Polymers with Individual Fibres of Cellulose. *Eur. J. Inorg. Chem.* **2007,** *35,* 5571–5577.
75. Stejskal, J.; Trchova, M.; Kovarova, J.; Prokes, J.; Omastova, M. Polyaniline-Coated Cellulose Fibers Decorate With Silver Nanoparticles. *Chem. Pap.* **2008,** *62,* 181–186.
76. Lee, B. H.; Kim, H. J.; Yang, H. S. Polymerization of Aniline on Bacterial Cellulose and Characterization of Bacterial Cellulose/Polyaniline Nanocomposite Films. *Curr. Appl. Phys.* **2012,** *12,* 75–80.
77. Merlini, C.; Barra, G. M. O.; Schmitz, D. P.; Ramôa, S. D. A. S.; Silveira, A.; Araujo, T. M.; Pegoretti, A. Polyaniline-Coated Coconutfibers: Structure, Properties and Their Use as Conductive Additives in Matrix of Polyurethane Derived from Castor Oil. *Polym. Test.* **2014,** *38,* 18–25.
78. Onar, N.; Aksit, A. C.;. Ebeoglugil, M. F; Birlik, I.; Celik, E.; Ozdemir, I. Structural, Electrical, and Electromagnetic Properties f Cotton Fabrics Coated with Polyaniline and Polypyrrole. *J. Appl. Polym. Sci.* **2009,** *114,* 2003–2010.
79. Araujo, J. R.; Adamo, C. B.; De Paoli, M. A. Conductive Composites of Nylon-6 with Polyaniline Coated Vegetal Fibre. *Chem. Eng. J.* **2011,** *174,* 425–31.
80. Dutta, D.; Sarma, T. K.; Chowdhury, D.; Chattopadhyay, A. A Polyaniline-Containing Filter Paper that Acts as a Sensor, Acid, Base, And Endpoint Indicator and also Filters Acids and Bases. J. *Colloid Interf. Sci.* **2005,** *283,* 153–159.
81. Xia, Y.; Lu, Y. Fabrication and Properties of Conductive Conjugated Polymers / Silk Fibroin Composite Fibers. *Compos. Sci. Technol.* **2008,** *68,* 1471–1479.
82. Chen, J.; Winther-Jensen, B.; Pornputtkul, Y.; West, K.; Kane-Maquire, L.; Wallace, G. G. Synthesis of Chiral Polyaniline Films Via Chemical Vapor Phase Polymerization. *Electrochem. Solid St. Lett.* **2006,** *9,* 9–11.
83. Kim, J. Y.; Lee, J. H.; Kwon, S. J. The Manufacture and Properties of Polyaniline Nano–Films Prepared Through Vapor-Phase Polymerization. *Synth. Met.* **2007,** *157,* 336–342.
84. Zhang, H.; Li, C. Chemical Synthesis of Transparent and Conducting Polyaniline-Poly(Ethylene Terephthalate) Composite Films. *Synth. Met.* **1991,** *44,* 143–146.
85. Gong, X.; Dai, L.; Mau, A. W. H.; Griesser, H. J. Plasma–Polymerized Polyaniline Films: Synthesis and Characterization. *J. Polym. Sci. Pol. Chem.* **1998,** *36,* 633–643.
86. Wang, J.; Neoh, K. G.; Zhao, L.; Kang, E. T. Plasma Polymerization of Aniline on Different Surface Functionalized Substrates. *J. Colloid. Interf. Sci.* **2002,** *251,* 214–224.
87. Yang, P.; Xie, J.; Yang, W. A Simple Method to Fabricate a Conductive Polymer Micropattern on an Organic Polymer Substrate. *Macromol. Rapid. Comm.* **2006,** *27,* 418–423.
88. Talbi, H.; Just, P. E.; Dao, L. H. Electropolymerization of Aniline on Carbonized Polyacrylonitrile Aerogel Electrodes: Applications for Supercapacitors. *J. Appl. Electrochem.* **2003,** *33,* 465–473.
89. Molina, J.; Río, A. I.; Bonastre, J.; Cases, F. Electrochemical Polymerisation of Aniline on Conducting Textiles of Polyester Covered with Polypyrrole/*AQSA. Eur. Polym. J.* **2009,** *45,* 1302–1315.
90. Molina, J.; Río, A. I.; Bonastre, J.; Cases, F. Influence of the Scan Rate on the Morphology of Polyaniline Grown on Conducting Fabrics. Centipede-Like Morphology. *Synth. Met.* **2010,** *160,* 99–107.
91. Huang, Z.; Wang, P. C.; Macdiarmid, A. G.; Xia, Y.; Whitesides, G. M. Selective Deposition of Conducting Polymers on Hydroxyl-Terminated Surfaces with Printed Monolayers of Alkylsiloxanes as Templates. *Langmuir.* **1997,** *13,* 6480–6484.

92. Huang, Z.; Wang, P. C.; Feng, J.; Macdiarmid, A. G.; Xia, Y.; Whitesides, G. M. Selective Deposition of Films of Polypyrrole, Polyaniline and Nickel on Hydrophobic/Hydrophilic Patterned Surfaces and Applications. *Synth. Met.* **1997,** *85,* 1375–1376.
93. Barros, R. A.; Martins, C. R.; Azevedo, W. M. Writing with Conducting Polymer. *Synth. Met.* **2005,** *155,* 35–38.
94. Souza Jr., F. G.; Oliveira, G. E.; Anzai, T.; Richa, P.; Cosme, T.; Nele, M.; Rodrigues, C. H. M.; Soares, B. G.; Pinto, J. C. A Sensor for Acid Concentration Based on Cellulose Paper Sheets Modified with Polyaniline Nanoparticles. *Macromol. Mater. Eng.* **2009,** *294,* 739–748.
95. Lee, U. J.; Lee, S. H.; Yoon, J. J.; Oh, S. J.; Lee, S. H.; Lee, J. K. Surface Interpenetration Between Conducting Polymer and PET Substrate for Mechanically Reinforced ITO–Free Flexible Organic Solar Cells. *Sol. Energ. Mat. Sol. C.* **2013,** *108,* 50–56.
96. Lee, K.; Cho, S.; Park, S. H.; Heeger, A. J.; Lee, C. W.; Lee, S. H. Metallic Transport in Polyaniline. *Nature.* **2006,** *441,* 65–68.
97. Lee, S. H.; Lee, D. H.; Lee, K.; Lee, C. W. High-Performance Polyaniline Prepared Via Polymerization in a Self–Stabilized Dispersion. *Adv. Func. Mater.* **2005,** *15,* 1495–1500.
98. Tang, Q.; Wu, J.; Sun, X.; Li, Q.; In, J. Layer–by–Layer Self-Assembly of Conducting Multilayer Film from Poly(Sodium Styrenesulfonate) and Polyaniline. *J. Colloid Interf. Sci.* **2009,** *337,* 155–161.
99. Cruz-Estrada, R. H.; Cupul-Manzano, C. V. Structure Formation in Polyaniline–Based Polymer Blends. *J. Mater. Sci.* **2005,** *40,* 6571–6579.
100. Eftekhari, A. Nanostructured Conductive Polymers, John Wiley & Sons Ltd., U.K., 2010; pp 776
101. Farrokhzad, H.; Van Gerven, T.; Van Der Bruggen, B. Preparation and Characterization of aonductive Polyaniline/Polysulfone Film and Evaluation of the Effect of Co-Solvent. *Eur. Polym. J.* **2013,** *49,* 3234–3243.
102. Stockton, W. B. Structure and Electrical Properties of Assemblies of Polyaniline: From Blends to Self-Assembled Multilayers., Doctoral Thesis, Doctor of Philosophy at the Massachusetts Institute of Technology, 1995.
103. Koul, S.; Chandra, R.; Dhawan, S. K. Conducting Polyaniline Composites: A Reusable Sensor Material for Aqueous Ammonia. *Sens. Actuators. B-Chem.* **2001,** *75,* 151–159.
104. Chen, C. H.; Kan, Y. T.; Mao, C. F.; Liao, W. T.; Hsieh, C. D. Fabrication and Characteriza-Tion of Water-Based Polyurethane/Polyaniline Conducting Blend Films. *Surf. Coat. Tech.* **2013,** *231,* 71–76.
105. Saïdi, S.; Mannaï, A.; Derouiche, H.; Mohamed, A. B. Effect of Drying Temperature on Structural and Electrical Properties of PANI : PVDF Composite Thin Films and Their Application as Buffer Layer for Organic Solar Cells. *Mat. Sci. Semicon. Proc.* **2014,** *19,* 130–135.
106. Hsu, C. F.; Kilmartin, P. A. Antioxidant Capacity of Robust Polyaniline–Ethyl Cellulose Films. *React. Funct. Polym.* **2012,** *72,* 814–822.
107. Hong, C. H.; Ki, S. J.; Jeon, J. H.; Che, H. L.; Park, I. K.; Kee, C. D.; Oh, I. K. Electroac-Tive Biocomposite Actuators Based on Cellulose Acetate Nanofibers with Specially Chopped Polyaniline Nanoparticles through Electrospinning. *Compos. Sci. Technol.* **2013,** *87,* 135–141.
108. Liu, D. Y.; Sui, G. X.; Bhattacharyya, D. Synthesis and Characterization of Nanocellulose–Based Polyaniline Conducting Films. *Compos. Sci. Technol.* **2014,** *30,* 31–36.

109. Chatzidaki, E. K.; Favvas, E. P.; Papageorgiou, S. K.; Kanellopoulos, N. K.; Theophilou, N. V. New Polyimide–Polyaniline Hollow Fibers: Synthesis, Characterization and Behavior in Gas Separation. *Eur. Polym. J.* **2007,** *43,* 5010–5016.
110. Hechavarrıa, L.; Hu, H.; Rincon, M. E. Polyaniline-Poly(2-Acrylamido-2-Methyl-1-Propanosulfonic Acid) Composite Thin Films: Structure and Properties. *Thin Solid Films.* **2003,** *441,* 56–62.
111. Martins, C. R.; De Paoli, M. A. Antistatic Thermoplastic Blend of Polyaniline and Polystyrene Prepared in a Double-Screw Extruder. *Eur. Polym. J.* **2005,** *41,* 2867–2873.
112. Nand, A. V.; Ray, S.; Travas-Sejdic, J.; Kilmartin, P. A. Characterization of Antioxidant Low Density Polyethylene/Polyaniline Blends Prepared Via Extrusion. *Mater. Chem. Phys.* **2012,** *135,* 903–911.
113. Kazimi, M. R.; Shah, T.; Jamari, S. B.; Faizal, C. M. Characterization of Functionalized Low Density Polyethylene/Polyaniline Nano Fiber Composite. *J. Med. Bioeng.* **2014,** *3,* 306–310.

CHAPTER 13

HYBRID NANOSYSTEMS BASED ON CONJUGATED POLYAMINOARENES DOPED BY FERRUM–CONTAINING COMPOUNDS

O. I. AKSIMENTYEVA[1], YU. YU. HORBENKO[1], and O. I. KONOPELNYK[2]

[1]*Department of Physical and Colloid Chemistry, Faculty of Chemistry, Ivan Franko National University of Lviv, 6 Kyryla & Mefodia Str., Lviv 79005, Ukraine. E-mail: aksimen@ukr.net*

[2]*Department of General Physics, Faculty of Physics, Ivan Franko National University of Lviv, 8 Kyryla & Mefodia Str., Lviv 79005, Ukraine. E-mail: konopel@ukr.net*

CONTENTS

ABSTRACT

The main objective of this chapter is to investigate of hybrid nanosystems based on conjugated polyaminearenes doped by ferrum containing compounds.

INTRODUCTION

One of the main tasks of modern physical and chemical investigations is the development of scientific principles and new approaches to the creation of polymeric and composite materials with present functional properties, namely electrical, optical, magnetic, an ability to the charge storage and its transfer, the catalysis of a number of reactions, sensorics, and so forth.[1–5] The actual problem of the science and technology at the present step is an investigation of the mechanism of formation and the development of new nanosized composite materials based on polymers doped or filled by inorganic clusters, in particular carbonic (graphene,[6,7] fullerene,[8,9] and nanotubes,[10,11]) silicium (silicium (IV) oxide,[12–14] nanocrystals of silicon,[1,15] porous silicon,[16–18] and silicium carbide[19]) as well as by the compounds of transition metals, in particular by ferrum-containing ones, namely clusters of ferrum and its oxide, magnetite, the complexes of ferrum, and so forth.[20–25] Due to the nanostructures of such composites (the size of the particles is to dozens nm) they have unique magnetic, spectral, and electrochemical properties. General scientific problem at the development of high dispersed and film composites of the polymers with inorganic nanoparticles are understanding of the character of components interaction. For this purpose, it is necessary to investigate the physical chemistry both of initial compounds and of the obtained composites; this gives the possibility to control the parameters of the synthesized materials. This knowledge can be obtained via the complex investigations of crystalline, molecular, supramolecular and electronic structure, the dispersion degree of the particles, their distribution, determination of the relationship between the structure and magnetic, electrical, and optical properties of nanomaterials. The investigations presented in this chapter are dedicated to the solution of the abovesaid tasks.

13.1 NANOSYSTEMS BASED ON CONJUGATED POLYMERS AND FERRUM-CONTAINING COMPOUNDS

The term "*nanosystem*" has in mind such objects, in which at least one component is nanodimensional.[26,27] When we consider the conjugated (electroconductive) polymers, we have deal with a number of nanoscale elements, which depending on the method of synthesis can be considered as meso- or nanometals with an average particles size 10–20 nm or less,[28] to form globular or fibrillar nanostructures (polymeric wires and nanotubes with the diameter 5–20 nm.[29,30]) Doping of conjugated polymers by ferrum-containing compounds is also associated with the formation of nanoclusters of the dopant, which then is a part of the conjugated polymeric chain. For example, ferrum (III) chloride due to the interaction with the polymer and solvent forms the clusters of ferrum oxide.[26] Colloidal clusters are formed in solutions via chemical reaction and may have the dimensions from 1 to 100 nm. The sizes of the magnetite nanoparticles in accordance with the X-ray investigations (**XRD**) is 10 nm.[24] Colloidal clusters can be for a long time in the liquid phase, do not settle and do not coagulate due to weak intercluster interactions, charge repulsion and surface passivation. Reaction of the hydrolysis of inorganic salts of metals can be as an example of the reaction of nanoclusters obtaining:

$$FeCl_3 + 3H_2O \leftrightarrow Fe\,(OH)_3 + 3HCl \tag{13.1}$$

Nanosystems based on conjugated polymers and ferrum-containing compounds may be the basis of the development of new polymeric materials with magnetic functions (paramagnetic, superparamagnetic, ferro- and ferrimagnetic). Polymeric magnetic materials cause the growing scientific interest due to their practical use in modern technologies. Flexibility, small losses during the magnetization, the ability to form the films, small relative density designate an important role of these materials in various fields, such as molecular electronics, recording media, sensors, biological researches, new diagnostic methods in medicine.[31,32] Composites based on electroconductive polymers and oxides of transition metal (Fe, V, Mo, and Mn) are promising for the manufacture of electrode materials of chemical power sources.[22] Hybrid materials based on polyaminoarenes (**PAAr**) doped by $[Fe(CN)_6]^{4-/3-}$ can have the potential applications in storage batteries.[33] Ferromagnetic composites of ferrum oxides (e. g., magnetite) with polymers are underlying in the production of toners for electronic printing,[34] recording and storing information systems,[35] and their combination with fluorescent and electroconductive polymers enables to get the composites with two functions, namely magnetic and luminescent,[32] or magnetic and electrical.[36,37] Understanding the nature of intermolecular interactions that determine the electrical, optical, and magnetic properties of electroconducting polymers is the fundamental basis for the creation of new nanoscale materials.

In order to obtain new knowledge about such objects, we investigated the conditions of synthesis, the structure and properties of hybrid nanocomposites which are consisted of conjugated polymeric matrices of various type, including polyaniline **(PAn)**, poly–*o*–methoxyaniline **(PoMA)**, poly–*o*–toluidine **(PoT)**, poly–*o*–anisidine **(PoA)**, polyaminothiazole **(PAT)** doped with magnetic ferrum-containing compounds ($FeCl_3$, $K_3Fe(CN)_6$) able to form the magnetic nanoclusters. Another direction was the study of magnetic nanoparticles based on ferrum oxides (Fe_xO_y, Fe_3O_4) encapsulated by polymeric shells. In this section, an influence of ferrum-containing compounds on the properties of polymers of polyaniline type is considered on the example of ferrum (III) chloride as a dopant. The chemical structure of the polymeric matrices and ferrum-containing compounds (doping additives (dopants)) are presented in Table 13.1.

TABLE 13.1 Chemical Structure of Polymeric Matrixes and Ferrum-Containing Compounds.

№	Polymer		Chemical structure	Doping additive
1	poly–*o*–toluidine	**PoT**	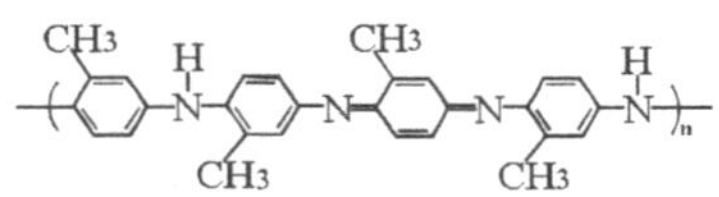	$FeCl_3$
2	poly–*o*–anisidine	**PoA**	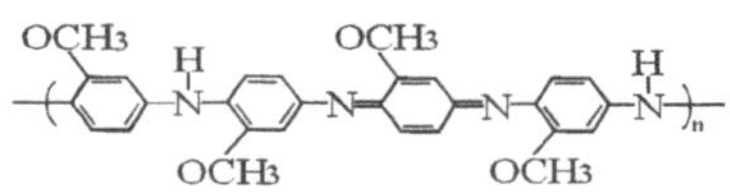	$FeCl_3$ $K_3Fe(CN)_6$
3	polyaniline	**PAn**	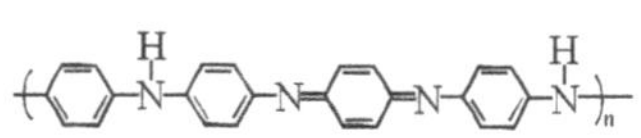	$FeCl_3$, Fe_3O_4 $K_3Fe(CN)_6$
4	polyaminothiazole	**PAT**	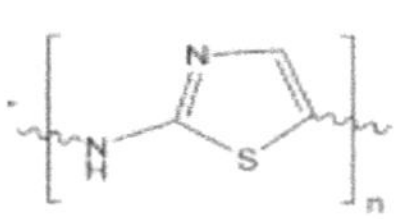	$FeCl_3$

13.2 SYNTHESIS AND STRUCTURAL CHARACTERISTICS OF CONJUGATED POLYMERS DOPED BY FERRUM-CONTAINING COMPOUNDS

The most simplest way to introduce the magnetic ions into the structure of **PAArs** is their doping by ferrum (III) chloride, which is carried out by the holding of polymer in the solution of dopant[38] or under chemical synthesis stage.[39] Ferrum (III) chloride is used for doping of many electroconductive polymers, including polyparaphenylene or polyacetylene since the 1980s.[38,40] It was found, that during the doping mixed redox and acid–base reactions take place. At this, a half of $FeCl_3$ molecules acts as an oxidizing agent leading to the formation of polycarbonium cations, and the second half of ferrum (III) chloride molecules incorporates the anions Cl^- forming the impurities $FeCl_4^-$ (Fig. 13.1).[40] The same type of the process is described for the polypyrrol, polythiophenes and their derivatives.[41]

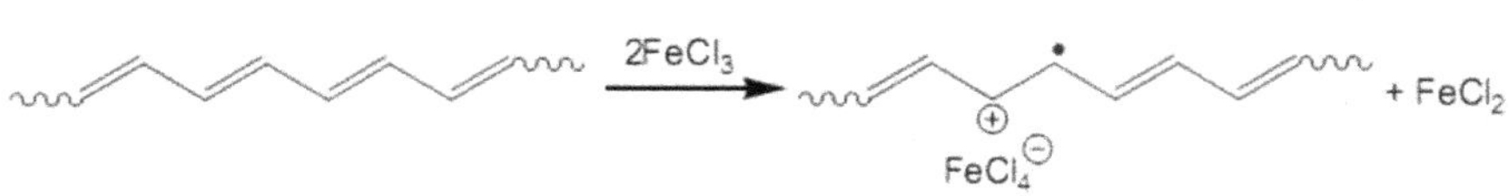

FIGURE 13.1 Doping of the polyacetylene under the action of $FeCl_3$.[40] (From: Sichel, E. K., Rubner, M. F., Georger, J. Jr., Papaefthymiou, G. C., Ofer, S., Frankel, R. B., "Magnetic Phase Transition, Aggregate Formation, and Electrical Conductivity in Fecl3-Doped Polyacetylene," in *Phys. Rev. B.* **1983,** *28*(11), 6588–6590. © 1983. Used with permission from the American Physical Society.)

Similar doping by redox type also is occurred into leucoemeraldine, namely in the most reduced form of **PAn**.

We have investigated the effect of doping by ferrum (III) chloride and potassium hexacyanoferrate on the structure and electronic properties (optical absorption, conductivity, and ESR spectra) of conjugated polymeric matrices of various types, but the main attention was accentuated on **PAArs**, namely **PAn** and its derivatives.

Synthesis of conjugated polymers doped with ferrum-containing compounds was performed by the method of oxidative polymerization of appropriate monomers in the presence in a reaction medium of ferrum (III) chloride, which acted as oxiding and doping agent, that is, ferrum-containing compound directly participates in the formation of the polymer and into its doping.[42] The process of conjugated polymers doping by potassium hexacyanoferrate was conducted under conditions of the contact exchange reaction accordingly to ref.[33]

Among conjugated **PAArs**, the *ortho*–substituted derivatives of **PAn** (namely, poly–*o*–anisidine and poly–*o*–toluidine) call an especial interest.[43,44] The presence of electron-donor substituents of the benzene ring into *ortho*–position to the amine group creates the local peculiarities of molecular structure of **PAArs** due to a decrease of the length of the polymer chain coupling and delocalization of electron in comparison with **PAn**.[45] Consequently, the electronic structure of polymers is changed, that determines the features of its doping, optical and ESR spectra, as well as parameters of the charge transfer.[39,46] One of the most interesting representative of the class of conjugated conductive **PAArs** is poly–*o*–anisidine (**PoA**), which has a very high electrical conductivity, reveals interesting and stable electrochemical and electrochromic properties, making it attractive for use in organic displays and sensors.[46,47]

Due to the presence of electron-donor methoxyl substituent (–OCH_3) of benzene ring in the *ortho*-position to the amine group, this polymer unlike to **PAn**, significantly is dissolved in acids and in organic solvents, namely chloroform, acetonitrile, **DMFA**, etc., which gives the possibility to obtain the ultrathin films of **PoA** on various surfaces.[43] At the same time, the physical and chemical properties of **PoA** doped with the compounds of transition metals, unlike to its analogue (**PAn**)[48,49] studied insufficiently.

We have studied an influence of doping by ferrum (III) chloride on the structure, optical spectra, and conductivity parameters of poly–*o*–anisidine.

In order to obtain the samples of **PoA**, it was used the method of chemical synthesis under the action of oxidants, namely ferrum (III) chloride and for the comparison ammonium persulfate (**APS**) as described in ref.[42] 0.2 M solution of *o*–anisidine (*Aldrich*) and 0.2 M solution of ferrum (III) chloride or **APS** were used for the synthesis; 1 M HCl was used as a solvent. The samples of polymers were ground into the porcelain mortar to obtain the dispersed powder. To confirm the molecular structure of polymers, the method of infrared spectroscopy with *Fourier* transforming (*FTIR*) was applied using the *FTIR* spectrometer "*Avatar*," the measurements were performed on the samples pressed into the tablets with KBr. The spectrum of **PoA** in the range 2000–400 cm^{-1} is shown in Figure 13.2.

According to the data of *IR* spectroscopy, the molecular structure of elementary link of **PoA** corresponds to the emeraldine salt: for the synthesized polymers the absorption bands at 3350 (N–H), 3000 (C–H), 1571 (C=C bond into quinoid ring), 1580 (C=N), 1505, 1360, 1100, 824, 744 (C–C bond into benzoquinoid ring), 1000 (C–O) cm^{-1} are observed.

Ferrum (III) chloride is not sufficiently strong oxidant in order to oxidize of doped emeraldine, namely half-oxidated form of **PAAr**. Instead, under considering of the $FeCl_3$ as *Lewis* acid, it can be postulated that in this case the acid–base doping via possible complexation reactions of *Lewis* acid with the main centers of the polymer and dopant is possible. Such type of doping by compound $FeCl_3$ for unsubstituted **PAn** firstly was described by F. Genoud et al.[48] Based on the data of *IR* spectroscopy as well as structures described in the references; we propose the following scheme of the **PoA** doping (Figure 13.3):

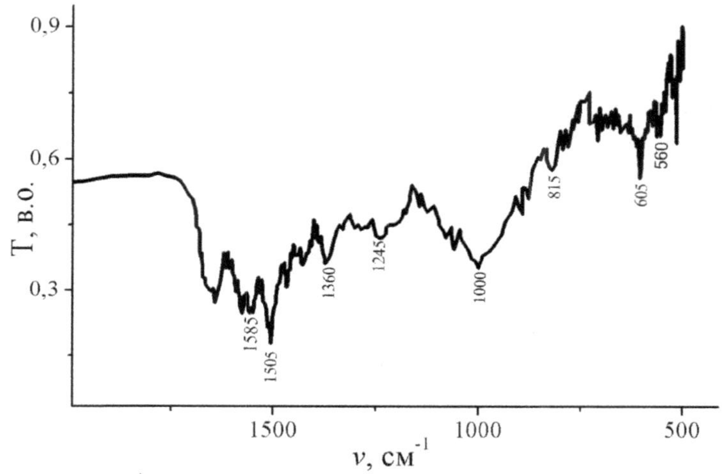

FIGURE 13.2 *IR* spectrum of poly–*o*–anisidine in the range υ = 2000–400 cm^{-1}.

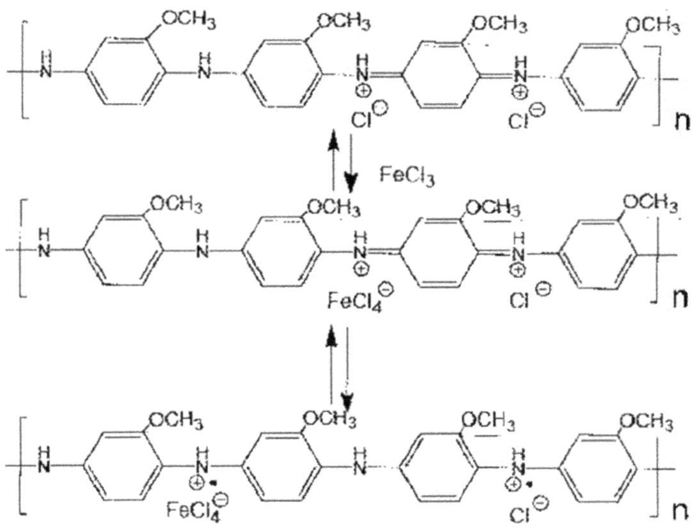

FIGURE 13.3 The scheme of the doping process of **PoA** by $FeCl_3$ compound.

Emeraldine doped with ferrum (III) chloride has otherness properties than it is set for the main (undoped) form of this polymer,[40,49] that can be conditioned both by the structural features of the obtained material, and by its electronic properties.

Unlike crystalline solids with ordered structures, the polymeric materials are characterized by the absence of the long-range order. Supramolecular structure of most polymers is consisting of amorphous and crystalline regions. It is considered that the "islands" of crystallinity surrounded by the amorphous matrix that creates some barrier for the electron transfer.

According to the results of the **XRD** (automatic diffractometer *STOE STADI*, *Cu Kα*1–irradiation) of the structure of obtained polymers the diffractogram of the **PoA** sample obtained in the presence of **APS** as oxidant contains the amorphous halo and some weak vague peaks at 2θ = 5.6 and 24.5° (Fig. 13.4 a), which indicates the formation of almost amorphous phase. For the pattern doped with ferrum (III) chloride an amplification of the peaks intensity is traced as well as the appearance of broad amorphous halo in the range of angles 2θ = 18–40 (Fig. 13.4 b) is observed. The average size of domains of the coherent scattering (under approximation average linear size of the particles of grains of phase – crystallites) defined for this amorphous halo accordingly to ref.[50,51] is 5.79 Å. These dimensions correspond to two elementary links in the polymeric chain that retain the linear (coplanar) orientation. Based on the fact that on the diffractograms doped with **PoA** (Fig. 13.4 b) there are not the diffraction peaks characteristic for ferrum (III) chloride,[50] and at the same time the significant change in *X*-ray diffraction spectrum is recorded compared with undoped **PoA** (Fig. 13.4 a), we can talk about the formation of hybrid structures **PoA**–$FeCl_3$. Difractogram of the sample **PoA**–$FeCl_3$ shows the formation practically amorphous phase polymer–ferrum (III) chloride.

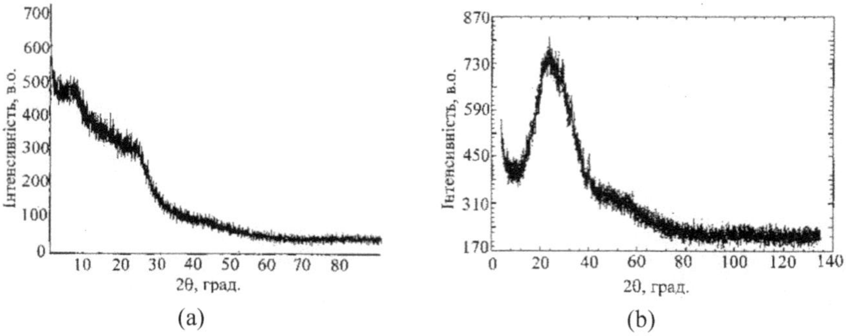

FIGURE 13.4 X-ray powder diffractograms of poly–*o*–anisidine obtained via chemical synthesis in the presence of **APS** (a) and ferrum (III) chloride (b).

As the structural studies showed, synthesized samples of **PAn** and **PoT** are characterized predominantly by amorphous–crystalline structure, and on the background of amorphous halo, there are diffuse peaks that indicate the existence of crystal structures (Fig. 13.5 b). For the sample **PoT**–$FeCl_3$, the diffuse peak at $2\theta = 5.932°$ corresponding to interplanar distance $d = 13.887$ Å is observed.

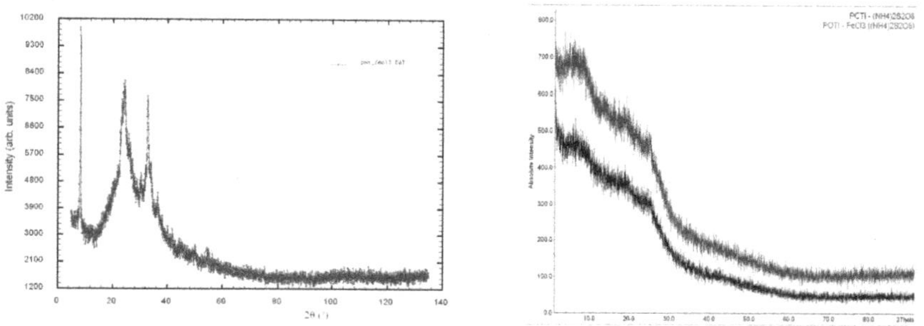

(a) (b)

FIGURE 13.5 (a) *XRD* powder diffractograms of samples **PAn**–$FeCl_3$ (diffractometer *DRON*–2.0 M, *Fe Kα*–irradiation); (b) *XRD* powder diffractograms of samples **PoT** and **PoT**–$FeCl_3$ (diffractometer *STOE STADI P, Cu Kα*1–irradiation).

At the comparison of the diffractograms of doped **PAAr**s samples (Fig. 13.4 and 13.5), one can argue that in the area of angles ~25°2θ (interplanar distance $d = 4.47$ Å) they are similar, namely on the diffractograms of the pattern **PAn**–$FeCl_3$ an amorphous halo also is observed. However, the most interesting is the existence of a group of narrow, well-formed peaks that correspond to interplanar distances $d \sim 13.46$ Å and 3.43 Å at the angles 8.25°2θ and 32.83°2θ, respectively, which indicates the formation of the phase with high crystallinity degree. It can be assumed that in a case of **PAn** the process of the complexes of type [**PAn** + $FeCl_4^-$] formation takes place; that phase gives own diffraction picture unlike to X-rays of compounds $FeCl_3$, $FeCl_2$, and others. Thus, the introduction ferrum-containing compound on the stage of the synthesis significantly alters the crystalline structure of polymers of polyaniline series, but the nature of an impact of dopant largely depend on the type of polymeric matrix, mechanism of chelation or acid–base interactions. This can lead to the formation of a completely amorphous hybrid structure (**PoA**–$FeCl_3$), and to creation of islands of crystalline phase and even the appearance of a new phase in the case of **PAn** doped with ferrum (III) chloride. A significant increase of the intensity at low angles 2θ-ray diffraction (< 2.0°) indicates the formation of the fractal structures.[52,53]

13.3 AN IMPACT OF DOPING BY FERRUM-CONTAINING COMPOUNDS ON THE PARAMETERS OF THE CONDUCTIVITY AND ELECTRONIC ABSORPTION SPECTRA OF CONJUGATED POLYAMINOARENES

The charge transfer in low-dimensional polymeric systems, which include the conjugated polymers, can be described in the "domain" or "granular" model of the conductivity. In accordance with the modern concepts[45,54–57] in the conjugated polymer there are well-ordered areas (domains or crystallites) with high conductivity, besides one polymer may be included in several crystallographic cells. Conjugation of electronic system of neighboring elementary links leads to the charge delocalization along the polymeric chain and to the transfer of the charge both in a one-dimensional (1*D*), and in three-dimensional (3*D*) direction through the interchain transitions. It is expected that within the crystalline domains there is a significant overlapping of the wave functions throughout the all volume of the domain.[56] Transport charge between domains occurs through a hopping mechanism via low conductive amorphous membranes, creating an energy barrier of the conductivity. A measure of this barrier may be an activation energy of the charge transport, an assessment of which is carried out using the activation equation.[45,57]

As same as typical organic semiconductors, the conjugated **PAAr**s doped with $FeCl_3$ show a decrease of specific resistivity with increasing of temperature.[42,57] The linear character of the dependence $lg(R/R_0) - 1/T$ (Fig. 13.4) in the temperature range $T = 294–404$ K makes it possible to calculate the activation energy of the conductivity (E).

As can be seen from the data presented in Figure 13.6 and in Table 13.2, for **PAn** and its derivatives doped with ferrum (III) chloride, the value of activation energy of conductivity does not change significantly, which may indicate a retaining of the mechanism of conductivity in systems where the ferrum-containing dopant is introduced at the stage of synthesis.

In order to determine the influence of the doping with magnetic ions on the electronic structure of conjugated **PAAr**, their optical absorption spectra were studied. The spectra were obtained using a spectrophotometer *SF–46* on film samples with a thickness of 540 ± 20 nm deposited on SnO_2 electrode by the method of electrochemical polymerization from 0.1 M solutions of monomers in 1 M HCl or 0.5 M H_2SO_4 at current density $i = 0.1$ mA/cm^2, electrolysis time 10 min and additionally doped with 0.01 M solution of ferrum (III) chloride for 60 min.[42,58]

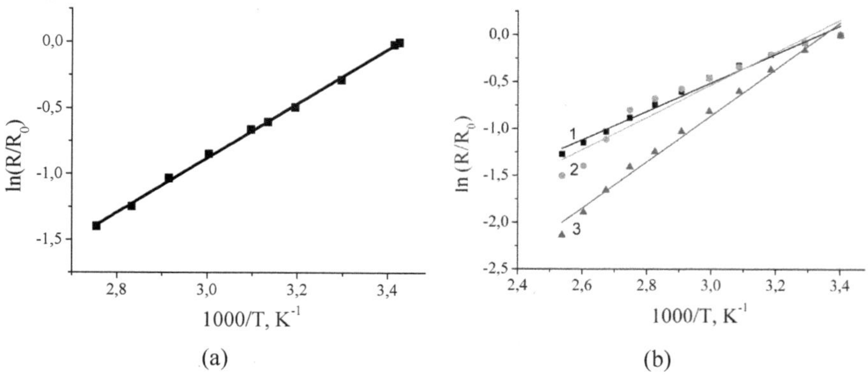

FIGURE 13.6 Temperature dependence of the specific resistance of polymers doped with magnetic particles: (a) poly–*o*–anisidine doped with ferrum (III) chloride; (b) poly–*o*–toluidine, doped with different dopants: **1** is sulfate acid, **2** is toluene sulfonic acid (**TSA**), **3** is ferrum (III) chloride.

TABLE 13.2 Parameters of Conductivity of Conjugated Polymers Doped by Ferrum (III) Chloride.

№ of sample	Polymer	Proton dopant	Ferrum-containing dopant	ρ(T = 293), Ohm*cm	E, eV
1	**PoT**	H_2SO_4	–	(4.18 ± 0.02) 10^2	0.27 ± 0.01
2	PoT	**TSA**	–	(0.13 ± 0.01) 10^2	0.29 ± 0.01
4	PoT	TSA	$FeCl_3$	(1.66 ± 0.04) 10^2	0.44 ± 0.02
5	**PoA**	**TSA**	–	(1.90 ± 0.01) 10^3	0.32 ± 0.01
6	**PoA**	**TSA**	$FeCl_3$	(0.21 ± 0.01)10^3	0.35 ± 0.02
8	**PAn**	**TSA**	–	2.5 ± 0.1	0.16 ± 0.01
9	**PAn**	**TSA**	$FeCl_3$	2.5 ± 0.1	0.16 ± 0.01
10	**PAT**	H_2SO_4	$FeCl_3$	–	0.80 ± 0.02

It was established that under the action of the ferrum salts as well as other heavy metals[58] the color of **PAArs** films is changed. These color changes correspond to the spectral transformations, fixed by us using the optical spectroscopy (Fig. 13.7).

The obtained spectra are typical for **PAArs** containing both highly localized and delocalized polarons.[58] As shown in Figure 13.7 a, regardless of the type of doping the optical spectrum of **PoA** consists of two bands, namely in the region of 400–440 nm (π–π* transition in bandgap) and broad band with a maximum at $\lambda = 780$ nm, which stretches into the near infrared area till $\lambda > 900$ nm.

Peaks at 400–440 nm can be attributed to the transition from the localized polaron level to π^*–orbital. An intense band in the region of 700–900 nm corresponds to the presence of the transition from π–orbitals on the localized polaron. Expansion of the absorption band in the near infrared region of the spectrum indicates greater length of the delocalization of electrons for polyanisidine doped with $FeCl_3$, compared with the acid-doped **PoA**. However, a longer exposure (30–60 min) of polymeric film in a solution of ferrum chloride causes a decrease in the intensity of the polaron absorption bands, perhaps because of the growing the share of pernigraniline fragments.

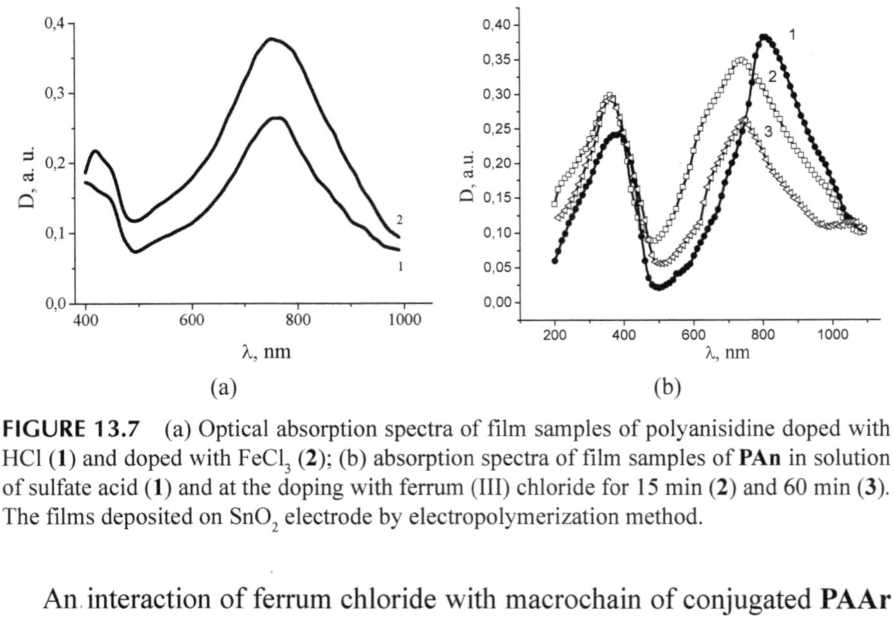

FIGURE 13.7 (a) Optical absorption spectra of film samples of polyanisidine doped with HCl (**1**) and doped with $FeCl_3$ (**2**); (b) absorption spectra of film samples of **PAn** in solution of sulfate acid (**1**) and at the doping with ferrum (III) chloride for 15 min (**2**) and 60 min (**3**). The films deposited on SnO_2 electrode by electropolymerization method.

An interaction of ferrum chloride with macrochain of conjugated **PAAr** may occur in accordance with the mechanism of acceptor doping,[43,55] where the ions Fe^{3+} serve as an acceptor, and they are easily reduced to Fe^{2+}, and can oxide the polymer chain. These processes cause an increase of the concentration of the charge carriers, namely polarons, and accordingly, an increasing of the optical absorption both in polaronic and in bipolaron band. This can form a complex with charge transfer by $[\mathbf{PAn}]^+[FeCl_4]^-$ type, which leads to the change both in the electronic structure of the polymer, and in its crystal structure, as it shown in Section 13.2.

Thus, the studies of conductivity, optical absorption, and structural parameters clearly show that the doping of conjugated **PAAr**s with ferrum-containing compounds can cause a significant change of their electronic properties. The action of this dopant is occurred as a result of the acceptor doping and formation of complexes with the charge transfer. For a more detailed study of the nature of doping particles in the systems **PAAr**–ferrum-containing dopant, we have used the method of paramagnetic resonance spectroscopy (EPR).

13.4 DYNAMICS OF MAGNETIC CENTERS IN NANOSYSTEMS BASED ON PAARS DOPED WITH FERRUM-CONTAINING COMPOUNDS

The basis of cooperative physical and chemical phenomena peculiar to substances, such as electrical conductivity, magnetic susceptibility, optical activity, is the interaction of electric and magnetic charges that make up the structure of substances. In the case of conjugated **PAAr**s and their composites with organic and inorganic compounds such elementary charged particles can be quasiparticles of polaron or soliton type[55–57] (ion-radicals) with the charge $q = 1$ and spin $s = ½$. An interaction of these charges in the material causes the presence of electroconductivity, magnetization, and the ability to convert the electrical and light energy, an absorption and emission of light in the visible spectrum, providing the interconnection between electrical, magnetic, and optical properties of the materials.

We have previously found that undoped matrixes of **PAn**, like as the most acid-doped **PAAr**s at room temperature preferably exhibit the diamagnetic properties.[57,59] However, after doping of such matrix with iodine or with strong proton acids[60] it is observed the paramagnetic behavior, namely an increase of the magnetic susceptibility in a magnetic field (Fig. 13.8), the temperature dependence of which can be described by the Curie–Weiss law at $T < 120$ K (Fig. 13.9).

Studies of the spectra of electron paramagnetic resonance (EPR) have shown that in the temperature range 4.2–300 K the acid-doped matrixes of **PAAr**s give sustained EPR signal over time even at room temperature with *g*-factor at 2.0003–2.0027.[21,38,57] However, little researched is the influence of the doping with magnetic ions on paramagnetic properties of polymers and also the impact of polymeric matrix on the nearest surroundings of the magnetic center, and magnetic behavior of polymeric composites. To solve this problem, we conducted a study of the temperature dependence of the ESR for nanosystems obtained based on matrixes of conjugated polymers doped with ferrum-containing compounds, namely ferrum (III) chloride.

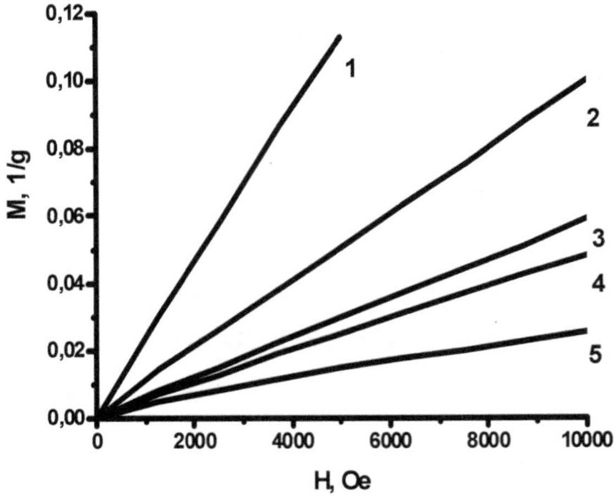

FIGURE 13.8 Dependence of the magnetization on field at $T = 4.2$ K for doped **PAn** with acids: **1** is 1.8–aminonaphthylsulfoacid (**ANSA**), **2** is 1.5-**ANSA**; **3** is HCl; **4** is **ANSA**, **5** is H_2SO_4.

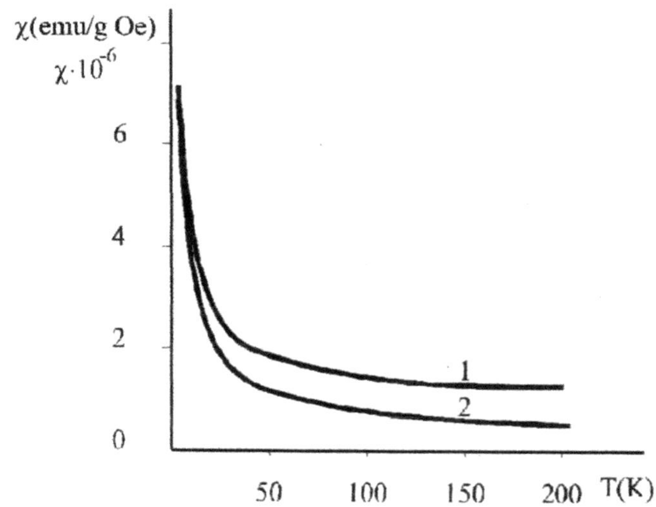

FIGURE 13.9 Temperature dependence of magnetic receptivity of **PAn** doped with iodine in the field 0.1 T (**1**) and 5 T (**2**).

Researches were carried out on powder samples of polymers sealed in quartz ampoules, using EPR spectrometer of *X*-ray diapason in the temperature range 462–300 K as described in ref.[38] The temperature dependence of magnetic susceptibility were studied using the vibration magnetometer in accordance with ref.[25]

Study of doping on the magnetic properties of polymers was carried out for dispersed samples of doped and undoped matrixes of **PAT** and **PoA** obtained by the method of chemical synthesis. Synthesis of samples was conducted by oxidative polymerization with the following treatment by ammonia solution to remove of inorganic impurities and to obtain a polymer in the form of emeraldine base.

For the obtaining of doped **PAT** the anhydrous ferum (III) chloride was used as an oxidizing agent at a ratio of aminothiazole: $FeCl_3$ = 1:2; acetone was used as a solvent. Synthesis and purification of products were performed according to the method described in ref.[61] By studying of the obtained polymer structure with the use of *IR* and electronic spectroscopy methods previously, it was found that the formation of π–conjugated chain of **PAT** proceeds via the participation of amino group and activated to electrophilic substitution positions of thiazole ring in accordance with the scheme presented in Figure 13.10.

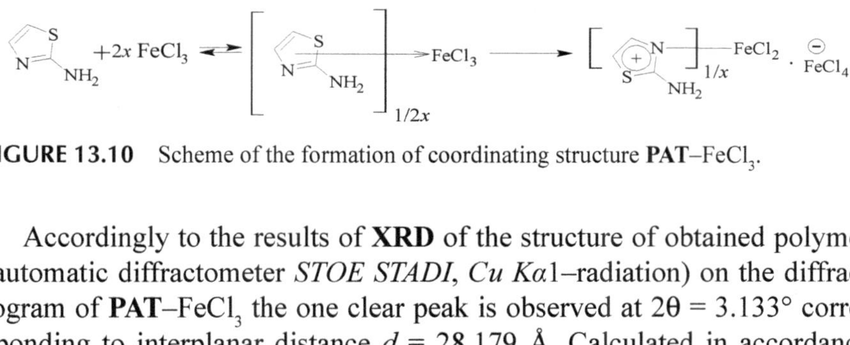

FIGURE 13.10 Scheme of the formation of coordinating structure **PAT**–$FeCl_3$.

Accordingly to the results of **XRD** of the structure of obtained polymer (automatic diffractometer *STOE STADI*, *Cu Kα*1–radiation) on the diffractogram of **PAT**–$FeCl_3$ the one clear peak is observed at $2\theta = 3.133°$ corresponding to interplanar distance $d = 28.179$ Å. Calculated in accordance with the *Ehrenfest* formula $2R_1 \mathrm{Sin}\theta = 1.23\,\lambda$ ($\lambda = 1.1540598$ Å) radius of the first coordination sphere is $R_1 = 34.66$ Å.

As the results of the studies have shown, the monomer compound, namely 2-aminothiazole gives a weak EPR signal. Instead, the obtaining polymer **PAT** has clearly expressed paramagnetic properties, as is evidenced by fairly strong ESR signal even at room temperature. The obtained spectra of EPR of the synthesized polymers (Fig. 13.11 b) are characterized by wide asymmetric line (singlet) without fine structure with the *g*-factor 2.010 ± 0.001 neared to the *g*-factor of a free electron.[63] These values of *g*-factor inherent to paramagnetic centers located in a chain of the conjugation and stabilized by resonant interaction with aromatic ring,[64] which confirms the structure of **PAT**.[61]

For **PAT** doped with ferrum (III) chloride in the range $T = 4.2\text{–}293$ K at lowering of the temperature the change in shape of the EPR spectrum is observed. In addition to the line of polymeric matrix that gives the signal with the value $g_1 = 2.01$ (Fig. 13.11 a, curve **2**; Fig. 13.9 b, curve **1**) at temperatures $T < 80$ K there is a new signal (Fig. 13.9 b, curve **2**).

In the EPR spectra the resonance line 2 with the effective value $g_2 = 4.09 \pm 0.02$, which is characteristic for ions[38,39] Fe^{3+} can be observed. With decreasing of temperature to 5 K to width of this line (ΔHpp) increases, while line 1 is hardly variable.

The temperature dependence of inverse magnetic susceptibility of **PAT**–$FeCl_3$ in the temperature range 80–300 K (Fig. 13.12) is ordered to the Curie–Weiss low, indicating the paramagnetic nature of magnetism in the studied systems.[25]

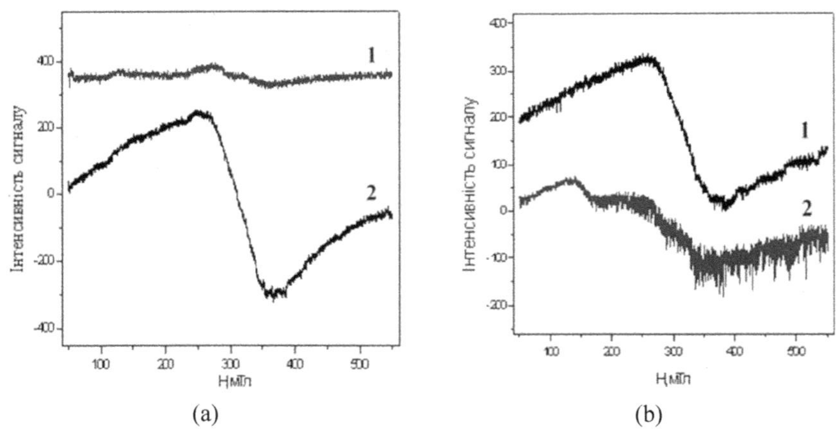

FIGURE 13.11 (a) EPR spectra of monomer (aminothiazole) (**1**) and polymer (**PAT**) (**2**) at $T = 295$ K; (b) EPR spectra of complex **PAT**–$FeCl_3$ at $T = 295$ (**1**) and 77 K (**2**).

The ions of $3d^n$ group of ferrum known as the main source of free radicals that cause of the disruption of cells of the nervous system, so the study of the dynamics of their behavior is essential for forecasting of the processes in biological objects. It was shown, that the ions Fe^{3+} can act as a paramagnetic probe,[38,62] since in the formed polymeric complex reveals its own, different from the polymer matrix, temperature behavior, allowing to observe its dynamics in conditions of similar molecular surroundings. This phenomenon can be used in biology and medicine to monitor the behavior of nerve cells.

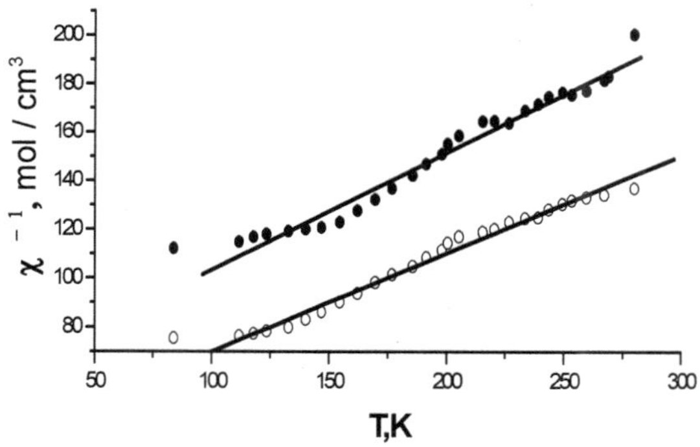

FIGURE 13.12 The temperature dependence of inverse magnetic susceptibility of **PAT**–$FeCl_3$ at the densities of magnetic field 5 T (**1**) and 0.1 T (**2**).

Another modeling object to study the dynamics of magnetic centers can be the molecule of **PoA**, which under the doping forms the complexes of ferrum wit heteroatoms of polymeric chain, namely nitrogen and oxygen.[39] In accordance with carried out quantum-chemical calculations by the method described in ref.,[38] it was proposed the structure of ferrum complex with molecule of **PoA** as shown in Figure 13.13.

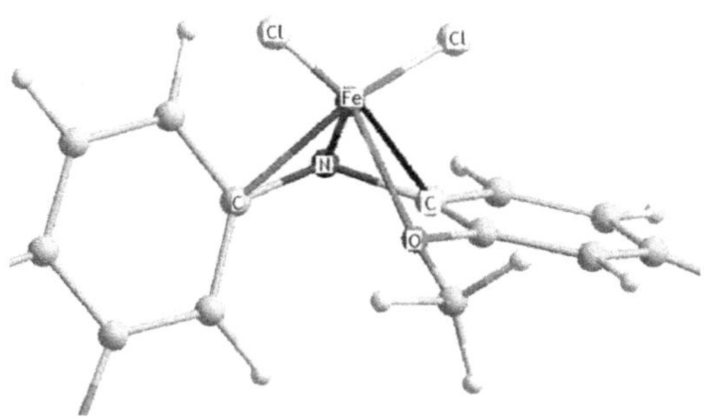

FIGURE 13.13 Molecular structure of the coordinating complex of ferrum ion in the **PoA**. Investigated ion of ferrum is surrounded by the following atoms: nitrogen, two carbon atoms, oxygen, and two chlorine atoms.

Quantum chemical calculations of molecular structure of **PoA** as a coordination polymer showed the presence of two structurally nonequivalent positions of magnetic samples Fe(**1**) and Fe(**2**) (Fig. 13.14). The presence of such structural nonequivalent positions of magnetic centers is apparented in some features of chemical and physical properties of polymers. Positions Fe(**1**) and Fe(**2**) of magnetic probes in a matrix of **PoA** indicated by arrows directed toward the 17 and 26 nitrogen atoms (Fig. 13.14). The resulting structure of the energy states is the sextette with the coordination number equal to six. This is agreed with the calculated distributions of a charge in the structures. With the use of the calculated methods, it were obtained the ionization potentials of polymer and ferric ion.[39]

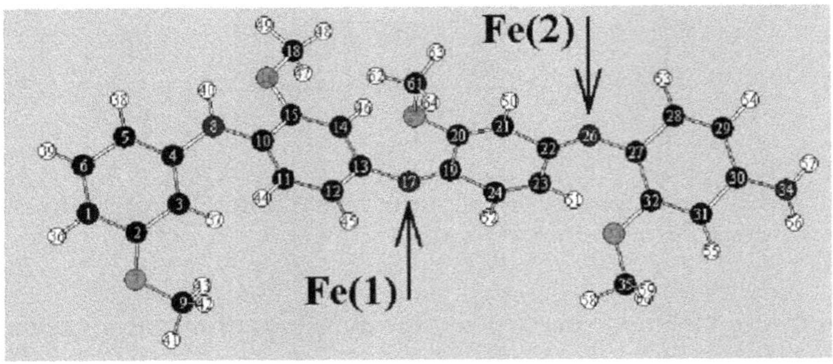

FIGURE 13.14 Molecular structure of the fragment of chain of **PoA**. Two structurally nonequivalent positions of Fe(**1**) and Fe(**2**) are denoted by arrows.

Using the quantum chemical calculations the distances from the equilibrium state of the ferrum ion to the atoms of the first coordination sphere has been determined. Investigated ferric ion is surrounded by such atoms as nitrogen N at a distance of 1.91 Å, two carbon atoms C at distances of 2.16 and 2.11 Å, oxygen O at a distance of 2.46 Å, and two chlorine atoms Cl at a distance of 1. 91 and 1.93 Å.

EPR spectrum of Fe^{3+} ions in the matrix of **PoA** is consisted of two lines, namely low-temperature, which is manifested at lower margins (Fig. 13.15, line on the left) and high-temperature (Fig. 13.15, line on the right).

As can be seen from the EPR spectra presented in Figure 13.15, the redistribution of the lines intensities takes place at the change of temperature. The temperature dependence of the intensity of these lines is shown in Figure 13.16.

EPR spectra of ferrum ions in the structures with long-range order of symmetry (monocrystals) and in the structure of short-range order symmetry (polymers) have a different number of lines that is explained by corresponding reasons. Complexes with ferrum ions in the polymer are structurally unequal. EPR spectrum of each complex with ferrum ion has one isotropic line. Such isotropic lines of EPR spectra from the all complexes of ferrum are imposed on each other. This results in a range of isotropic ESR line (the line 2 with the *g*-factor = 2.0).

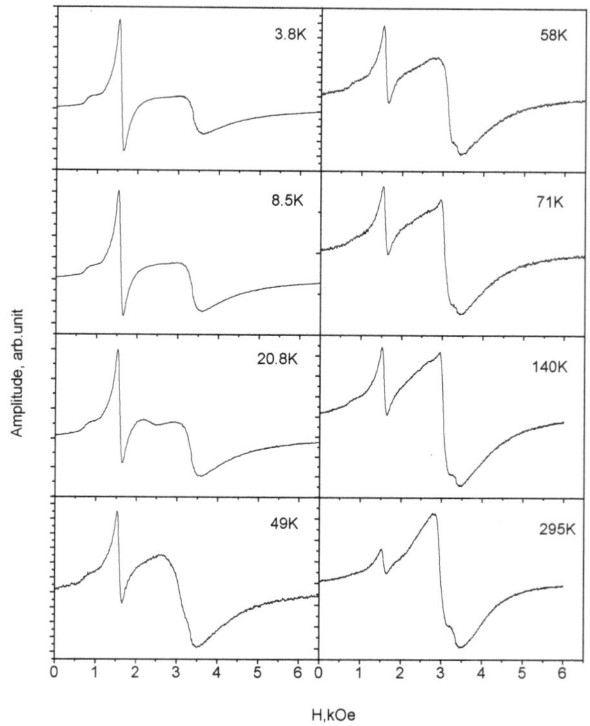

FIGURE 13.15 EPR spectra of ferrum ions into the matrix of PoA at different temperatures in the range 4.2–295 K.

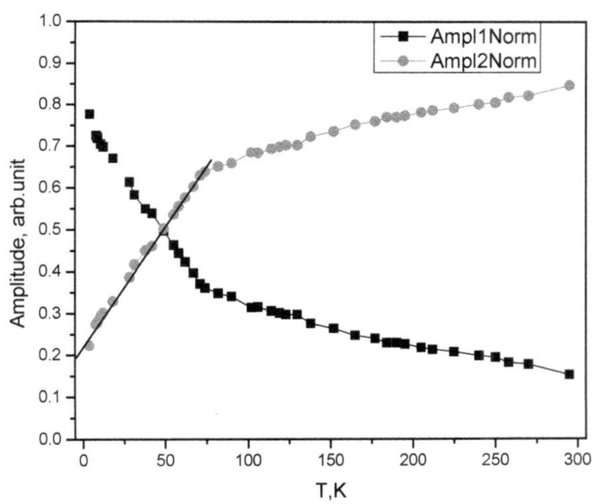

FIGURE 13.16 Temperature transformation of the 1 and 2 lines intensities of EPR spectrum for the ferrum ions in **PoA** for the temperature range $T = 4.2–295$ K.

Anisotropic lines in the EPR spectra of ferrum ions in the polymer from the all complexes are overlapping. This results in a line of anisotropic EPR spectrum lines (the line 1 with the *g*-factor 4). The reason lies in the mixed energy states of ferrum ions.[39] Such mixing of the energy states are the result of structural nonequivalence of complexes with the ferrum ions as described in ref.[38,64]

Thus, with the change of temperature the redistribution of the lines 1 and 2 intensities takes place in the EPR spectrum of Fe^{3+} ions in **PoA**. In this case, we can speak about a new effect in the temperature behavior of the magnetic centers. This effect is found in the case of structural non-equivalence of ferrum complexes, which is manifested in an orientational statistical averaging according to symmetry axes of ferrum complexes. The observed phenomenon is the basis for development of a new method of nanomaterials researches.[39]

13.5 CONCLUSIONS

So, the use of EPR methodology for the study of polymeric and bioorganic nanoobjects, such as metalloproteins, is promising for the development of nanomedicine,[65,66] since the EPR signal is sensitive to the smallest amounts of paramagnetic particles, namely organic free radicals and metal ions.[64]

KEYWORDS

- **conjugated polymers**
- **polyaminoarenes**
- **ferrum-containing compounds**

REFERENCES

1. Svrcek, V. Hybrid Optoelectronic and Photovoltaic Materials based on Silicon Nanocrystals and Conjugated Polymers. In *Optoelectronic Devices and Properties;* Sergiyenko, O., Eds.; National Institute of Advanced Industrial Science and Technology (AIST): Tsukuba, Japan, 2011; pp 197.
2. Martin, J.; Maiz, J.; Sacristan, J.; Mijangos, C. Tailored polymer-based nanorods and nanotubes by "template synthesis": From Preparation to Applications. *Polymer.* **2012,** *53,* 1149–1166.

3. Bai, H.; Shi, G. Gas Sensors Based on Conducting Polymers. *Sensors*. **2007,** *7,* 267–307.
4. Gong, X. Toward High Performance Inverted Polymer Solar Cells. *Polymer*. **2012,** *53,* 5437–5448.
5. Saxena, V.; Malhotra, B. D. Prospects of Conducting Polymers in Molecular Electronics. *Curr. Appl. Phys.* **2003,** *3,* 293–305.
6. Dhand, V.; Rhee, K. Y.; Kim, H. Ju.; Jung, D. Ho. A Comprehensive Review of Graphene Nanocomposites: Research Status and Trends. *J. Nanomater.* **2013,** Article ID 7**63953,** 14, http://Dx. Doi.Org/10.1155/2013/763953.
7. Das, T. K.; Prusty, S. Graphene-Based Polymer Composites and Their Applications. *Poly. Plast. Technol. Eng.* **2013,** *52,* 319–331.
8. Kamat, P. V. Meeting the Clean Energy Demand: Nanostructure Architectures for Solar Energy Conversion. *J. Phys. Chem. C.* **2007,** *111,* 2834–2860.
9. Aksimentyeva, O. I.; Konopelnyk, O. I.; Dyakonov, V. P.; Shapovalov, V. A.; Horbenko, Yu. Yu. Charge Separation in Polyphenylacetylene – Fullerene Nanostructures/In Book: "*Fullerenes and Nanostructures in Condenced Matter*". Edit. Center of BSU: Minsk, Brlarus, **2011**; pp 178–183.
10. Herandez, J. J. ; García-Gutierrez, M. C. ; Nogales, A. ; Rueda, D. R.; Kwiatkowska, M.; Szymczyk, A.; Roslaniec, Z.; Concheso, A.; Guinea, I.; Ezquerra, T. A. Influence of Preparation Procedure on the Conductivity and Transparency of SWCNT-Polymer Nanocomposites. *Compos. Sci. Technol.* **2009,** *69,* 1867–1872.
11. Alig, I. ; Pötschke, P. ; Lellinger, D. ; Skipa, T. ; Pegel, S. ; Kasaliwal G. R. ; Villmow, T. ; Gaurav, R. Establishment, Morphology and Properties of Carbon Nanotube Networks in Polymer Melts. *Polymer*. **2012,** *53,* 4–28.
12. Liu, P. Preparation and Characterization of Conducting Polyaniline/Silica Nanosheet Composites. *Curr. Opin. Solid State Mater. Sci.* **2008,** *12,* 9–13.
13. Li, X.; Wang, G.; Lib, X. Surface Modification of Nano-Sio_2 Particles Using Polyaniline. *Surf. Coat. Technol.* **2005,** *197,* 56–60.
14. Arancis, R.; Joy, N.; Aparna P.; Vijayan, R. Polymer Grafted Inorganic Nanoparticles, Preparation, Properties, and Applications: A Review. *Polym. Rev.* **2014,** *54,* 268–347.
15. The New Frontiers of Organic and Composite Nanotechnology, V. Erokhin, M. K. Ram, Ö. Yavuz, Eds.; Elsevier: Boston, 2008.
16. Aksimentyeva O. I.; Monastyrskyi L. S.; Savchyn B. M.; Stakhira, P.; Vertsimakha, Ya.; Tsizh, B. Electronic Processes in Conducting Polymer-Porous Silicon Interface. *Molec. Cryst. Liq. Cryst.* **2007,** *467,* 73–83.
17. Monastyrskii, L.; Aksimentyeva, O. I.; Olenych, I.; Yarytska, L. Transmission Infrared Spectra of Low Dimension Hybrid Composites on a Porous Silicon Base. Visnyk Lviv Univ, *Ser. Phys.* **2009,** *44,* 271–276.
18. Nguyen, T. P.; Le Rendu, P.; Tran, V. H. Electrical and Optical Properties of Conducting Polymer/Porous Silicon Structures. *J. Porous Mater.* **2000,** *7,* 393–396.
19. Davies, J.; Hamada, H. Flexural Properties of a Hybrid Polymer Matrix Composite Containing Carbon and Silicon Carbide Fibres, *Adv. Composite Maters.* **2001,** *10*(1), 77–96.
20. Zhang, Z.; Wan, M. Nanostructures of Polyaniline Composites Containing Nano-Magnet. *Synth. Met.* **2003,** *132,* 205–212.
21. Vasyukov, V. N.; Dyakonov, V. P.; Shapovalov, V. A.; Aksimentyeva, O. I.; Szymczak, H.; Piehota, S. Temperature-Induced Change in the ESR Spectrum of the Fe^{3+} Ion in Polyaniline. *J. Low Temp. Phys.* **2000,** *26*(4), 265–269.

22. Bencsik, G.; Janaky, C.; Krivan, E.; Lukacs, Z.; Endrıdi, B.; Visy, C. Conducting Polymer Based Multifunctional Composite Electrodes. React. *Kinet. Catal. Lett.* **2009,** *96*(2), 421–428.
23. Sawada, H.; Yoshioka, H.; Kawase, T.; Ueno, K.; Hamazaki, K. Preparation of Magnetic Nanoparticles by yhe Use of Self-Assembled Fluorinated Oligomeric Aggregates – A New Approach to the Dispersion of Magnetic Particles on Poly(Methyl Methacrylate) Film Surface. *J. Fluorine Chem.* **2005,** *126,* 914–917.
24. Opalnych, I.; Aksimentyeva, O. I.; Dyakonov, V.; Piechota, S.; Ułański, J.; Demchenko, P.; Ukrainets, A. Structure and Thermodeformation Properties of Polymer-Magnetite Hybrid Composites. *Mater. Sci.* **2012,** *48,* 95–100.
25. Dyakonov, V. P.; Zubov, E.; Aksimentyeva,O. I.; Dyakonov, S.; Piechota ; Szymczak, H. Low-Temperature Magnetic Behavior of the Organic-Based Magnet $Na[Feo_6(C_{10}H_6N)_3]$. *Low Temp. Phys.* **2014,** *40,* 835–841.
26. Suzdaliev, I. P. Nanotechnology: Physics and Chemistry of Nano-Clusters, Nanostructures and Nanomaterials, Komkniga: Moscow, Russia, 2006. (27) *New Materials;* Karabasov, Yu. S., Eds., MISIS: Mscow, Russia, 2002.
28. Srinivasan, D.; Natarajan, T. S.; Rangarajan, G.; Bhat, S. V.; Wessing, B. Electron Spin Resonance Absorption in Organic Metal Polyaniline and its Blend with PMMA. *Solid State Comm.* **1999,** *110,* 503–508.
29. Tran H. D.; Wang, Y.; Arcy, J. M. D.; Karner,B. R. Toward an Understanding of the Formation of Conducting Polymer Nanofibers. *ACS Nano.* **2008,** *2,* 1841–1848.
30. Wang, X.; Shao, M.; Shao G.; Wu, Z.; Wang, S. A Facile Route to Ultra-Long Polyaniline Nanowires and the Fabrication of Photoswitch. *J. Colloid Interface Sci.* **2009,** *332,* 74–77.
31. Challa S. S. R. Nanomaterials for Medical Diagnosis and Therapy, In *Nanotechnologies For The Life Sciences;* Kumar, Eds.; Wiley-VCH Verlag Gmbh,Germany, 2007; Vol. 10.
32. Zebli, B.; Susha A. S.; Sukhorukov, G. B.; Rogach, A. L.; Parak, W. J. Magnetic Targeting and Cellular Uptake of Polymer Microcapsules Simultaneously Functionalized with Magnetic and Luminescent Nanocrystals. *Langmuir.***2005,** *21,* 4262–4265.
33. Gomez-Romero, P.; Torres-Gomez, G. Molecular Batteries: Harnessing $Fe(CN)_6^{3-}$ Electroactivity in Hybrid Polyaniline–Hexacyanoferrate Electrodes. *Adv. Mater.* **2000,** 12(19), 1454–1465.
34. Tiberto, P.; Barrera, G.; Celegato, F.; Coïsson, M.; Chiolerio, A.; Martino, P.; Pandolfi, P.; Allia, P. Magnetic Properties of Jet-Printer Inks Containing Dispersed Magnetite Nanoparticles. *Eur. Phys. J. B.* **2013,** *86,* 173; DOI: 10.1140/Epjb/E2013-30983-8.
35. Netto, C. G. C. M. ; Toma, H. E.; Andrade, L. H. Superparamagnetic Nanoparticles as Versatile Carriers and Supporting Materials for Enzymes. *J. Mol. Catal. B: Enzym.* **2013,** *85–86,* 71–92.
36. Jacobo, S. E.; Aphesteguy, J. C.; Anton, R. L.;Schegoleva, N. N.; Kurlyandskaya, G. V. Influence of the Preparation Procedure on the Properties of Polyaniline Based Magnetic Composites. *Eur. Polym. J.* **2007,** *43,* 1333–1346.
37. Aphesteguy, J. C.; Jacobo, S. E. Composite of Polyaniline Containing Iron Oxides. *Physica. B.* **2004,** *354,* 224–227.
38. Vasyukov, V. N.; Shapovalov, V. A.; Dyakonov, V. P.; Dmitruk, A. F.; Aksimentjeva,O. I.; Szymczak, H.; Piechota, S. Investigation of Structure of Fe^{3+} Magnetic Center in Polyparaphenylene. *Int. J. Quantum Chem.* **2002,** *88,* 425–529.
39. Shapovalov, V. A.; Shapovalov, V. V.; Rafailovich, M.; Piechota, S.; Dmitruk, A.; Aksimentyeva, E. I..; Mazur, A. Dynamic Characteristic of Molecular S Tructure

of Poly-Ortho-Methoxyaniline with Magnetic Probes. *J. Phys. Chem.* **2013,** *117,* 7830−7834.

40. Sichel, E. K.; Rubner, M. F.; Georger, J. Jr.; Papaefthymiou, G. C.; Ofer, S.; Frankel, R. B. Magnetic Phase Transition, Aggregate Formation, And Electrical Conductivity in Fecl_3-Doped Polyacetylene *Phys. Rev. B.* **1983,** *28*(11), 6588–6590.
41. Pron, A.; Kucharski, Z.; Budrowski, C.; Zagórska, M.; Kirchene, S.; Suwalski, J.; Dehe, G.; Lefrant, S. Mössbauer Spectroscopy Studies Of Selected Conducting Polypyrroles. *J. Chem. Phys.* **1985,** *83*, 5923–5927.
42. Horbenko, Yu.; Aksimentyeva, O. I. Structure and Physical-Chemical Properties of Poly-Ortho-Anisidine Doped with Ferric (III) Chloride. Bulletin of L'viv University, Ukraine, *Chem.* **2013,** *54*(II), 353–357.
43. Fonseca, L. H. M.; Rinaldi, A. W.; Rubira, A. F.; Cótica, L. F.; De Medeiros, S. N.; Paesano, Jr. A.; Santos, I. A.; Girotto, E. M. Structural, Magnetic, and Electrochemical Properties of Poly(*O*-Anisidine)/Maghemite Thin Films. *Mater. Chem. Phys.* **2006,** *97,* 252–255.
44. Aksimentyeva, O. I.; Konopelnyk, O. I.; Grytsiv; Ya. M.; Martyniuk, G. V. Charge Transport in Electrochromic Films of Polyorthotoluidine. *Funct. Mater.* **2004,** *11*(2), 300–304.
45. Kohlman, R. S.; Joo, J.; Epstein, A. J. Conducting Polymers: Electrical Conductivity. In *Physical Properties of Polymers Handbook;* Mark, J. E.; Eds.; Amer. Inst. Phys. Woodbury: New York, 1996.
46. Cherpak, V. ; Stakhira, P. ; Aksimentyeva, O. I. ; Hotra, Z.; Tsizh, B.; Volynyuk, D.; Bordun, I. ; Vacuum-Deposited Poly(*O*-Methoxyaniline) Thin Films: Structure and Electronic Properties. *J. Non-Crys. Solids.* **2008,** *354,* 4273–4277.
47. Aksimentyeva, O. I.; Konopelnyk, O. I.; Cherpak, V.; Stakhira, P.; Fechan, A.; Hlushyk, I. Conjugated Polyaminoarenes as an Electrochromic Layer For Non-Emissive Displays. *Ukr. J. Phys. Opt.* **2005,** *6*(1), 27–32.
48. Genoud, F.; Kulszewicz-Bajer, I.; Bedel, A.; Oddou, J. L.; Jeandey, C.; Pron. Lewis, A. Acid Doped Polyaniline. Part II: Spectroscopic Studies of Emeraldine Base and Emeraldine Hydrochloride Complexation with $Fecl_3$ *Chem. Mater.* **2000,** *12*(3), 744–749.
49. Gosk, J. B.; Kulszewicz-Bajer, I.; Twardowski, A. Magnetic Properties of Polyaniline Doped with $Fecl_3$. *Synth. Met.* **2006,** *156,* 773–778.
50. Roisnel, T.; Rodriguez-Carvajal, J. Winplotr: A Windows Tool for Powder Diffraction Patterns Analysis. *Mater. Sci. Forum.* **2001,** *378–381,* 118–123.
51. De Keijser, Th. H.; Langford, J. I.; Mittemeijer, E. J.; Vogels, A. B. P. Use of the Voigt Function in a Single-Line Method for the Analysis of X-Ray Diffraction Line Broadening. *J. Appl. Cryst.* **1982,** *15,* 308–313.
52. Silova, O. A.; Shilov, V. V. Nanocomposite Oxide and Organic-Inorganic Hybrid Materials Produced y Sol-Gel Method. Synthesis. Properties. Application. Nanosystems, Nanomaterials and Nanotechnologies, 2003, *1,* 9–83.
53. Shpak, A. P.; Kunitskiy, Yu. A.; Gomza, Yu. P.; Nesin, S. D.; Suhuy, K. M.; Sperkach, S. O.; Leonov, D. S. Influence of the Synthesis Conditions of Organic-Inorganic Polyonencontaining Sol-Gel – Nanocomposites on Their Fractal Structure and Proton Conductivity. *Nanosystems Nanomater. Nanotechnol.* **2005,** *3*(4), 973–984.
54. Bisquertz, J.; Garcia-Belmontez, G. Interpretation of AC Conductivity of Lightly Doped Conducting Polymers in Terms of Hopping Conduction, *Synth. Met.* **2004,** *40,* 396–402.
55. Heeger, A. J. Semiconducting and Metallic Polymers: The Fourth Generation of Polymeric Materials. *Synth. Met.* **2002,** *123,* 23–42.

56. Prigodin, V. N.; Epstein, A. J. Nature of Insulator-Metal Transition and Novel Mechanism of Charge Transport in the Metallic State of Highly Doped Electronic Polymers. *Synth. Met.* **2002,** *125,* 43–53.
57. Aksimentyeva, O. I. *Electrochemical Methods of Synthesis and Conductivity of Conjugated Polymers;* Svit: L'viv, Ukraine, 1998.
58. Konopelnyk, O. I.; Aksimentyeva, O. I.; Demchenko, P. Yu. Influence of Heavy Metal Ions on Absorption Spectra and Structure of Polyaminoarene. Reports XIII Int. Conf. *Physics and Technology of Thin Films and Nanosystems;* Ivano-Frankivsk, Ukraine, 16–21 May, 2011; 30.
59. Aksimentyeva,O. I.; Baran, M.; Dyakonov, V. P.; Magnetic Properies of Dopped Polyaniline. *Solid State Phys.* **1996,** *38,* 2266–2278.
60. Aksimentyeva, O. I.; Dyakonov, V. P. Chapter 9. Effect of Aminonaphthalene Sulfonic Acid Nature on the Structure and Physical Properties of Their Copolymers with Aniline, In *Functional Polymer Blends and Nanocomposites;* a Practical Engineering Approach; Zaikov, G. E., Bazylak, L. I.; Haghi, A. K.;, Eds.; Apple Academic Press Ink.: New Jersey, 2013; pp 217.
61. Dubrovskii, R.; Aksimentyeva, O. I. Oxidative Polymerization of Aminothiazole in the Presence of Ferric Chloride. *Tekh.* **2007,** *43,* 91−95.
62. Horbenko, Yu. Yu.; Aksimentyeva, O. I.; Shapovalov, V. A. ESR Study of Model Magnetic Centers in Metalloproteins. Reports II Int. Conf. *Nanobiophysics: Fundamental and Applied Aspects*; Kyiv, Ukraine, 6–9 October, 2011; 48.
63. Hedvig, P. *Applied Quantum Chemistry;* Mir: Moscow, Russia, 1977.
64. Klyava, Y. G. EPR Spectroscopy of Disordered Solids; Zinatne: Riga, Latvia, 1988.
65. Popp, S.; Packschies, L.; Radzwill, N.; Vogel, K. P.; Steinhoff, H. J.; Reinstein, J. Structural Dynamics of the Dnak-Peptide Complex. *J. Mol. Biol.* **2005,** *347,* 1039–1052.
66. Aksimentyeva, O. I.; Savchyn, V. P.; Dyakonov, V. P.; Piechota, S.; Horbenko, Yu. Yu.; Opaynych, I. Ye.; Demchenko, P. Yu.; Popov, A.; Szymczak, H. Modification of Polymer-Magnetic Nanoparticles by Luminescent and Conducting Substances. *Mol. Cryst. Liq. Cryst.* **2014,** *590,* 35–42.

CHAPTER 14

STRUCTURE AND THERMAL STABILITY OF SILICA–GLAUCONITE / POLYANILINE COMPOSITE

M. M. YATSYSHYN[1], V. M. MAKOGON[1], O. V. RESHETNYAK[1], and J. BŁAŻEJOWSKI[2]

[1]*Department of Physical and Colloid Chemistry, Faculty of Chemistry, Ivan Franko National University of Lviv, 6 Kyryla & Mefodia Str., Lviv 79005, Ukraine.*
E-mail: m_yatsyshyn@franko.lviv.ua

[2]*Department of Physical Chemistry, Faculty of Chemistry, University of Gdańsk, 18 J. Sobieskiego Str., Gdańsk 80952, Poland*

CONTENTS

ABSTRACT

Silica-glauconite/polyaniline (**Si-Gl/PAn**) composites were synthesized by *in situ* polymerization in the presence of 0.5 M H_2SO_4 as dopant by adding of Si-Gl microparticles into aniline (**An**) aqueous solution. The composites were characterized by X-ray diffraction (**XRD**) analysis; Fourier transform infrared (**FTIR**) spectroscopy, Raman spectroscopy, thermogravimetry (**TG**), differential thermogravimetry (**DTG**), and differential scanning calorimetry (**DSC**). Using the **XRD** analysis and **FTIR** spectroscopy, it was shown that the structure of **Si-Gl/PAn** composite is amorphous–crystalline. In addition, an analysis of the physical–chemical properties confirms that the synthesized samples of **Si-Gl/PAn** represent by themselves the composite, but not a mechanical mixture of the components. At the same time, the **PAn** in composite is characterized by higher degree of crystallinity compared with a pure **PAn**. Using the **FTIR** and Raman spectroscopy it was showed that between the macromolecules of **PAn** and the surface of Si-Gl particles available interphase interaction through the formation of hydrogen bond. It was established that the thermal destruction of pure **PAn** and **Si-Gl/PAn** into inert atmosphere (argon) is a multistage process; it was described some stages of this process. Using the **FTIR** spectroscopy and the mass spectrometry (**MS**) methods, the nature of the gaseous products of the thermolysis of individual components (Si-Gl and **PAn**) and **Si-Gl/PAn** composite in the temperature range 20–1000 °C have been determined. It was found, that the introduction of natural mineral (**NM**) Si-Gl into the **PAn** insignificantly accelerates the thermal destruction of **PAn**. Based on the results of **TG**–analysis it was evaluated the composition of the obtained composite: at the mass ratio of Si-Gl:**An** in the reaction mixture, which was ~1:1, the mass ratio of Si-Gl:**PAn** in the obtained composite was about 1:2.

14.1 INTRODUCTION

Synthesis of composite materials based on polyaniline (**PAn**) and inorganic substances, including the natural or synthetic minerals, can be performed by chemical oxidation of aniline (**An**) in the presence of dispersed particles of mineral filler into reaction solution.[1] Combination of properties of the **PAn** and inorganic substances of micro-, and especially of nano-sized dispersion degree leads to the formation of composites with physical and chemical properties that are much better compared to individual properties both of polymer and inorganic oxide or mineral.[2] In many cases the thermal stability of **PAn** into obtained compo-sites is increased,[3] and the composites acquire electromagnetic,[4,5] magnetic,[6,7] catalytic,[8–10] adsorptive[11], and other properties.

In recent decades, the researchers actively develop the various hybrid mineral-polymeric composites, among which a special attention is given to the materials based on **PAn** and natural minerals (**NMs**), including montmorillonite (**MMT**) [**MMT** (*SWy*–1, *China*),[12] Na^+**MMT** (*SWa*–1, *Wyoming, USA*), Na^+**MMT** (*Korea*),[13,14] **MMT**–KSF (*Aldrich*),[15] Na**MMT** clay (*China*)[16,17] modified by inorganic carbonates **MMT** (*China*),[18] modified by organic compounds **MMT** (so-called organo-montmorillonite, *Gonzales, Texas*)[19]], mordenite,[20,21] halloysite (*China*),[22] vermiculite (*China*),[23] kaolinite (*Turkey*),[24] bentonite (*India*),[25] layered perofskite,[26] porous zeolite *Y* (*NTN, Greece*),[27,28] diatomite (*China*),[29] Cu^{2+}-exchanged zeolite *Y* (13*X*, *Sig-ma–Aldrich*),[30] pumice (*Turkey*),[31] smectite—a type of calcium montmorillonite ("*Grey Fuller's Earth*" ™, *United Kingdom*),[32] silica-glauconite, glauconite-silica and glauconite (*Ukraine*).[33–36] Composites based on **PAn** and various **NMs** combine the properties both of the polymer and **NM**, thus showing the synergism effect.[29]

An important characteristic among the acquired physical and chemical properties is already above-mentioned increased thermal stability of **PAn** in different composite materials with the **NM**-fillers.[12–34] Combination of thermal or thermogravimetric (**TG**) analyzer with the mass-spectrometer,[37] gas chromatograph[38], or **FTIR**-spectrometer allows not only to investigate the thermal stability of the material, but also to determine the nature of the gaseous products of the thermolysis and to control the kinetics of their formation during the gradual heating of the investigated samples. In particular, in *ref.*[39] it is shown that the products of thermal destruction of the composite NiO/**PAn** (synthesized by chemical oxidation of **An** by $(NH_4)_2S_2O_8$ in aqueous HCl in the presence of the oxide nanoparticles) in air are H_2O, N_2, NO, CO_2, and NO_2, to which the ions with a mass to charge ratio *m*/*z* (electron ionization) of 18, 28, 30, 44, and 46 Da correspond, respectively, on the mass-spectra. At the study of pyrolysis of the **PAn** doped with HCl and HNO_3, in addition to the above listed fragments others fragments with *m*/*z* from 36 to 548 Da were identified,[39] which correspond to C_5H_6, C_4H_3NH, and S_2H_2 fragments. At the analysis of mass-spectra of the products of thermal destruction of **PAn**, synthesized in the presence of 1.0 M H_2SO_4, the fragments with *m*/*z* from 17 to 98 Da were identified.[40]

Based on the thermogravimetry (**TG**) results it was found, that the process of **PAn** decomposition is characterized by two or three stages, namely: a stage of the eliminating of water, a stage of a loss of the doping substance and, really, a stage of the destruction of **PAn**.[41–43] The results of the **PAn** thermolysis in the form of emeraldine base can confirm of this fact.[19] In this case, only two stages of the mass loss were observed on the thermograms, namely at temperatures < 120 °C (eliminating of absorbed water; the loss ~ 3.27 *mass.* %) and in the temperature range 320–620 °C (thermal destruction of polymer; the loss ~96.73 *mass.* %). In the temperature range 120–320 °C the loss of the mass was not fixed.[19] In a case of the emeraldine salts of the **PAn** between these two stages there is another stage of the loss of mass caused by the eliminating of doping component (inorganic or organic acid).[41–43] As to the interpretation of the nature of the second stage, different researchers are divided in opinion on the reasons for reducing of the mass of the sample in the range 110–320 °C. In particular, the authors of *ref.*[15] affirm about the simultaneous eliminating both of water and of HCl (as doping component) in the temperature range 50–240 °C. At the same time, it can be clearly argued that in the macromolecular chains of doped **PAn** there are three types of water, an eliminating of which will be observed on thermograms at different temperatures, namely: 1) absorption water or superficial water, which is evaporated in the range from ~60 to 120 °C; 2) water that enters into macromolecular chain in the form of the hydration shell of the doping ion and is eliminated at 120–320 °C; 3) water combined with the amino/imino-groups of the macromolecular chains of **PAn** via hydrogen bonds and is eliminated practically in the same temperature range.[37,44,45] Obviously, that the part of the water molecules of the hydration shell also are involved in the formation of hydrogen bonds with the nitrogen atoms of macromolecular chains. So, the evaporation of doping component can occur simultaneously with the evaporation of water of the hydration shell and water combined by hydrogen bonds with the macromolecular chains of **PAn**.

At the same time, as the analysis of the results shows, the nature of the filler can significantly impact not only on the thermal stability of the polymer in the composite, but also on change the nature of the products of its thermolysis and the kinetics of their elimination. Therefore, continuing the previous studies[33,45] we have investigated the thermal destruction of **PAn** and **Si-Gl/PAn** composite using the methods of mass- and **FTIR**-spectrometry to identify the products of this process.

14.2 EXPERIMENTAL DETAILS

14.2.1 MATERIALS

An (95%, *Sigma–Aldrich*) was purified by vacuum distillation under reduced pressure of four Torr and then was stored under argon. Water was distilled twice. Other used in the work chemicals was used without additional purification.

Sample of glauconite–silica is derived from Adamivs'ke-2 deposit occurrence (Khmelnytskyi *region, Ukraine*). The chemical composition of produced silica–glauconite (Si-Gl) sample was ~70% SiO_2 and ~30% glauconite (in oxides terms ± 0.5 mass.%): SiO_2 – 56.6; Al_2O_3 – 11.4; Fe_2O_3 and FeO – 14.2; MgO – 3.8; TiO_2 – 1.8; K_2O – 4.0; CaO – 1.6; Na_2O 0.6; H_2O – ~5.0–6.0. The fraction of Gl–Si dispersion with particles size < 20 μm was used in this work. Before an investigation the initial Si-Gl was washed with distilled water, dried at 150 °C, was rubbing in a porcelain mortar and was sieved through a nylon sieve with holes 20 μm. Only the fraction of Gl-Si dispersion with particles size < 20 μm was used in this work. Si-Gl is an insulator and non-inflammable. It is also insoluble in water, and does not react with other substances in the air.

14.2.2 SYNTHESIS OF PAn AND Si-Gl/PAn COMPOSITE

Samples of **PAn** and **Si-Gl/PAn** composites were synthesized by *in situ* polymerization, which was similar to our previous work.[46] Firstly, 0.93 g of **An** was dissolved in 50 mL 0.5 M H_2SO_4 aqueous solution. At the synthesis of **PAn** sample, the 50 mL of cooled to 275 ± 1.0 K 1.0 M solution of oxidant (sodium peroxydisulfate, ≥ 99.0%, *Sigma–Aldrich*) in 0.5 M H_2SO_4 aqueous solution was added once to this solution (cooled also to 275 ± 1.0 K) and the reaction mixture was stirred for 1.0 h. The synthesis of examples of **Si-Gl/PAn** composites was performed similarly. The difference was only that 1 g of Si-Gl was added into the initial solution of **An** and the reaction mixture was ultrasonificated for 10 min for disaggregation of mineral particles. After this, the reaction mixture was left at room temperature for 24 h.

Obtained samples of **PAn** and **Si-Gl/PAn** were then filtered, washed thoroughly with distilled water to neutral pH value and dried in vacuum at 60 °C for 24 h.

14.2.3 INSTRUMENTAL

XRD, FTIR & Raman Spectra. The **XRD** patterns of **Si-Gl/PAn** composite, individual Si-Gl, and **PAn** were recorded on a diffractometer *DRON*–3 fitted with *CuK*α radiation (λ = 1.5404 Å) at 40 kV and 40 mA, with a scanning speed of 10 min^{-1}. Fourier transform infrared (**FTIR**) absorption spectra of samples were performed on a *Brüker IFS* 66 **FTIR** spectrophotometer in the wavelength range of 4000–400 cm^{-1} at scanning rate 30 cm^{-1}. Specimens were compacted into pellets with *KBr*. Raman scattering spectra (2 cm^{-1}) were registered in the wavelength range of 3500–0 cm^{-1} using a *Brüker IFS* 66 with additional module *FRA* 106 *FT* with cooled liquid nitrogen MCT detector. Nd:YAG laser (λ = 1064 nm) was used as the excitation source. The power of laser was ranged from 40 to 130 mW depending on the nature of the samples.

Thermal Properties & Thermogravimetry/Mass Spectrometry. Thermogravimetric (**TG**) and differential thermogravimetric (**DTG**) studies of the samples Si-Gl, **PAn,** and **Si-Gl/PAn** were performed under argon with a volume rate of gas flow s_V = 20 mL·min^{-1} using the **TG** analyzer *NETZSCH TG*209. The scanning speed of temperature (s_T) was 15 °C·min^{-1} under investigated temperature range from 293 to 1000 K. The evolution of gaseous products degradation was studied by *IR* spectroscopy and by mass-spectroscopy using a spectrophotometer *BRUKER IFS*66 **FTIR** and *Jupiter STA*449*F*3 mass spectrometer (electron ionization), respectively, which were coupled with microbalance.

14.3 STRUCTURE CHARACTERIZATION OF Si-Gl/PAn COMPOSITE

XRD patterns of **Si-Gl/PAn** composite and individual components (Si-Gl and **PAn**) are shown in Figure 14.1. Three intense peaks at 2θ angles around 20.6°, 26.5°, 68.0°, and other low-intensity diffraction peaks at Si-Gl (see Fig. 14.1a) indicate the polycrystalline state of the mineral component. In addition, two intense peaks at 2θ angles[47,48] around 20.6° and 26.5° also confirm the high content of SiO_2 in the composition of Si-Gl.

It can be seen, that the individual **PAn** exhibited two broad peaks at 2θ angles around 21.4° and 24.9° (see Fig. 14.1b), which indicated that the individual **PAn** had crystallinity to a certain extent. The peak centered at 2θ = 21.4° can be ascribed to the periodicity parallel to the polymer chain, while the peak at 2θ = 24.9° may be caused by the periodicity perpendicular to the polymer chains of **PAn**.[49] In addition, an availability of the diffraction peak 2θ angles around 24.9° (see Fig. 14.1b) shows that the sample of individual **PAn** is partly doped with H_2SO_4, that is in the form of the emeraldine base and emeraldine salt.[50]

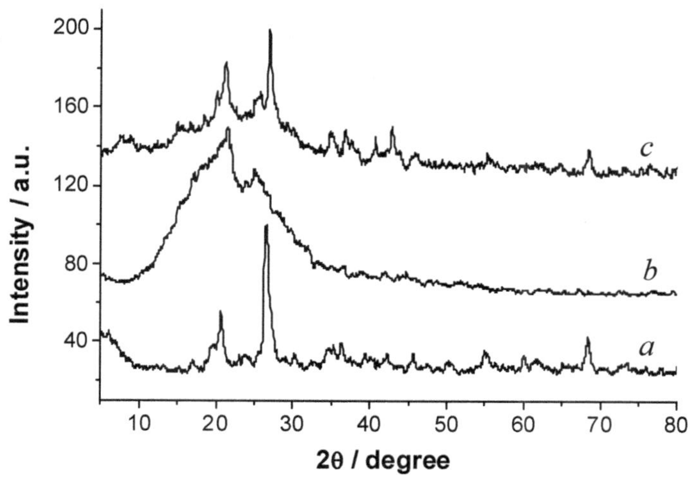

FIGURE 14.1 XRD patterns of individual samples of Si-Gl (a) and **PAn** (b), and **Si-Gl/PAn** composite (c).

The **XRD** pattern of **Si-Gl/PAn** composite (see Fig. 14.1c) integrated the characteristics of **PAn** and Si-Gl individual samples. It means that the Si-Gl particles existed in the composite. While, compared with individual **PAn** and Si-Gl microparticles, the peak intensity for the composite decreased, which showed that the Si-Gl microparticles addition decreased the crystallinity degree of **PAn**. The **XRD** pattern of the composite retains all peaks present in the Si-Gl (2θ = 20.6°, 26.5°, 68.0°, 34.8°, 36.7°, 40.7°, 42.8°, 55.4°, and 68.4° degrees) with decrease in intensity, showing that the basic Si-Gl structure is retained even after the composite formation.[15] **PAn** in **Si-Gl/PAn** composite has a higher degree of crystallinity of the sample concerning to sample of individual **PAn**, as it is evidenced by higher intensity of diffraction peaks of **PAn** at 2θ angles around 21.4° and 24.9° in **XRD** pattern of **Si-Gl/PAn** composite. Overall, the supramolecular structure of the samples of **PAn** and **Si-Gl/PAn** composite is characterized by the coexistence of amorphous and crystalline regions.

Figure 14.2 represents the **FTIR** spectra of (a) Si-Gl particles, (b) individual **PAn,** and (c) **Si-Gl/PAn** composite. Generalized data as for the interpretation of the characteristic absorption bands of individual components and of the composite are shown in Table 14.1. For **FTIR** spectra of Si-Gl particles (see Fig. 14.2a) typical are intense bands at 3424, 1027, and 461 cm^{-1} and less intense bands at 2933, 1632, 781, and 689 cm^{-1}. The form of *IR*-spectrum for Si-Gl (see Fig. 14.2a) corresponds to the **FTIR**-spectra of similar materials.[23,32,51] Their main feature is the presence of highly intense broad characteristic band at ~1027 cm^{-1}. The broad intense band on the **FTIR**-spectrum of Si-Gl within the 1510–825 cm^{-1} is a combination of bands that correspond[22,31] to the vibrations of Si–O, stretching vibrations Si–O–Si and deformation vibrations Me–Me–OH, as for example, Al–Al–OH, Al–Fe–OH, Al–Mg–OH, etc.[15,32,51] In Figure 14.2, the absorption peaks 3556 cm^{-1} is attributed to the H–O–H stretching and peak 1632 cm^{-1} is for H–O–H in absorbed water.

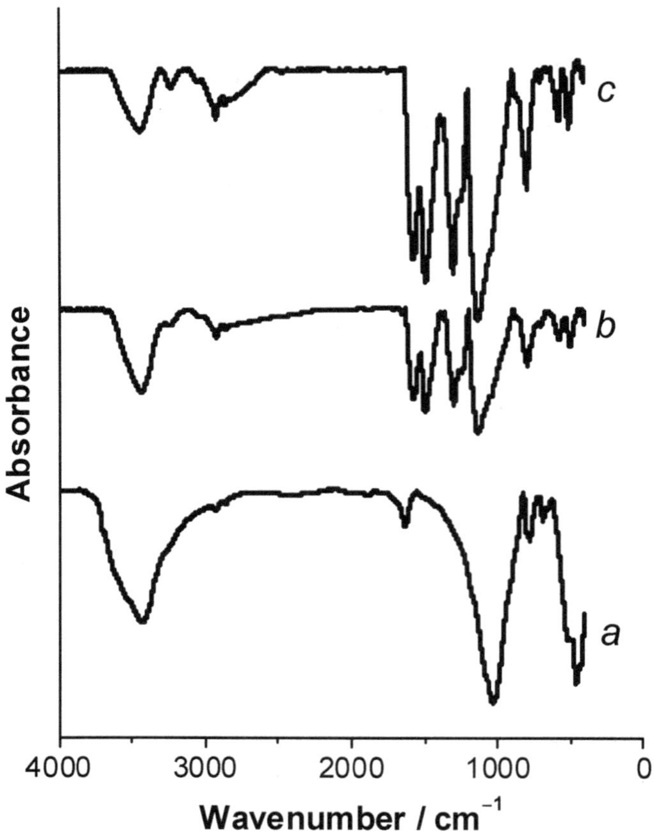

FIGURE 14.2 FTIR spectra of individual samples of Si-Gl (a) and **PAn** (b), and **Si-Gl/PAn** composite (c).

TABLE 14.1 FTIR Vibrational Modes for Individual Samples of Si-Gl and **PAn**, and **Si-Gl/PAn** Composite.

Sample	Interatomic bond	Wave number/cm^{-1}
Si-Gl	Si–OH	3674
	O–H	3431
	H–O–H	1632
	Si–O	1032
	Al–Al–OH, Al–Fe–OH, Al–Mg–OH (etc.)	779
	–O–H	692
	Si–O–Si, Si–O–Al	510–465
PAn	O–H	3433
	N–H	3252
	C–H	2922
	C=C of quinoid ring	1574
	C=C of benzene ring	1495
	C–N	1298
	C–H of quinoid ring	1138
	benzene and quinoid ring C–H	797
	aliphatic C–H	577
	C–N–C	503
Si-Gl/PAn	O–H	3441
	N–H	3229
	C–H	2922
	C=C of quinoid ring	1572
	C=C of benzene ring	1491
	C–N	1300
	C–H of quinoid ring	1138
	benzene and quinoid ring C–H	798
	aliphatic C–H	580
	C–N–C	509

FTIR spectrum of individual **PAn** (see Fig. 14.2b) is in good agreement with those reported in the *ref.*[52–54]. There are a number of the intense bands at 3436, 1575, 1490, 1300, 1134, and 795 cm^{-1}, which indicate that the investigated substance is **PAn** (see Table 14.1). The characteristic peak within the 340–2880 cm^{-1} corresponds to the so-called "H–peak", which is a demonstration of the formation of hydrogen bond that takes place at the interaction of (–NN–) amine and protonated ($=NH^{+}$–) imine groups of the macromolecules of **PAn** between themselves or with the surface –O–H groups of Si-Gl. Low-intensity absorption peak of **PAn** at 3222 cm^{-1} is attributed to the stretching vibrations of $-NH_2-$ groups[52] in protonated **PAn**. A weak peak at 2918 cm^{-1} is attributed to the valence vibrations of C–H groups of aromatic cycle.[53] The characteristic peaks at 1575 and 1490 cm^{-1} for emeraldine salt form of **PAn** can be clearly seen, which are ascribed to C=C stretching vibration of quinoid rings and benzenoid rings, respectively.[15] The presence of absorption band at 1300 cm^{-1} is corresponding to the C–N stretching mode for benzenoid units, while the band at 1134 cm^{-1} is attributed to C–H in plane bending vibration of quinonoid units. And the band at 795 cm^{-1} is associated with C–H out–of–plane bending vibration of benzenoid unit, which indicates the polymer formation in 1,4-positions.[19,55,56] An intense band at 1300 cm^{-1} indicates a state of doped **PAn** and is evidence of its high electroconductivity.[54]

The **FTIR** absorption spectrum of **Si-Gl/PAn** composite (see Fig. 14.2c and Table 14.1) occurs with the peaks at 3442, 2927, 1571, 1489, 1297, 1139, 798, 580, and 509 cm^{-1} that are similar with the individual **PAn**, but all bands are shifted to higher wave number. As it is shown in Figure 14.2(c), for the characteristic bands of **Si-Gl/PAn** composite a small displacement is typical, which is a sign of weak interfacial interactions through the formation of hydrogen bonds.[52]

Figure 14.3 indicates that Raman spectra of **PAn** (see Fig. 14.3a) and **Si-Gl/PAn** (see Fig. 14.3b) composite are obviously similar in the range of 400–2000 cm^{-1} wave number region. The peak observed at 1598 cm^{-1} for individual **PAn** and at 1590 cm^{-1} for **Si-Gl/PAn** is due to C=C stretching of *p*-disubstituted benzene.[57,58] Other main peak at 1374 cm^{-1} is due to C=N stretching of a quinoid structure **PAn**. The peak at 1503 cm^{-1} on the spectrum of individual **PAn** and at 1501 cm^{-1} in the spectrum of **Si-Gl/PAn** was attributed to C=C stretching of a quinoid structure, as well as *p*-disubstition of benzene rings.[57,59–61]

Similarly, **Si-Gl/PAn** composite (see Fig. 14.2c) had the characteristic bands of **PAn**, which indicated the presence of **PAn** in the composite. The intensity of the peaks in the composite spectrum (see Fig. 14.2b) increased relative to the intensity of the peaks for sample of individual **PAn**. Nevertheless, comparison of the Figure 14.2b and c seen that the presence of Si-Gl leads to the shifts of some peaks in **PAn** or the changes of relative intensity. Analogous effect is observed also for the Raman spectra. Several specific bands for **PAn** are observed at 1598, 1503, 1374, 1172, and 811 cm^{-1}, while in the spectrum of **Si-Gl/PAn** these bands are shifted to 1590, 1501, 1374, 1172, and 811 cm^{-1}, respectively. These changes confirm that synthesized sample is a composite, but not simple blend of the two components.[27] Besides, it indicates on interfacial interactions between the particles of mineral filler and Si-Gl conductive polymer matrix (protonated emeraldine form of **PAn**), which should be associated with hydrogen bonds formed between components of the composite.[28,62–64]

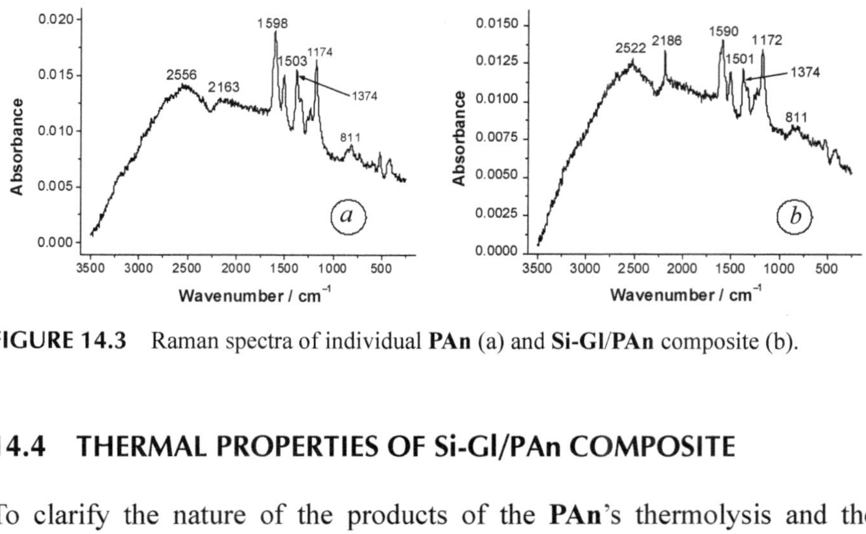

FIGURE 14.3 Raman spectra of individual **PAn** (a) and **Si-Gl/PAn** composite (b).

14.4 THERMAL PROPERTIES OF Si-Gl/PAn COMPOSITE

To clarify the nature of the products of the **PAn**'s thermolysis and the impact of mineral filler on thermal stability of the polymeric matrix we have conducted in an inert atmosphere the **TG** researches of the samples of individual components and of the composite with the fixation of mass-spectra (**MS**) and **FTIR**-spectra of the gaseous products of the thermolysis. The results of **TG–MS** investigations are represented on Figure 14.4.

The **TG–MS** diagram for the sample of individual Si-Gl (see Fig. 14.4a) is a relatively poor because in this case it was recorded the signals of only three fragments, namely with the *m*/*z* of 12, 17, and 18 Da. The fragments with *m*/z = 12 Da were observed for sample in the 120–580 and 600–810 °C temperature intervals and correspond to carbon (C), which is formed in the results of the thermolysis of organic residues in this mineral material. Proof of this can be the fact that, depending on the sample, the mass loss under combustion of the samples Si-Gl can reach 10.5%.[34] The fragments of the mass to charge ratio *m/z*, which accounted 17 and 18 are, respectively, the particles OH and molecules H_2O, whose appearance in the gas phase is two-stage. The first stage is observed in the temperature range 70−280 °C and is associated with the evaporation of superficial water; molecules physically linked to the surface of Si-Gl or are in their pores.

The second stage of water desorption is occurred at temperatures from 280 to 700 °C and corresponds to the elimination of chemisorbed and crystallization water from Si-Gl.

During the thermolysis of individual sample of **PAn** in a gas phase the fragments with *m/z* of 12, 14, 16, 17, 18, 28, 66, 78, and 93 Da (see Fig. 14.4b) were fixed, to which the eliminating of C, N (or CH_2, more likely) O (or NH_2, more likely), NH_3 (or OH), H_2O (or $NH4^+$), CO (or N_2), C_5H_6 (or C_4H_3NH and S_2H_2), C_6N_6, and $C_6N_5NH_2$ correspond, respectively.[33,40,65] An eliminating of C from the sample of individual **PAn** proceeds in two stages at the temperatures 520–710 and 710–900 °C, and an eliminating of CH_2 proceeds into three stages, to which the temperatures 80–220, 220–230, and 680–860 °C correspond, respectively. An eliminating of NH_2 proceeds in four stages at temperatures 80−200, 200−500, 500−740, and 740−900 °C. An eliminating of OH and H_2O (or NH_3) also occurs in four stages at temperatures 70−200, 200−450, 450580, and 580−950 °C. An eliminating of CO or N_2 occurs mainly in 680−900 °C temperature range, but it can be also selected three little intense stages at lower temperatures, namely at 100−220, 220−350, and 520−620 °C, respectively. Intensive eliminating of fragments with m/z = 66 Da is observed in the temperature ranges 190−450 and 450−900 °C. The destruction of the polymeric chain from the eliminating of C_6H_6 is two-stage intensive process proceeding at 350−490 and 490−900 °C, while for the formation of $C_6H_5NH_2$ fragments there is only one step (intensive peak), which corresponds to the temperature range 410−800 °C. The products of the **PAn** thermal destruction, which were identified by us, correspond to those described in *ref.*[33,40,45,65].

At the **TG–MS** diagram of **Si-Gl/PAn** composite (see Fig. 14.4c), there are signals of identical as same as for the thermolysis of individual **PAn** fragments. At the same time, in the temperature range 400−700 °C along with the fragments of m/z = 78 and 93 Da an eliminating of a new particle with m/z = 54 Da was recorded, the formation of which can be caused only by thermal destruction of aromatic or quinoid rings. Such fragment was not identified by the authors of *ref.*[37,39] though they traced the peaks that correspond to the fragments with *m/z* less than 78 Da. This difference shows the influence of the filler on thermal stability of the polymeric matrix, namely the catalytic activity of some metal ions, which is a part of glauconite, as to destruction of benzenoid/quinoid rings of **PAn**.

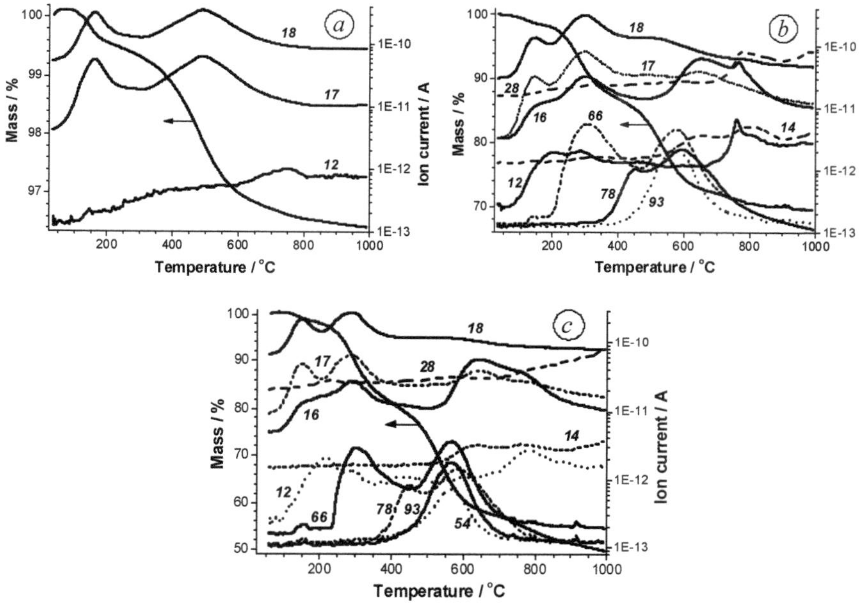

FIGURE 14.4 **TG–MS** diagrams for the samples of individual Si-Gl (a), **PAn** (b), and **Si-Gl/PAn** composite (c). Numerals over curves corresponds to the values of *m/z* for the fragments.

Detailed analysis of the references shows that the results obtained by us of **TG–MS** analysis somewhat are differed from the results of other researchers. In particular, the authors of *ref.*[39] did not identify the signals of the fragments with $m/z < 78$ Da, while the authors of *ref.*[37] did not identify the fragment with $m/z = 12$ Da. At the same time, the identified by us products both in the case of **PAn** and in the case of **Si-Gl/PAn** composite, practically, correspond to the products identified under thermal destruction of **PAn** doped with H_2SO_4.[40] The presence of fragment with $m/z = 66$ Da among the products of **PAn** decomposition (at the temperature 190−460 °C) and **Si-Gl/PAn** composite (225−460 °C) is often attributed to the formation of particles S_2H_2 as the product of the doping agent **PAn**−H_2SO_4 reducing, which in the temperature ranges is eliminated from the samples. However, from our point of view, in the investigated system no so strong reductant, which could recover the sulfur (+6) to the disulfide. Therefore, more likely, in this case, it is appeared the formation of fragments C_5H_6 or C_4H_3NH as products of decomposition of benzenoid and quinoid cycles of **PAn**. In addition, there is the question about the source for the fragments with *m/z* = 28 Da. However, taking into account practically the independence of the intensity signal on temperature, it can be suggested that its appearance may be due to the presence of the adsorbed air by **PAn** and **Si-Gl/PAn** composite.

The **TG–MS** dependence for the fragments with *m/z* = 17 and 18 Da (see Fig. 14.4 b,c) requires also more detailed consideration. This makes it possible to analyze the processes of water eliminating from the samples during their heating because the authors of *ref.*[37,44] do not specify the description of the H_2O eliminating stages within 160–500 °C. The analysis shows, that the eliminating of water from the sample Si-Gl takes place in two stages: the first is in the temperature range 75–300 °C at maximum speed of the eliminating at 180 °C, and the second is in the temperature range from 300 to 700 °C (wide intense peak at 490 °C) (see Fig. 14.4a). At the same time, for the eliminating of water (the fragments with *m/z* = 17 and 18 Da) from the samples of **PAn** and **Si-Gl/PAn** typical are four main stages: the first is in the temperature range 70–190 °C (intense peak at 150 °C); the second is in the temperature range 200–400 °C (highly intense peak at 275 °C); the third is within 450–700 °C (weak peak at ~525 °C) and the fourth is within 720–850 °C (weak peak at 790 °C) (see Fig. 14.4b,c). The intensity and, therefore, the amount of H_2O, which is eliminated in the second stage (within the temperature 200–400 °C) compared to the first stage for the sample **Si-Gl/PAn** is slightly higher due to, in our opinion, an additional eliminating of water from the particles Si-Gl. Even more detailed analysis of **TG–MS** diagrams within 200–350 °C temperatures shows that the loss of H_2O by the samples of **PAn** and **Si-Gl/PAn** at the second stage of the process is actually additive value of the two parallel processes, the maximum speed of which is observed at temperatures 250 and 300 °C, respectively. The presence of two sources of water eliminating can be explained by the fact that water that introduces in the hydration shell of doping **PAn** ion (HSO_4^-), and the molecules H_2O bounded with the imine and amine nitrogen atoms of **PAn** macromolecules is eliminated at once.

Presented in Figure 14.5 total results of **TG**, differential **TG** analysis, and differential scanning calorimetry (**TG, DTG,** and **DSC** curves, respectively) for the samples Si-Gl, **PAn,** and **Si-Gl/PAn** illustrate the correlation of the processes that are occurred during their heating, such as weight loss by the samples and the sign of thermal effects accompanying the thermal degradation. **DTG** curves confirm the stages of the loss of mass by the samples, and **DSC** curves illustrate the presence of thermal effects accompanying thermal conversion of the samples. As it's illustrated by Figure 14.5(b), during the heating of the individual sample of **PAn** an eliminating of water and doping component is began at 151 °C, while in the case of composite **Si-Gl/PAn** at achieving of the significantly higher temperature 162 °C. At the same, the actual thermal destruction of **PAn** in the samples of individual **PAn** and composite **Si-Gl/PAn** actually begins at the same values of temperature, namely at 410 and 412 °C, respectively.

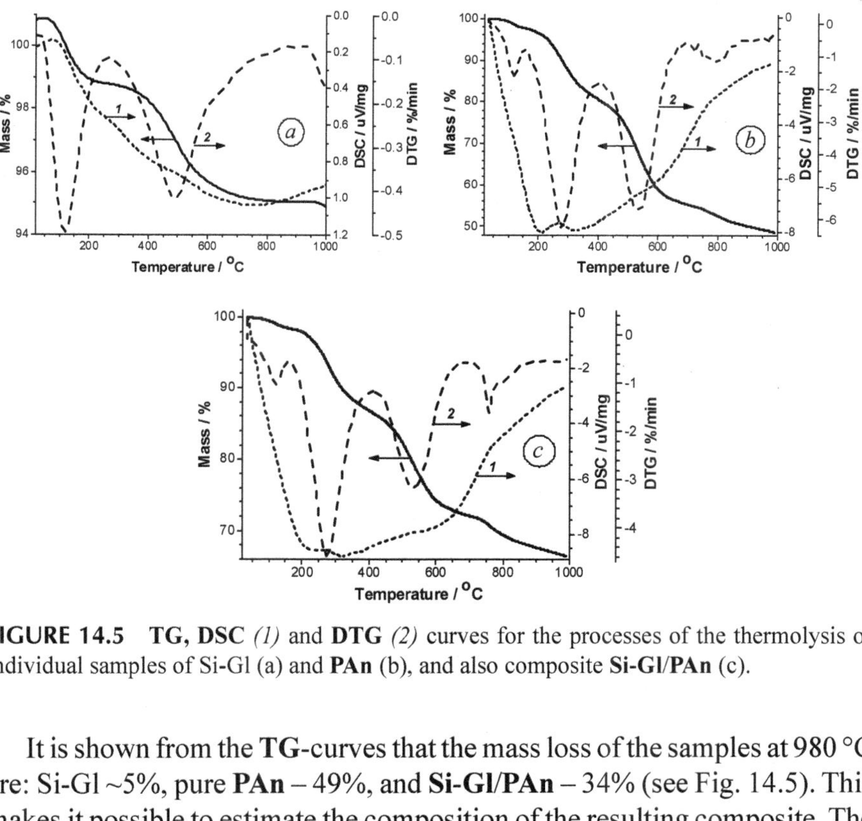

FIGURE 14.5 **TG, DSC** *(1)* and **DTG** *(2)* curves for the processes of the thermolysis of individual samples of Si-Gl (a) and **PAn** (b), and also composite **Si-Gl/PAn** (c).

It is shown from the **TG**-curves that the mass loss of the samples at 980 °C are: Si-Gl ~5%, pure **PAn** – 49%, and **Si-Gl/PAn** – 34% (see Fig. 14.5). This makes it possible to estimate the composition of the resulting composite. The results show that at the mass ratio of Si-Gl:**An** in the reaction mixture, which was 1:0.93, the mass ratio of **PAn:Si-Gl** in the composite was about 33:67, that is, 1:2. One can assume that the reason for this is the sedimentation of the most coarse fractions of the mineral component during the synthesis of the composite or *vice versa* – the most fine particles were not eliminated during filtration.

The results of **TG–DSC** analysis of the samples of individual **PAn** and composite **Si-Gl/PAn** confirm the presence of the fourth stages of loss of mass and give the possibility to estimate their energy (see Fig. 14.5b,c).[66,67] Determined values of the loss of mass by the samples on each stages (1–4) and their temperature ranges are listed in Table 14.2. Temperatures of peaks on **DTG**-curves (see Fig. 14.5b,c) that correspond to the eliminating of HSO_4^- are 277 and 274 °C for individual **PAn** and **Si-Gl/PAn** composite, respectively. The temperature of **DTG** peak, which corresponds to the stage of active decomposition of pure **PAn** is 420 °C, and the temperature peak of the **PAn** decomposition in **Si-Gl/PAn** is 410 °C. Differences in temperatures of these peaks may indicate the influence of Si-Gl on a slight decrease in thermal stability of **PAn** in composite **Si-Gl/PAn**. Character both **DTG** and **DSC** dependencies for thermal destruction of **PAn** and **Si-Gl/PAn** composite is similar.

TABLE 14.2 The Loss of Mass by Investigated Samples.

Sample	The stages of the loss of mass				Total loss of mass, *mass.* %
	1	2	3	4	
	Temperature ranges of the loss mass stages/ °C (The loss of mass / *mass.* %)				
Si-Gl	60–150 (0.57)	150–810 (2.90)	–	–	3.47
PAn	60–160 (1.39)	160–410 (20.20)	410–675 (23.54)	675–910 (6.03)	51.16
Si-Gl/PAn	60–170 (1.02)	170–410 (14.84)	410–725 (14.47)	725–810 (2.42)	32.75

In Figure 14.6, there is an evolution of **FTIR**-spectra of gaseous products decomposition for investigated samples obtained within 20–1000 °C during thermolysis 0–3800 c. Their comparison shows that thermal degradation processes that occur in these samples have some differences. In particular, peaks that correspond to the products of sample **Si-Gl/PAn** decomposition are characterized by much higher intensity compared with the peaks that correspond to the products of individual **PAn** decomposition. Obviously, the particles of mineral filler Si-Gl accelerate the thermal destruction of **PAn**, deposited on the particles of Si-Gl. The differences in the spectra of gaseous products of thermal destruction of **PAn** and **Si-Gl/PAn** composite are observed primarily in the spectral ranges 3900–3400 and 1800–1250 cm^{-1}, starting from the ~600 *s* ($T \approx 150$ °C) from the beginning of the samples heating (see Fig. 14.6 b,c). The presence of Si-Gl in **Si-Gl/PAn** composite affects the process of thermal destruction of **PAn** that is reflected on the character of **FTIR**-spectra evolution (see Fig. 14.6c). In particular, FTIR-spectra of gaseous products of thermolysis of the sample of composite are more intense (see Fig. 14.6c).

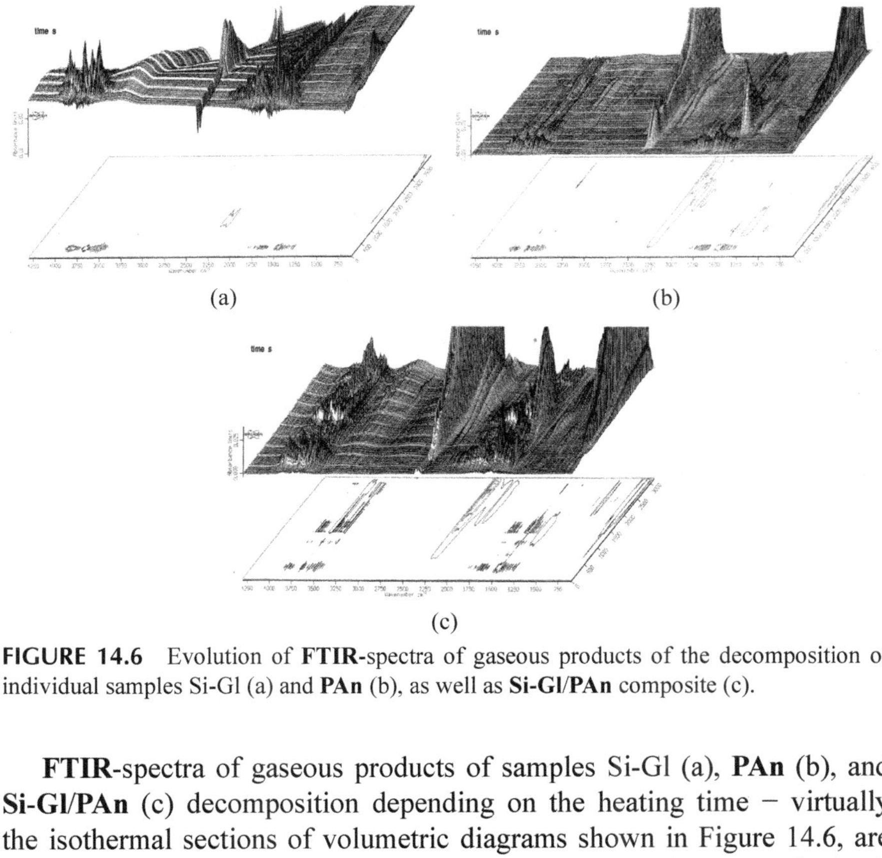

FIGURE 14.6 Evolution of **FTIR**-spectra of gaseous products of the decomposition of individual samples Si-Gl (a) and **PAn** (b), as well as **Si-Gl/PAn** composite (c).

FTIR-spectra of gaseous products of samples Si-Gl (a), **PAn** (b), and **Si-Gl/PAn** (c) decomposition depending on the heating time – virtually the isothermal sections of volumetric diagrams shown in Figure 14.6, are represented in Figure 14.7. This makes it possible to analyze more in detail the impact of the products of thermolysis on the kinetics of the process. In particular, the characteristic bands in the ranges 4000–3400 and 2100–1250 cm^{-1} that are occurred after ~360 c of heating[68] (at the achieving of T ~90 °C) corresponds to the eliminating of water, which is agreed with the spectra shown in Figure 14.2, and also with the results of **TG–MS** and **DSC** analysis (see Figs. 14.3 and 14.4). In addition, the obtained **FTIR**-spectra confirm the different nature of the processes of water eliminating from the investigated samples: if the sample Si-Gl loses water almost on the initial stages of heating to 120 °C, then an eliminating of water from the samples **PAn** and **Si-Gl/PAn** composite continues practically throughout the whole process of the thermolysis.

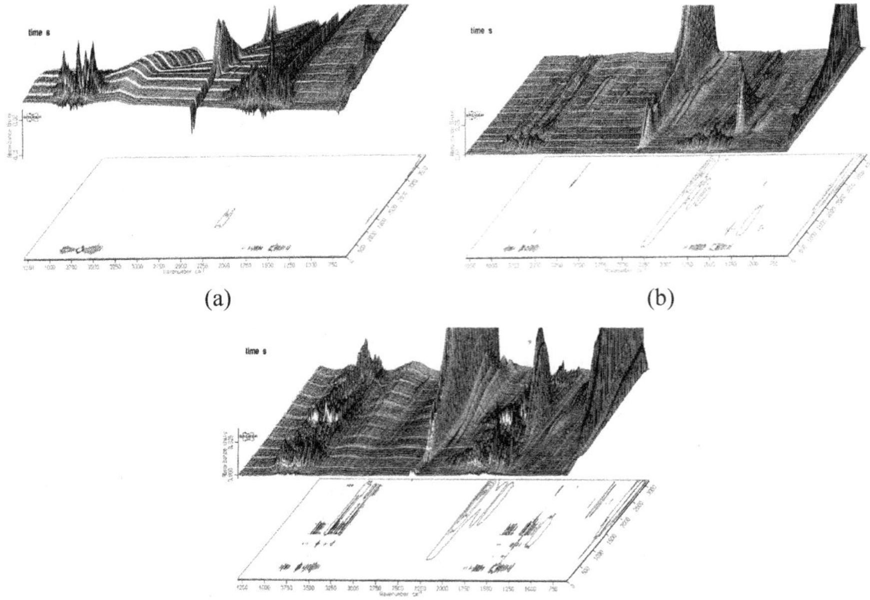

(a) (b)

(c)

FIGURE 14.7 The dynamics of the intensities change of characteristic bands of **FTIR**-spectra of gaseous products of the decomposition of samples Si-Gl (a), **PAn** (b), and **Si-Gl/PAn** (c) depending on the heating time (time from the start of the thermolysis is represented directly on figures).

Characteristic band (doublet), which can be traced at 2400–2250 cm^{-1} arises after ~1400 c from the beginning of the thermolysis (T = 350 °C) and corresponds to eliminating of CO.[68] It is significant, that this band was recorded not only during the thermolysis of **PAn** samples and composite, but also individual sample Si-Gl (see Fig. 14.7a). At the same time, a fragment with m/z = 28 absent on Figure 14.4. This seemingly contradictory result is due to the fact that the possibility of CO eliminating during the **TG–MS** analysis of this sample was not considered and, appropriate signal was not detected. Intensive narrow band, which can be traced on the **FTIR**-spectra (see Figs. 14.6b,c and 14.7b,c) within the 740–600 cm^{-1} also belongs to CO. This band is appeared along with the band at 2350–2250 cm^{-1} and is almost concomitant to it.

As it was shown above, thermal destruction of **PAn** is accompanied by the eliminating of gaseous products, among which the main ones are benzene and **An** (see Fig. 14.4b,c). These products are eliminated at temperatures higher than 350 °C. On **FTIR**-spectra the band, which correspond to

C_6H_6 and $C_6H_5NH_2$, namely an intense peak within 2250–2350 cm^{-1} (see Fig. 14.7b,c) arises after 1440 and 1560 *s* ($T \approx 360$ and $T \approx 390$ °C), respectively, from the beginning of the heating of a sample **PAn**, and **Si-Gl/PAn** composite.

14.5 CONCLUSIONS

Silica-glauconite/polyaniline (**Si-Gl/PAn**) composites with 1:2 mass ratios of components were synthesized by *in situ* polymerization at the presence of 0.5 M H_2SO_4 as dopant by introduction of Si-Gl microparticles into reaction mixture (**An** water solution). The results of **XRD** analysis showed that the structure of the samples of **PAn** and **Si-Gl/PAn** composite is amorphous–crystalline. At this, **PAn** into the composite possesses by a higher crystallinity compared with a sample of individual **PAn**. With the use of **FTIR** and Raman spectroscopy, it was showed that between the macromolecules of **PAn** and the surface of Si-Gl particles there are interfacial interactions through the hydrogen bonding, and it was confirmed that the obtained samples represent by themselves the **Si-Gl/PAn** composite, but not a mechanical mixture of the initial components.

It was established, that the thermal destruction of pure **PAn** and **Si-Gl/PAn** composite is a multistage process. With the use of **TG,** analysis within 20–1000 °C with the mass-spectroscopic fixation of gaseous products of thermal decomposition, which were determined the products of Si-Gl, **PAn,** and **Si-Gl/PAn** composite decomposition in an inert atmosphere. It was found that the thermal destruction of samples of **PAn** and **Si-Gl/PAn** composite proceeds with the formation of practically identical products. The difference is only in the formation during the **Si-Gl/PAn** thermal degradation of the fragments with $m/z = 54$ Da, which from our point of view is a testament of the catalytic destruction of benzenoid/quinoid cycles of **PAn** at temperatures over 450 °C with the formation of fragments C_5H_6 or C_4H_3NH. The results of **TG**, **DSC** studies show that **Si-Gl/PAn** composite has slightly lower thermal stability compared with individual **PAn**, which can be caused also by the catalytic influence of the metal ions – components of Si-Gl.

Volumetric diagrams of the evolution over time of the **FTIR**-spectra of gaseous products of thermolysis for the samples Si-Gl, pure **PAn,** and **Si-Gl/PAn** composite have been constructed. On their basis, it has been confirmed the nature of the gaseous products of the decomposition of investigated samples previously determined using the **TG–MS** analysis.

KEYWORDS

- **polyaniline**
- **silica–glauconite**
- **composite**
- **structure**
- **thermogravimetry**
- **mass spectrometry**
- **differential scanning calorimetry**

REFERENCES

1. Malinauskas, A. Chemical Deposition of Conducting Polymers. *Polymer.* **2001,** *42,* 3957–3972.
2. Wan, M. *Conducting Polymers with Micro or Nanometer Structure*. Springer, Berlin, 2010.
3. Qi, Y. N.; Xu, F.; Sun, L. X.; Zeng, J. L.; Liu, Y. Y. Thermal Stability and Glass Transition Behaviour of PANi/α-Al_2O_3 Composites. *J. Therm. Anal. Calorim.* **2008,** *94,* 553–557.
4. Reddy, K. R.; Lee, K. P.; Gopalan, A. I. Self-Assembly Approach for the Synthesis of Electromagnetic Functionalized Fe_3O_4/Polyaniline Nanocomposites: Effect of Dopant on the Properties. *Colloids Surf. A: Physicochem. Eng. Aspects.* **2008,** *320,* 49–56.
5. Hsieh, T. H.; Ho, K. S.; Bi, X. T.; Han, Y. K.; Chen, Z. L.; Hsu, C. H.; Chang, Y. C. Synthesis and Electromagnetic Properties of Polyaniline-Coated Silica/Maghemite Nanoparticles. *Eur. Polym. J.* **2009,** *45,* 613–620.
6. Lee, G.; Kim, J.; Lee, J. H. Development of Magnetically Separable Polyaniline Nanofibers for Enzyme Immobilization and Recovery. *Enz. Microbial Techn.* **2008,** *42,* 466–472.
7. Singh, K.; Ohlan, A.; Kotnal, R. K.; Bakhshi, A. K.; Dhawan, S. K. Dielectric and Magnetic Properties of Conducting Ferromagnetic Composite of Polyaniline with γ-Fe_2O_3 Nanoparticles. *Mater. Chem. Phys.* **2008,** *112,* 651–658.
8. Anunziata, O. A.; Gomez, C. M. B.; Martınez, M. L. Interaction of Water and Aniline Adsorbed onto Na-AlMCM-41 and Na-AlSBA-15 Catalysts as Hosts Materials. *Catalysis Today*. **2008,** *133,* 897–905.
9. Wang, X.; Tang, S.; Zhou, C.; Liu, J.; Feng, W. Uniform TiO_2–PANI Composite Capsules and Hollow Spheres. *Synth. Met.* **2009,** *159,* 1865–1869.
10. Li, X.; Wang, D.; Cheng, G.; Luo, Q.; An, J.; Wang, Y. Preparation of Polyaniline-Modified TiO_2 Nanoparticles and their Photocatalytic Activity under Visible Light Illumination. *Appl. Catalysis. B: Environment.* **2008,** *81,* 267–273.
11. Yang, C.; Li, H.; Xiong, D.; Cao, Z. Hollow Polyaniline/Fe_3O_4 Microsphere Composites: Preparation, Characterization, and Applications in Microwave Absorption. *React. Funct. Polym.* **2009,** *69,* 137–144.

12. Feng, B.; Su, Y.; Song, J.; Kong, K. Electropolymerization of Polyaniline/Montmorillonite Nanocomposite. *J. Mater. Sci. Lett.* **2001,** *20,* 293–294.
13. Kim, B. H.; Jung, J. H.; Kim, J. W. et al. Effect of Dopant and Clay on Nanocomposites of Polyaniline (PAN) Intercalated into Na+-Montmorillonite (Na+-MMT). *Synth. Met.* **2001,** *121,* 1311–1312.
14. Lee, D.; Char, K. Thermal Degradation Behavior of Polyaniline in Polyaniline/Na+-Montmorillonite Nanocomposites. *Polym. Degr. Stab.* **2002,** *75,* 555–560.
15. Binitha, N. N.; Sugunan, S. Polyaniline/Pillared Montmorillonite Clay Composite Nanofibers. *J. Appl. Polym. Sci.* **2008,** *107,* 3367–3372.
16. Sun, F.; PAn, Y.; Wang, J.; Wang, Z.; Hu, C.; Dong, Q. Synthesis of Conducting Polyaniline-Montmorillonite Nanocomposites via Inverse Emulsion Polymerization in Supercritical Carbondioxide. *Polym. Composite.* **2010,** *31,* 163–172.
17. Wu, Q.; Xue, Z.; Qi, Z.; Wang, F. Synthesis and Characterization of PAn/Clay Nanocomposite with Extended Chain Conformation of Polyaniline. *Polymer.* **2000,** *41,* 2029–2032.
18. Mo, Z.; Zhang, P.; Zuo, D.; Sun Y.; Chen, H. Synthesis and Characterization of Polyaniline Nanorods/$Ce(OH)_3$-PrO_3/Montmorillonite Composites Through Reverse Micelle Template. *Mater. Res. Bull.* **2008,** *43,* 1664–1669.
19. Salahuddin, N.; Ayad, M. M.; Ali, M. Synthesis and Characterization of Polyaniline–Organoclay Nanocomposites. *J. Appl. Polymer Sci.* **2008,** *107,* 1981–1989.
20. Enzel, P.; Bein, T. Inclusion of Polyaniline Filaments in Zeolite Molecular Sieves. *J. Phys. Chem.* **1989,** *93,* 6270–6272.
21. Do Nascimento, G. M.; Temperini, M. L. A. Structure of Polyaniline Formed in Different Inorganic Porous Materials: A Spectroscopic Study. *Eur. Polym. J.* **2008,** *44,* 3501–3511.
22. Zhang, L.; Wang, T.; Liu, P. Polyaniline-Coated Halloysite Nanotubes via *In-Situ* Chemical Polymerization. *Appl. Surf. Sci.* **2008,** *255,* 2091–2097.
23. Liu, P. Preparation and Characterization of Conducting Polyaniline/Silica Nanosheet Composites. *Curr. Op. Sol. Stat. Mater. Sci.* **2008,** *12,* 9–13.
24. Duran, N. G.; Karakısla, M.; Aksu, L.; Sacak, M. Conducting Polyaniline/Kaolinite Composite: Synthesis, Characterization and Temperature Sensing Properties. *Mater. Chem. Phys.* **2009,** *118,* 93–98.
25. Sudha, J. D.; Reena, V. L. Structure–Directing Effect of Renewable Resource Based Amphiphilic Dopants on the Formation of Conducting Polyaniline-Clay Nanocomposite. *Macromol. Symp.* **2007,** *254,* 274–283.
26. Uma, S.; Gopalakrishnan, J. Polymerization of Aniline in Layered Perovskites. *Mater. Sci. Eng.* **1995,** *34,* 175–179.
27. Vitoratos, E.; Sakkopoulos, S.; Dalas, E.; et al. D. C. Conductivity and Thermal Aging of Conducting Zeolite/Polyaniline and Zeolite/Polypyrrole Blends. *Curr. Appl. Phys.* **2007,** *7,* 578–581.
28. Li, X.; Li, X.; Wang, G. Fibrillar Polyaniline/Diatomite Composite Synthesized by One-Step *In Situ* Polymerization Method. *Appl. Surf. Sci.* **2005,** *249,* 266–270.
29. Li, X.; Li, X.; Dai, N.; Wang, G. Large-Area Fibrous Network of Polyaniline Formed on the Surface of Diatomite. *Appl. Surf. Sci.* **2009,** *255,* 8276–8280.
30. Densakulprasert, N.; Ladawan, W.; Datchanee, C.; Hiamtup, P.; Sirivat, A.; Schwank, J. Electrical Conductivity of Polyaniline/Zeolite Composites and Synergetic Interaction with CO. *Mater. Sci. Eng. B.* **2005,** *117,* 276–282.

31. Gok, A.; Gode, F.; Turkaslan, B. E. Synthesis and Characterization of Polyaniline/Pumice (PAn/Pmc) Composite. *Mater. Sci. Eng. B.* **2006,** *133,* 20–25.
32. Rajapakse, R. M. G.; Krishantha, D. M. M.; Tennakoon, D. T. B.; Dias, H. V. R. Mixed-Conducting Polyaniline-Fuller's Earth Nanocomposites Prepared by Stepwise Intercalation. *Electrochim. Acta.* **2006,** *51,* 2483–2490.
33. Yatsyshyn, M.; Grynda, Yu.; Kun'ko, A.; Dumanchuk, N. Thermal Stability of the Polyaniline/Silica–Glauconite Composites. *Visnyk Lviv Univ. Ser. Khim.* **2011,** *52,* 268–276 (In Ukrainian).
34. Yatsyshyn, M. M.; Reshetnyak, O. V.; Dumanchuk, N. Ya.; Kulyk, Yu. O.; Fartushok, N. V.; Stadnyk, Yu. V. Hybrid Mineral-Polymeric Composite Materials on the Basis of the Polyaniline and Glauconite-Silica. *Ch&ChT.* **2013,** *7,* 441–444.
35. Yatsyshyn, M. M.; Grynda, Yu. M.; Kun'ko, A. C. et al. *Conductive Magnetic Composite Material Based on Polyaniline.* Pat. Ukr. 62888, Publ. Sept. 26, 2011.
36. Yatsyshyn, M. M.; Koval'chuk, E. P.; Turba, Z. V.; et al. *Magnetic, Conductive, Composite Material Based on Polyaniline and Glauconite-Silica.* Pat. Ukr. 78462, Publ. March. 25, 2013.
37. Hacaloglu, J.; Argin, E.; Kucukyavuz, Z. Characterization of Polyaniline via Pyrolysis Mass Spectrometry. *J. Appl. Pol. Sci.* **2008,** *108,* 400–405.
38. Borros, S.; Munoz, E.; Folch, I. Study of Some Pyrolysis–Gas Chromatography Indexes for the Differentiation among Oxidation States of Polyaniline. *J. Chromatography A.* **1997,** *837,* 273–279.
39. Qi, Y.; Zhang, J.; Qiu, S.; Sun, L.; Xu, F.; Zhu, M.; Ouyang, L.; Sun, D. Thermal Stability, Decomposition and Glass Transition Behaviour of PANI/NiO Composites. *J. Therm. Anal. Calorim.* **2009,** *98,* 533–537.
40. Sreedhar, B.; Sairam, M.; Chattopadhyay, D. K.; Mitra, P. P.; Rao, D. V. M. Thermal and XPS Studies on Polyaniline Salts Prepared by Inverted Emulsion Polymerization. *J. Appl. Polymer Sci.* **2006,** *101,* 499–508.
41. Pielichowski, K. Kinetic Analysis of Thermal Decomposition of Polyaniline. *Solid State Ion.* **1997,** *104,* 123–132.
42. Tsocheva, D.; Zlatkov, T.; Terlemezyan, L. Thermoanalitycal Studies of Polyaniline Emeraldine Base. *J. Therm. Anal.* **1998,** *53,* 895–904.
43. Rodrigues, P. C.; de Souza, G. P.; Neto, J. D. D. M.; Akcelrud, L. Thermal Treatment and Dynamic Mechanical Thermal Properties of Polyaniline. *Polymer.* **2002,** *43,* 5493–5499.
44. Tsocheva, D.; Mokreva, P.; Terlemezyan, L. Copolymers of Aniline and o-Methoxyaniline. II. Thermoanalytical Studies. *J. Appl. Polymer Sci.* **2007,** *104,* 2729–2734.
45. Grynda, Yu. M.; Luhodid, A. S.; Yatsyshyn, M. M. Thermal Destruction of Polyaniline. The Second All-Ukrainian Scientific Conference of Students and Graduate Students "Karazin chemicals readings – 2010", Kharkiv, April 19–22, 2010. Abstracts. Kharkiv, Operative polygraphy, 2010; pp 189–190 (In Ukrainian).
46. Yatsyshyn, M.; Grynda, Yu.; Kun'ko, A.; Kulyk, Yu. The Polymerization of Anyline in Presence of Glauconite. *Visnyk Lviv Univ. Ser. Khim.* **2010,** *51,* 395–406 (In Ukrainian).
47. Li, X.; Wang, G.; Li, X. Surface Modification of Nano-SiO2 Particles using Polyaniline. *Surf. Coat. Technol.* **2005,** *197,* 56–60.
48. Yatsyshyn, M. M.; Il'kiv, Z. V.; Halamay, R. I.; Struk, V. M.; Reshetnyak, O. V. *Method of Cleaning Glauconite from Silica and Other Admixtures.* Pat. Ukr. 86632, Publ. Jan. 10, 2014.

49. Cheng, F.; Tang, W.; Li, C.; Chen, J.; Liu, H.; Shen, P.; Dou, S. Conducting Poly(aniline) Nanotubes and Nanofibers: Controlled Synthesis and Application in Lithium/Poly(aniline) Rechargeable Batteries. *Chem. Eur. J.* **2006,** *12,* 3082–3088.
50. He, B. L.; Dong, B.; Wang, W.; Li, H. L. Performance of Polyaniline/Multi-Walled Carbon Nanotubes Composites as Cathode for Rechargeable Lithium Batteries. *Mater. Chem. Phys.* **2009,** *114,* 371–375.
51. Lin, J.; Tang, Q.; Wu, J.; Sun, H. Synthesis, Characterization and Properties of Polyaniline/Expanded Vermiculite Intercalated Nanocomposite. *Sci. Technol. Adv. Mater.* **2008,** *9,* 1–6.
52. Dhawale, D. S.; Salunkhe, R. R.; Jamadade, V. S.; Dubal, D. P.; Pawar, S. M.; Lokhande, C. D. Hydrophilic Polyaniline Nanofibrous Architecture using Electrosynthesis Method for Supercapacitor Application. *Curr. Appl. Phys.* **2010,** *10,* 904–909.
53. Binh, P. T. Electrochemical Polymerization of Aniline by Current Pulse Method in the Presence of m-Aminobenzoic Acid in Chlorhydric Acid Solution. *Macromol. Symp.* **2007,** *228,* 249–250.
54. Xu, J. C.; Liu, W. M.; Li, H. L. Titanium Dioxide Doped Polyaniline. *Mater. Sci. Engineer. C.* **2005,** *25,* 444–447.
55. Gu, Y.; Chen, C. C.; Ruan, Z. W. Enzymatic Synthesis of Conductive Polyaniline using Linear BSA as the Template in the Presence of Sodium Dodecyl Sulfate. *Synth. Met.* **2009,** *159,* 2091–2096.
56. Tang, Q.; Wu, J.; Sun, X.; Li, Q.; Lin, J. Layer-by-Layer Self-Assembly of Conducting Multilayer Film from Poly (Sodium Styrenesulfonate) and Polyaniline. *J. Colloid Interf. Sci.* **2009,** *337,* 155–161.
57. Wang, L.; Jing, X.; Wang, F. On the Iodine-Doping of Polyaniline and Poly-Ortho-Methylaniline. *Synth. Met.* **1991,** *41,* 739–744.
58. Quillard, S.; Louarn, G.; Buisson, J. P.; Lefrant, S.; Masters, J.; MacDiarmid, A. G. Vibrational Analysis of Reduced and Oxidized Forms of Polyaniline. *Synth. Met.* **1993,** *55,* 475–480.
59. Berrada, K.; Quillard, S.; Louarn, G.; Lefrant, S. Polyanilines and Substituted Polyanilines: A Comparative Study of the Raman Spectra of Leucoemeraldine, Emeraldine and Pernigraniline. *Synth. Met.* **1995,** *69,* 201–204.
60. Sacak, M.; Akbulut, U.; Batchelder, D. N. Batchelder Monitoring of Electroinitiated Polymerization of Aniline by Raman Microprobe Spectroscopy. *Polymer.* **1998,** *40,* 21–26.
61. Ciric-Marjanovic, G.; Trchova, M.; Stejskal, J. The Chemical Oxidative Polymerization of Aniline in Water: Raman Spectroscopy. *J. Raman Spectrosc.* **2008,** *39,* 1375–1387.
62. Niu, Z. W.; Yang, Z. Z.; Hu, Z. B.; Lu, Y. F.; Han, C. C. Polyaniline–Silica Composite Conductive Capsules and Hollow Spheres. *Adv. Funct. Mater.* **2003,** *13,* 949–954.
63. Grzeszczuk, M.; Szostak, R. Electrochemical and Raman Studies on the Redox Switching Hysteresis of Polyaniline. *Solid State Ion.* **2003,** *157,* 257–262.
64. Yuan, P.; Wu, D. Q.; He, H. P.; Lin, Z. Y. The Hydroxyl Species and Acid Sites on Diatomite Surface: A Combined IR and Raman Study. *Appl. Surf. Sci.* **2004,** *227,* 30–39.
65. Yatsyshyn, M. M.; Grynda, Yu. M.; Reshetnyak, O. V.; et al. Physico-Chemical Properties of the Polyaniline-Mineral Composites. XVIth International Seminar on Physics and Chemistry of Solids. Abstract. ISPCS'10. Lviv, June, 2010, p 151.
66. Han, Y. G.; Kusunose, T.; Sekino, T. One-Step Reverse Micelle Polymerization of Organic Dispersible Polyaniline Nanoparticles. *Synth. Met.* **2009,** *159,* 123–131.

67. Doca, N.; Vlase, G.; Vlase, T.; Pert, M.; Ilia, G.; Plesu, N. TG, EGA and Kinetic Study by Non-Isothermal Decomposition of a Polyaniline with Different Dispersion Degree. *J. Therm. Anal. Calorim.* **2009,** *97,* 479–484.
68. NIST Chemistry WebBook. NIST Standard Reference Database Number 69. http://webbook.nist.gov/chemistry.

INDEX

A

Q

R

S

T

V

W

X

Lightning Source UK Ltd.
Milton Keynes UK
UKHW011559090221
378470UK00005B/65

9 781771 883429